Mathematics in Cyber Research

Mathematics in Cyber Research

Edited by
Paul L. Goethals
Department of Mathematical Sciences, United States Military Academy, USA

Natalie M. Scala
College of Business and Economics, Towson University, USA

Daniel T. Bennett
United States Military Academy, USA

CRC Press is an imprint of the
Taylor & Francis Group, an **informa** business

First edition published 2022
by CRC Press
6000 Broken Sound Parkway NW, Suite 300, Boca Raton, FL 33487-2742

and by CRC Press
4 Park Square, Milton Park, Abingdon, Oxon, OX14 4RN

© 2022 Taylor & Francis Group, LLC

CRC Press is an imprint of Taylor & Francis Group, LLC

Reasonable efforts have been made to publish reliable data and information, but the author and publisher cannot assume responsibility for the validity of all materials or the consequences of their use. The authors and publishers have attempted to trace the copyright holders of all material reproduced in this publication and apologize to copyright holders if permission to publish in this form has not been obtained. If any copyright material has not been acknowledged, please write and let us know so we may rectify it in any future reprint.

Except as permitted under U.S. Copyright Law, no part of this book may be reprinted, reproduced, transmitted, or utilized in any form by any electronic, mechanical, or other means, now known or hereafter invented, including photocopying, microfilming, and recording, or in any information storage or retrieval system, without written permission from the publishers.

For permission to photocopy or use material electronically from this work, access www.copyright.com or contact the Copyright Clearance Center, Inc. (CCC), 222 Rosewood Drive, Danvers, MA 01923, 978-750-8400. For works that are not available on CCC, please contact mpkbookspermissions@tandf.co.uk

Trademark notice: Product or corporate names may be trademarks or registered trademarks and are used only for identification and explanation without intent to infringe.

Library of Congress Cataloging-in-Publication Data

Names: Goethals, Paul L., editor. | Scala, Natalie M. (Natalie Michele), editor. | Bennett, Daniel T., editor.
Title: Mathematics in cyber research / edited by Paul L. Goethals, Department of Mathematical Sciences, United States Military Academy, USA, Natalie M. Scala, College of Business and Economics, Towson University, USA, Daniel T. Bennett, National Renewable Energy Laboratory (NREL), USA.
Description: First edition. | London ; Boca Raton : C&H/CRC Press, 2022. | Includes bibliographical references and index.
Identifiers: LCCN 2021043509 (print) | LCCN 2021043510 (ebook) | ISBN 9780367374679 (hardback) | ISBN 9781032208046 (paperback) | ISBN 9780429354649 (ebook)
Subjects: LCSH: Computer science--Mathematics. | Cyberspace--Mathematics. | Operations research--Mathematics.
Classification: LCC QA76.9.M35 M389 2022 (print) | LCC QA76.9.M35 (ebook) | DDC 004.01/51--dc23
LC record available at https://lccn.loc.gov/2021043509
LC ebook record available at https://lccn.loc.gov/2021043510

ISBN: 978-0-367-37467-9 (hbk)
ISBN: 978-1-032-20804-6 (pbk)
ISBN: 978-0-429-35464-9 (ebk)

DOI: 10.1201/9780429354649

Typeset in CMR10
by KnowledgeWorks Global Ltd.

Editor's Note: The views expressed herein do not necessarily represent the views of the U.S. Department of Defense or the U.S. Government.

To Dr. B. Rae Cho, Professor Emeritus, and my doctoral advisor at Clemson University. Thank you for initiating and inspiring my curiosity for research. I would not be capable of this achievement without your example, your advice, and your consistent thoughtfulness.

—Paul L. Goethals

To the Mathematics and Computer Science faculty at John Carroll University, especially Patrick Chen, Barbara D'Ambrosia, Jerry Moreno, Doug Norris, Carl Spitznagel, and David Stenson. Your love of mathematics always inspired me, and your mentorship will never be forgotten.

—Natalie M. Scala

Contents

Preface	ix
List of Contributors	xvii
CHAPTER 1 ■ Combinatorics	1
CHEYNE HOMBERGER	
CHAPTER 2 ■ Cryptography	53
GRETCHEN L. MATTHEWS, AIDAN W. MURPHY	
CHAPTER 3 ■ Algebraic Geometry	97
LUBJANA BESHAJ	
CHAPTER 4 ■ Topology	133
STEVE HUNTSMAN, JIMMY PALLADINO AND MICHAEL ROBINSON	
CHAPTER 5 ■ Differential Equations	171
PARISA FATHEDDIN	
CHAPTER 6 ■ Network Science	207
ELIE ALHAJJAR	
CHAPTER 7 ■ Operations Research	233
PAUL L. GOETHALS, NATALIE M. SCALA AND NATHANIEL D. BASTIAN	
CHAPTER 8 ■ Data Analysis	267
RAYMOND R. HILL, DARRYL K. AHNER	
CHAPTER 9 ■ Statistics	303
NITA YODO, MELVIN RAFI	
CHAPTER 10 ■ Probability Theory	335
DAVID M. RUTH	

CHAPTER 11 ▪ Game Theory   363
    ANDREW FIELDER

CHAPTER 12 ▪ Number Theory   393
    DANE SKABELUND

CHAPTER 13 ▪ Quantum Theory   421
    TRAVIS B. RUSSELL

CHAPTER 14 ▪ Group Theory   453
    WILLIAM COCKE, MENG-CHE 'TURBO' HO

CHAPTER 15 ▪ Ring Theory   475
    LINDSEY-KAY LAUDERDALE

Index   497

Preface

INTRODUCTION

Since the dawn of the computer, we have observed the continued growth of cyberspace. The Internet of Things via users, mobile devices, and new technologies, as well as the need for greater interconnectivity among our social, economic, and political systems continue to fuel its expansion. Its importance to business, matters of security, and networking is clear, enabling real-time communication at a global scale. Today's emphasis on automated systems and robotics, as well as emerging fields such as artificial intelligence, suggest that this space will only continue to increase its role and significance to society in future years.

Of critical importance to the international community today are cyber topics such as the security of information and data, the protection of infrastructure, and the defense of networks or automated processes. The growth of cybercrime along with the perceived increase in the threat of an attack or breach of a system has created an environment of general uncertainty and concern. To complicate matters, protecting information is frequently characterized by dichotomies such as security versus privacy or offensive versus defensive action. Moreover, the precise advantages and benefits that cyberspace offers—connectivity, speed of transmission, and open access are in many ways the factors that produce the greatest disadvantages and vulnerabilities in many of our systems and processes.

In the last decade, both scholars and practitioners have sought novel ways to address the problem of cybersecurity. Innovative outcomes have included applications such as blockchain as well as creative methods for cyber forensics, software development, and intrusion prevention. Accompanying these technological advancements, discussion on cyber matters at national and international levels has focused primarily on the topics of law, policy, and strategy. The objective of these efforts is typically to promote security by establishing agreements among stakeholders on regulatory activities. Varying levels of investment in cyberspace, however, come with varying levels of risk; in some ways, this can translate directly to the degree of emphasis for pushing substantial change.

At the very foundation or root of cyberspace systems and processes are tenets and rules governed by principles in *mathematics*. Topics such as encrypting or decrypting file transmissions, modeling networks, performing data analysis, quantifying uncertainty, measuring risk, and weighing decisions or adversarial courses of action represent a very small subset of activities highlighted by mathematics. To facilitate education and a greater awareness of the role of mathematics in cyber systems and processes, a description of research in this area is needed. This ties directly to the objective of this book: to familiarize educators and young researchers with the breadth of mathematics in cyber-related research.

ORGANIZATION AND INTENT

To showcase areas where practitioners and researchers may contribute mathematically to the cyber domain, the chapters in this book are defined by sub-fields of mathematics. Some of these fields are theoretical, covering a foundation of the mathematical proofs and corollaries needed for further study. Some of the fields are more applied, offering solution techniques widely used in this domain of research. The intent of this book is to present a foundation of topics in both applied and pure mathematics in order to give the reader an appreciation of the breadth of research. Each chapter will introduce a mathematical sub-field, describe relevant work in this field associated with the cyber domain, and provide methods and tools or examples of how it is used. Implications of the research and extensions where further work is needed may also be described within each chapter.

The sub-fields outlined in this book are certainly not all inclusive. There are a wide number of other mathematical sub-fields where cyber research is currently being performed: complex analysis, linear algebra, numerical methods, and set theory, to name just a few. The chapters presented in this book are meant only to provide a representative display of the available topics where research is performed today. Emphasis is placed on the relationship of these topics to cyber systems, processes, or applications where security is of paramount interest. Naturally, topics such as cryptography and network defense are of immense importance to security and so may be highlighted in multiple mathematical sub-fields throughout the book.

The target audience for *Mathematics in Cyber Research* is the college undergraduate student or educator that is either interested in learning about cyber-related mathematics or intends to perform research within the cyber domain. The book may also appeal to practitioners within the commercial industry or government sectors, where techniques are applied first-hand to strengthen the security of cyber systems and processes.

ABOUT THE EDITORS

The editors of this book possess more than 70 years of combined experience in security and defense-related matters and represent several different perspectives in academia, government, and industry related to cyber research. A brief introduction of each editor is provided below.

Dr. Paul L. Goethals is an Academy Professor and the Director of Research for the Department of Mathematical Sciences at the United States Military Academy. He earned an M.S. degree in Mathematics from Florida State University in 2002 and a Ph.D. in Industrial Engineering from Clemson University in 2011. Dr. Goethals has served 29 years of active duty service within the United States Army, most recently as a Research Scientist at the Army Cyber Institute and as a member of the Commander's Action Group within the United States Cyber Command. He currently serves on the editorial board for the *Cyber Defense Review* and has served as the Editor-in-Chief of the *International Journal of Experimental Design and Process Optimization* since 2012.

Dr. Natalie M. Scala is an Associate Professor and Director of the graduate programs in Supply Chain Management in the College of Business and Economics at Towson University. She earned Ph.D. and M.S. degrees in Industrial Engineering from the University of Pittsburgh. Her primary research is in decision analysis, with foci on military applications and cybersecurity. Specific projects include army unit readiness, risk in voting systems, cybersecurity metrics and best practices, naval seabasing, nuclear power plants, and workforce planning. Dr. Scala frequently consults to government clients and has extensive professional experience, to include positions with Innovative Decisions, Inc., the United States Department of Defense, and the RAND Corporation. Her first book, a co-edited volume titled *The Handbook of Military and Defense Operations Research*, was released by CRC Press in early 2020.

Dr. Daniel T. Bennett is the Senior Research and Operations Advisor for Energy Security and Resilience at the National Renewable Energy Laboratory (NREL). He has a Ph.D. and M.S. in Electrical Engineering from the University of Colorado at Boulder. Dan came to NREL from the Army Cyber Institute at the United States Military Academy, where he led research for the Army's cyber think tank. While at West Point, he was also an Associate Professor in the Department of Electrical Engineering and Computer Science, where he was the Academy's recognized expert in analog, digital, and wireless communications systems and networks. He also recently served a yearlong tour as Technical Director Advisor to the 2-star commander of United States Cyber Command's Cyber National Mission Force at Ft. Meade, MD. Previous experiences include 15 months as Director of the Joint Network Operations & Security Center for Combined Joint Task Force-101 in Bagram, Afghanistan. Currently, he is a technical lead on the Department of Energy's '5G Catalogue'.

ABOUT THE AUTHORS

The authors of this book represent a diverse group of experts at various levels of academia, government, and industry. Some of the authors are scientists specializing in their given field, while other authors are practitioners relying on their experience and knowledge of real-world systems or processes. This collection of perspectives provides a unique vantage point for the current state of research in the cyber domain. To showcase the authors, a brief biography of each contributor, by chapter, is provided in the paragraphs that follow.

Chapter 1: *Combinatorics*

Dr. Cheyne Homberger earned his Ph.D. in Mathematics from the University of Florida, focusing on using analytic combinatorics to analyze structural properties of permutations. He currently leads a team of operations researchers and data scientists within the Department of Defense, leveraging advanced analytical techniques against critical challenges in national security.

Chapter 2: *Cryptography*

Dr. Gretchen L. Matthews is a Professor of Mathematics at Virginia Tech and Director of the Commonwealth Cyber Initiative Southwest Virginia. She is affiliated faculty with Virginia Tech's Hume Center for National Security & Technology and Computational Modeling & Data Analytics program. Her research is in applied algebra, with foci on coding theory and cryptography and applications in data storage, protection, and privacy. She is passionate about diversity and inclusion in STEM and has developed a number of programs to broaden participation. She earned a B.S. from Oklahoma State University and a Ph.D. from Louisiana State University, both in Mathematics. She spent time as a postdoc at the University of Tennessee and on the faculty at Clemson University before joining Virginia Tech.

Aidan W. Murphy is a Ph.D. student in the Virginia Tech Department of Mathematics and a student of Gretchen L. Matthews. His research interest is algebraic coding theory, particularly in locally recoverable codes and code-based cryptography. Outside of research, he serves on the Virginia Tech Graduate Student Assembly as Parliamentarian and chairs the Standing Committee on Internal and Judicial Affairs as well as the special Subcommittee on Constitutional Overhaul. He also has interest in mathematics education at and above the undergraduate level. He earned a B.S. in Mathematics from the State University of New York at Geneseo and an M.S. in Mathematics at Virginia Tech.

Chapter 3: *Algebraic Geometry*

Dr. Lubjana Beshaj is a Cyber Fellow of Mathematics at the Army Cyber Institute and an Assistant Professor in the Department of Mathematical Sciences at West Point. She is a member of the American Mathematical Society and Women in Mathematics. Her research interests include cryptography, elliptic and hyperelliptic curve cryptography, post-quantum cryptography, and more specifically, isogeny-based cryptography, Jacobian varieties, and arithmetic of algebraic curves.

Chapter 4: *Topology*

Steve Huntsman is an applied mathematician working on cyber-oriented problems. He earned a bachelor's degree from Arizona State University in 1997 and a master's degree in Mathematics from New York University in 1999 before leaving to work at the Institute for Defense Analyses. He later completed coursework and qualifying exams for a Ph.D. in Physics at the Naval Postgraduate School in 2006 before leaving to start a network security company. Since then, he has continued to work in defense research and development apart from a brief stint as a data scientist. His research interests have included statistical physics, bioinformatics, quantum computing, cybersecurity, networking, autonomy, data science, and machine learning.

Jimmy Palladino earned a combined bachelor's degree in Applied Mathematics from American University in 2016 and Mechanical Engineering from Columbia University in 2018. Between 2015 and 2016, he researched applications of algebraic topology for the analysis of wireless networks. His most recent work was as a bioinformatics engineering intern at the Simons Foundation in 2020. His interests include network analysis, data science, computational mathematics, and statistical modeling.

Dr. Michael Robinson is an applied mathematician working as an Associate Professor at American University. His interests include signal processing, dynamics, and applications of topology. He earned bachelor's and master's degrees from Rensselaer Polytechnic Institute in 2002 and 2003, respectively. From that time, he worked on projects involving radio propagation and network planning, bistatic radar processing, and advanced radar simulation. In 2008, he earned a Ph.D. in Applied Mathematics at Cornell University, in which he developed topological methods for studying the dynamics of parabolic equations. His more recent efforts follow an emerging trend started during postdoctoral work at the University of Pennsylvania of topologically-motivated signal processing techniques. His current projects involve topological approaches to tracking, communication network analysis, sonar target recognition, and data fusion.

Chapter 5: *Differential Equations*

Dr. Parisa Fatheddin is a Senior Lecturer at Ohio State University, Marion, with her research focus being stochastic partial differential equations. She has published research articles in *Stochastic Processes and Applications, Journal of Stochastic Analysis, Journal of Applied Probability, and Stochastic Analysis and Applications*. She observes the importance of finding applications of her area of study to other fields, such as cybersecurity and optics. From fall 2015 to spring 2016, Dr. Fatheddin served as a Postdoctoral Fellow at an optics laboratory at the Air Force Institute of Technology, where she applied her knowledge of probability theory to problems connected to laser technology and published her work with Dr. Jonathan Gustafsson in *Optics Communications*. Dr. Fatheddin continues to conduct research on both theoretical and applied probability and stochastic analysis to help the growth of mathematics in other fields of study.

Chapter 6: *Network Science*

Dr. Elie Alhajjar is a Research Scientist at the Army Cyber Institute and an Assistant Professor in the Department of Mathematical Sciences at the United States Military Academy, where he teaches and mentors cadets from all academic disciplines. His research interests include network science and mathematical modeling. He has presented his research work at international meetings in North America, Europe, and Asia. Before coming to West Point, Dr. Elie Alhajjar had a research appointment at the National Institute of Standards and Technology in Gaithersburg, Maryland. He holds M.S. and Ph.D. degrees in Mathematics from George Mason University as well as master's and bachelor's degrees from Notre Dame University.

Chapter 7: *Operations Research*

Dr. Paul L. Goethals (biography provided earlier).

Dr. Natalie M. Scala (biography provided earlier).

Dr. Nathaniel D. Bastian is a decision analytics professional with expertise in the scientific discovery and translation of actionable insights into effective decisions using algorithms, techniques, tools, and technologies from operations research, data science, artificial intelligence (AI), systems engineering, and economics to research, design, develop, and deploy intelligent decision-support models, tools, and systems for descriptive, predictive, and prescriptive analytics. His disciplinary expertise spans optimization, simulation, statistical computing, machine learning, intelligent systems, big data analytics, decision science, econometrics, and engineering management. His primary research interest is computational stochastic optimization and learning for making inferences and decisions under uncertainty, with projects in generative machine learning for cyber, data science, and artificial intelligence (AI) for cyber network security decision-support, AI system assurance, and AI-enabled command and control for multi-domain operations.

Chapter 8: *Data Analysis*

Dr. Raymond R. Hill is a Professor of Operations Research in the Department of Operational Sciences at the Air Force Institute of Technology. He has a Ph.D. in Industrial and Systems Engineering from The Ohio State University and has research interests in applied statistical models, modeling and simulation, applications of optimization and search heuristics, and military operations research applications. He is the Director of the Test and Evaluation Certificate Program, a Principal Investigator in the Science of Test Research Consortium, and Editor-in-Chief of the *Military Operations Research* journal and the *Journal of Defense Analytics and Logistics*.

Dr. Darryl K. Ahner is a Professor of Operations Research in the Department of Operational Sciences at the Air Force Institute of Technology. He is the founding Director of the Office of the Secretary of Defense Scientific Test and Analysis Techniques Center of Excellence, established in 2012, and the founding Director of the Department of Homeland Security Center of T&E Best Practices, established in 2020. He has a Ph.D. in Systems Engineering from Boston University, is a licensed Professional Engineer, and has research interests in stochastic models, modeling and simulation, stochastic dynamic programming, optimization of test and evaluation, autonomous systems, and military operations research applications.

Chapter 9: *Statistics*

Dr. Nita Yodo is an Assistant Professor in the Department of Industrial and Manufacturing Engineering at North Dakota State University (NDSU). She earned her Ph.D. degree in Industrial Engineering from Wichita State University. Her research interests include designing resilient, secure, and sustainable engineering systems. She is also a research faculty advisor of the Advanced System Engineering Laboratory and Center of Quality, Reliability, and Maintainability Engineering research groups at NDSU.

Dr. Melvin Rafi earned his Ph.D. degree in Aerospace Engineering from Wichita State University. He is currently a Postdoctoral Researcher at the General Aviation Flight Laboratory at Wichita State University. His research interests include intelligent resilient adaptive control, loss-of-control prediction systems for aircraft, augmented-reality pilot advisory displays, and the development of aircraft flight simulation systems.

Chapter 10: *Probability Theory*

Dr. David M. Ruth is an Associate Professor of Mathematics at the United States Naval Academy (USNA). He was commissioned in the U.S. Navy in 1991 and served in operational assignments involving submarine warfare, surface warfare, and military planning. Dr. Ruth joined the USNA faculty in 2009 as a Permanent Military Professor, where he has served in a variety of leadership roles. In 2020 he was awarded the USNA Military Professor Teaching Excellence Award. His research interests include nonparametric statistics, graph-theoretic statistical methods, multivariate change detection, and statistical learning.

Chapter 11: *Game Theory*

Dr. Andrew Fielder has a background in agent-based systems within computer science primarily focused on security problems. He received a Ph.D. from the University of Birmingham on the subject of intelligent agents for the evaluation of wholesale electricity production designs. Following this, he moved to Imperial College London, where he worked on the topic of games for cybersecurity, focusing on the economics and socio-technical aspects of security decision making. He was granted a National Technical Authority fellowship and worked on mathematical models for cybersecurity as well as approaches to utilizing AI and machine learning for supporting security with a focus on national infrastructure.

Chapter 12: *Number Theory*

Dr. Dane Skabelund studied Mathematics at Brigham Young University and then at University of Illinois at Urbana-Champaign, where he received his Ph.D. He held a subsequent position as a Postdoctoral Researcher at Virginia Tech. He now works for the Department of Defense.

Chapter 13: *Quantum Theory*

Dr. Travis Russell is a native of Jacksonville, Texas. He received his B.S. in Physics from Stephen F. Austin State University in 2009, his M.S. in Mathematics from the University of Texas at Tyler in 2011, and his Ph.D. in Mathematics from the University of Nebraska in 2017. He is currently a Research Assistant Professor at the Army Cyber Institute, an interdisciplinary research institute serving the cyber branch of the U.S. Army based at West Point, New York. He also teaches in the Department of Mathematics at the United States Military Academy.

Chapter 14: *Group Theory*

Dr. William Cocke received his Ph.D in Mathematics from the University of Wisconsin-Madison in 2019. He is currently serving as a Lieutenant in the Cyber Branch of the United States Army and continues to learn and utilize mathematics. His research interests are primarily focused on finite group theory and computations about finite groups. He enjoys collaborating with coauthors throughout the United States and the world. Besides mathematics, William enjoys computer science and linguistics. Depending on the conversation, he is conversant in Urdu, Hindi, and Mandarin. Before working on his doctorate, William received a M.S. and B.S. in Mathematics from Brigham Young University in Provo, Utah.

Dr. Turbo Ho is an Assistant Professor of Mathematics at California State University Northridge. He received his Ph.D. from the University of Wisconsin-Madison in 2017 under the guidance of Uri Andrews and Tullia Dymarz. Before that, he received his M.A. from the University of Wisconsin-Madison and B.S. from National Taiwan University. Turbo has held visiting positions at Purdue University, the Mathematical Sciences Research Institute, and Hausdorff Research Institute for Mathematics. His research interests lie in group theory and logic with a focus on the computability, complexity, and model theory of finitely generated groups.

Chapter 15: *Ring Theory*

In 2009, Lindsey-Kay Lauderdale received her B.S. in Mathematics from the University of Illinois at Urbana-Champaign; she received her Ph.D. in Mathematics from the University of Florida in 2014. The focus of her Ph.D. thesis was on group theory, and in particular, she studied maximal subgroups of finite groups. Since graduation, she has expanded her research program to include graph theory, combinatorics, number theory, and their applications throughout the sciences. Currently, she is an Assistant Professor in the Department of Mathematics at Towson University in Maryland.

ACKNOWLEDGMENTS

We would like to thank the authors for the many hours of work that they sacrificed in completing this project. In terms of both content and contribution, we do not find any comparative product in the world of literature. Special thanks are also given to our publisher representatives at Taylor & Francis / CRC Press, Mansi Kabra and Callum Fraser, for their guidance and direction throughout the process.

To the reader of *Mathematics in Cyber Research*, we hope that it inspires you to further investigate and solve problems in cyber-related work. As long as our world depends on technology for its existence, there will be a need for attention and further contribution in this research space.

—Paul L. Goethals, Natalie M. Scala, and Daniel T. Bennett

List of Contributors

Darryl K. Ahner
Air Force Institute of Technology
Wright-Patterson Air Force Base, Ohio, USA

Elie Alhajjar
Army Cyber Institute
West Point, New York, USA

Nathaniel D. Bastian
Army Cyber Institute
West Point, New York, USA

Lubjana Beshaj
Army Cyber Institute
West Point, New York, USA

William Cocke
Army Cyber Command
Fort Gordon, Georgia, USA

Parisa Fatheddin
Ohio State University
Marion, Ohio, USA

Andrew Fielder
Imperial College London
London, United Kingdom

Paul L. Goethals
United States Military Academy
West Point, New York, USA

Raymond R. Hill
Air Force Institute of Technology
Wright-Patterson Air Force Base, Ohio, USA

Meng-Che 'Turbo' Ho
California State University Northridge
Northridge, California, USA

Cheyne Homberger
Department of Defense
USA

Steve Huntsman
New York University
New York City, New York, USA

Lindsey-Kay Lauderdale
Towson University
Towson, Maryland, USA

Gretchen L. Matthews
Virginia Tech
Blacksburg, Virginia, USA

Aidan W. Murphy
Virginia Tech
Blacksburg, Virginia, USA

Jimmy Palladino
American University
Washington, D.C., USA

Melvin Rafi
Wichita State University
Wichita, Kansas, USA

Michael Robinson
American University
Washington, D.C., USA

David M. Ruth
United States Naval Academy
Annapolis, Maryland, USA

Travis B. Russell
Army Cyber Institute
West Point, New York, USA

Dane Skabelund
Department of Defense
USA

Natalie M. Scala
Towson University
Towson, Maryland, USA

Nita Yodo
North Dakota State University
Fargo, North Dakota, USA

CHAPTER 1

Combinatorics

Cheyne Homberger

CONTENTS

1.1	Introduction	2
1.2	Basic Enumeration and Brute Force	2
	1.2.1 Sequences	4
	1.2.2 Permutations	4
	1.2.3 Subsets	5
	1.2.4 Multisets	7
	1.2.5 Password Strength	8
	1.2.6 The Binomial Theorem	9
1.3	Discrete Probability and Hash Functions	12
	1.3.1 Probability and Expectation	12
	1.3.2 Hash Functions	15
	1.3.3 Collisions and Cyber Attacks	16
	1.3.4 Birthday Attacks	17
	1.3.5 The Coupon Collector's Problem	18
1.4	Inclusion-Exclusion and Random Mappings	20
	1.4.1 Principle of Inclusion–Exclusion	20
	1.4.2 Random Functions and Stirling Numbers	23
	1.4.3 Cycles in Permutations	25
	1.4.4 Random Mapping Statistics and Applications	26
1.5	Generating Functions and Asymptotic Enumeration	29
	1.5.1 Ordinary Generating Functions	30
	1.5.2 Recurrences	31
	1.5.3 Combining Classes	33
	1.5.4 Recursive Structures	36
	1.5.5 Asymptotic Analysis	39
	1.5.6 Bivariate Generating Functions and Combinatorial Statistics	42
1.6	Combinatorial Designs and Software Testing	45
	1.6.1 Latin Squares	46
	1.6.2 Black-Box Testing	47
	1.6.3 Orthogonal Arrays	48
	1.6.4 Test Generation	49
1.7	Conclusion	50

DOI: 10.1201/9780429354649-1

1.1 INTRODUCTION

Combinatorics can be defined as the study of finite structures, and combinatorial problems arise in all areas of research. These problems are often handled using ad hoc methods, and so the field of combinatorics can appear to an outsider to be a disjointed mix of topics and techniques. Studying the field directly—rather than simply to solve the problem at hand—reveals deep connections and equips any researcher with the tools and perspective needed to tackle a variety of challenges.

Combinatorialists are typically concerned with the existence, the enumeration, and the properties of structures of finite sets and their combinations. In discussing the cultures of problem-solving and theory-building mathematicians, Gowers wrote that he uses the word 'combinatorics' to refer to 'problems that are reasonable to attack more or less from first principles,' and that this definition overlaps considerably with the traditional one (Gowers, 2000). Subfields are named after the types of problems to which they are applied (e.g., combinatorial optimization, geometry, and data analysis) as well as the types of methods used to solve problems (e.g., probabilistic, analytic, and algebraic combinatorics). Graph theory, one of the oldest branches of combinatorics, often stands alone.

Cybersecurity research intersects this field in a variety of ways. In this chapter we showcase a selection of useful techniques alongside applications. There are a number of topics *not* included in this chapter. We note in particular that we do not cover graph theory, but do make use of some of its basic vocabulary. We also do not cover combinatorial optimization or the analysis of algorithms, though these fields have considerable overlap to the material here. Finally, the fields of information theory, coding theory, cryptography (the development of encryption systems), and cryptanalysis (the breaking of encryption systems) are full of problems of a combinatorial nature; the statements of many of these problems are beyond the scope of this chapter, but the methods presented here can be readily applied.

For readers interested in continuing to learn more about the field of combinatorics, we recommend the accessible and more comprehensive overview given by Bóna (2017). For graph theory, West (1996) provides a gentle introduction while Diestel (2018) gives a more advanced treatment. Stanley's two-volume textbook (2012, 1999) remains the definitive reference for enumeration. Finally, Bóna (2015) highlights current research directions.

The cyber problems we'll focus on are broad, but include password strength, hash functions, statistics of random mappings, and software testing. The methods we'll use will include enumeration, discrete probability, the principle of inclusion-exclusion, generating functions, and combinatorial designs.

1.2 BASIC ENUMERATION AND BRUTE FORCE

Enumeration—counting the number of objects in a set—is the foundation of combinatorics, and already has clear applications to cybersecurity. Assessing the strength of a password, for example, amounts to counting the size of the set of all passwords: there are $10^4 = 10,000$ different four-digit sequences, while there are $26^4 = 456,976$

different four-lowercase-letter sequences. It is intuitive that a password chosen from a larger set is inherently stronger than one chosen from a smaller set; the techniques developed in this chapter will allow us to be more precise.

First, a note on notation: Since we're primarily dealing with integers, it will be convenient to set a few conventions. We denote by $\mathbb{Z}$, $\mathbb{N}$, and $\mathbb{P}$ the sets of integers, non-negative integers, and positive integers, respectively. Further, for $n \in \mathbb{P}$, we denote by $[n]$ the set of positive integers less than or equal to n. That is,

$$\mathbb{Z} = \{\ldots -1, 0, 1, \ldots\}, \quad \mathbb{N} = \{0, 1, 2, \ldots\}, \quad \mathbb{P} = \{1, 2, 3, \ldots\}, \quad [n] = \{1, 2, \ldots n\}.$$

We'll start with a motivating example:

Example 1.1. An attacker is trying to break into two phones, each of which uses a 4-digit passcode to unlock. By looking closely at the physical screen, they see smudges indicating which of the numbers are used in the passcodes. See Figure 1.1: Phone A shows smudges on the 2, 4, 5, and 9, while phone B only shows smudges on the 0, 5, and 7. Which of the phones is easiest to crack?

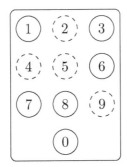 

FIGURE 1.1 Finger smudges on two phones with 4-digit passcodes.

Phone A uses four distinct numbers as a passcode, while phone B uses only 3 distinct numbers, so one of them must be repeated in its 4-digit code. It turns out that the 3-smudge phone is 50% harder to break into than the 4-smudge phone.[1] In this section, we'll develop the techniques and vocabulary to explain why.

The sets we'll be interested in most commonly arise from other, smaller sets. To see what we mean, let B be the 2-element set $\{0, 1\}$. A more interesting set is the set of ordered sequences using elements of set B. In this section we'll consider four basic constructions—sequences, permutations, sets, and multisets— each of which creates a family of finite sets from a single set. We call these families *classes*.

Definition 1.1 (Combinatorial Classes). A *combinatorial class* is a (possibly infinite) set $\mathcal{C}$ equipped with a *size* function $|\cdot| : \mathcal{C} \to \mathbb{N}$, with the property that for each $k \in \mathbb{N}$, there are finitely many elements of $\mathcal{C}$ having size k.

Let $\mathcal{C}$ be a combinatorial class. For each $k \in \mathbb{N}$, denote by $\mathcal{C}_k$ the subset of $\mathcal{C}$ having size k. That is,

$$\mathcal{C}_k = \{c \in \mathcal{C} \ : \ |c| = k\}.$$

---

[1]Phone A uses one of $4! = 24$ possible passcodes, while phone B uses one of $3/2 \cdot 4! = 36$.

Finally, for each class $\mathcal{C}$ we associate a sequence of non-negative integers $(c_0, c_1, c_2, \ldots)$ in which $c_k = |\mathcal{C}_k|$. This sequence is called the *enumeration* of $\mathcal{C}$.

1.2.1 Sequences

Definition 1.2 (Sequences). For a finite set S, a *sequence of length k* from S is an ordered sequence $(a_1, a_2, \ldots a_k)$ of length k in which $a_i \in S$ for each i. We denote by $\text{SEQ}(S)$ the set of all sequences with elements from S. It will be helpful later to decide that the *empty sequence*, denoted by $()$ or ϵ, is an element of $\text{SEQ}(S)$.

We'll sometimes write these elements without commas and parentheses to simplify notation. That is, using 10010 to denote $(1, 0, 0, 1, 0)$.

For example, we have

$$\text{SEQ}(B) = \{(), (0), (1), (0,0), (0,1), (1,0), (1,1), (0,0,0), \ldots\},$$

which has the enumeration sequence

$$(2^0, 2^1, 2^2, 2^3, 2^4, 2^5, \ldots) = (1, 2, 4, 8, 16, 32, \ldots).$$

Of course, there is nothing special about the set $\{0, 1\}$—we could have just as easily made sequences from any other set with two distinct elements (e.g., $\{a, b\}$, $\{\text{orange}, \text{blue}\}$, $\{\heartsuit, \diamondsuit\}$). Further, if our underlying set had n elements, we would have n^k unique sequences of size k. We capture this in the following theorem:

Theorem 1.1 (Sequence Enumeration). *If S is a finite set with n elements, then the number of sequences of length k using elements of S is n^k. In other words, the enumeration sequence of $\text{SEQ}(S)$ is given by*

$$(n^0, n^1, n^2, n^3, n^4, n^5, \ldots).$$

A note on notation: Sequences arise in a number of different mathematical and computer science contexts, leading to a number of different notations. The elements of the sequence are sometimes called *letters*, the underlying set is the *alphabet*, and the sequences (written without commas and parentheses) are called *words*. The empty sequence is denoted by ϵ. For example, aacba is a word of length 5 on the alphabet $\{a, b, c\}$.

1.2.2 Permutations

Definition 1.3 (Permutations). A *permutation* of size k (also denoted k-permutations) from S is a sequence of size k drawn from S in which none of the entries are repeated. The class of all permutations of S is denoted $\text{PERM}(S)$.

For example,

$$\text{PERM}(\{1, 2, 3\}) = \{\epsilon, 1, 2, 3, 12, 13, 21, 23, 31, 32,$$
$$123, 132, 213, 231, 312, 321\}.$$

By direct enumeration,[2] we see that the enumeration sequence for permutations of a 3-element set is $(1, 3, 6, 6, 0, 0, 0, \ldots)$.

Note that $\text{PERM}(S)$ is a finite set as well as a combinatorial class. To count the number of k-permutations of an n-element set, note that we have n choices for the first element, $n-1$ choices for the second element (any element except the one used for the first), $n-2$ choices for the third, and $n-k+1$ for the kth. Since each combination of choices leads to a unique permutation, we have the following theorem.

Theorem 1.2 (Permutation Enumeration). *Let $n, k \in \mathbb{N}$. The number of k-permutations of an n-element set is equal to $n(n-1)(n-2)\cdots(n-k+1)$. This quantity is denoted $n^{\underline{k}}$ and is read n falling factorial k. In the special case, when $n = k$, we have n factorial: $n! := n^{\underline{n}} = n(n-1)(n-2)\cdots 1$.*

Note what happens when $k > n$: intuitively, we can't pick out more than n distinct elements from an n-elements set. Set $n = 4$ and $k = 7$ and see what happens:
$$n^{\underline{k}} = 4^{\underline{7}} = 4 \cdot 3 \cdot 2 \cdot 1 \cdot 0 \cdot (-1) \cdot (-2) = 0.$$

The formula still works! When presented with a new enumeration formula, you should ask what happens in the so-called *degenerate cases*. Formulas which handle degenerate cases gracefully will be useful, as they will be used as the building blocks for more complicated enumerations.

The n-permutations of an n-element set are a special case that arises frequently enough to deserve their own notation. When we refer to an *n-permutation* or a *permutation of length n* without specifying the set S, we are referring to an n-permutation of the set $[n]$, and the set of all such permutations is denoted $\mathfrak{S}_n$. The ubiquity of these permutations is due in part to the fact that they are equivalent to *bijections* (invertible functions) from an n-element set to itself. To see this, note that we can associate each element $(a_1, a_2, \ldots, a_n)$ of $\text{PERM}([n])$ with a function $\pi \in \mathfrak{S}_n$ by $\pi(k) = a_k$, and this association is easily reversible. We'll treat n-permutations as both sequences and functions in the coming sections.

Our two constructions so far, SEQ and PERM have relied on the *ordering* of a collection of elements drawn from a finite set. For our next two, we'll consider *unordered* collections.

1.2.3 Subsets

Definition 1.4 (Subsets). *Let S be a finite set. A *subset* of S is an unordered collection of non-repeated elements of S. That is, a subset is a collection $R \subseteq S$ with $R = \{r_1, r_2, \ldots r_k\}$ with each element distinct and from S. We denote by $\text{SET}(S)$ the set of all subsets of S.*

For example, let $S = \{1, 2, 3, 4\}$ and count the number of subsets of size 3. Rather than counting them directly, we start by counting the permutations of length 3, and then figure out which are identical if we ignore their order.

---
[2]'Direct Enumeration' is combinatorialist-speak for 'counting by hand.'

$$\text{SET}(\{1,2,3,4\})_3 = \{123, 132, 213, 231, 312, 321,$$
$$124, 142, 214, 241, 412, 421,$$
$$134, 143, 314, 341, 413, 431,$$
$$234, 243, 324, 342, 423, 432\}.$$

The only difference between $\text{PERM}(\{1,2,3,4\})_3$ and $\text{SET}(\{1,2,3,4\})_3$ is whether or not the order of the elements matters. There are 24 elements in $\text{PERM}(\{1,2,3,4\})$. The listing above splits into 4 subsets—the first line shows all orderings of the entries $\{1,2,3\}$, the second shows all orderings of $\{1,2,4\}$, the third shows orderings of $\{1,3,4\}$, and the fourth $\{2,3,4\}$. In each case, we know that the number of orderings is the number of permutations of a 3-element set: $3 \cdot 2 \cdot 1 = 6$. This tells us two things:

- There are 4 different subsets of $\{1,2,3,4\}$ of size 3.

- The number of subsets is equal to the number of 3-permutations of the 4-element set divided by the number of permutations of a 3-element set: $(4 \cdot 3 \cdot 2)/3! = 4$ subsets.

Any given *unordered* collection leads to a certain number of *ordered* collections. More precisely, an unordered collection of size k leads to $k!$ different ordered collections. Thus, if we denote by $\binom{n}{k}$ the number of unordered k-sets from an n-element set, then we have that

$$k!\binom{n}{k} = n^{\underline{k}} = n(n-1)(n-2)\cdots(n-k+1).$$

By dividing both sides by $k!$, we've just proved the following theorem.

Theorem 1.3 (Subset Enumeration). *The number of k-subsets of an n element set is denoted by $\binom{n}{k}$, which is equal to*

$$\binom{n}{k} = \frac{n^{\underline{k}}}{k!} = \frac{n!}{k!(n-k)!} = \frac{n(n-1)(n-2)\cdots(n-k+1)}{k!}.$$

Thus, for an n-element set S, the enumeration sequence for $\text{SET}(S)$ is equal to

$$\left(\binom{n}{0}, \binom{n}{1}, \binom{n}{2}, \binom{n}{3}, \ldots\right) = \left(1, n, \frac{n^2-n}{2}, \frac{n^3-2n^2-n}{6}, \ldots\right).$$

The numbers $\binom{n}{k}$ are known as the *binomial coefficients* and will be used extensively. Note first that if $k > n$ or $k < 0$, then the numerator is 0, and we define $\binom{n}{k}$ to be zero. Next, we note that for $0 \le k \le n$, the expression $n!/(k!(n-k)!)$ is not obviously an *integer*, let alone one that counts something useful. Viewed a different way, the counting argument can be viewed as a *combinatorial proof* that $n!$ is divisible by $k!(n-k)!$.

It's also worth noting that the formula exhibits a certain symmetry: replacing k with $n-k$ gives the same result. This implies that, for fixed n, there is the same number of k-subsets as $(n-k)$-subsets. We've just proven this fact—it's a simple consequence of the formula—but here is a more combinatorial proof: if A is a k-element subset of the n-element set n, then $S \setminus A$ (the elements of S not in A) is an $(n-k)$-element subset of S, and taking the complement again gets us back to A. Thus, we have an invertible mapping between the k- and $(n-k)$-element subsets of S, showing that they must be equal in number.

1.2.4 Multisets

Definition 1.5 (Multisets)**.** Let S be a finite set. A *multiset* of S is an unordered collection of (possibly repeated) elements of S. That is, a multiset is a collection $M = \{m_1, m_2, \ldots m_k\}$ with $m_i \in S$ for each i. Since the order is unimportant, two multisets are identical if and only if they have the same elements with the same multiplicity. The size of a multiset is the total number of elements, including repeats. The (infinite) set of all multisets of S is denoted by $\text{MSET}(S)$.

For example, if $S = \{1, 2, 3, 4\}$, then $\{1, 1, 1, 1\}$, $\{1, 2, 2, 3\}$, $\{1, 2, 3, 4\}$, and $\{1, 1, 3, 4\}$ are all distinct size-4 multisets of S, but $\{1, 1, 2, 2\}$ and $\{2, 1, 1, 2\}$ denote the same multiset. Since the order doesn't matter, we typically write the elements in non-decreasing order for convenience.

To count multisets of size k drawn from a set of size n, it may seem that we could follow the strategy for subsets: first count the sequences of length k, then divide by the different possible orderings. This won't work, however, since the multiset $\{1, 2, 3, 4\}$ can be rearranged into $4! = 24$ different sequences, while the multiset $\{1, 1, 1, 1\}$ leads to only a single sequence. Since each multiset corresponds to a potentially different number of sequences, there is no single number by which we can divide.

We use a classic method for enumerating these multisets known as *stars and bars*, which will use our newly developed enumerations of both subsets and sequences. Let $S = \{1, 2, 3, 4\}$ as above and let M be the multiset $\{1, 1, 2, 3, 3, 3, 4\}$. We can represent M as a sequence of *stars* (representing elements) and *bars* separating the type of element. That is, we can represent the two 1s as two stars followed by a bar, then one star for the single 2, then a bar and 3 stars for the 3s, then a bar and a star for the single 4:

$$\{1,1,2,3,3,3,4\} \mapsto \underbrace{**}_{1s} | \underbrace{*}_{2s} | \underbrace{***}_{3s} | \underbrace{*}_{4s}.$$

If a term never appears, we represent this by a pair of bars with no stars in between. Further, if we replace each star with a 0 and each bar with a 1, we have a mapping between multisets and binary sequences. Some additional examples:

$$\begin{aligned}
\{1\} &\mapsto *||| &&\mapsto 0111 \\
\{1,1,1,1,4,4,4\} &\mapsto ****|||*** &&\mapsto 0000111000 \\
\{2,3,4\} &\mapsto |*|*|* &&\mapsto 101010
\end{aligned}$$

We enumerate multisets by enumerating these binary strings. The number of zeros in the sequence corresponds to the size of the multiset since each zero corresponds to an element. Note that each binary sequence above has exactly 3 ones—this is because there are four distinct elements in the underlying set, leading $4 - 1 = 3$ bars to separate these. Thus, in general, the number of k-multisets of an n element set is equal to the number of binary sequences with k zeros and $n - 1$ ones.

How many such sequences are there? The total number of binary sequences of length $n+k-1$ is given by 2^{n+k-1}, but this allows any number of 1s and 0s. In our case, from our $n + k - 1$ positions, we need only to *choose* which k-subset of this $(n+k-1)$-element set will be assigned 0s and let the remainder be 1s. Alternatively, we could choose which $n-1$ positions will be 1s, and let the remainder be 0s. Either way, we arrive at the following theorem.

Theorem 1.4 (Multiset Enumeration). *The number of k-multisets of an n-element set S is given by*

$$\text{MSET}(S)_k = \binom{n+k-1}{k} = \binom{n+k-1}{n-1}.$$

The enumeration of these four constructions will form the building blocks for a variety of more complicated problems. We summarize the results of this section in Table 1.1.

TABLE 1.1 Enumerations of the four basic constructions for an n-element set S

$\lvert \text{SEQ}(S)_k \rvert$	$=$	n^k	
$\lvert \text{PERM}(S)_k \rvert$	$=$	$n^{\underline{k}}$	$= \frac{n!}{k!}$
$\lvert \text{SET}(S)_k \rvert$	$=$	$\binom{n}{k}$	$= \frac{n!}{k!(n-k)!}$
$\lvert \text{MSET}(S)_k \rvert$	$=$	$\binom{n+k-1}{k}$	$= \binom{n+k-1}{n-1}$.

1.2.5 Password Strength

We know intuitively that the password `1234` is easier to break than the password `p@ssw04d`. The concept of *information entropy* is commonly used to measure the strength of a password. We say that a given password has *n bits of entropy* if an attacker would need to try 2^n different passwords to guarantee that they find the correct one. Thus, assessing the strength of a password boils down to enumerating the set of possible passwords, known as the *password* (or *key*) *space*.

Note that password strength is an attribute of the password space rather than the password itself. In general, we measure the strength of a password by enumerating the smallest reasonable space containing it—for example, all lowercase letters, letters and special characters, or dictionary words followed by digits.

For example, the password `01101101` chosen at random from all binary strings of length 8 has 8 bits of entropy since an attacker would need to try 2^8 passwords to be sure of finding the right one. A 4-digit password with decimal digits has roughly

13.3 bits of entropy, since $10^4 \approx 2^{13.3}$. A six-digit passcode consisting of the numbers 1-4 has the same strength as a twelve-digit binary string because $4^6 = 2^{12}$. If we choose a random binary string of length between 1 and 10, we have 11 bits of entropy, since $\log_2(2^1 + 2^2 + \cdots 2^{10}) \approx 11$.

It can generally be assumed that a password uses some combination of upper and lowercase letters, digits, and some set of special characters. This leads to an alphabet size of around 90 (depending on the special characters allowed). It can also be assumed that passwords are generally between 4 and 20 characters long. Does this mean that every password has strength equal to

$$\log_2\left(90^4 + 90^5 + \cdots 90^{20}\right) \approx 1324?$$

Of course not! Attackers employ more clever strategies than simply trying every possible combination of digits. They may start with short passwords on smaller alphabets (say, just lowercase letters), then try common dictionary words, then try lowercase words followed by digits, and so on. Thus, a password of **1234** will be cracked much more quickly than one which uses the full space of characters.

While an attacker may not necessarily know the requirements of a password, it is good practice that any security system should be secure even if the attacker knows everything about the system except the password. This is concept is known as *Kerckhoff's Principle*.

Example 1.2 (Digits in Passwords). A system's password policy requires a 10 digit password consisting of lowercase letters and digits.

- A randomly chosen string from this space has strength $\log_2(36^{10}) \approx 51.7$.

- A random password known to have exactly three numbers has strength $\log_2\left(\binom{10}{3} \cdot 26^7 \cdot 10^3\right) \approx 49.7$.

- A password known to have seven letters followed by three digits at the end has strength $\log_2\left(26^7 \cdot 10^3\right) \approx 42.9$.

Note that adding a single bit of entropy makes the password *twice as hard* to crack. Thus, the first case is four times as secure as the second, and roughly 500 times more secure than the third.

1.2.6 The Binomial Theorem

If anything deserves to be called the fundamental theorem of combinatorics, it is the binomial theorem. It connects enumeration with algebra and analysis, and this connection will be further explored in section 1.5.

Theorem 1.5 (Binomial Theorem). *Let $x, y \in \mathbb{R}$, and $n \in \mathbb{N}$. Then*

$$(x+y)^n = \sum_{k=0}^{n} \binom{n}{k} x^k y^{n-k}.$$

Proof: Expand the left-hand side:
$$\underbrace{(x+y)(x+y)\cdots(x+y)}_{n \text{ terms}}.$$

By the distributive property, each term in the expansion arises from a choice of one summand from each factor. To get the term $x^k y^{n-k}$, we take an x from k factors and a y from the remaining ones. The number of ways to choose which k of the n factors to pull an x from is precisely $\binom{n}{k}$, completing the proof. $\square$

It will be useful to generalize this to non-integer values of n. First, we need to generalize the binomial coefficients: if $r \in \mathbb{R}$ and $k \in \mathbb{N}$, define $\binom{r}{k} = r^{\underline{k}}/k!$. Note that this definition reduces to the usual one if $r \in \mathbb{N}$.

Theorem 1.6 (Generalized Binomial Theorem). *Let $x, r \in \mathbb{R}$, with $|x| < 1$. Then*
$$(1+x)^r = \sum_{k \geq 0} \binom{r}{k} x^k.$$

Proof: Letting $f(x) = (1+x)^r$, we calculate the Taylor series for f centered at $x = 0$ (also known as the Maclaurin series), using methods from calculus. That is, taking derivatives of both sides and plugging in 0 shows that the two sides represent the same function when $|x| < 1$. $\square$

Why would a combinatorialist care about binomial coefficients with non-integer values? Surprisingly, these will turn up several times throughout the course of this chapter. For example, consider the expression $1/\sqrt{1-4x}$. Let's see what happens when we apply the theorem:
$$\frac{1}{\sqrt{1-4x}} = (1-4x)^{-1/2} = \sum_{k \geq 0} \binom{-1/2}{k} (-4)^k x^k.$$

Expanding the summand, we see that
$$\binom{-1/2}{k} 4^k x^k = \frac{(-1/2)(-3/2)\cdots(-(2k-1)/2)}{k!}(-4)^k x^k$$
$$= (-1)^k \frac{(1 \cdot 3 \cdot \cdots \cdot (2k-1))}{2^k k!}(-4)^k x^k$$
$$= \frac{1 \cdot 2 \cdot 3 \cdot \cdots \cdot (2k-1) \cdot 2k}{(2 \cdot 4 \cdot \cdots \cdot 2k) 2^k k!} 4^k x^k$$
$$= \frac{2k!}{2^k \cdot 2^k \cdot k! \cdot k!} 4^k x^k = \binom{2k}{k} x^k.$$

Thus, we have that
$$\frac{1}{\sqrt{1-4x}} = \sum_{k \geq 0} \binom{2k}{k} x^k.$$

The numbers $\binom{2k}{k}$ are called the *central binomial coefficients*, and though it may seem like a fluke that they arise in such a strange expression, we'll see many more examples of this in later sections.

The binomial theorem enables us to apply deep ideas from algebra and analysis to tackle enumerative problems. To see what we mean, we'll prove two propositions on finite sets in two contrasting ways: first, a direct, *combinatorial proof* using an elementary counting argument, followed by an indirect, *analytic proof* using the binomial theorem.

Proposition 1.1. Let $n \geq 0$. There are a total of 2^n subsets of $[n]$.

Combinatorial Proof: Identify each subset $S \subset [n]$ with the vector $\vec{v} = \langle v_1, v_2, \ldots v_n \rangle$ where $v_i = 1$ if $i \in S$ and $v_i = 0$ otherwise. Each binary vector of length n corresponds to a unique subset, and so the set of all binary vectors of length n is in bijection with the subsets of $[n]$. There are 2^n such binary vectors, and so this is the number of subsets. $\square$

Analytic Proof: Since $\binom{n}{k}$ counts the number of subsets of size k, the total number of sets is $\sum_k \binom{n}{k}$. By the binomial theorem, we have

$$\sum_{k=0}^{n} \binom{n}{k} = \sum_{k=0}^{n} \binom{n}{k} 1^k = (1+1)^n = 2^n.$$

$\square$

Proposition 1.2. For all integers $n \geq 1$, we have

$$n 2^{n-1} = \sum_{k=0}^{n} k \binom{n}{k}.$$

Combinatorial Proof: Claim that both sides count the same thing: the number of ways of forming a (possibly empty) committee from a group of n people, and electing one of those people as the committee president. The left-hand side counts the ways of first choosing a president, then choosing the committee members as a subset of the remaining $n - 1$ entries. The right-hand side counts this by the size of the committee, first forming a committee by choosing a subset of size k, then electing a president from within. $\square$

Analytic Proof: From the binomial theorem, we have $(x+1)^n = \sum_{k=0}^{n} \binom{n}{k} x^k$. Taking the derivative of both sides with respect to x, then substituting $x = 1$ completes the proof. $\square$

These examples give some insight into the power and general way of thinking about analytic methods in combinatorics. In each of the problems above, the combinatorial proofs rely on clever but disparate ideas, and there are no obvious generalizations. On the other hand, the analytic proofs follow a simple pattern: encode the numbers of interest (in this case, the numbers $\binom{n}{k}$) into an analytic object with some indeterminate x, then use analytic operations—substitute in values, take derivatives—and interpret the results on the numbers of interest.

In fact, analytic proofs often prove much stronger theorems than their combinatorial counterparts. In the last theorem above, note that the analytic proof actually holds for all real values of n, while the combinatorial proof is only valid when n is a positive integer.

1.3 DISCRETE PROBABILITY AND HASH FUNCTIONS

Enumeration forms the foundation of any introductory probability sequence, and, conversely, ideas from probability can be productively applied to purely combinatorial problems. Randomness is central to cybersecurity (and all real-world processes), and by incorporating concepts from probability, we open ourselves to a range of new combinatorial applications.

In this section we'll also cover hash functions, which are both a fundamental concept underlying much of cybersecurity and an excellent source of combinatorics and probability problems. We'll touch on birthday attacks, a cryptanalytic attack that rests on some simple combinatorial ideas, and the classic coupon collector's problem.

1.3.1 Probability and Expectation

This chapter isn't intended to provide an introduction to probability, but we'll quickly define a few concepts to standardize notation.

Denote by Ω a finite set of outcomes (e.g., the result of a rolled die), and, for an event $A \subset \Omega$, let $\mathbb{P}[A]$ denote the probability of this event occurring. If all events in Ω are equally likely, then for any $A \in \Omega$, we have $\mathbb{P}[A] = |A|/|\Omega|$. Two events $A, B \subset \Omega$ are *independent* if $\mathbb{P}[A \cap B] = \mathbb{P}[A] \cdot \mathbb{P}[B]$. A random variable X is a function $X : \Omega \to \mathbb{N}$. The *expectation* (or average) of a random variable is defined as

$$\mathbb{E}[X] = \sum_{A \in \Omega} X(A)\mathbb{P}[A] = \sum_{x \in \mathbb{N}} x\mathbb{P}[X = x].$$

To see that these summations are equivalent, note that they both include all possible values of the random variable and all events in the sample space: the first is indexed by the sample space, while the second is indexed by the output of the variable.

For example, let $\Omega = \{(d_1, d_2) : d_1, d_2 \in [6]\}$ be the set of outcomes resulting from rolling two dice, and let $X : \Omega \to \mathbb{N}$ be the random variable equal to the sum of the two dice. Then, since Ω has 36 equally likely outcomes, we have that

$$\begin{aligned}\mathbb{E}[X] &= 1 \cdot \mathbb{P}[X = 1] + 2 \cdot \mathbb{P}[X = 2] + 3 \cdot \mathbb{P}[X = 3] + \cdots \\ &\quad + 12 \cdot \mathbb{P}[X = 12] + 13 \cdot \mathbb{P}[X = 13] + \cdots \\ &= 1 \cdot 0 + 2 \cdot \frac{1}{36} + 3 \cdot \frac{2}{36} + \cdots + 12 \cdot \frac{1}{36} + 13 \cdot 0 + \cdots \\ &= \frac{252}{36} = 7.\end{aligned}$$

Let's return to brute force attacks. We know that if a password has strength k, then an attacker must attempt 2^k different passwords to *guarantee* that they find the right one. But how many must they try on average?

Proposition 1.3 (Expected Time for Brute Force Attacks). Let p be a randomly chosen password from a space of size n (i.e., a password with strength $\log_2(n)$). The expected number of passwords that need to be attempted before the attack succeeds is $(n-1)/2$.

Proof: Let $P = \{p_1, p_2, \ldots, p_n\}$ be the set of all possible passwords, in random order, and let X be the random variable indicating the number of passwords an attacker must try before finding the correct one. We know that $p = p_i$ for some unknown i. Since we've put the passwords in random order, i is equally likely to be any number in $[n]$. An attacker, trying passwords from P in order, must make exactly i attempts before succeeding, so $X = i$.

Therefore, we have

$$\mathbb{E}[X] = 1 \cdot \mathbb{P}[X=1] + 2 \cdot \mathbb{P}[X=2] + \cdots + n \cdot \mathbb{P}[X=n]$$
$$= 1 \cdot \frac{1}{n} + 2 \cdot \frac{1}{n} + \cdots + n \cdot \frac{1}{n}$$
$$= \frac{1}{n}(1 + 2 + \cdots + n) = \frac{n(n-1)}{2n} = \frac{n-1}{2}.$$

$\square$

Calculating probabilities directly often boils down to simply enumerating sets. Expectation, however, has an important property that greatly simplifies many problems.

Theorem 1.7 (Linearity of Expectation). *Let X, Y be two random variables from the same sample space, and let $c \in \mathbb{R}$. Then*

$$\mathbb{E}[X+Y] = \mathbb{E}[X] + \mathbb{E}[Y] \quad \text{and} \quad \mathbb{E}[c \cdot X] = c \cdot \mathbb{E}[X].$$

Proof: This proof follows directly from the definition. For the first claim, we have:

$$\mathbb{E}[X+Y] = \sum_{A \in \Omega} (X(A) + Y(A)) \mathbb{P}[A]$$
$$= \sum_{A \in \Omega} X(A) \mathbb{P}[A] + \sum_{A \in \Omega} Y(A) \mathbb{P}[A]$$
$$= \mathbb{E}[X] + \mathbb{E}[Y].$$

The proof of the second claim follows analogously. $\square$

Linearity of expectation is a powerful tool in any probabilist's or combinatorialist's toolbox, greatly simplifying many calculations.

We'll start with a classic example. A *fixed point* of a permutation $\pi : [n] \to [n]$ is a value $i \in [n]$ such that $\pi(i) = i$. How many fixed points do we expect in a random permutation of length n? Remarkably, we find that the answer doesn't depend on n.

Theorem 1.8 (Fixed Point Expectation). *Let X be the number of fixed points in a randomly chosen permutation of length $n \in \mathbb{P}$. Then $\mathbb{E}[X] = 1$.*

Proof: Let $\pi \in \mathfrak{S}_n$, and define a new set of random variables $\{X_i\}_{i \in [n]}$, defined as

$$X_i(\pi) = \begin{cases} 1 & i \text{ is a fixed point of } \pi \\ 0 & i \text{ is not a fixed point of } \pi \end{cases}.$$

Note that since X is the total number of fixed points, we have that $X = \sum_{i \in [n]} X_i$. These variables are not independent—for example, if the first $(n-1)$ entries are fixed points, then the final one must be as well—but we can still apply linearity of expectation:

$$\mathbb{E}[X] = \sum_{i \in [n]} \mathbb{E}[X_i].$$

We can calculate $\mathbb{E}[X_i]$ directly:

$$\mathbb{E}[X_i] = 0 \cdot \mathbb{P}[i \text{ is not a fixed point}] + 1 \cdot \mathbb{P}[i \text{ is a fixed point}].$$

Note that there are $(n-1)!$ permutations in which i is fixed since we can permute the remaining entries in any way. Thus the probability that i is fixed is $(n-1)!/n! = 1/n$.

Putting this all together,

$$\mathbb{E}[X] = \sum_{i \in [n]} \mathbb{E}[X_i] = \sum_{i \in [n]} \frac{1}{n} = 1.$$

□

This style of proof—decomposing a random variable into a sum of indicator random variables and applying linearity of expectation—is a common one and is a versatile method for addressing complex expectation problems. Calculating the above expectation directly would require enumerating permutations that have exactly k fixed points for each $k \in [n]$. This is a difficult problem and one for which we have not yet developed the tools (see Section 1.4). Linearity allows us to calculate the overall expectation without needing to worry about these intermediate enumerations.

Permutations without fixed points are called *derangements* and are a classic object of study. They are relevant to cybersecurity as well: consider a substitution cipher, in which letters from an alphabet are permuted. No matter the size of the alphabet, under a random mapping, we would expect exactly one letter to be mapped to itself.

Let's try another example, this time on password reuse.

Example 1.3. Suppose that a company randomly generates computer passwords for each employee from some set of possible passwords. How many employees do we expect to have the same password as someone else? That is, letting n be the number of possible passwords, k the number of employees, and X the number of employees who don't share a password with someone else, what is $\mathbb{E}[X]$?

To calculate this directly we would need to find $\mathbb{P}[X = i]$ for all i, which quickly gets out of hand. As you might expect, the linearity of expectation greatly simplifies the problem.

Let person i's password be p_i, and let X_i be the variable which takes the value 1 if person i doesn't share a password with anyone else, and 0 otherwise. Then $X = \sum_i X_i$, and the expectation of X_i is the probability that everyone else is assigned a password other than p_i, which is equal to $\left(\frac{n-1}{n}\right)^{k-1}$. Thus we have

$$\mathbb{E}[X] = \sum_{i \in [k]} \mathbb{E}[X_i] = \sum_{i \in [k]} \left(\frac{n-1}{n}\right)^{k-1} = \frac{kn}{n-1}\left(1 - \frac{1}{n}\right)^k.$$

Note that, since $(1 - 1/n)^n \to e^{-1}$ as n approaches infinity,[3] if $n = k$ then as n gets large, this value approaches n/e. It follows then that, if we give out n passwords to n people, the proportion of people with unique passwords approaches $1/e \approx 36.8\%$.

1.3.2 Hash Functions

A *hash function* is any function which maps from some domain to a fixed range of values. That is, a function $f : M \mapsto [n]$, where M is typically understood to be some large space (e.g., binary strings of any length) and n is some positive integer (often a power of 2).

Hash functions are a critical component of many areas of computer science research, and form the basis for several commonly used data structures including hash tables and bloom filters. Hash functions come in all shapes and sizes, and various properties may be useful for different purposes. In this chapter we'll consider *cryptographic hash functions*.

Definition 1.6 (Cryptographic Hash Function). A *cryptographic hash function* (or, for simplicity, a *hash*) of *size* n is a function $f : M \mapsto [n]$ for some $n \in \mathbb{P}$ with the following properties:

- For a given $m \in M$, it is easy to compute $f(m)$.

- For a given $k \in [n]$, it is difficult to compute $f^{-1}(k)$.

- For $m_1 \neq m_2 \in M$, there should be no discernible relationship between $f(m_1)$ and $f(m_2)$.

- The function f should appear to an outside observer to be random: for any input m, $f(m)$ appears equally likely to take on any value from $[n]$ with equal probability.

The definition above is informal. A hash f is *not* a random variable: on a given input, f will always return the same output. However, we want f to *look* random,

---
[3]This limit is proven in an undergraduate calculus class using logarithms.

in that there should be no way of predicting the output of a given input. Said another way, to identify an input m that produces a hash value of k, our only option should be a *brute-force* approach: try every possible value of m until we get a match. Designing and implementing hash functions is a major topic of research that we won't cover in this chapter.

Example 1.4 (Probability of Hash Collision). For a given hash function f of size n, a *collision* is an instance of two inputs m_1 and m_2 such that $m_1 \neq m_2$ and $f(m_1) = f(m_2)$.

Given a pair m_1 and m_2, what is the probability that they collide? Suppose $f(m_1) = k$. Since we assume that each output is equally likely, it follows that the probability that $f(m_2) = k$ is equal to $1/n$.

Recall that in Example 1.3, we calculated the expected number of unique passwords when randomly assigning passwords to people. This can easily be translated to calculating expected number of unique values when a set of inputs is hashed.

Real-world hash functions have size on the order of 2^{256}, which means that a given pair has a 2^{-256} chance of collision. Note, however, that since the domain is (in general) much larger than the range, collisions are guaranteed to exist by the *pigeonhole principle*.[4] For a good hash function, finding these collisions should be difficult.

1.3.3 Collisions and Cyber Attacks

Hashes are used in a variety of ways to secure information, and finding and avoiding hash collisions is an important aspect of cyber attacks and defenses. We won't go over all the details of how hashes are used, but we'll present a few simplified examples to give the flavor of the area.

Example 1.5 (Passwords). A well-designed system will never store your actual password but rather a *hash* of your password. That is, if your password is p, then the system actually stores $f(p)$ alongside your username, for some hash f. When attempting to log in with password m, it computes $f(m)$ and compares to the stored hash. If these match, you're allowed to log in; if not, you may be prompted to try again. Thus, if an attacker wants to break into your account, they don't necessarily need *your* password—any password which has the same hash as yours will do.

Example 1.6 (File Integrity). When transferring files or messages, it is often useful to verify that the contents match what is expected. Hashes are used to provide some level of assurance that the contents haven't been altered (maliciously or accidentally) during transfer. The sender computes the hash of the message and transmits this alongside the larger message. The receiver can then also calculate the hash and compare this to the one sent. If the message has been altered, then with high probability, the hashes will not match.

Note, a *checksum* is a simple mechanism that can be used to identify transmission errors. An additional value is transmitted alongside the message; this value

---

[4] If n pigeons nest in k holes, with $n > k$, then two pigeons must reside in the same hole.

can be as simple as the sum of the digits transmitted, modulo some fixed number. The recipient can quickly verify this value (by *checking* the *sum*), and if it doesn't match, something must have gone wrong. Note, however, that this is *not* a cryptographic hash, and an adversary may be able to defeat this mechanism. Checksums are valuable for ensuring data *integrity*, but not data *authenticity*.

Hashes are used as building blocks for more complex security mechanisms, and the *collision resistance*—the inability to easily find hash collisions—is a major factor in the level of security provided. We showed that the probability of a collision is inversely proportional to the size of the hash, and so larger hashes will be more collision resistant. Larger hashes are also more resource intensive to compute, so applications must balance computational efficiency with security.

1.3.4 Birthday Attacks

Let f be a hash function of size n. Suppose we want to find two messages m_1 and m_2 which collide. We've shown that two different messages collide with probability $1/n$, which means that, at random, we would need to generate n different pairs of messages before we would expect to see a collision. The *birthday attack* allows one to find two messages m_1 and m_2 in roughly $\sqrt{n}$ time.

This attack rests on the following simple idea: if one generates a number of messages, there is a much larger number of *pairs* of these messages, any of which may lead to a collision. The name comes from the classical "Birthday Paradox," which states that if there are 23 people in a room, there is a greater than 50% chance that two of them share a birthday.

Let's see how this works: a given pair collides with probability $1/n$. If I generate k messages, there are $\binom{k}{2}$ pairs of messages, and so we would expect to see $\frac{1}{n}\binom{k}{2}$ collisions. A quick calculation shows that when $k \approx \sqrt{2n}$, this expectation is close to 1. Note: this *doesn't* mean that we would expect to see a collision as often as not, but it shows that the average number of collisions is 1.

To be more precise, consider the probability that there are no hash collisions amongst k messages $\{m_1, m_2, \ldots m_k\}$. This means that the values $\{f(m_1), f(m_2), \ldots f(m_k)\}$ are all distinct. There are $n^{\underline{k}}$ different distinct sequences amongst the n^k sequences, so the probability that there is *at least one* hash collision is equal to

$$\mathbb{P}[f(m_i) = f(m_j) \text{ for some } i,j] = 1 - \frac{n(n-1)(n-2)\cdots(n-k+1)}{n^k}.$$

Some messy analytic approximation which we won't cover here can be used to show that we have a 50% probability of a hash collision when $k \approx \sqrt{2n\ln(2)} \approx 1.17\sqrt{n}$. [5]

So how is this useful? How can we turn this into an *attack*? Suppose you want to trick someone into signing a fraudulent contract. If you can draft a plausible

---

[5] Using $n = 365$ gives $k \approx 22.5$, solving the classical birthday problem.

contract that has the same hash as the fraudulent one, you can have them sign it. Then, this valid signature can then be applied to the fraudulent contract.

Say I have a real message m_1 and a false message m_2, and I'd like them to have the same hash value. I could introduce some minor alterations to m_2 (add some noise, flip some bits, etc.) and check the hashes. If I keep trying altered versions, I would expect to eventually find one which matches the hash of m_1 after roughly n tries. *However*, if I can simultaneously alter m_1 and m_2, I need only produce roughly $\sqrt{n}$ different messages for each before I expect to find a hash collision.

For example, suppose I'm using a hash function of size 2^{128}, and I have two emails m_1 and m_2: one real and one false. If I can find 64 tweaks to each message (e.g., add a space after a sentence, add a line of white space, change the punctuation, alter a word), then I can build a set of 2^{64} almost-real emails and 2^{64} false emails. Since there are $2^{64} \cdot 2^{64} = 2^{128}$ pairs of emails, there is a good chance of finding a collision that can be used to defeat a verification. An attack using this idea requires roughly $1.17 \cdot 2^{64} \approx 2.2 \times 10^{19}$ operations, while a naive collision attack would use $2^{128} \approx 3.4 \times 10^{38}$.

Note that this heuristic can be applied to a variety of problems and can be rephrased as follows: if you're trying to find a connection between two items, it's often more efficient to explore from both ends simultaneously than to search from a single point.

1.3.5 The Coupon Collector's Problem

The Birthday Problem asks, essentially, how many random samples must I draw from a finite set until I see the first repeat. An inverse problem is: how many random samples must I draw until *every* draw will be a repeat? Or, equivalently, how many draws until I've seen every possible value? This problem gets its name from a classic game, in which random coupons are distributed in sealed boxes, and the goal is to purchase boxes until you've collected them all.

We model the problem as follows: let $(a_1, a_2, a_3, \ldots)$ be a sequence of values where each a_i is drawn independently and uniformly from $[n]$. Let X be the smallest value such that $(a_1, a_2, \ldots, a_X)$ contains all the values from 1 through n. What is the expectation of X?

It's easy to see that $X \geq n$, since we can't possibly see all the values before the nth element, and $X = n$ if and only if every value is distinct. There is no upper bound on X, however, since a given value may not occur until very late in the sequence. For example, if $n = 4$, the sequence $(3, 3, 1, 4, 1, 3, 4, 2)$ has $X = 8$ (see Figure 1.2). We'll assume that the sequence ends when the final element is seen since we won't care what happens after that.

How long do we expect such a sequence to be? Again, the linearity of expectation will be the key to the solution. First, we need a lemma.

Lemma 1.1 (Expected Waiting Time). *Let X be a random variable equal to 1 with probability $p \in [0, 1]$ and 0 otherwise. Consider a sequence of $(x_1, x_2, x_3, \ldots)$ of independent draws from x. If $k \in \mathbb{P}$ is the smallest value such that $x_k = 1$, then $\mathbb{E}[k] = 1/p$.*

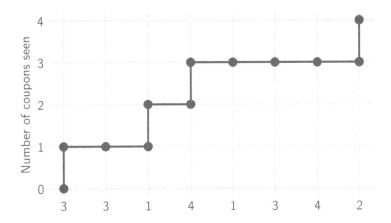

FIGURE 1.2 A coupon sequence of length 8 with 4 coupons.

Proof: We calculate the expectation from the definition. The probability of seeing a 1 on the first draw is p. If the first 1 is in the kth position, each item before this is a zero—this happens with probability $(1-p)^{k-1}p$. Putting this together, we have

$$\mathbb{E}[k] = 1 \cdot p + 2 \cdot (1-p)p + 3 \cdot (1-p)^2 p + \ldots = p \sum_{i \in \mathbb{N}} (i+1)(1-p)^i.$$

By manipulating binomial coefficients, we can show that $(i+1) = (-1)^i \binom{-2}{i}$. Using this fact, we apply the Binomial Theorem:

$$\mathbb{E}[k] = p \sum_{i \in \mathbb{N}} \binom{-2}{i} (1-p)^i = p(1-(1-p))^{-2} = p/p^2 = 1/p.$$

Thus, the expected position of the first 1 is the reciprocal of the probability p. □

Theorem 1.9 (Coupon Collecting Expectation). *Let $(a_1, a_2, \ldots)$ be a sequence drawn uniformly at random from $[n]$. Let t be the smallest integer such that the sequence $(a_1, \ldots a_t)$ contains all the values from $[n]$.*

Letting $h_n = \sum_{i \in [n]} 1/i$ denote the nth harmonic number, we have that

$$\mathbb{E}[t] = n \cdot h_n.$$

As n grows large, $h_n / \ln n$ approaches 1, so the expectation can be approximated by $n \ln n$.

Proof: We decompose t into smaller variables. For each $i \in [n]$, let t_i be the number of draws between seeing the first occurrences of the $i-1$st unique element and the ith (with $t_1 = 1$). In Figure 1.2, we have that $t_1 = 1$, $t_2 = 2$, $t_3 = 1$, and $t_4 = 4$.

It follows then that $\sum_{i \in [n]} t_i = t$, and so

$$\mathbb{E}[t] = \sum_{i \in [n]} \mathbb{E}[t_i].$$

We calculate $\mathbb{E}[t_i]$ directly. Once we've seen $i-1$ entries, the probability that the next entry is *new* is $(n-i+1)/n$. Thus, by the lemma, the expected waiting time is the reciprocal: $\mathbb{E}[t_i] = \frac{n}{n-i+1}$.

Putting it all together:

$$\mathbb{E}[t] = \sum_{i \in [n]} \frac{n}{n-i+1} = n\left(1 + \frac{1}{2} + \frac{1}{3} + \ldots + \frac{1}{n}\right).$$

And the proof is complete. □

For example, if there are 100 coupons, $\ln(100) \approx 4.6$, so we would expect to need ≈ 460 random draws before we've seen them all.

Problems of this type appear frequently in computer science. Consider, for example, the problem in reverse: suppose we are observing random numbers emitted by some unknown process, and we are trying to estimate the total number of unique outputs possible. Suppose that after 500 iterations, we have seen 400 unique random numbers. How many more possible outputs might we expect?

Consider the 501st draw. We estimate that it has a $\approx 80\% = 400/500$ chance of being unique. The actual probability is $\frac{n-400}{n}$, where n is the (unknown) total number of possible outputs. Setting this equal to .8 and solving for n gives us an estimate that $n = 2000$.

Remarkably, a variant of this idea was used by the Allies in World War II to estimate the number of German tanks based on the set of unique serial numbers which had been observed. [6] Though this application, known as the *German Tank Problem*, is not *technically* in the realm of cybersecurity, we hope the reader will agree that it is certainly in the same spirit.

1.4 INCLUSION-EXCLUSION AND RANDOM MAPPINGS

In the discussion of hash functions above, we mostly ignored the set of inputs. In this section we consider functions from one finite set to another or from a set to itself. Functions on finite sets are a foundational object of study, and understanding their properties will be helpful in various cybersecurity applications. We'll develop new enumerative methods to answer questions about these functions and extend our toolbox, and we'll present several applications to cryptanalysis.

1.4.1 Principle of Inclusion–Exclusion

Recall that a fixed point of a permutation is a value that maps to itself. A *derangement* is a permutation with no fixed points. In Theorem 1.8, we calculated that the expected number of fixed points in a random permutation is one. How many derangements are there of length n?

The counting techniques we covered in Section 1.2 aren't flexible enough to answer this question directly. This problem motivates a powerful and general counting

---

[6] Note that in this example, the sampling was done *without replacement* since each serial number was used only once. It is more complicated to solve this case, but the basic idea is similar.

method: the *principle of inclusion-exclusion*. The general problem is this: given a finite set S and a set of subsets $A_1, A_2, \ldots A_n \subseteq S$, how many elements are in the union of these subsets?

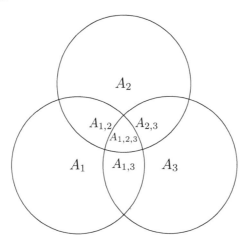

FIGURE 1.3 A Venn Diagram with three sets, with $\bigcap_{i \in I} A_i$ denoted by A_I.

For an intuitive example, consider the case of $n = 3$, and see Figure 1.3. To count the number of elements in $\{A_1 \cup A_2 \cup A_3\}$, we can start by adding the sizes of each of A_1, A_2, and A_3:
$$|A_1| + |A_2| + |A_3|.$$

Note that elements which are in more than one of the sets A_i are counted more than once. We correct this by subtracting each of the intersections:
$$|A_1| + |A_2| + |A_3| - |A_1 \cap A_2| - |A_1 \cap A_3| - |A_2 \cap A_3|.$$

Finally, note that elements in $|A_1 \cap A_2 \cap A_3|$ are counted three times, then subtracted three times. We correct this by adding these back in once more, leading to
$$|\cup_{i=1}^3 A_i| = |A_1| + |A_2| + |A_3|$$
$$- |A_1 \cap A_2| - |A_1 \cap A_3| - |A_2 \cap A_3|$$
$$+ |A_1 \cap A_2 \cap A_3|.$$

This idea generalizes to the following theorem.

Theorem 1.10 (Principle of Inclusion-Exclusion). *Let $\{A_i\}_{i=1}^n$ be a set of finite sets. The size of the union of all the sets is given by:*
$$\left| \bigcup_{1 \leq i \leq n} A_i \right| = \sum_{1 \leq k \leq n} (-1)^{k+1} \left(\sum_{1 \leq i_1 < \cdots i_k \leq n} |A_{i_1} \cap \cdots \cap A_{i_k}| \right).$$

Proof: We prove that any element of $\cup_{1 \leq i \leq n} A_i$ is counted precisely once on the right-hand side. Let x be any such element, and let $I \subseteq [n]$ be such that $x \in A_i$ if and only if $i \in I$. We see that x is counted exactly I times by the first term, $-\binom{|I|}{2}$ times by the second term, $\binom{|I|}{3}$ times by the third, and so on.

Thus, we need only show that

$$\binom{n}{1} - \binom{n}{2} + \binom{n}{3} - \cdots = 1.$$

If we multiply both sides by -1 and add 1 to both sides, this is equivalent to the statement that

$$\sum_{k \geq 0} (-1)^k \binom{n}{k} = 0,$$

which follows immediately from the binomial theorem applied to $(1-1)^n$. $\square$

Let's apply this theorem to permutations.

Proposition 1.4 (Counting Derangements). The number of derangements of length n is equal to

$$n! - \frac{n!}{2} + \frac{n!}{6} - \frac{n!}{24} + \cdots + (-1)^k \frac{n!}{k!} = n! \sum_{0 \leq k \leq n} (-1)^k/k!.$$

It follows that for large n, the probability that a random permutation is a derangement is $1/e$.

Proof: We count permutations that fix at least one element, then subtract these from the number of all permutations. Let $n \in \mathbb{P}$, and for each $i \in [n]$ let A_i be the set of permutations that fix *at least i*, though they may fix more elements as well. For ease of notation, for $I \subseteq [n]$, denote by A_I the intersection $\cap_{i \in I} A_i$. The set A_I is the set of permutations that fix at least the elements of I.

We see that $|A_I| = (n - |I|)!$, since every permutation of the non-fixed entries leads to a unique permutation. Noting that $\cup_{i \in [n]} A_i$ is the set of all permutations which fix at least one element of $[n]$, we apply the theorem:

$$\left| \bigcup_{i \in [n]} A_i \right| = \sum_{k \in [n]} (-1)^{k+1} \sum_{\substack{I \subseteq [n] \\ |I| = k}} |A_I|$$

$$= \left| \bigcup_{i \in [n]} A_i \right| \sum_{k \in [n]} (-1)^{k+1} \binom{n}{k} (n-k)!$$

$$= n! \sum_{1 \leq k \leq n} \frac{(-1)^{k+1}}{k!}.$$

The number of permutations *without* a fixed point is $n!$ minus this quantity, which is equal to $n! \sum_{0 \leq k \leq n} (-1)^k / k!$. Note that this sum approaches e^{-1} as $n \to \infty$, and so the proportion of permutations that have no fixed points approaches $1/e$. $\square$

1.4.2 Random Functions and Stirling Numbers

Random (or pseudorandom) functions arise in all areas of cybersecurity research. Analyzing the properties of these functions helps to both understand the security and to identify vulnerabilities within encryption systems. The principle of inclusion-exclusion, combined with the basic enumerations, allows us to enumerate various classes of finite functions.

Let $r, s \in \mathbb{P}$, and let R, S be r- and s-element sets. Recall that a function is *injective* (or *one-to-one*) if each element of the domain maps to a unique element of the range, *surjective* (or *onto*) if every element of the range has a preimage, and *bijective* if it is both injective and surjective.

Consider the set of all functions $f : R \to S$ under various conditions. Assume, for simplicity, that $R = [r]$ and $S = [s]$. We'll enumerate each case:

- **No Restrictions:** We can specify a function by assigning $(f(1), f(2), \ldots f(r))$. Each of these can take on a value from $[s]$. Thus, these functions correspond to sequences and so are enumerated by s^r.

- **Injective** f: If f is injective, we can still specify it by assigning the values in sequence—as above—except that the same element cannot be repeated. Thus, there are $s(s-1)(s-2)\cdots(s-k+1) = s^{\underline{r}}$ injective functions.

- **Bijective** f: A function f can only be bijective if the domain and range have the same size, that is, $r = s$. In this case, any injective function is necessarily surjective, and hence bijective. So, there are $r! = s!$ bijective functions when $r = s$, and zero otherwise.

- **Surjective** f: This one requires a bit more work:

Theorem 1.11 (Enumerating Surjections). *The number of surjections from an r-element set to an s-element set is*

$$\sum_{0 \leq k \leq s} (-1)^k \binom{s}{k} (s-k)^r.$$

Proof: We'll use inclusion-exclusion to enumerate those functions which are *not* surjections. Let $f : R \to S$ be a function, and say that an element $y \in S$ is *missed* if there is no $x \in R$ such that $f(x) = y$. For each $i \in S$, let A_i be the set of functions that miss i, and for $I \subset S$, A_I the functions that miss every element of I. Note, a function in A_I may miss other elements as well.

The size of A_I is $(s - |I|)^r$ since we need only specify to which of the $S \setminus I$ elements each $x \in R$ is sent. Also, we know that there are $\binom{s}{k}$ subsets $I \subset S$ of size k.

By Theorem 1.10 we have

$$\left| \bigcup_{i \in S} A_i \right| = \sum_{1 \leq k \leq s} (-1)^{k+1} \sum_{\substack{I \subseteq S \\ |I|=k}} |A_I|$$

$$= \sum_{1 \leq k \leq s} (-1)^{k+1} \binom{s}{k} (s-k)^r.$$

Subtracting this from the number of all functions s^r gives the number of surjections. □

If $f : R \to [s]$ is a surjection, then each element of S has at least one preimage. Thus, R can be written as a disjoint union as follows:

$$R = f^{-1}(1) \bigsqcup f^{-1}(2) \bigsqcup \cdots \bigsqcup f^{-1}(s).$$

A decomposition of a finite set into the union of disjoint subsets is called a *set partition*. Since each of these subsets above is associated with a specific element of $[s]$, we say that this is a *labeled set partition* of the set R.

Set partitions arise often enough in combinatorics that the numbers which enumerate them are called the *Stirling Numbers of the Second Kind* and are given their own symbol: $\{{r \atop s}\}$ is the number of partitions of an r-element set into s blocks. We've just seen that the number of surjections from an r-element set to an s-element set is equal to $s!\{{r \atop s}\}$ since there are $s!$ ways of labeling the s blocks. We capture this in the following theorem.

Theorem 1.12 (Stirling Numbers of the Second Kind). *Let $r, s \in \mathbb{P}$, and let $\{{r \atop s}\}$ denote the Stirling number of the second kind, equal to the number of set partitions of an r-element set into s-parts. Then*

$$\left\{ {r \atop s} \right\} = \frac{1}{s!} \sum_{k=0}^{s} (-1)^k \binom{s}{k} (s-k)^r.$$

Also, the number of surjections from an r-element set to an s-element set can be written as $s!\{{r \atop s}\}$.

Note that *any* function between finite sets is surjective onto its image. This simple fact can be used to prove a complicated identity.

Corollary 1.1. *Let $r, s \in \mathbb{P}$. Then*

$$s^r = \sum_{m=0}^{s} m! \binom{s}{m} \left\{ {r \atop m} \right\} = \sum_{m=0}^{s} \binom{s}{m} \sum_{k=0}^{m} (-1)^k \binom{m}{k} (m-k)^r.$$

Proof: Claim that each term counts the set of all functions from an r-element set R to an s-element set S. The left-hand term is clear. The middle term counts these functions by the size m of their image, first choosing the image in $\binom{s}{m}$ ways and then choosing a surjection onto this image in $m!\{{r \atop m}\}$ ways. The right-hand term is the same, with the Stirling number expanded. □

Good luck finding a non-combinatorial proof of this identity.

1.4.3 Cycles in Permutations

Denote by $\mathfrak{S}_n$ the set of all n-permutations (considered as bijections from $[n]$ to itself), and by $\mathfrak{S}$ the set of all permutations of any length. For a function $\pi \in \mathfrak{S}_n$, we can represent π as the sequence $\pi(1), \pi(2), \ldots, \pi(n)$. This representation is known as the *one-line notation* for the permutation. For example:

$$\mathfrak{S}_1 = \{1\}$$
$$\mathfrak{S}_2 = \{12, 21\}$$
$$\mathfrak{S}_3 = \{123, 132, 213, 231, 312, 321\}$$

There is another common representation that uncovers a different sort of structure. Consider what happens when the function is applied repeatedly to an element. That is, let π be a permutation consider the sequence

$$(1, \pi(1), \pi(\pi(1)), \pi(\pi(\pi(1))), \cdots.$$

Since there are only n possible elements in the images, this sequence must eventually repeat itself. Once it does, it will continue to repeat. We can run the following experiment: start with 1, apply the permutation until we see a repeat, and record this sequence. Then, pick an element we haven't seen yet, and start applying the function until we repeat. Then pick a new element, and continue until we've seen everything. This shows that every element lies in a *cycle*, and that each permutation can be expressed as a collection of disjoint cycles.

For example, let $\pi = 43572681$. By repeatedly applying π, we find that it decomposes into 3 cycles as shown in Figure 1.4.

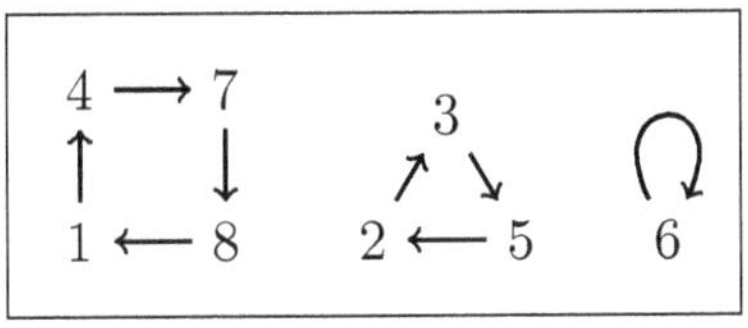

FIGURE 1.4 A visualization of the permutation $\pi = 43572681 = (523)(6)(8147)$.

This leads to an alternative representation for permutations, the so-called *cycle notation*. By writing out the cycles, we can uniquely specify the permutation. In this case, we can write $\pi = (523)(6)(8147)$—writing each cycle (in order) inside parentheses. Note, there are multiple equivalent ways of writing this down: the order we write the cycles doesn't matter, and the elements within each cycle can be rotated cyclically.

Each permutation has a unique *canonical cycle representation*, which starts each cycle with its largest entry and orders the cycles in increasing order of its first entry. The representation above is canonical. The canonical representation is more than just convenient: it can actually be used to prove some useful facts, including the following proposition.

Proposition 1.5. Let π be a random n-permutation, and let i and j be any elements of $[n]$. The probability that i and j lie in the same cycle is $1/2$.

Proof: Assume without loss of generality that $i = 1$ and $j = n$, and write the permutation in canonical order. We know (because of the properties of the canonical order) that everything written to the right of n lies in the same cycle as n, so 1 is in the same cycle as n if and only if 1 lies to the right of n. Since every ordering leads to a unique permutation, we see that this occurs exactly half of the time. □

A cycle of length 1 is a fixed point, and so we know that the number of cycles of length 1 in a random permutation is expected to be one. An *involution* is a function that is its own inverse. That is, a function $f : R \to S$ such that $f(f(x)) = x$ for all $x \in R$. An involution is a permutation (since it must be both surjective and injective). It follows from the cycle representation that involutions are precisely those permutations that have no cycles of length greater than two.

To count involutions, we split them into cases based on the number of fixed points. If a permutation of length $2k$ has no fixed points, the remaining entries must be grouped in pairs. We have $2k - 1$ options for the first image, $2k - 3$ options for the next, and so on, leading to $(2k-1)(2k-3)\cdots 3 \cdot 1$. This expression is called the *double factorial* of $2k - 1$, and is denoted $(2k-1)!!$. Summing over all possible numbers of fixed points proves the following proposition.

Proposition 1.6 (Counting Involutions). The number of involutions on a set of size n is

$$\sum_{k=0}^{\lfloor n/2 \rfloor} \binom{n}{2k}(2k-1)!!.$$

Preimage attacks are a class of cyber attacks that rely on finding a preimage for a value of a function. That is, for a function f and a value y, we want to find a value x such that $f(x) = y$. If we know that the function is a permutation, we can solve this easily. Simply start with the value y and iterate, and consider the sequence:

$$(y, f(y), f(f(y)), \ldots).$$

Since I know that f is a permutation, all of these values are in the same cycle as y, which means that I must eventually come back around to y. The value just before y is my preimage.

1.4.4 Random Mapping Statistics and Applications

Let S be a finite set and $f : S \to S$. Consider the *iteration* of f on any starting point. That is, the sequence $\{a_i\}_{i \geq 0}$ where $a_0 \in S$ and $a_{i+1} = f(a_i)$. Since we have a finite domain and range, this sequence must eventually repeat. For example, letting $S = \{0, 1, \ldots, 34\}$ and $f(x) = x^2 + 1 \pmod{35}$, Figure 1.5 shows the action of this function. Starting at $a_0 = 0$, we have the sequence:

$$(0, 1, 2, 5, 26, 12, 5, 26, 12, 5, 26, \ldots).$$

This last part of the sequence that repeats—$(5, 26, 12)$—is called a *cycle* of length three. This graph is called an *iterated function graph*, and it follows that any such graph must be composed of paths leading to cycles.

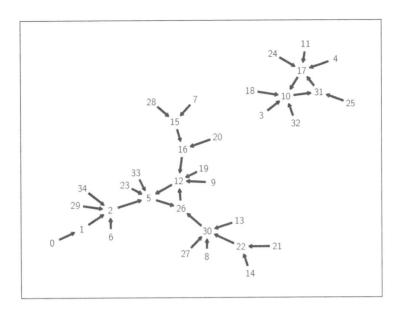

FIGURE 1.5 A graph of the mapping f given by $f(x) = (x^2 + 1) \pmod{35}$.

If we have a function f (say, a hash function), this principle can be used to find values x, y such that $f(x) = f(y)$ (a hash collision). Simply pick a random value and iterate, and hope that you run into a non-trivial cycle. (In the example above, we found that $f(2) = f(12) = 5$.)

A large number of attacks and algorithms are based on finding cycles and, therefore, the structure of random mappings. For example, how many different cycles do we expect? How long must we go until we enter a cycle? How big is the average cycle? Each of these have implications to cybersecurity research. A foundational paper in this area (Flajolet & Odlyzko, 1990) applies techniques that we'll develop in Section 1.5 to solve some of these problems.

A classic application of random mapping statistics is Pollard's rho algorithm, which is used to efficiently factor numbers that are the product of two primes. Numbers of this form are used extensively in cryptology and factoring them efficiently is a major topic of research.

The algorithm relies on the fact that calculating the greatest common divisor (the gcd) of two integers is very fast computationally while factoring algorithms are typically much slower. It also uses Floyd's cycle-finding algorithm, also known as the tortoise and hare algorithm: a simple method for identifying cycles in a random map.

Proposition 1.7 (Floyd's Cycle-Finding Algorithm). Let f be a function from a finite set to itself, and let $\{x_n\}_{n=0}^{\infty}$ be a sequence given by $x_{k+1} = f(x_k)$, and x_0 any element of the set. We define two related sequences: the *tortoise* $t_k = x_k$ and the *hare* $h_k = x_{2k}$. That is, at each step, the tortoise takes one step along the sequence, while the hare takes two.

Then there exist infinitely many values $m \in \mathbb{N}$ such that $t_m = h_m$, and each of these values lie in the same cycle of the original sequence.

Proof: We'll sketch the idea of the proof. The reader is encouraged to draw a random mapping graph and follow the journeys of the tortoise and hare.

Since the function is a mapping on a finite set, it must eventually reach a cycle. Once both the tortoise and hare sequences have entered this cycle, they chase each other around and must eventually land on the same vertex—and will continue to do so. □

Example 1.7 (Pollard's Rho Algorithm[7]). Let n be an integer that is known to be the product of two unknown primes p and q. Pollard's rho algorithm is an efficient algorithm for finding these factors.

If we can find a positive integer $a < n$ such that $\gcd(a, n) = d \neq 1$, then d is a nontrivial divisor of n and so must be equal to either p or q. The remaining factor can then be obtained by calculating n/d. One *could* simply generate random values of a and calculate $\gcd(a, n)$, but our odds of finding a nontrivial factor would be low: equal to $2/n$ if a is chosen uniformly at random from $\{0, 1, \ldots, n-1\}$. The algorithm is a more efficient process for finding such an integer a.

Let f be any function from $\{0, 1, \ldots, n-1\}$ to itself, and consider the sequence $\{a_k\}_{k=0}^{\infty}$ defined by $a_0 = 0$ and $a_{k+1} = f(a_k)$ for all k. Since f is a function from a finite set to itself, this sequence must eventually enter a cycle, and Floyd's algorithm can be used to find a value m such that $x_m = x_{2m}$.

Though we can't compute them, there are two related sequences given by $b_k = a_k \pmod{p}$ and $c_k = a_k \pmod{q}$. These sequences must also enter cycles, and since they are drawn from smaller sets of values, they will typically enter into cycles much faster than the original sequence. Floyd's algorithm produces values i, j for which $b_i - b_{2i} \equiv 0 \pmod{p}$ and $c_j - c_{2j} \equiv 0 \pmod{q}$, and these values will (typically) be much smaller than the index m obtained for the original sequence.

Consider what happens when $b_i - b_{2i} \equiv 0 \pmod{p}$: this means that p divides $b_i - b_{2i}$, and so p also divides $a_i - a_{2i}$. Therefore, this difference has a non-trivial common divisor with n. While we can't calculate the sequence $\{b_k\}$ directly (since we don't know p), we *can* calculate $a_k - a_{2k}$ and $\gcd(a_k - a_{2k}, n)$ for each k. Pollard's rho algorithm simply computes these differences and calculates these gcds to find a non-trivial factor.

Note: this algorithm may fail if the underlying function graph has certain undesirable properties—namely, if the sequences $\{b_k\}$ and $\{c_k\}$ cycle at the same time and frequency. Typically, $f(x) = x^2 + 1 \pmod{n}$ is chosen as the function, as this leads to a 'random-looking' functional graph, and the algorithmic complexity is estimated by analyzing the properties of a random mapping. Typical implementations will try a new function after failing to find a factor after a certain number of iterations. This algorithm has motivated research into the average behavior of

---

[7] This algorithm, developed in 1975, is named for the fact that an iterated sequence drawn from a finite set will eventually loop back on itself in a cycle. The graph of this trajectory looks somewhat like the Greek letter rho: ρ.

random mappings, including analyzing the average number of iterations before we can expect to find a cycle.

For further reading on the principle of inclusion-exclusion and generalizations, see Chapters 2 and 3 of Stanley (2012).

The combinatorics of permutation is a large topic, of which cycle structure is one small part. Bóna (2012) provides an overview of this area. An introduction to *permutation patterns*—a particularly active area of study which has applications to computer science, statistics, and genomics—is given by Vatter (2015).

1.5 GENERATING FUNCTIONS AND ASYMPTOTIC ENUMERATION

How many leaves does a random binary tree have? How many cycles does a random permutation have? How long is the longest string of 0s in a random binary string? How many steps should we expect Pollard's rho algorithm to take? Each of these questions is relevant to cybersecurity, and each asks for the *expected* behavior of a random combinatorial structure. In this section we cover *generating functions*, which provide a framework for answering these questions and many more.

A generating function is a way of encoding a sequence $\{a_n\}_{n \geq 0}$ as a single algebraic object—an infinite power series—which can then be added, multiplied, and otherwise combined with other sequences. Generating functions come in *ordinary* and *exponential* forms (among others), and each will be useful for different situations.

This section will also cover *asymptotic analysis*. In many problems, an approximate solution may be more useful than an exact solution. To illustrate, consider the sequences $\{a_n\}$ and $\{b_n\}$ given by

$$a_n = \frac{(2n)!}{n! \cdot n!} - \frac{(2n)!}{(n-1)!(n+1)!}$$
$$b_n = \frac{4^n}{\sqrt{\pi n^3}}.$$

Which one is easier to work with? Which one is easier to compare to other sequences? The sequence $\{b_n\}$ is simply an exponential function divided by a power of n, with a constant factor, while $\{a_n\}$ is the difference of divided factorials. Factorials and summations are often difficult to work with, both technically and intuitively.

It turns out that sequence b_n is an *approximation* of sequence a_n, in the sense as n gets large the ratio of these sequences tends towards 1. This means that for large n, we can—in some sense—treat these sequences as the same. Approximations can be more useful than exact solutions. We'll present more examples and revisit the sequence $\{a_n\}$ later in Examples 1.13 and 1.14. See Wilf (1982) for an enlightening discussion on characterizing a 'useful answer' to a problem.

The definition of a generating function is easy enough.

Definition 1.7 (Ordinary and Exponential Generating Functions). Let $\{a_n\}_{n \geq 0}$ be a sequence of real numbers. The *ordinary generating function* (OGF) in *indeterminate* z for the sequence is given by

$$A(z) = a_0 + a_1 z + a_2 z^2 + \cdots = \sum_{n \geq 0} a_n z^n.$$

The *exponential generating function* (EGF) for the sequence is

$$A(z) = a_0 + a_1 z + a_2 \frac{z^2}{2} + a_3 \frac{z^3}{6} + \cdots = \sum_{n \geq 0} a_n \frac{z^n}{n!}.$$

Note: these are considered *formal power series in* z, rather than expansions of functions of z. This will allow us to brush aside questions of convergence.

Stringing coefficients along a series might seem like a strange thing to do. It's worth pointing out, however, that we've already encountered this idea: recall the binomial theorem, which states that

$$(1+x)^n = \sum_{k \geq 0} \binom{n}{k} x^k.$$

This means that the expression $(1+x)^n$ is the ordinary generating function for the sequence $\left\{\binom{n}{k}\right\}_{k \geq 0}$. (Note that when n is a positive integer this sequence has finitely many nonzero terms, so the expression is sometimes called a *generating polynomial*.)

1.5.1 Ordinary Generating Functions

For ease of notation, if $\mathcal{A}$ is a combinatorial class, we use a_n for the number of objects of size n and $A(z)$ for the generating function of the enumeration sequence. In this section, all generating functions will be ordinary.

Example 1.8. Let $\mathcal{B}$ be the class of binary strings. Then the ordinary generating function is given by

$$B(z) = \sum_{n \geq 0} 2^n z^n = 1 + 2z + 4z^2 + 8z^3 + \cdots = \frac{1}{1-2z}.$$

This last equality follows from the fact that $(1 + 2z + 4z^2 + \ldots)(1 - 2z) = 1$, or from the Binomial Theorem.

Letting $\mathcal{P}$ be the class of permutations, we have

$$P(z) = \sum_{n \geq 0} n! z^n = 1 + z + 2z^2 + 6z^3 + 24z^4 + \cdots.$$

In the example above, the generating function $B(z)$ can be expressed simply, without any summation signs, as $1/(1-2z)$. This is a *closed form expression*, defined as one which can be evaluated in a finite number steps. Finding closed form expressions for infinite series is not always possible—$\sum n! z^n$ has none, for

TABLE 1.2 Common series and their closed form representations.

$$\frac{1}{1-z} = \sum_{n\geq 0} z^n \qquad e^z = \sum_{n\geq 0} \frac{z^n}{n!}$$

$$\frac{1}{1-az} = \sum_{n\geq 0} a^n z^n \qquad e^{az} = \sum_{n\geq 0} a^n \frac{z^n}{n!}$$

$$\frac{z}{(1-z)^2} = \sum_{n\geq 0} n z^n \qquad z e^z = \sum_{n\geq 0} n \frac{z^n}{n!}$$

$$\frac{z^k}{(1-z)^k} = \sum_{n\geq 0} n^{\underline{k}} z^n \qquad z^k e^z = \sum_{n\geq 0} n^{\underline{k}} \frac{z^n}{n!}$$

example—but can help us unlock facts about the series. Note that the binomial theorem itself simply provides the closed form for a series of binomial coefficients.

See Table 1.2 for some commonly used series and their closed forms; many of these follow directly from the binomial theorem with a bit of calculus. Each of these is understood to be an identity within the ring of formal power series over the reals with indeterminate z, denoted $\mathbb{R}[[z]]$. Since we're not treating these as functions of z, we don't need to worry about convergence.

1.5.2 Recurrences

Generating functions can be used for solving recurrences, as seen in the following example:

Example 1.9. Define a sequence $\{a_n\}_{n\geq 0}$ by the following recurrence:

$$a_0 = 0, \quad \text{and for all } n \geq 1, \quad a_n = 2a_{n-1} + 1.$$

This recurrence uniquely defines a sequence, and if one wants to know the value of a_n for any n, they need only calculate the sequence term by term. But is there a way of finding a closed-form representation for a_n for any n?

Let's start by setting $A(z) = \sum_{n\geq 0} a_n z^n$. The recurrence equation $a_n = 2a_{n-1} + 1$ isn't a single equation, it's actually infinitely many of them: one for every n. This makes it difficult to handle analytically. However, as we've seen, generating functions help us to turn infinitely many things into a single one.

Let's take this recurrence relation, multiply both sides by z^n, and sum over all values of n for which the equation holds:

$$\sum_{n\geq 1} a_n z^n = \sum_{n\geq 1} (2a_{n-1} + 1) z^n.$$

The left-hand side is precisely the generating function for the sequence, minus the a_0 term. We can expand the right-hand side into two different summations

leading to
$$A(z) - a_0 = 2\sum_{n \geq 1} a_{n-1}z^n + \sum_{n \geq 1} 1z^n.$$

The first term on the right-hand side is just $A(z)$ multiplied by a factor of $2z$, and the second term can be expressed as $z/(1-z)$. This gives
$$A(z) - a_0 = 2zA(z) + \frac{z}{1-z}.$$

This is now a finite expression that fully captures the recurrence relation. Further, we can *solve* this expression algebraically for $A(z)$ to obtain
$$A(z) = \frac{1}{1-2z}\left(a_0 + \frac{z}{1-z}\right).$$

Plugging in our initial value of $a_0 = 0$ gives
$$A(z) = \frac{z}{(1-2z)(1-z)}.$$

Thus, we've determined a closed form representation for the generating function of the sequence $\{a_n\}_{n \geq 0}$. So what? It turns out that a bit more algebra can get us to an explicit formula for the nth term. The method of partial fractions[8] allows us to break this rational function into simpler terms, which can then be unrolled back into series:

$$\begin{aligned}
A(z) = \sum_{n \geq 0} a_n z^n &= \frac{z}{(1-2z)(1-z)} \\
&= \frac{1}{1-2z} - \frac{1}{1-z} \\
&= \sum_{n \geq 0} 2^n z^n - \sum_{n \geq 0} z^n \\
&= \sum_{n \geq 0} (2^n - 1)z^n.
\end{aligned}$$

This shows that $a_n = 2^n - 1$. We can quickly verify that this satisfies both the initial condition and the recurrence.

Generating functions can be used to solve many types of more complicated recurrences, and the solutions follow a similar pattern: first encode the recurrence equation into a relation on infinite series, then find a closed form solution for the generating function, and finally analyze this result to find a closed form solution for the coefficients.

We'll sketch this process out for another example: finding a formula for the nth Fibonacci number.

---

[8]The method of partial fractions is typically taught in an undergrad calculus course as a way of evaluating integrals.

Example 1.10 (Fibonacci Numbers). Let $F(z) = \sum_{n \geq 0} f_n z^n$ be the generating function for the *Fibonacci numbers*, which are defined as $f_0 = f_1 = 1$ and $f_{n+2} = f_{n+1} + f_n$ for all $n \geq 0$.

Multiplying both sides of the recurrence relation by z^n and summing over $n \geq 0$, we find that
$$F(z) = zF(z) + z^2 F(z) + 1.$$
Solving for $F(z)$ yields:
$$F(z) = \frac{1}{1 - z - z^2}.$$

The partial fraction decomposition expresses the coefficients in terms of the roots of the polynomial $1 - z - z^2$. The details are tedious but are not conceptually difficult and are the sort of thing that your favorite computer algebra system will happily solve for you. This leads to
$$f_n = \frac{1}{\sqrt{5}} \left(\frac{1 + \sqrt{5}}{2} \right)^n - \frac{1}{\sqrt{5}} \left(\frac{1 - \sqrt{5}}{2} \right)^n.$$

Good luck finding a combinatorial proof of this identity.

1.5.3 Combining Classes

Recurrences occur often in real problems, and being able to find closed solutions can be helpful. However, the true utility of generating functions lies in their ability to represent *classes of objects*, rather than just sequences. To see this, we need to understand how generating functions and classes can be combined.

First, we define two simple classes which will prove useful: let $\mathcal{E}$ denote the *null class* containing a single element of size zero, and let $\mathcal{Z}$ denote the *atomic class*, which contains a single element of size 1. The generating functions for these classes are given by 1 and z, respectively.

Combinatorial classes can be roughly classified as *labeled* or *unlabeled*. In a class, each object of size n can be viewed as some structure built upon n different *atoms*—in some classes these atoms each carry a unique *label* from some n element set (typically $[n]$), while in other classes these atoms are indistinguishable. For example, we can have labeled and unlabeled graphs and trees.

Definition 1.8 (Combining Classes). Let $\mathcal{A}$ and $\mathcal{B}$ be combinatorial classes which are either both labeled or both unlabeled. We define two new constructions based on these classes.

- Let $\mathcal{A} \otimes \mathcal{B}$ denote the class whose elements are
$$\{(\alpha, \beta) : \alpha \in \mathcal{A},\ \beta \in \mathcal{B}\},$$
and where the size of the element (α, β) is equal to the sum of the sizes of α and β. This class is called the *cartesian product* of the two classes.

- Let $\mathcal{A} \oplus \mathcal{B}$ denote the class whose elements are given by the disjoint union[9] of the elements of $\mathcal{A}$ and $\mathcal{B}$, and the size of an element is the size in its parent class.

Recall the construction $\text{SEQ}(\mathcal{A})$, which denotes the class whose elements are (possibly empty) sequences of elements from $\mathcal{A}$, and the size of such a sequence is given by the sum of the sizes of the elements in the sequence. We can describe this with the $\otimes$ and $\oplus$ constructions as follows:

$$\text{SEQ}(\mathcal{A}) = \mathcal{E} \oplus \mathcal{A} \oplus (\mathcal{A} \otimes \mathcal{A}) \oplus (\mathcal{A} \otimes \mathcal{A} \otimes \mathcal{A}) \oplus \cdots.$$

In each case, if the classes are labeled, then the new class is labeled as well, and the labels on the individual objects are inherited from the originals. In particular, if $(\alpha, \beta) \in \mathcal{A} \otimes \mathcal{B}$ is labeled, then the labels are distributed across the objects α and β in such a way that the order of the labels in the original objects is respected.

Theorem 1.13 (Operations on Ordinary Generating Functions). *Let $\mathcal{A}$ and $\mathcal{B}$ denote unlabeled classes with ordinary generating functions $A(z)$ and $B(z)$.*

- *The generating function of $\mathcal{A} \oplus \mathcal{B}$ is $A(z) + B(z)$*
- *The generating function of $\mathcal{A} \otimes \mathcal{B}$ is $A(z) \cdot B(z)$.*
- *The generating function of $\text{SEQ}(\mathcal{A})$ is given by $\frac{1}{1-A(z)}$.*

Proof: To prove the first claim, note that the number of elements of size n in $\mathcal{A} \oplus \mathcal{B}$ is precisely the sum of the number of elements of this size in each of the classes. This number is $a_n + b_n$, which is the coefficient of z^n in $A(z) + B(z)$.

To prove the second, again we consider the number of elements of size n. Such an element is a pair (α, β), where $|\alpha| = k$ and $|\beta| = n - k$ for some integer k with $0 \leq k \leq n$. For a fixed value of k, there are $a_k \cdot b_{n-k}$ such pairs since we have a_k choices for the object from $\mathcal{A}$ and b_{n-k} choices for the object from $\mathcal{B}$. The product of the generating functions is given by

$$A(z) \cdot B(z) = (a_0 + a_1 z + a_2 z^2 + \cdots)(b_0 + b_1 z + b_2 z^2 + \cdots)$$
$$= \sum_{n \geq 0} \left(\sum_{k=0}^{n} a_k b_{n-k} \right) z^n,$$

so the coefficient of z^n is precisely the number of elements of size n.

For the final claim, we apply the previous results and note that if $\mathcal{C} = \text{SEQ}(\mathcal{A})$, then

$$\mathcal{C} = \mathcal{E} \oplus \mathcal{A} \oplus (\mathcal{A} \otimes \mathcal{A}) \oplus (\mathcal{A} \otimes \mathcal{A} \otimes \mathcal{A}) \oplus \cdots,$$

and so

$$C(z) = 1 + A(z) + A(z)^2 + A(z)^3 + \cdots = \frac{1}{1 - A(z)},$$

completing the proof. $\square$

---

[9]That is, we assume that the parent class of each element can be determined, even if they appear identical.

Labeled constructions go well with exponential generating functions, as seen by the following theorem.

Theorem 1.14 (Operations on Exponential Generating Functions). *Let $\mathcal{A}$ and $\mathcal{B}$ denote labeled classes with exponential generating functions $A(z)$ and $B(z)$.*

- *The generating function of $\mathcal{A} \oplus \mathcal{B}$ is $A(z) + B(z)$.*
- *The generating function of $\mathcal{A} \otimes \mathcal{B}$ is $A(z)B(z)$.*
- *The generating function of $\text{Seq}(\mathcal{A})$ is given by $\frac{1}{1-A(z)}$.*

Proof: The first and third claims are analogous to the proof of Theorem 1.13. To prove the second, consider the number of elements of size n in class $\mathcal{A} \otimes \mathcal{B}$. Such an element is a pair (α, β) with $|\alpha| = k$ and $|\beta| = n - k$ for some k. For a fixed value of k, there are $\binom{n}{k} a_k b_{n-k}$ such pairs, since we need to choose the labels in $\binom{n}{k}$ ways and choose the pair of objects in $a_k b_{n-k}$ ways. Summing over all k, we have that there are $\sum_k \binom{n}{k} a_k b_{n-k}$ objects of size n in the class, and we claim that this is precisely the coefficient of $z^n/n!$ in the product.

The product of the generating functions is equal to

$$A(z) \cdot B(z) = \sum_{n \geq 0} a_n \frac{z^n}{n!} \cdot \sum_{n \geq 0} b_n \frac{z^n}{n!}$$

$$= \sum_{n \geq 0} \left(\sum_{k=0}^{n} \frac{a_k}{k!} \cdot \frac{b_{n-k}}{(n-k)!} \right) z^n$$

$$= \sum_{n \geq 0} \left(\sum_{k=0}^{n} \frac{n! \cdot a_k b_{n-k}}{k!(n-k)!} \right) \frac{z^n}{n!}$$

This proves the claim and completes the proof. $\square$

There are two other operations on labeled classes that will prove useful: the *cycle* and the *set* operations. For a labeled class $\mathcal{A}$, define $\text{Cyc}(\mathcal{A})$ to be the class of nonempty sequences of $\text{Seq}(\mathcal{A})$ in which sequences that are rotationally equivalent are considered equal. More precisely, $(\alpha_1, \alpha_2, \ldots, \alpha_n) = (\alpha_k, \alpha_{k+1}, \ldots, \alpha_{k-1})$ for all $k \in [n]$. The class $\text{Set}(\mathcal{A})$ is similar: its objects are *unordered sets* of objects from $\mathcal{A}$.

Theorem 1.15 (The Cycle and Set Operators). *If $A(z)$ is the exponential generating function for the labeled class $\mathcal{A}$, then the exponential generating function for $\text{Cyc}(\mathcal{A})$ is*

$$A(z) + A(z)^2/2 + A(z)^3/3 + \ldots = \sum_{n=1}^{\infty} \frac{A(z)^n}{n} = \ln\left(\frac{1}{1 - A(z)}\right),$$

and the exponential generating function for $\text{Set}(\mathcal{A})$ is

$$1 + A(z) + A(z)^2/2 + A(z)^3/3! + \ldots = \sum_{n=1}^{\infty} \frac{A(z)^n}{n!} = e^{A(z)}.$$

Proof: Since $\mathcal{A}$ is a labeled class, we can represent each object as a sequence starting with its lowest element. Each cycle of length k then leads to exactly k different representations. Thus, the number of unique cycles of length k is $A(z)^k/k$. Summing over all k proves the first claim. The proof of the second claim is analogous. □

Many classes can be built up from smaller ones. By understanding the above operations, one can quickly and easily create generating functions for various classes. Let's warm up with some easy ones:

Example 1.11. A binary digit is a class with two elements of size 1, representing a zero and a one. Call this class $\mathcal{D}$, and so $D(z) = 2z$. The class $\mathcal{B}$ of binary strings is then equal to $\text{Seq}(\mathcal{D})$. Thus, the ordinary generating function is given by

$$B(z) = \frac{1}{1 - D(z)} = \frac{1}{1 - 2z} = \sum_{n \geq 0} 2^n z^n.$$

Example 1.12. A permutation is a sequence in which each entry has a unique label. Thus, if $\mathcal{P}$ is the class of permutations, we see that $\mathcal{P} = \text{Seq}(\mathcal{Z})$ and so the exponential generating function is

$$P(z) = \frac{1}{1 - z} = \sum_{n \geq 0} n! \frac{z^n}{n!}.$$

1.5.4 Recursive Structures

When each element of a sequence is described by previous elements, we say that this is a recursive formula for the sequence. When an *object* is defined by smaller objects of the same type, we say that we have a recursive description for the *object*.

Rooted trees, a common topic of study across computer science and combinatorics, serve as a good example. Some vocabulary will be helpful.

Definition 1.9 (Rooted Trees). A *rooted tree* is a tree in which one node is identified as the *root*. The distance from any node to the root is called the *level* of the node. For a node v at level k, a neighbor of v at level $k - 1$ is the *parent* of v, and any neighbors at level $k + 1$ are the *children* of v. For any vertex w, there is a unique path to the root—if this path passes through v, then w is a *descendant* of v, while v is an *ancestor* of w.

Example 1.13 (Binary Trees). A *binary tree* is a rooted tree in which each node has two (possibly empty) children, referred to as the *left* and the *right* child. For example, there are five different binary trees on four nodes. How many binary trees are there on n nodes?

Given a binary tree, note that removing the root results in two binary trees. Conversely, given any ordered pair of binary trees, we can create a new tree by creating a new root as a parent of the two existing roots.

Letting $\mathcal{B}$ be the class of binary trees, claim that

$$\mathcal{B} = \mathcal{B} \otimes \mathcal{Z} \otimes \mathcal{B} \oplus \mathcal{E}.$$

In this expression, $\mathcal{Z}$ represents the class that contains a single root, and $\mathcal{E}$ represents an empty tree. Thus, the expression simply states the following: a binary tree is either a pair of binary trees joined by a root, or it is empty.

It follows then that, if $B(z)$ is the generating function for binary trees, then

$$B(z) = zB(z)^2 + 1.$$

To solve this for $B(z)$, we can use the quadratic formula.[10] The quadratic formula gives two solutions; we choose the one which has non-negative integer coefficients:[11]

$$B(z) = \frac{1 - \sqrt{1 - 4z}}{2z}.$$

So what? Recall that we saw the term $\sqrt{1-4z}$ back in Section 1.2.6. We can apply the Binomial Theorem to see that

$$\sqrt{1-4z} = \sum_{n \geq 0} \binom{1/2}{n} 4^n z^n.$$

We expand the binomial coefficient as follows:

$$\binom{1/2}{n} 4^n = \frac{(1/2)(-1/2)(-3/2) \cdots ((-2n+3)/2)}{n!} \cdot 4^n$$

$$= (-1)^{n-1} \cdot \frac{1 \cdot 1 \cdot 3 \cdot \cdots \cdot (-2n+3)}{2^n n!} \cdot 4^n$$

$$= (-1)^{n-1} \cdot \frac{(2n-2)!}{2^n \cdot 2^n \cdot (n-2)! \cdot n!} \cdot 4^n$$

$$= (-1)^{n-1} \cdot 2 \cdot \binom{2n-2}{n}.$$

With a bit more series manipulation, we find that

$$B(z) = \sum_{n \geq 0} \frac{1}{n+1} \binom{2n}{n},$$

and so the number of binary trees on n nodes is $\frac{1}{n+1}\binom{2n}{n}$.

Notice that, to enumerate binary trees, *we didn't need to do any direct counting*—we simply found a recursive description for the objects themselves and let the generating functions do all the work (along with some tedious but straightforward algebra). Let's try again.

---

[10]This is allowed because the ring of formal power series is an *integral domain*, which means that the quadratic formula can obtained by the method of completing the square.

[11]We know a priori that one must exist, because there are a non-negative-integer number of trees of a given size.

Example 1.14 (Catalan Trees). A *Catalan Tree* is a rooted tree in which every node has some number *ordered* children. For example, there are five Catalan trees on four nodes. Letting $\mathcal{C}$ be the class of Catalan tree, we have the following recursive description:

$$\mathcal{C} = \mathcal{Z} \otimes (\mathcal{E} \oplus \mathcal{C} \oplus \mathcal{C} \otimes \mathcal{C} \oplus \cdots)$$
$$= \mathcal{Z} \otimes \text{SEQ}(\mathcal{C}).$$

Thus, we have

$$C(z) = \frac{z}{1 - C(z)}, \quad \Longrightarrow \quad C(z) = z + C(z)^2.$$

Letting $B(z)$ be the generating function for binary trees above, we find that, remarkably, $C(z) = zB(z)$. Therefore, we have that

$$c_{n+1} = b_n = \frac{1}{n+1}\binom{2n}{n}.$$

Remarkably, Catalan trees and binary trees are *both enumerated by the same numbers*; the numbers $\frac{1}{n+1}\binom{2n}{n}$ are called the *Catalan numbers*. At a high level, the two recurrences above can be described as: an object is either empty or is an atom with two smaller objects, or, an object is an atom with some (possibly zero) ordered smaller objects, respectively. These two ideas are ubiquitous within combinatorial structures, and where we find them, we find the Catalan numbers.

In the case of labeled structures, we have similar recurrences. Recall that a permutation can be written as either a sequence of entries or a set of cycles. It follows then that, if $\mathcal{P}$ is the class of permutations, we have

$$\mathcal{P} = \text{SEQ}(\mathcal{Z}) = \text{SET}(\text{CYC}(\mathcal{Z})).$$

This leads to several equivalent expressions for the generating function:

$$P(z) = \frac{1}{1-z} = e^{\ln\left(\frac{1}{1-z}\right)} = e^{\underbrace{z}_{\text{1-cycles}} + \underbrace{z^2/2}_{\text{2-cycles}} + \underbrace{z^3/3}_{\text{3-cycles}} + \underbrace{z^4/4}_{\text{4-cycles}} + \cdots}.$$

From this, we can quickly develop the generating function for various special cases of permutations. For example, recalling that involutions are those permutations whose cycles have lengths 1 and 2, we see that the generating function for involutions is equal to

$$e^{x+z^2/2} = 1 + \frac{z}{1!} + \frac{2z^2}{2!} + \frac{4z^3}{3!} + \frac{10z^4}{4!} + \frac{26z^5}{5!} + \cdots.$$

1.5.5 Asymptotic Analysis

How big is n factorial? Is it bigger or smaller than $(n/2)^n$ for large n? The *asymptotic behavior* of a sequence is often more enlightening than an exact formula. We first need to make precise the notion of *asymptotic growth*.

Definition 1.10 (Asymptotic Growth). Let $\{a_n\}_{n \geq 0}$ and $\{b_n\}_{n \geq 0}$ be two sequences. We say that the sequences have the same *asymptotic growth* or are *asymptotically equivalent*, which we denote by $a_n \sim b_n$, if

$$\lim_{n \to \infty} \frac{a_n}{b_n} = 1.$$

A famous example of asymptotic growth is *Stirling's Approximation* for the factorials. We won't present a proof, but this can be derived by approximating the value of $\ln(n!)$ using integrals and a bit of analysis.

Theorem 1.16 (Stirling's Approximation). *The following two sequences are asymptotically equivalent:*

$$n! \sim \sqrt{2\pi n} \left(\frac{n}{e}\right)^n.$$

With some algebraic manipulation, we find that the Catalan numbers can be written as

$$c_n = \frac{1}{n+1}\binom{2n}{n} = \binom{2n}{n} - \binom{2n}{n-1}.$$

These numbers are the same as the numbers presented at the start of this section, where we claimed that the numbers c_n are asymptotically equivalent to the numbers $b_n = \frac{4^n}{\sqrt{\pi n^3}}$. Stirling's Approximation can be used to prove this.

In Example 1.10 we used the partial fraction decomposition to break apart a rational function to find an exact formula for the coefficients. This idea can be generalized, allowing us to find exact (or asymptotic) formulas for the coefficients of any rational function, which is expressible as a power series about $z = 0$.

Lemma 1.2. Let $a, b, c, d \in \mathbb{R}$. Then

$$F(z) = \frac{a \cdot z^b}{(1 - cz)^d} \quad \Longleftrightarrow \quad [z^n]F(z) = a \cdot c^{n-b}\binom{n+d-b-1}{d-1}.$$

In particular, we have that

$$[z^n]F(z) \sim a \cdot \frac{n^{d-1}}{(d-1)!} \cdot c^{n-b}.$$

Proof: By the binomial theorem, we have that

$$\frac{1}{(1-cz)^d} = \sum_{n \geq 0} c^n \binom{n+d-1}{d-1} z^n.$$

To prove the first claim we note that multiplying a sequence by az^b multiplies each coefficient by a and shifts the coefficient of z^{n-b} to be the coefficient of z^n. The asymptotic bound then follows by expanding the binomial coefficient and identifying the highest power of n. □

Theorem 1.17 (Coefficients of Rational Generating Functions). *Let $F(z)$ be a rational function which is expressible as a power series about $z = 0$, having poles at $\{\alpha_1, \alpha_2, \ldots, \alpha_n\}$ with multiplicities $\{m_1, m_2, \ldots, m_n\}$. Suppose that $|\alpha_1| < |\alpha_i|$ for all $i \in [n]$; in this case we call α_1 the* dominant pole.

Letting f_n be the coefficient of z^n in $F(z)$, we have

$$f_n \sim c \cdot \frac{n^{m_1-1}}{(m_1-1)!} \left(\frac{1}{\alpha_1}\right)^n,$$

where

$$c = \lim_{z \to \alpha_1} (1 - \alpha_1 z)^{m_1} F(z).$$

Proof: By the method of partial fractions, $F(z)$ can be broken down to a sum of terms of the form $\frac{a \cdot z^b}{(1-cz)^d}$. The proof then follows by applying the lemma and noting that the term with the highest exponent will dominate as n goes to infinity. □

To see how this can be used in practice, we first apply it to a contrived example. Suppose a system's password policy requires that a password be made up of: a sequence of at least one lowercase letter, followed by a (possibly empty) sequence of digits, followed by another (possibly empty) sequence of lowercase letters, followed by a sequence of at most three digits, followed by a final sequence of at least one uppercase letter.

For example, `abcd23ef456GH` is an allowed password under this scheme. How many possible length-n passwords are allowed? How does this compare to the 26^n passwords, which only consist of lowercase letters?

Let $\mathcal{F}$ be the class of these passwords, with generating function $F(z)$. By reading the rules left-to-right, and letting $\mathcal{U}, \mathcal{L}, \mathcal{D}$ be the classes of uppercase letters, lowercase letters, and digits, respectively, we have

$$\mathcal{F} = \text{SEQ}_{\geq 1}(\mathcal{L}) \otimes \text{SEQ}(\mathcal{D}) \otimes \text{SEQ}(\mathcal{L}) \otimes \text{SEQ}_{\leq 3}(\mathcal{D}) \otimes \text{SEQ}_{\geq 1}(\mathcal{U}),$$

$$F(z) = \frac{26z}{1-26z} \cdot \frac{1}{1-10z} \cdot \frac{1}{1-26z} \cdot (1 + 10z + 10^2 z^2 + 10^3 z^3) \cdot \frac{26z}{1-26z}$$

$$= \frac{676 z^2 (1 + 10z + 100z^2 + 1000z^3)}{(1-26z)^3 (1-10z)}.$$

For a sanity check, we turn to Python and find that

$$F(z) = 676z^2 + 66248z^3 + 3999216z^4 + 192173280z^5 + \cdots.$$

This function has two poles, one at $1/26$ with multiplicity 3 and one at $1/10$ with multiplicity 1. The pole at $1/26$ dominates, so the asymptotic formula will have a 26^n in it. We can rewrite this function as $F(z) = \frac{1}{(1-26z)^3} \cdot R(z)$, and so $R(z) = (1-26z)^3 F(z)$. Taking the limit, we find that

$$\lim_{z \to 1/26} R(z) = \frac{873}{338} \approx 2.58,$$

and so, near $z = 1/26$, $F(z)$ behaves like $2.58/(1-26z)^3$. Therefore, we have that the number of these passwords is given approximately by

$$f_n \sim \frac{873}{338} \cdot \frac{n^2}{2!} 26^n \approx 1.29 \cdot n^2 \cdot 26^n.$$

We present a few more examples—the calculations are analogous to those above.

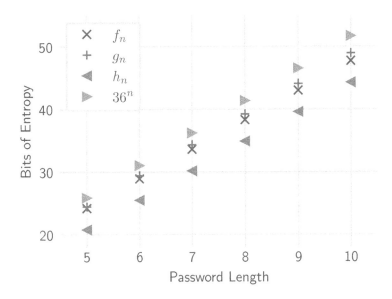

FIGURE 1.6 Asymptotic growth of the size of password spaces from Example 1.15.

Example 1.15 (Password Asymptotics). We can use Theorem 1.17 to quickly compare the strengths of various password formats. Figure 1.6 shows the graphs of these asymptotic growths.

- The ordinary generating function $F(z) = (1-26z)^{-1}(1-10z)^{-1}$ counts passwords having some number of lowercase letters followed by some number of digits and has poles at $1/26$ and $1/10$, each with multiplicity one. Letting f_n be the nth coefficient, we can calculate that

$$f_n \sim \frac{13 \cdot 26^n}{8} = 1.625 \cdot 26^n.$$

This shows that, on average, such a password is 62.5% more secure than one consisting of only letters. However, this is still much less secure than using the 36^n passwords, which include mixed letters and numbers.

We could go further and find an exact formula for the nth coefficient by using the partial fraction decomposition, but the asymptotics are more enlightening.

- The ordinary generating function $G(z) = (1-26z)^{-1} \cdot 10z \cdot (1-26z)^{-1}$ counts passwords consisting of some string of lowercase letters, a single digit, and then some other string of lowercase letters. We see that $G(z)$ has a single pole at $z = 1/26$ with multiplicity two. Applying the theorem tells us that

$$g_n \sim \frac{5 \cdot n \cdot 26^n}{13} \approx 0.38 \cdot n \cdot 26^n.$$

For short passwords, this is weaker than simply doing all lowercase digits, but quickly becomes stronger as n gets large.

- Finally, consider passwords that have some number of lowercase letters followed by exactly two digits. The generating function is given by $H(z) = (1 - 26z)^{-1} \cdot (10z)^2$, and

$$h_n \sim \frac{10^2 \cdot 26^n}{26^2} = \frac{25}{169} \cdot 26^n \approx 0.15 \cdot 26^n.$$

Of course, a simple counting argument shows that the above asymptotic relationship is actually an *equality*—to create a password of length n, one has 26^{n-2} choices for the letters, and 10^2 choices for the digits. This tells us that such a password is actually only $\approx 15\%$ as strong as a password which is all lowercase letters.

1.5.6 Bivariate Generating Functions and Combinatorial Statistics

How many leaves do we expect to find in a random binary tree of size n? How many cycles do we expect in a random n-permutation? How long must we iterate a random mapping before we expect to find a cycle? Each of these questions asks for some *statistic* on a random combinatorial object. We've seen that we can use generating functions to enumerate these objects; with a slight addition, we can also use them to calculate various statistics of interest.

We'll illustrate with an example. The class $\mathcal{B}$ of binary trees—this time *not* including the empty tree—can be decomposed as

$$\mathcal{B} = \mathcal{Z} \oplus (\mathcal{Z} \otimes \mathcal{B}) \oplus (\mathcal{B} \otimes \mathcal{Z}) \oplus (\mathcal{B} \otimes \mathcal{Z} \otimes \mathcal{B}).$$

The first term on the right-hand side represents the case of a node with no children, the second two represent a node with a single right- and left-child, and the final term represents a node having two children. The class $\mathcal{Z}$ has only a single element of size one, and as trees are recursively constructed, the number of these elements is precisely the number of nodes of the tree.

If we want to keep track of leaves, we can use a *new object* which we attach to each leaf. Formally, we equip the class $\mathcal{B}$ with a *second size function*—for an element $\beta \in \mathcal{B}$, we say that $|\beta|_\mathcal{Z}$ is the number of nodes in β while $|\beta|_\mathcal{U}$ is the number of leaves.

Let $\mathcal{U} = \{\mu\}$ be a class containing only a single element for which $|\mu|_\mathcal{Z} = 0$ and $|\mu|_\mathcal{U} = 1$, and $\mathcal{Z}$ be the usual atomic class, containing a single element ζ with $|\zeta|_\mathcal{Z} = 1$ and $|\zeta|_\mathcal{U} = 0$.

Now, we can rework our recursion, incorporating class $\mathcal{U}$ to track leaves:

$$\mathcal{B} = \underbrace{\mathcal{Z} \otimes \mathcal{U}}_{\substack{\text{term representing} \\ \text{a leaf}}} \oplus (\mathcal{Z} \otimes \mathcal{B}) \oplus (\mathcal{B} \otimes \mathcal{Z}) \oplus (\mathcal{B} \otimes \mathcal{Z} \otimes \mathcal{B}).$$

This expression translates to a relationship between infinite power series in two indeterminates z and u, letting z be the generating function for the class $\mathcal{Z}$ and u be the generating function for the class $\mathcal{U}$:

$$B(z, u) = uz + 2zB(z, u) + zB(z, u)^2.$$

Solving, again with the quadratic formula, gives

$$B(z, u) = \sum_{\beta \in \mathcal{B}} z^{|\beta|_z} u^{|\beta|_u} = \frac{1 - 2z - \sqrt{1 - 4z + 4z^2 - 4uz^2}}{2z}.$$

So what? Bivariate generating functions are considerably more difficult to work with than the single variable version. We first note that plugging in $u = 1$ gives us the single-variable generating function for the Catalan numbers (minus the constant term), as expected. We can compute some coefficients grouped by powers of z using a computer algebra system:

$$B(z, u) = \frac{1 - 2z - \sqrt{1 - 4z + 4z^2 - 4uz^2}}{2z} \tag{1.1}$$
$$= uz + 2uz^2 + (4u + u^2)z^3 + (8u + 6u^2)z^4 + (16u + 24u^2 + 2u^3)z^5 + \dots.$$

This expansion yields insight into the structure of these trees. Examining the coefficient of z^5, this tells us that there are a total of $16 + 24 + 2 = 42$ binary trees on 5 nodes, and that 16 of them have only a single leaf, 24 have two leaves, and only 2 have three leaves. Thus, remarkably, this Equation 1.1 captures the full distribution of leaves across all binary trees of all sizes. We need only figure out how to extract useful information.

Consider, for example, if we want to know the *expected* number of leaves in a random binary tree on n nodes. Since we already know the total number of trees, we need only compute the *total number* of leaves across all binary trees on n nodes.

Considering Equation 1.1 once again, we see that plugging in $u = 1$ yields the single-variable generating function enumerating binary trees. However, if we *differentiate* with respect to u, and *then* plug in $u = 1$, we see that the coefficient of z^5, for example, is $(16 + 2 \cdot 24 + 3 \cdot 2) = 70$, which is the total number of leaves in all binary trees on 5 nodes. Thus, the single-variable generating function for the total number of leaves in all n-node binary trees is:

$$\partial_u B(z, u)|_{u=1} = \frac{z}{\sqrt{1 - 4z}}$$
$$= \sum_{n \geq 0} \binom{2n - 2}{n - 1} z^n$$
$$= z + 2z^2 + 6z^3 + 20z^4 + 70z^5 + \dots.$$

We summarize with the following proposition.

Proposition 1.8. The total number of leaves in the set of all binary trees on n nodes is $\binom{2n-2}{n-1}$.

Corollary 1.2. The expected number of leaves in a random binary tree is equal to
$$\frac{n(n+1)}{4n-2}.$$
For large n, the expectation that a randomly chosen node is a leaf is approximately $1/4$.

Proof: The first claim follows from dividing the total number of leaves by the total number of trees (the nth Catalan number). The second claim follows from dividing the total number of nodes, n, by the expected number of leaves and taking the limit as $n \to \infty$. □

This strategy can be used for any object with a recursive decomposition. By inserting a new variable u to keep track of a statistic of interest, we can create a bivariate generating function that encodes the distribution of this statistic. The differentiation trick we used above extracts the total value of the statistic, which led to the expectation. To calculate higher moments (e.g., the *variance*) of a statistic, we need only take additional derivatives.

Let's apply this same strategy to the permutation cycle structure. Recall that if $\mathcal{P}$ is the class of permutations, then we have

$$P(z) = e^{\underbrace{z}_{\text{1-cycles}} + \underbrace{z^2/2}_{\text{2-cycles}} + \underbrace{z^3/3}_{\text{3-cycles}} + \underbrace{z^4/4}_{\text{4-cycles}} + \ldots}.$$

To calculate statistics on the number of r-cycles (for any $r \in \mathbb{P}$), we tack a u on the corresponding term. Note that since a 1-cycle is just a fixed point, we could calculate the fixed point expectation using this method. Using a computer algebra system, we see, for example, that

$$e^{uz+z^2/2+z^3/3+\cdots} = uz + (u^2+1)z^2 + (2+3u+u^3)z^3 + \cdots.$$

This tells us, among other things, that of the permutations of length 3, exactly 2 have no fixed points, 3 have one, and 1 has three. We could calculate expectation by differentiating, but let's prove a more general fact.

Proposition 1.9. The expected number of r-cycles in a random n-permutation is equal to $1/r$.

Proof: Let $\mathfrak{S}$ be the class of all permutations, and for $\pi \in \mathfrak{S}$, let $|\pi|_l$ be the length and $|\pi|_c$ the number of r-cycles of π. Letting $p_{n,k}$ be the number of permutations of length n which have precisely k r-cycles, we have that

$$P(z,u) = \sum_{\pi \in \mathfrak{S}} z^{|\pi|_l} u^{|\pi|_c} = \sum_{n,k \geq 0} p_{n,k} u^k z^n$$
$$= e^{z + \frac{z^2}{2} + \cdots + \frac{z^{r-1}}{(r-1)} + \frac{uz^r}{r} + \frac{z^{r+1}}{(r+1)} + \cdots}$$
$$= e^{\ln\left(\frac{1}{1-z}\right) - \frac{z^r}{r} + \frac{uz^r}{r}}$$
$$= \frac{e^{((uz^r - z^r)/r)}}{1-z}.$$

Taking the derivative with respect to u, and setting $u = 1$ gives

$$\partial_u P(z,u)|_{u=1} = \frac{z^r}{r(1-z)} = \sum_{n=r} \frac{z^n}{r}.$$

Since this is an exponential generating function, this means that the total number of r-cycles in all n-permutations (with $n \geq r$) is $n!/r$, and so the expected number in a random permutation is $1/r$. □

Again, we could use this same method to calculate higher moments of the distribution by taking additional derivatives.

Discrete structures are fundamental to computer science, and the behavior of these structures is essential to understanding algorithmic complexity and performance. Understanding the average properties of a structure influenced by random processes can help design better algorithms and identify flaws within encryption systems.

In the field of cybersecurity, encryption systems are based on assumptions of true randomness—a pseudorandom process may be susceptible to attacks if it exhibits non-random characteristics. For example, a hash function that produces cycles after a few iterations may be vulnerable to specific attacks aimed at finding collisions.

The subfield of *analytic combinatorics* is primarily concerned with applying generating functions and complex analytic methods to solve combinatorial problems. Flajolet and Sedgewick (2009) provide a comprehensive introduction to this area, including both the combinatorial tools used to construct the generating function and the methods from complex analysis used to analyze them. This book focuses mostly on single-variable generating functions, while Pemantle and Wilson (2013) covers the multi-variate case.

1.6 COMBINATORIAL DESIGNS AND SOFTWARE TESTING

We end this chapter with a brief introduction to *combinatorial designs*, a topic with strong applications to statistics and to system testing. Designs are families of finite sets satisfying specific properties, and the study of these objects is a classic research topic in combinatorics.

We focus here on a specific application to software testing. Adequate testing leads to more reliable and more secure software. Bugs and errors can be exploited by adversaries to gain access and disrupt systems. As software becomes more complex, testing becomes more difficult; the theory of combinatorial designs can be applied to more effectively and efficiently identify vulnerabilities in software and hardware systems.

1.6.1 Latin Squares

A classic example of a combinatorial design is the *Latin square*. For $n \in \mathbb{P}$, a Latin square of size n is an $n \times n$ array filled with elements from some n-element set in a way such that each element appears precisely once in each row and column.

If $A = (a_{ij})_{i,j \in [n]}, B = (b_{i,j})_{i,j \in [n]}$ are two Latin squares, then we define the *product* (denoted $A \otimes B$) to be the array $((a_{ij}, b_{ij}))_{i,j \in [n]}$. For example, if

$$A = \begin{pmatrix} 1 & 3 & 2 \\ 3 & 2 & 1 \\ 2 & 1 & 3 \end{pmatrix} \text{ and } B = \begin{pmatrix} 3 & 1 & 2 \\ 2 & 3 & 1 \\ 1 & 2 & 3 \end{pmatrix}, \text{ then } A \otimes B = \begin{pmatrix} (1,3) & (3,1) & (2,2) \\ (3,2) & (2,3) & (1,1) \\ (2,1) & (1,2) & (3,3) \end{pmatrix}.$$

Note that the original squares can be recovered from the product by simply reading off the first and second elements of each entry separately.

Definition 1.11 (Orthogonal Arrays). Two Latin squares A and B of the same size n are said to be *orthogonal* if each ordered pair (i, j) appears precisely once in the array $A \otimes B$.

$$\begin{pmatrix} (1,1) & (2,6) & (3,4) & (4,5) & (5,3) & (6,2) \\ (2,2) & (3,1) & (6,5) & (5,4) & (1,6) & (4,3) \\ (3,3) & (4,5) & (1,2) & (2,6) & (6,4) & (5,1) \\ (4,4) & (6,3) & (5,6) & (3,2) & (2,1) & (1,5) \\ (5,5) & (1,4) & (2,3) & (6,1) & (4,2) & (3,6) \\ (6,6) & (5,2) & (4,1) & (1,3) & (3,5) & (2,4) \end{pmatrix}$$

FIGURE 1.7 The product of two squares that are not orthogonal.

For example, the Latin squares A, B above are orthogonal because each of the nine pairs appears exactly once in the product. Figure 1.7 shows the product of two *non-orthogonal* Latin squares of size six—to see this, note that the pairs $(2, 6)$ and $(4, 5)$ occur twice, while the pairs $(2, 5)$ and $(4, 6)$ never appear.

The study of orthogonal Latin squares goes back over 200 years. The '36 Officers Problem,' posed by Leonard Euler in 1782, asks to find a pair of orthogonal Latin squares of size 6. Euler referred to these as *Graeco-Latin squares* since he used Greek letters for the first entry and Latin for the second. He conjectured that there could be no such pair of size 6, nor of any size congruent to 2 (mod 4). While it is true that there is no pair of size 2 or 6, it turns out that pairs of orthogonal Latin squares exist for all other sizes, disproving Euler's conjecture. Klyve and Stemkoski (2006) give an excellent and accessible survey of this problem's two-century history and the variety of methods involved.

1.6.2 Black-Box Testing

Suppose we have an office printer, which has the following settings: paper size, which can be set to A4, Letter, or Legal; single-sided or double-sided options; color and black and white modes; modes for text, pictures, or both; portrait and landscape settings.

We say that this printer has five *parameters*, each of which can take on two or three different *values*. A full assignment of values to parameters—say, printing a single-sided color picture on legal paper in landscape mode—is called a *configuration*. Finally, a set of configurations to be tested is called a *test set*.

Trying out all possible configurations to check for errors is called *exhaustive testing*. If we conduct exhaustive testing on this printer, we would need to print $3 \cdot 2 \cdot 2 \cdot 3 \cdot 2 = 72$ different sheets of paper. Real software systems may have hundreds of parameters, each with dozens of values, which makes exhaustive testing prohibitively expensive. Even a relatively simple system with thirty binary parameters has $2^{30} = 1{,}073{,}741{,}824$ different configurations.

Software testing is a vast and complicated subject; here we consider only a simple version of it called *black-box* testing, in which the system is treated as opaque in that we only interact with it by plugging in configurations and observing the output. In fact, we are really only engaging in the form of black-box testing, sometimes called *smoke testing*—run the system, and look (or smell) for fire—which is often the first step in a more robust testing plan.

For a given system, our goal is to determine if there is any configuration that leads to an error. Full exhaustive testing is required to be 100% sure that a system is free of errors of this form. How many tests are then required to be 99% sure? or 90% sure? It turns out, with a simple and intuitive idea (backed by experimental evidence), we can detect errors with high confidence using only a small number of tests.

To illustrate, consider the printer described above. If the printer breaks while printing a double-sided color text on letter paper in landscape mode, we might expect the error to be caused by some subset of these features. For example, it might be that any page printed on letter paper in landscape mode, or any color text, or even any double-sided page breaks the printer. The key idea is that *most errors are caused by a bad interaction between a relatively small number of parameters*. This intuitive idea is supported by empirical data (Kuhn, Kacker, & Lei, 2010), which found that 60% to 90% of errors are caused by at most two parameters interacting, and were unable to find any errors caused by more than six parameters.

It follows then that if we want to detect errors in a small number of tests, we should design a test set that includes as many different parameter interactions as possible. For $t \in \mathbb{P}$, a *t-way interaction* is an assignment of t parameters to specific values. We say that a test set is *t-covering* if every possible t-way interaction appears somewhere in the test set. For example, Table 1.3 shows a two-way covering set for the printer—note that for every pair of parameters, every possible combination of values is included in at least one test.

TABLE 1.3 A two-way covering test set of size nine for a printer

test	paper	duplex	color	content	orientation
1	a4	false	false	text	landscape
2	a4	true	true	pictures	portrait
3	a4	false	true	both	landscape
4	letter	true	false	text	portrait
5	letter	false	true	pictures	landscape
6	letter	true	false	both	portrait
7	legal	false	true	text	portrait
8	legal	true	false	pictures	landscape
9	legal	false	false	both	portrait

For a specified system and a specified t, our goal is to construct the shortest possible set of tests which is t-covering.

1.6.3 Orthogonal Arrays

Leaving Latin squares aside for a moment, we consider a different design: the *orthogonal array*.

Definition 1.12 (Orthogonal Arrays). Let t, v, k, λ be positive integers. A $t - (v, k, \lambda)$ orthogonal array is a $k \times \lambda v^t$ array containing entries from the set $[v]$ satisfying the property that for every sequence in $\text{SEQ}([v])_t$ and every choice of columns from $\text{SET}([k])_t$, there are exactly λ rows such that the entries in the chosen columns form precisely that sequence.

This definition needs a little unpacking. For example, the array depicted in Figure 1.8 is a $2 - (3, 4, 1)$ orthogonal array. To see this, note that: the entries are taken from the set $[3]$, there are four columns, and for every sequence in the set $\text{SEQ}([3])_2 = \{(1,1), (1,2), \ldots (3,3)\}$ and for every pair of columns, that sequence appears in those columns in exactly one row. The parameter t is the *strength* of the array, and an array can be checked using the *t-finger test*: put t of your fingers on different columns, and run them down the array—regardless of the columns chosen, we should see each t-tuple appearing exactly λ times.

Returning to the language of system testing, a $t - (v, k, \lambda)$ array is a test set for a system with k parameters, each having v values, in which every t-way interaction appears precisely λ times. For testing purposes, we (typically) only care that each t-way interaction appears *at least once*—an array with this property is called a *covering array*.

It turns out that orthogonal arrays are closely related to sets of mutually orthogonal Latin squares, as seen by the following theorem.

Theorem 1.18. *Let $v, k \in \mathbb{P}$ with $v \geq 2$ and $k \geq 3$. The set of orthogonal arrays with parameters $2 - (v, k, 1)$ is in bijection with the set of ordered sequences of $k - 2$ mutually orthogonal Latin squares of size v.*

$$\begin{pmatrix} 1 & 1 & 1 & 1 \\ 1 & 2 & 2 & 2 \\ 1 & 3 & 3 & 3 \\ 2 & 1 & 2 & 3 \\ 2 & 2 & 3 & 1 \\ 2 & 3 & 1 & 2 \\ 3 & 1 & 3 & 2 \\ 3 & 2 & 1 & 3 \\ 3 & 3 & 2 & 1 \end{pmatrix} \qquad \begin{pmatrix} 1 & 2 & 3 \\ 2 & 3 & 1 \\ 3 & 1 & 2 \end{pmatrix} \quad \begin{pmatrix} 1 & 2 & 3 \\ 3 & 1 & 2 \\ 2 & 3 & 1 \end{pmatrix}$$

FIGURE 1.8 A $2-(3,4,1)$ orthogonal array and a pair of orthogonal Latin squares of size three.

Proof: Let A be a $2-(v,k,1)$ orthogonal array. We construct a sequence of size n Latin squares, and then claim that they are orthogonal. The first two columns of A will be the *index* columns, and each remaining column will lead to a new Latin square.

First, create a set of $(k-2)$ $n \times n$ arrays whose entries are initially blank; we'll go down the rows of A and fill in entries one-by-one. Let $(a_1, a_2, a_3, a_4, \ldots a_k)$ be a row of A, and consider the a_1 row and a_2 column of each of the squares: insert a_3 into this position in the first square, a_4 in the second square, and so on. Claim that this operation is well defined and that the resulting set of squares are both Latin and mutually orthogonal.

This follows largely by definition. Since each possible a_1, a_2 appears precisely once, this process is well defined. Since each tuple (a_1, a_k) and (a_2, a_k) appears precisely once for each k, we see exactly one of each value per row and column, so these are Latin squares. Finally, a lack of orthogonality would lead to the same pair appearing twice in array A, so we see that the resulting squares must be mutually orthogonal.

The above construction can also be reversed, showing that any set of mutually orthogonal Latin squares can be used to generate an orthogonal array. □

Figure 1.8 shows an orthogonal array along with the ordered pair of orthogonal Latin squares corresponding to this array. The first two columns form the index, the third leads to the first Latin square, and the fourth column to the second Latin square.

1.6.4 Test Generation

Given a system S with k parameters each having at most v values, full exhaustive testing requires v^k tests. An orthogonal array, in which each t-way interaction is included exactly once, requires only v^t.

Given such a system, we could simply construct a set of mutually orthogonal Latin squares and use these to construct an orthogonal array. There are many methods for constructing sets of mutually orthogonal Latin squares, including techniques drawn from fields as diverse as finite algebra, projective geometry, and number theory. This strategy, however, requires that each parameter in the system has the same number of values. Further, these methods may be computationally difficult and fragile: if we change the system slightly, we have to start over from scratch.

Generating covering arrays for realistic systems is a topic of current research, which we won't get into here. A variety of algorithms exist, including the flexible In-Parameter-Order (IPO) algorithm and its generalizations, and a t-covering test set can be generated with roughly $v^t \ln k$ tests (Kuhn et al., 2010).

For example, a small system with 30 binary parameters requires over a billion tests for exhaustive testing, while 6-way testing can be accomplished with ≈ 218 tests and 2-way testing can be accomplished with only 14. These methods allow for system developers to more effectively and efficiently design tests, ultimately leading to more robust and secure systems.

1.7 CONCLUSION

Combinatorics is a broad field, and its tools and techniques can be applied to a variety of problems. The topics and techniques covered in this chapter give insight into the combinatorialist's toolbox and way of thinking about new problems. In particular, these examples are intended to show some of the connections between seemingly different problems.

A hallmark of a combinatorial solution is that, when viewed from a certain angle, it may seem simple or even obvious. A number of famous theorems in the field spent decades as open problems until they were approached from a new point of view and solved succinctly. When approaching a new problem—whether a major research question or a smaller issue that arises in other work—it pays to attack it from multiple perspectives.

Each of the sections in this chapter leads to new areas of active research, both in the applications to cybersecurity topics and in the study of combinatorial objects themselves. In addition to the examples presented here, algorithm design, network analysis, and high-performance computing are all active areas of research in which combinatorics plays a key role.

Cybersecurity is a rapidly growing field of research and encompasses a vast array of important areas ranging from deeply technical mathematics and computer science research to sociological and policy questions. All of these areas involve the analysis of finite structures, and all of these areas can benefit from combinatorial methodology—though it may not be obvious at first. A working knowledge of these methods equips a researcher to effectively tackle new problems. We hope that this chapter encourages further study and application.

ACKNOWLEDGMENTS

This chapter has benefited substantially from discussions with Jenny Zito and Vince Vatter and from corrections and feedback given by the referee and the editors.

REFERENCES

Bóna, M. (2012). *Combinatorics of permutations* (Second ed.). CRC Press, Boca Raton, FL. Retrieved from https://doi.org/10.1201/b12210

Bóna, M. (Ed.). (2015). *Handbook of enumerative combinatorics*. CRC Press, Boca Raton, FL.

Bóna, M. (2017). *A walk through combinatorics*. World Scientific Publishing Co. Pte. Ltd., Hackensack, NJ.

Diestel, R. (2018). *Graph theory* (Fifth ed., Vol. 173). Springer, Berlin.

Flajolet, P., & Odlyzko, A. M. (1990). Random mapping statistics. In *Advances in cryptology—EUROCRYPT '89* Vol. 434, pp. 329–354. Springer, Berlin. Retrieved from https://mathscinet.ams.org/mathscinet-getitem?mr=1083961.

Flajolet, P., & Sedgewick, R. (2009). *Analytic combinatorics*. Cambridge University Press, Cambridge. Retrieved from https://doi.org/10.1017/CBO9780511801655 doi: 10.1017/CBO9780511801655.

Gowers, W. T. (2000). The two cultures of mathematics. In *Mathematics: Frontiers and perspectives* (pp. 65–78). Amer. Math. Soc., Providence, RI.

Klyve, D., & Stemkoski, L. (2006). Graeco-Latin squares and a mistaken conjecture of Euler. *College Math. J.*, *37*(1), 2–15. doi: 10.2307/27646265

Kuhn, D. R., Kacker, R. N., & Lei, Y. (2010). *Practical combinatorial testing* (Tech. Rep. Sp 800-142). Gaithersburg, MD, USA.

Pemantle, R., & Wilson, M. C. (2013). *Analytic combinatorics in several variables* (Vol. 140). Cambridge University Press, Cambridge. Retrieved from https://doi.org/10.1017/CBO9781139381864.

Stanley, R. P. (1999). *Enumerative combinatorics, Volume 2*. Cambridge University Press, Cambridge. Retrieved from https://doi.org/10.1017/CBO9780511609589.

Stanley, R. P. (2012). *Enumerative combinatorics, Volume 1* (Second ed.). Cambridge University Press, Cambridge.

Vatter, V. (2015). Permutation classes. In *Handbook of enumerative combinatorics* (pp. 753–833). CRC Press, Boca Raton, FL.

West, D. B. (1996). *Introduction to graph theory*. Prentice Hall, Inc., Upper Saddle River, NJ. Retrieved from https://mathscinet.ams.org/mathscinet-getitem?mr=1367739

Wilf, H. S. (1982). What is an answer? *Amer. Math. Monthly*, 89(5), 289–292, doi: 10.2307/2321713.

CHAPTER 2

Cryptography

Gretchen L. Matthews

Aidan W. Murphy

CONTENTS

2.1	Introduction		53
2.2	Coding Theory Prerequisites		56
	2.2.1	Basic Concepts and Terminology	56
	2.2.2	Code Constructions	67
		2.2.2.1 Reed-Solomon Codes	67
		2.2.2.2 Goppa Codes	70
		2.2.2.3 Parity-Check Codes	70
	2.2.3	Decoding	74
2.3	Code-Based Cryptosystems		77
	2.3.1	Public-Key Cryptography	77
	2.3.2	The McEliece Public-Key Cryptosystem	78
	2.3.3	The Niederreiter Public-Key Cryptosystem	79
2.4	Attacks and Potential Repairs		81
	2.4.1	Sidelnikov-Shestakov Attack	82
	2.4.2	Distinguisher Attacks	84
	2.4.3	Information Set Decoding	85
	2.4.4	Variants	86
		2.4.4.1 Berger-Loidreau	86
		2.4.4.2 Random Linear Code Encryption	88
		2.4.4.3 LDPC and MDPC	90
2.5	Conclusion		92

2.1 INTRODUCTION

Cryptography is an important element of cybersecurity, as it allows for the secure communication and storage of data. It protects data from tampering and allows for authentication so that unauthorized parties may not access, manipulate, or falsify data. Mathematics has played a key role in the development of cryptography, especially over the last century. The focus of this chapter is code-based cryptography,

DOI: 10.1201/9780429354649-2

which is a relatively new branch of public-key cryptography highly relevant to current and future cybersecurity challenges.

Coding theory was introduced to protect information from distortion and degradation, as error-correcting codes incorporate redundancy to guarantee the recovery of information. It is still employed in that capacity, and the codes themselves have evolved over time to address coding for flash memories, distributed storage, and other applications. At the same time, public-key cryptosystems have developed to support wide-ranging applications including defense, secure messaging, e-commerce, and blockchain. The fields of coding theory and cryptography have mostly developed in parallel, with the primary point of intersection discovered in the late 1970s by Robert McEliece (McEliece, 1978) and early 1980s by Harald Niederreiter (Niederreiter, 1986). While this marked the advent of code-based cryptography, the area lay mostly dormant for a number of years while RSA and elliptic curve cryptography flourished and saw widespread implementation. All are public-key cryptosystems, but RSA (Rivest, Shamir, and Adelman, 1978) and elliptic curve cryptosystems (Koblitz, 1987) have smaller public key sizes than code-based systems with similar security guarantees. For that reason, little attention was paid to the practical application of error-correcting codes to cryptography at that time. Shor's Algorithm (Shor, 1994), a quantum algorithm for solving the mathematical problems on which the security of RSA and elliptic curve cryptography is based, was announced in 1994, prompting researchers to consider alternatives. A timeline of key developments related to the topics in this chapter is found in Table 2.1.

At present time, we do not have large-scale quantum computers capable of running Shor's Algorithm on reasonably sized problems. However, more and more attention is being paid to the impact of quantum algorithms on cryptographic protocols, especially as we enter an era in which some entities may have access to powerful quantum computing before others. In this timeframe, most (if not all) communications will be conducted via classical methods, while some more financially potent or dominant parties would have the power to intercept and decipher messages meant for others. It is also the case that large amounts of information communicated or generated today may be stored in anticipation of the ability to decrypt when quantum computing is more viable, in what is sometimes termed a download now, decrypt later attack. Post-quantum cryptography is a way of securing classical information, meaning strings of elements from a finite alphabet, that is believed to be robust even in the presence of quantum algorithms. A distinction must be made between post-quantum cryptography and quantum cryptography. Quantum cryptography uses quantum mechanics to securely communicate. It comes with the promise of provably secure communications and the ability to detect eavesdropping. This would obviously be a major scientific advance, but it is not yet within reach. For those reasons, we focus on post-quantum cryptography. This is portrayed in Figure 2.1.

This chapter is centered on post-quantum cryptography. It does not include systems such as RSA and elliptic curve cryptography which are presently known to be vulnerable to quantum algorithms and are already detailed in a number of accessible references including Hoffstein, Pipher, and Silverman (2008), Smart

TABLE 2.1 Key developments in coding theory and cryptography relevant to code-based cryptography

Year	Development
1948	Shannon's Theorem
1950	Hamming codes
1960	Reed-Solomon codes
1963	LDPC codes discovered by Gallagher
1970	Goppa codes
1973*	RSA cryptosystem developed by Cocks (but this work was classified until 1997)
1976	Diffie-Hellman key exchange
1977	RSA cryptosystem
1978	McEliece cryptosystem
1985	Elliptic-curve cryptography
1986	Niederreiter cryptosystem
1994	Shor's Algorithm
1996	LDPC codes rediscovered by MacKay and Neal

(2015), and Stinson and Paterson (2018). Both RSA, whose security is based on factoring integers, and elliptic curve cryptography, which is based on the elliptic curve discrete log problem, can be attacked in polynomial time via Shor's Algorithm. Post-quantum cryptography is often divided into a few major types: lattice-based, multivariate polynomial-based, supersingular elliptic curve isogeny-based, and code-based. Error-correcting codes are front and center in code-based cryptography, which is considered a primary alternative public-key cryptosystem that is more robust to quantum algorithms. In this chapter, we provide an overview of code-based cryptography relying mostly on undergraduate linear algebra.

This chapter is organized as follows. Section 2.2 contains necessary background in the theory of error-correcting codes, including basic concepts and terminology in Subsection 2.2.1, standard code constructions relevant to code-based cryptography

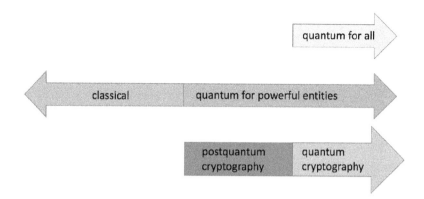

FIGURE 2.1 Projected evolution of computation and cryptography.

in Subsection 2.2.2, and a discussion of general decoding algorithms in Subsection 2.2.3. Section 2.3 is the heart of the chapter, with Subsection 2.3.1 providing a brief overview of public-key cryptography as preparation for Subsection 2.3.2, which introduces the McEliece public-key cryptosystem, and Subsection 2.3.3 containing the Niederreiter cryptosystem. Section 2.4 includes some attacks as well as some variants of McEliece and Niederreiter using codes introduced in earlier sections.

2.2 CODING THEORY PREREQUISITES

2.2.1 Basic Concepts and Terminology

The study of error-correcting codes dates back to the late 1940s with early work by Claude Shannon, whose seminal paper (Shannon, 2001) laid the groundwork for information theory, and Richard Hamming, who defined the first sophisticated family of error-corrected codes (Hamming, 1950). The basic idea is to ensure the reliable transmission of information from source to receiver by adding redundancy to messages, rather than sending the raw information itself. Indeed, we often think of the model depicted in Figure 2.2 below.

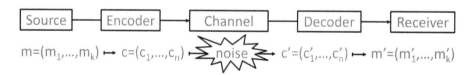

FIGURE 2.2 Error-correcting codes at work.

Here, the message m is encoded as a codeword c. Then c is sent across the channel. The channel output is denoted c', which may reflect noise ν added to the codeword c; hence, $c' = c + \nu$. The decoder takes c' as input and determines from it the most likely message sent, denoted m'. A goal of coding theory is to design and implement efficient encoders and decoders such that $m = m'$. This process is formalized below, following a summary of necessary background and notation.

The alphabet used for messages is typically taken to be a finite field. The symbols 0 and 1 (or strings of them) are often used to suggest digital information or data being communicated. Finite fields formalize and generalize this notion. Informally, a field is a set closed under addition, subtraction, multiplication, and division by nonzero elements in which these operations 'behave nicely.' More precisely, a field is a nonempty set S closed under addition and multiplication satisfying the following conditions:

1. addition and multiplication are both associative, meaning $a + (b + c) = (a + b) + c$ and $a \cdot (b \cdot c) = (a \cdot b) \cdot c$ for all $a, b, c \in S$;

2. addition and multiplication are both commutative, meaning $a + b = b + a$ and $a \cdot b = b \cdot a$ for all $a, b \in S$;

3. distributive laws hold, meaning $a \cdot (b + c) = a \cdot b + a \cdot c$ for all $a, b, c \in S$;

4. there is an additive identity in S, meaning there exists $0 \in S$ such that $a + 0 = a$ for all $a \in S$;

5. there is a multiplicative identity in S, meaning there exists $1 \in S$ such that $a \cdot 1 = a$ for all $a \in S$;

6. every element of S has an additive inverse in S, meaning for all $a \in S$ there exists $b \in S$ such that $a + b = 0$;

7. every nonzero element of S has a multiplicative inverse in S, meaning for all $a \in S \setminus \{0\}$ there exists $b \in S$ such that $a \cdot b = 1$.

For instance, the set $\mathbb{Q}$ of rational numbers is a field, whereas the set of integers $\mathbb{Z}$ is not, since the number 2 does not have multiplicative inverse, which is an integer. Other examples of fields include the set $\mathbb{R}$ of real numbers and the set $\mathbb{C}$ of complex numbers. Notice that the sum of field elements $a, b \in S$ is written $a + b$, and their product is expressed as $a \cdot b$ or sometimes simply ab.

Fields are typically introduced in abstract algebra. The curious reader is encouraged to consult any standard undergraduate text (such as Fraleigh and Brand (2020) and Gallian (2016)) to learn more about fields in general. For purposes of error-correction, we require that the field be finite so that there are only finitely many possibilities to check when repairing an incorrect symbol.

The theory of finite fields is quite rich and certainly goes beyond our needs in this chapter; more comprehensive and rigorous treatments of fields appropriate for the undergraduate reader can be found in Mullen and Mummert (2007). The goal here is to quickly equip the reader with a working knowledge of finite fields. We hope the more experienced reader will appreciate the necessary sacrificing of some details that are found in a typical abstract algebra course in service of the goal of covering highly relevant topics in code-based cryptography.

Fields in this chapter are usually denoted by $\mathbb{F}$. It turns out that a finite field $\mathbb{F}$ of size q exists if and only if $q = p^l$ where p is prime and l is a positive integer. Moreover, there is essentially one field with q elements, which will be denoted $\mathbb{F}_q$; here, 'essentially' means up to isomorphism (meaning up to a relabeling of the elements which respects the field operations). If $q = p$, $\mathbb{F}_p = \mathbb{Z}/p\mathbb{Z}$, the integers modulo p which is described below. In particular, we may write $\mathbb{F}_p = \{0, 1, \ldots, p-1\}$. According to the Division Algorithm, given an integer n there are unique integers Q and R so that $n = Qp + R$ and $0 \leq R < p$. Using this decomposition of n in terms of p, we define $n \mod p := R$; observe also that $n \mod p = n - \left\lfloor \frac{n}{p} \right\rfloor p$. The operations $\oplus$ and $\odot$ on $\mathbb{F}_p$ are defined as follows. For $a, b \in \mathbb{F}_p$, $a \oplus b = (a+b) \mod p$ and $a \odot b = (ab) \mod p$.

Example 2.1. Consider $p = 2$. Then $\mathbb{F}_2 = \{0, 1\}$ with addition and multiplication as in the table below.

$\oplus$	0	1
0	0	1
1	1	0

$\odot$	0	1
0	0	0
1	0	1

Notice that $1 \oplus 1 = 0$, since $1 + 1 = 2 = 1 \cdot 2 + 0$. The field $\mathbb{F}_2$ is called a binary field, as is any field that contains it. Care must be taken when working with fields to understand that elements in different fields may have the same name or label but have different meanings. For instance, the number 1 behaves differently as an integer than the element $1 \in \mathbb{F}_2$ does: in particular, as an integer, $1+1 = 2$, whereas as an element of $\mathbb{F}_2$, $1 \oplus 1 = 0$. Similarly, $1 \in \mathbb{F}_2$ is not the same as $1 \in \mathbb{F}_3$, as we will see below. Even so, the context should make it clear which field is being considered, and no confusion should arise. It is worth noting that binary fields all have even cardinality, though the proof goes beyond the scope of this chapter.

For another example, take $p = 3$. We can see that $\mathbb{F}_3 = \{0, 1, 2\}$ with operations as shown below.

$\oplus$	0	1	2
0	0	1	2
1	1	2	0
2	2	0	1

$\odot$	0	1	2
0	0	0	0
1	0	1	2
2	0	2	1

Here, we can see that $1 \oplus 2 = 0$; indeed $1 + 2 = 3 = 1 \cdot 3 + 0$, which means $(1+2) \mod 3 = 0$. Furthermore, $2 \oplus 2 = 1$, since $2 + 2 = 4 = 1 \cdot 3 + 1$ so that $(2+2) \mod 3 = 1$. Similarly, $2 \odot 2 = 1$.

For a final example, take $p = 5$. We can see that $\mathbb{F}_5 = \{0, 1, 2, 3, 4\}$ with operations as shown below.

$\oplus$	0	1	2	3	4
0	0	1	2	3	4
1	1	2	3	4	0
2	2	3	4	0	1
3	3	4	0	1	2
4	4	0	1	2	3

$\odot$	0	1	2	3	4
0	0	0	0	0	0
1	0	1	2	3	4
2	0	2	4	1	3
3	0	3	1	4	2
4	0	4	3	2	1

Here, we can see that $2 \oplus 2 = 4$ since $2 + 2 = 0 \cdot 5 + 4$, whereas $2 \oplus 4 = 1$ since $2 + 4 = 6 = 1 \cdot 5 + 1$.

It is worth noting that $\{0, 1, 2, 3\}$, the set of integers mod 4, with the mod 4 operations is not a field. While the set is closed under $\oplus$ and $\odot$ and satisfies conditions 1. - 6. as shown below, condition 7. fails. Not every nonzero element has a multiplicative inverse.

$\oplus$	0	1	2	3
0	0	1	2	3
1	1	2	3	0
2	2	3	0	1
3	3	0	1	2

$\odot$	0	1	2	3
0	0	0	0	0
1	0	1	2	3
2	0	2	0	2
3	0	3	2	1

In particular, there is no element $b \in \{0, 1, 2, 3\}$ such that $2 \odot b = 1$.

Fields of prime cardinality are called prime fields. All prime fields, up to isomorphism (meaning a relabeling of elements that preserves the field operations), are of the form $\mathbb{F}_p$ with mod p arithmetic as described above. They are the building blocks for non-prime fields, meaning those of cardinality p^l, where p is prime and $l > 1$. Before more background on finite fields is outlined, let us pause to develop some notation and see the role they play in coding theory.

The set of nonzero elements of a field $\mathbb{F}$ is denoted by $\mathbb{F}^* := \mathbb{F} \setminus \{0\}$. We use $\mathbb{Z}^+$ to denote the set of positive integers and $\mathbb{N}$ to mean the set of nonnegative integers. The set of all $m \times n$ matrices over a field $\mathbb{F}$ is denoted by $\mathbb{F}^{m \times n}$. The nullspace of a matrix $A \in \mathbb{F}^{m \times n}$ is $NS(A) := \{x \in \mathbb{F}^{n \times 1} : Ax = 0\}$. In a set S, the symbol 0 will be used to denote the element (if it exists) that acts as the additive identity, which might have different forms depending on S itself. For instance, $0 \in \mathbb{F}^{m \times n}$ means the $m \times n$ matrix whose entries are all $0 \in \mathbb{F}$, whereas $0 \in \mathbb{Z}$ means the integer 0. The particular form that 0 takes should be clear from the context. Given $M \in \mathbb{F}^{m \times n}$, $Row_i M$ denotes the ith row of M, and $Col_j M$ denotes the jth column of M. Given $\mathcal{I} := \{j_1, \ldots, j_t\} \subseteq \{1, \ldots, n\}$, the matrix obtained from M by taking only columns with indices in $\mathcal{I}$ is $M \mid_{\mathcal{I}} := [Col_{j_1} M \ldots Col_{j_t} M] \in \mathbb{F}^{m \times t}$. The entry in the ith row and j^{th} column of M will be written as M_{ij}. We sometimes use a permutation matrix, meaning a matrix that has exactly one '1' entry in each row and in each column and 0 entries elsewhere; such a matrix is nonsingular.

Vectors are generally considered as row vectors, so that the transpose of $u \in \mathbb{F}^k = \mathbb{F}^{1 \times k}$ is $u^T \in \mathbb{F}^{k \times 1}$. We use $[n] := \{1, \ldots, n\}$ to denote the set of indices of the coordinates of a word $w \in \mathbb{F}^n$. Given $w \in \mathbb{F}^n$ and $i \in [n]$, w_i denotes the ith coordinate of w. The standard basis vectors e_i of $\mathbb{F}^n$, where $i \in [n]$, have

$$e_{ij} = \begin{cases} 1 & i = j \\ 0 & otherwise. \end{cases}$$

Given $\mathcal{I} = \{i_1, \ldots, i_j\} \subseteq [n]$ and $u \in \mathbb{F}^n$, the vector $(u_i)_{i \in \mathcal{I}} := (u_{i_1}, \ldots, u_{i_j}) \in \mathbb{F}^{|\mathcal{I}|}$ is a vector formed by eliminating all coordinates whose indices are not in $\mathcal{I}$. The support of a vector $v \in \mathbb{F}^n$ is $\mathrm{Supp}(v) := \{i \in [n] : v_i \neq 0\}$. Given a set of vectors $S = \{u_1, \ldots, u_s\} \subseteq \mathbb{F}^n$, the vector space over $\mathbb{F}$ generated by S is denoted by

$$\langle S \rangle := \left\{ \sum_{i=1}^{s} a_i u_i : a_i \in \mathbb{F} \right\} \subseteq \mathbb{F}^n.$$

When convenient, we may write $\langle u_1, \ldots, u_s \rangle$ to mean $\langle S \rangle$.

Given vector spaces X and Y, we sometimes write

$$\begin{aligned} f : X &\to Y \\ x &\mapsto y \end{aligned}$$

to mean that f is a map from X to Y and $f(x) = y$. The image, or range, of the map f, is written as $f(X)$; that is, $f(X) = \{f(x) : x \in X\}$. The kernel, or nullspace, of the map f is

$$ker(f) := \{x \in X : f(x) = 0\}.$$

If f is a linear transformation, the Rank-Nullity Theorem from linear algebra can be phrased as $\dim f(X) = \dim X - \dim ker(f)$. Furthermore, f is injective if and only if $ker(f) = \{0\}$.

The set of polynomials in the indeterminate x with coefficients in $\mathbb{F}$ is denoted by $\mathbb{F}[x]$, meaning $\mathbb{F}[x] = \{\sum_{i=0}^{m} a_i x^i : a_i \in \mathbb{F}, m \in \mathbb{N}\}$. This set naturally inherits the addition and multiplication of $\mathbb{F}$. To be more precise, consider polynomials $a(x) = \sum_{i=0}^{s} a_i x^i, b(x) = \sum_{i=0}^{t} b_i x^i \in \mathbb{F}[x]$ which means $a_i, b_j \in \mathbb{F}$ for all $i, 0 \leq i \leq s$ and for all $j, 0 \leq j \leq t$. Without loss of generality, we may assume that $s \geq t$. We can then write $b(x) = \sum_{i=0}^{s} b_i x^i$ by taking $b_j = 0$ for all $j, t+1 \leq j \leq s$. Then

$$a(x) + b(x) = \sum_{i=0}^{s} (a_i + b_i) x^i$$

and

$$a(x)b(x) = \sum_{k=0}^{s+t} \left(\sum_{i=0}^{k} (a_i \cdot b_{k-i}) \right) x^k$$

where the sums $a_i + b_i$ and $\sum_{i=0}^{k} (a_i \cdot b_{k-i})$ as well as the products $a_i b_{k-i}$ are taken in $\mathbb{F}$. A polynomial $h(x) \in \mathbb{F}[x]$ is irreducible if and only if $h(x) \neq a(x)b(x)$ for any nonconstant polynomials $a(x), b(x) \in \mathbb{F}[x]$. If $f(x) = \sum_{i=0}^{m} a_i x^i \in \mathbb{F}[x]$ and $a \in \mathbb{F}$, then the evaluation of f at a is

$$f(a) = \sum_{i=0}^{m} a_i a^i \in \mathbb{F}, \qquad (2.1)$$

since $\mathbb{F}$ is closed under addition and multiplication.

We are now ready to define the key concept in this section, a linear code.

Definition 2.1. Given $n \in \mathbb{Z}^+$, a linear code C of length n over $\mathbb{F}_q$ is an $\mathbb{F}_q$-subspace of $\mathbb{F}_q^n$. Elements of C are called codewords.

All codes considered in this chapter will be linear codes; hence, we use the term code to mean linear code throughout. Hence, given $c, c' \in C$ where C is a code over $\mathbb{F}_q$, $ac + bc' \in C$ for all $a, b \in \mathbb{F}_q$. Note also that $0 \in C$. If $q = 2$, C is said to be a binary code; otherwise, C is called nonbinary.

A code C has three important classical parameters: length (Definition 2.1 above); dimension (Definition 2.2 below); and minimum distance (Definition 2.3 below).

Definition 2.2. The dimension k of a code C over $\mathbb{F}_q$ is its dimension as a vector space over $\mathbb{F}_q$, meaning

$$k := \dim_{\mathbb{F}_q} C.$$

Given vectors $u, v \in \mathbb{F}_q^n$, the Hamming distance from u to v is

$$d(u, v) := |\{i \in [n] : u_i \neq v_i\}|.$$

Notice that $d(u, v) = d(v, u)$, so we typically say the Hamming distance between u and v.

Definition 2.3. The minimum distance d of a code C is

$$d := \min \{d(u, v) : u, v \in C, u \neq v\}.$$

The weight of a vector $u \in \mathbb{F}_q^n$ is $wt(u) := d(u, 0)$, meaning

$$wt(u) = |\{i \in [n] : u_i \neq 0\}|$$

is the number of nonzero coordinates of u. Notice that $d(u, v) = wt(u - v)$ Because C is a vector space, $u - v \in C$ for all $u, v \in C$. As a result, the minimum distance of a code may be thought of as the minimum weight among the nonzero codewords; more precisely,

$$\begin{aligned} d &= \min\{wt(u) : u \in C, u \neq 0\} \\ &= \min |\{i \in [n] : c_i \neq 0, c \in C \setminus \{0\}\}|. \end{aligned}$$

We say that C is an $[n, k, d]$ code to mean C has length n, dimension k, and minimum distance d. If the minimum distance is not relevant to the discussion, we sometimes write $[n, k]$ code instead.

The information rate of an $[n, k, d]$ code is the ratio $r := \frac{k}{n}$, which gives a measure of efficiency of the code, in that each codeword of length n transmits a message of length k as demonstrated in Figure 2.2.

A code with minimum distance d can correct any $t := \lfloor \frac{d-1}{2} \rfloor$ errors. To see this, consider an $[n, k, d]$ code C and a word $w \in \mathbb{F}_q^n$. We claim that there is at most one codeword of C within distance t of w. Suppose there exist $c, c' \in C$ such that $d(c, w) \leq t$ and $d(c', w) \leq t$. Then $wt(c - w) \leq t$ and $wt(w - c') \leq t$, which implies

$$wt(c - c') \leq wt(c - w) + wt(w - c') \leq 2t.$$

Since C is a vector space and $c, c' \in C$, $c - c' \in C$. Hence, $c = c'$ as the minimum weight nonzero codeword of C is of weight $d > 2t$. It follows that each word in $\mathbb{F}_q^n$ has at most a single codeword within distance t of it. Thus, a received word w containing at most t errors, meaning $w = c + e$ where $wt(e) \leq t$, is decoded to the unique codeword c that is within distance t of it.

The relative distance of an $[n, k, d]$ code is $\frac{d}{n}$. Since $0 \leq k \leq n$ and $0 \leq d \leq n$, $0 \leq \frac{k}{n} \leq 1$ and $0 \leq \frac{d}{n} \leq 1$; thus, the information rate and relative distance are real numbers between 0 and 1. A code is said to have good error-correcting capability if its relative distance is near 1 while it is very efficient if its information rate is near 1. As one might suspect, these two quantities are linked, as demonstrated by the next result.

Theorem 2.1. *(Singleton Bound)* If C is an $[n, k, d]$ code, then

$$k + d \leq n + 1.$$

A code is called maximum distance separable (or MDS) if the minimum distance of the code meets the Singleton bound with equality.

A generator matrix of an $[n, k]$ code C is a $k \times n$ matrix $G \in \mathbb{F}_q^{k \times n}$ such that the set of rows of G is a basis for C as an $\mathbb{F}_q$-vector space. If $G = [I_k | A]$ for some $A \in \mathbb{F}_q^{k \times (n-k)}$, then G is said to be in systematic form. Encoding of a word $m \in \mathbb{F}_q^k$ may be accomplished by matrix multiplication to obtain the associated codeword. To see this, fix a generator matrix M for C. Then

$$c = mG.$$

If G is in systematic form, then

$$c = mG = m[I_k | A] = (m_1, \ldots, m_k, c_{k+1}, \ldots, c_n)$$

for some $c_{k+1}, \ldots, c_n \in \mathbb{F}_q$, and the first k symbols (or positions) are called information symbols (or positions).

A parity-check matrix of C over $\mathbb{F}_q$ is any matrix H whose nullspace is C; that is, H is a parity-check of C if and only if $Hc^T = 0$ for all $c \in C$ and $Hx^T \neq 0$ for all $x \in \mathbb{F}_q^n \setminus C$. A code necessarily has many different generator matrices and many different parity-check matrices. Notice that if C has generator matrix

$$G = [I_k | A],$$

then
$$H = \left[-A^T | I_{n-k}\right] \quad (2.2)$$
is a parity-check matrix for C. To see this, it may be verified that
$$HG^T = \left[-A^T | I_{n-k}\right] [I_k | A]^T = -A^T I_k^T + I_{n-k} A^T = 0 \in \mathbb{F}_q^{(n-k) \times n}. \quad (2.3)$$

The minimum distance d of the code C equals the minimum number of linearly dependent columns of a parity-check matrix H for C; that is,
$$d = \min \left\{ l : \{Col_{j_1} H, \ldots, Col_{j_l} H\} \text{ is linearly dependent} \right\}. \quad (2.4)$$
To see this, suppose that H has l linearly dependent columns $Col_{j_1} H, \ldots, Col_{j_l} H$. Then there exist $a_1, \ldots, a_l \in \mathbb{F}_q$, not all 0, such that
$$\sum_{i=1}^{l} a_i Col_{j_i} H = 0.$$
Let $u = \sum_{i=1}^{l} a_i e_{j_i}$. Then $Hu^T = 0$, so $u \in C \setminus \{0\}$. Hence, the minimum codeword weight among the nonzero codewords is the smallest number of linearly dependent columns of H.

The most basic examples of codes are simple parity-check codes and repetition codes which are presented in the next example.

Example 2.2. It is straightforward to check that
$$C = \left\{ (c_1, \ldots, c_n) : c_i \in \mathbb{F}_q, c_n = \sum_{i=1}^{n-1} c_i \right\} \subseteq \mathbb{F}_q^n$$
is an $[n, n-1, 2]$ code with generator matrix $G = [I_{n-1} | 1_{n-1}]$, where $1_{n-1} \in \mathbb{F}^{(n-1) \times 1}$ is the all-one vector. The message word $m \in \mathbb{F}_q^{n-1}$ is encoded as the codeword $\left(m_1, \ldots, m_{n-1}, \sum_{i=1}^{n-1} m_i\right)$. The code C is called a simple parity-check code. It is efficient in that the information rate is $\frac{n-1}{n}$ but is not able to correct any errors.

One may also verify that if $s, l \in \mathbb{Z}^+$, then
$$C = \{(c_1, \ldots, c_s, c_1, \ldots, c_s, \ldots, c_1, \ldots, c_s) : c_i \in \mathbb{F}_q\} \in \mathbb{F}_q^{sl}$$
is an $[sl, s, l]$ code with generator matrix
$$\begin{bmatrix} I_s & I_s & \cdots & I_s \end{bmatrix} \in \mathbb{F}_q^{s \times sl}.$$

The code C is called a repetition code. If l is large, then C, having information rate $\frac{s}{sl} = \frac{1}{l}$, is not considered to be efficient; however, it does have good error-correcting capability since it can correct $\lfloor \frac{l-1}{2} \rfloor$ errors.

The next example features an important family of codes, called Hamming codes, which are among the oldest error-correcting codes. They were invented by Richard Hamming to automatically correct errors in computations introduced by punched card readers.

Example 2.3. Let $m \in \mathbb{Z}^+$. Notice that there are 2^m vectors in $\mathbb{F}_2^m$, one of which is the all-zero vector $0 \in \mathbb{F}_2^m$. Consider a matrix $H \in \mathbb{F}_2^{m \times (2^m - 1)}$ whose columns are precisely the nonzero vectors in $\mathbb{F}_2^m$. Let $C = NS(H)$ be the binary code with H as a parity-check matrix. The length of C is then $2^m - 1$, and the dimension of C is $2^m - m - 1$. Observe that the standard basis vectors e_1 and e_2 are among the columns of H as is $e_1 + e_2$. Hence, H has three linearly dependent columns. According to (2.4), $d = 3$.

Taking $m = 3$, we obtain a $[7, 4, 3]$ code which can be described by the parity-check matrix

$$H = \begin{bmatrix} 1 & 1 & 0 & 1 & 1 & 0 & 0 \\ 1 & 0 & 1 & 1 & 0 & 1 & 0 \\ 0 & 1 & 1 & 1 & 0 & 0 & 1 \end{bmatrix} \in \mathbb{F}_2^{3 \times 7}.$$

Using (2.2), we can see that

$$G = \begin{bmatrix} 1 & 1 & 0 & 1 & 0 & 0 & 0 \\ 1 & 0 & 1 & 0 & 1 & 0 & 0 \\ 0 & 1 & 1 & 0 & 0 & 1 & 0 \\ 1 & 1 & 1 & 0 & 0 & 0 & 1 \end{bmatrix} \in \mathbb{F}_2^{4 \times 7}$$

is a generator matrix for C. Hence, a message word $m = (m_1, m_2, m_3, m_4) \in \mathbb{F}_2^4$ is encoded as the codeword

$$mG = (m_1 + m_2 + m_4, m_1 + m_3 + m_4, m_2 + m_3 + m_4, m_1, m_2, m_3, m_4).$$

The $[7, 4, 3]$ Hamming code described above is especially famous, as it was showcased in the first paper on the construction of error-correcting codes (Hamming, 1950).

More examples of codes will be considered in Subsection 2.2.2. To conclude this subsection, we consider ways in which existing codes give rise to new ones.

Definition 2.4. The dual of an $[n, k]$ code C over $\mathbb{F}_q$ is

$$C^\perp = \{w \in \mathbb{F}_q^n : w \cdot c = 0 \; \forall c \in C\}$$

where $w \cdot c := \sum_{i=1}^n w_i c_i$ is the usual dot product.

If C is an $[n, k]$ code over $\mathbb{F}_q$, then $C^\perp$ is an $[n, n - k]$ code over $\mathbb{F}_q$. Moreover, if H is a parity-check matrix for C, then H is a generator matrix for $C^\perp$ and vice versa.

If $C' \subseteq C$ where C and C' are both codes, then we say that C' is a subcode of C. Clearly, C' has the same length as C, dimension bounded above by that of C, and minimum distance bounded below by that of C; that is, if C is an $[n, k, d]$ code and C' is an $[n', k', d']$ code with $C' \subseteq C$, then $n' = n$, $k' \leq k$, and $d' \geq d$.

Definition 2.5. Let C be an $[n, k, d]$ code over $\mathbb{F}_q$ and $\mathcal{I} \subseteq [n]$. The puncturing of C at $\mathcal{I}$ is
$$P_\mathcal{I}(C) := \{(c_i)_{i \in [n] \setminus \mathcal{I}} : c \in C\} \subseteq \mathbb{F}_q^{n-|\mathcal{I}|}.$$
The shortening of C at $\mathcal{I}$ is
$$S_\mathcal{I}(C) := \{(c_i)_{i \in [n] \setminus \mathcal{I}} : c \in C, c_i = 0 \; \forall i \in \mathcal{I}\} \subseteq \mathbb{F}_q^{n-|\mathcal{I}|}.$$

Notice that $S_\mathcal{I}(C)$ is a subcode of $P_\mathcal{I}(C)$. Shortening and puncturing will be utilized in Section 2.4 to provide attacks on code-based cryptosystems.

There is one additional modification to codes that will be relevant to the discussion in this chapter. It applies to codes over alphabets which are non-prime fields. For that reason, we take a brief interlude to explore field extensions and fields whose cardinalities are not prime.

Suppose that $q = p^l$ where p is prime as above and l is an integer such that $l > 1$. Consider the set $\mathbb{F}_q := \{f(x) \in \mathbb{F}_p[x] : \deg f \leq l - 1\}$. Notice that $|\mathbb{F}_q| = p^l$; indeed, there are p^l polynomials of the form $\sum_{i=0}^{l-1} a_i x^i$ with $a_i \in \mathbb{F}_p$ as there are p choices for each of the l coefficients $a_0, \ldots, a_{l-1}$. We wish to define operations of addition and multiplication so that this set is a field. To do so, we employ an irreducible polynomial $h(x) \in \mathbb{F}_p[x]$ of degree l. Given any polynomial $f(x) \in \mathbb{F}_p[x]$, the Division Algorithm states that there exist unique polynomials $Q(x), R(x) \in \mathbb{F}_p[x]$ such that $f(x) = Q(x)h(x) + R(x)$ with either $R(x) = 0$ or $0 \leq \deg R(x) \leq l - 1$. We can define $f(x) \mod h(x) := R(x)$. Using this, we define the operations $\oplus$ and $\odot$ on $\mathbb{F}_q[x]$ as follows. For $a(x), b(x) \in \mathbb{F}_q$, $a(x) \oplus b(x) = (a(x) + b(x)) \mod h(x)$ and $a(x) \odot b(x) = (a(x)b(x)) \mod h(x)$; here, the sum $a(x) + b(x)$ and the product $a(x)b(x)$ are taken in $\mathbb{F}_p[x]$. It turns out that $\mathbb{F}_q$ with these operations is a field of cardinality p^l, and we write $\mathbb{F}_q = \mathbb{F}_p[x]/\langle h(x) \rangle$. When the context is clear, we usually write $+$ instead of $\oplus$ and $a(x)b(x)$ instead of $a(x) \odot b(x)$. Moreover, $\mathbb{F}_p \subseteq \mathbb{F}_q$, because elements of the set $\{0, \ldots, p-1\}$ are constant polynomials, meaning of degree 0. For that reason, we consider $\mathbb{F}_q$ as an extension of $\mathbb{F}_p$ and sometimes express this by writing

$$\begin{array}{c} \mathbb{F}_{p^l} \\ |l \\ \mathbb{F}_p. \end{array}$$

With this in mind, we provide an example.

Example 2.4. Take $p = 2$ and $l = 2$ so that $q = 2^2 = 4$. Notice that $\mathbb{F}_2[x] = \{\sum_{i=0}^m a_i x_i : m \in \mathbb{N}, a_i = \{0, 1\}\}$. In particular, the only possible coefficients of polynomials in this set are 0 and 1. One can check that $h(x) := x^2 + x + 1 \in \mathbb{F}_2[x]$ is irreducible. Indeed, if $h(x) = a(x)b(x)$ with $a(x)$ and $b(x)$ nonconstant polynomials, then $\deg a(x) = \deg b(x) = 1$. This implies $a(x), b(x) \in \{x, x+1\}$. Notice that $x \cdot x = x^2 \neq h(x)$, $x \cdot (x+1) = x^2 + x \neq h(x)$, and $(x+1) \cdot (x+1) = x^2 + 1 \neq h(x)$; the last equality follows from the facts that $(x+1)(x+1) = x^2 + 2x + 1$ and $2 \mod 2 = 0$. This confirms that $h(x)$ is irreducible. Hence, $\mathbb{F}_4 = \{0, 1, x, x+1\}$ with addition and multiplication mod $(x^2 + x + 1)$ is a field, meaning $\mathbb{F}_4 = \mathbb{F}_2[x]/\langle x^2 + x + 1\rangle$. One may verify this by consulting the tables below.

$\oplus$	0	1	x	$x+1$
0	0	1	x	$x+1$
1	1	$x+1$	0	x
x	x	$x+1$	0	1
$x+1$	$x+1$	x	1	0

$\odot$	0	1	x	$x+1$
0	0	0	0	0
1	0	1	x	$x+1$
x	0	x	$x+1$	1
$x+1$	0	$x+1$	1	0

Additional explanation of some entries may be helpful. For instance, $(x+1) \oplus 1 = x$, $x \oplus x = 0$, $x \oplus (x+1) = 1$ and $(x+1) \oplus (x+1) = 0$ all follow from the fact that $2 \mod 2 = 0$. Notice that $x \cdot x = x^2 \mod (x^2 + x + 1) = (x+1)$, because $x^2 = 1 \cdot (x^2 + x + 1) + (x+1)$; the last equality follows from the fact that $(x+1)+(x+1) = (x+x)+(1+1) = 0$. Also, $x \cdot (x+1) = x^2 + x \mod (x^2+x+1) = 1$, because $x^2 + x = 1 \cdot (x^2 + x + 1) + 1$; the last equality follows from the fact that $1 + 1 = 0$ in $\mathbb{F}_2$. Finally, $(x+1) \cdot (x+1) = x^2 + 2x + 1 \mod (x^2 + x + 1) = 0$, because $x^2 + x + 1 = 1 \cdot (x^2 + x + 1) + 0$.

Example 2.5. To obtain a field of cardinality 8, one must use an irreducible polynomial $h(x) \in \mathbb{F}_2[x]$ of degree 3. The reader may verify that both $x^3 + x^2 + 1$ and $x^3 + x + 1$ are irreducible polynomials in $\mathbb{F}_2[x]$. We may use either one to define the field of size 8. In this example, we take $h(x) = x^3 + x + 1$ and consider $\mathbb{F}_8 = \{ax^2 + bx + c : a, b, c \in \{0, 1\}\}$ with the operations $\oplus$ and $\odot$ given $\mod h(x)$, meaning $\mathbb{F}_8 = \mathbb{F}_2[x]/\langle x^3 + x + 1\rangle$. Partial addition and multiplication tables are given below. Here, in the interest of space, we suppress the rows and columns in which a summand or factor is 0 as $0 \oplus a = a = a \oplus 0$ and $0 \odot a = 0 = a \odot 0$ for all $a \in \mathbb{F}_8$.

$\oplus$	1	x	$x+1$	x^2	x^2+1	x^2+x	x^2+x+1
1	0	$x+1$	x	x^2+1	x^2	x^2+x+1	x^2+x
x	$x+1$	0	1	x^2+x	x^2+x+1	x^2	x^2+1
$x+1$	x	1	0	x^2+x+1	x^2+x	x^2+1	x^2
x^2	x^2+1	x^2+x	x^2+x+1	0	x	1	$x+1$
x^2+1	x^2	x^2+x+1	x^2+x	1	0	$x+1$	x
x^2+1	x^2+x+1	x^2	x^2+1	x	$x+1$	0	1
x^2+x+1	x^2+x	x^2+1	x^2	$x+1$	x	1	0

$\odot$	1	x	$x+1$	x^2	x^2+1	x^2+x	x^2+x+1
1	1	x	$x+1$	x^2	x^2+1	x^2+x	x^2+x+1
x	x	x^2	x^2+x	$x+1$	1	x^2+x+1	x^2+1
$x+1$	$x+1$	x^2+x	x^2+1	x^2+x+1	x^2	1	x
x^2	x^2	$x+1$	x^2+x+1	x^2+x	x	x^2+1	1
x^2+1	x^2+1	1	x^2	x	x^2+x+1	$x+1$	x^2+x
x^2+x	x^2+x	x^2+x+1	1	x^2+1	$x+1$	x	x^2
x^2+x+1	x^2+x+1	x^2+1	x	1	x^2+x	x^2	$x+1$

We leave it as an exercise to construct the field of cardinality 8 using the irreducible polynomial $h(x) = x^3 + x^2 + 1$. By writing its addition and multiplication tables explicitly, one can see that there is a bijection $\phi : \mathbb{F}_2[x]/\langle x^3 + x^2 + 1\rangle \to \mathbb{F}_2[x]/\langle x^3 + x + 1\rangle$ such that $\phi(f(x)+g(x)) = \phi(f(x))+\phi(g(x))$ and $\phi(f(x)g(x)) =$

$\phi(f(x))\phi(g(x))$ for all $f(x), g(x) \in \mathbb{F}_2[x]/\langle x^3 + x^2 + 1\rangle$. Such a map is called an isomorphism and confirms that these two representations are equivalent up to re-labeling.

It is also noting that while $\mathbb{F}_2 \subseteq \mathbb{F}_4$ and $\mathbb{F}_2 \subseteq \mathbb{F}_8$, $\mathbb{F}_4 \not\subseteq \mathbb{F}_8$. While elements labeled x and $x+1$ exist in both $\mathbb{F}_4$ and $\mathbb{F}_8$, their behavior is different in these two contexts. In particular, in $\mathbb{F}_4$, $x \cdot (x+1) = x^2 + x = 1$, whereas in $\mathbb{F}_8$, $x \cdot (x+1) = x^2 + x \neq 1$.

One can also consider the setting where q is a power of a prime that is not prime itself, $l > 1$, and
$$\begin{array}{c} \mathbb{F}_{q^l} \\ |\,l \\ \mathbb{F}_q, \end{array}$$
meaning $\mathbb{F}_q \subseteq \mathbb{F}_{q^l}$. For our purposes in this situation, it will be sufficient to consider both $\mathbb{F}_q$ and $\mathbb{F}_{q^l}$ as constructed as above beginning with $\mathbb{F}_p$ where $q = p^s$ for some s; that is, $\mathbb{F}_q = \mathbb{F}_p[x]/\langle h(x) \rangle$ where $h(x) \in \mathbb{F}_p[x]$ is irreducible with $\deg h(x) = s$ and $\mathbb{F}_{q^l} = \mathbb{F}_p[x]/\langle g(x) \rangle$ where $g(x) \in \mathbb{F}_p[x]$ is irreducible with $\deg g(x) = sl$.

In general, it can be shown that $\mathbb{F}_q \subseteq \mathbb{F}_{q'}$ if and only if $q' = q^l$ for some $l \in \mathbb{Z}^+$. Looking back at the end of Example 2.5, we note that the smallest field containing both $\mathbb{F}_4$ and $\mathbb{F}_8$ is $\mathbb{F}_{64}$.

The next procedure defines a code over the base field $\mathbb{F}_q$ from that over the extension $\mathbb{F}_{q^l}$.

Definition 2.6. A subfield subcode of an $[n, k, d]$ code $C \subseteq \mathbb{F}_{q^l}^n$ is of the form
$$C|_{\mathbb{F}_q} := C \cap \mathbb{F}_q^n.$$

If C is an $[n, k, d]$ code over $\mathbb{F}_{q^l}$, then $C|_{\mathbb{F}_q}$ is an $[n, \geq n - l(n-k), \geq d]$ code over $\mathbb{F}_q$.

There are many excellent textbooks on coding theory including Huffman and Pless (2003) and MacWilliams and Sloane (1977), and more details may be found there. Here, we focus on those code families and concepts frequently considered in code-based cryptography.

2.2.2 Code Constructions

2.2.2.1 Reed-Solomon Codes

Reed-Solomon codes are a type of evaluation code where codewords are obtained by evaluating univariate polynomials of bounded degree at field elements. They were first defined in 1960 by Irving Reed and Gustave Solomon (Reed and Solomon, 1960) and have seen extensive commercial use beginning with compact discs in 1982. They aid in the digital transfer of information for video broadcasting, satellite communication, and broadband as well as supporting distributed storage.

Fix an alphabet $\mathbb{F}_q$ and positive integers k and n with $k < n \leq q$. Write $\mathbb{F}_q = \{\alpha_1, \ldots, \alpha_q\}$. Set
$$L_k := \{f \in \mathbb{F}_q[x] : \deg f < k\}.$$

Recall from (2.1) that $f \in \mathbb{F}_q[x]$ and $\alpha \in \mathbb{F}_q$ implies $f(\alpha) \in \mathbb{F}_q$; hence, $f \in L_k$ and $\alpha \in \mathbb{F}_q$ implies $f(\alpha) \in \mathbb{F}_q$. We can extend this notion to define a map from the set L_k to $\mathbb{F}_q^n$ by evaluating $f \in L_k$ at n distinct field elements $\alpha_1, \ldots, \alpha_n$ as follows. Given $f \in L_k$, notice that $(f(\alpha_1), f(\alpha_2), \ldots, f(\alpha_n)) \in \mathbb{F}_q^n$. This map is called an evaluation map, and we write

$$ev(f) := (f(\alpha_1), f(\alpha_2), \ldots, f(\alpha_n))$$

to reflect this. This idea is captured as

$$ev: \begin{array}{rcl} L_k & \to & \mathbb{F}_q^n \\ f & \mapsto & (f(\alpha_1), f(\alpha_2), \ldots, f(\alpha_n)). \end{array}$$

The kernel of ev is

$$\begin{aligned} ker(ev) &= \{f \in L_k : ev(f) = 0\} \\ &= \{f \in L_k : f(\alpha_1) = f(\alpha_2) = \cdots = f(\alpha_n) = 0\}. \end{aligned}$$

Notice that if there exists nonzero $f \in ker(ev)$, then f has at least n distinct roots $\alpha_1, \ldots, \alpha_n$, because $f(\alpha_j) = 0$ for all $j \in [n]$. However, $\deg f < k < n$ which makes this impossible. Hence, there are no nonzero elements of $ker(ev)$. Therefore, $ker(ev) = \{0\}$. As a result, ev is injective; thus, each codeword $c \in C_k$ has a unique preimage which will be denoted f_c.

The image of the map ev is a subspace of $\mathbb{F}_q^n$, and it is equal to the Reed-Solomon code C_k; that is $C_k := ev(L_k)$. Said differently, the Reed-Solomon code C_k is

$$C_k = \{(f(\alpha_1), f(\alpha_2), \ldots, f(\alpha_n)) : f \in L_k\} \subseteq \mathbb{F}_q^n.$$

It is immediate that the length of C_k is n. Moreover, the Rank-Nullity Theorem indicates that the dimension of C_k is

$$\dim C_k = \dim ev(L_k) = \dim L_k - \dim ker(ev) = k - 0 = k.$$

Next, we determine the minimum distance d of C_k. By definition, $d = wt(ev(f))$ for some $f \in L_k \setminus \{0\}$. Notice that f has at most $k-1$ distinct roots among $\alpha_1, \ldots, \alpha_n$, meaning there are at most $k-1$ values among $\alpha_1, \ldots, \alpha_n$ with $f(\alpha_j) = 0$. Because the weight of $ev(f)$ is $wt(ev(f)) = n - |\{j : f(\alpha_j) = 0\}|$, $d \geq n - (k-1)$. By the Singleton Bound (Theorem 2.1), $d \geq n - k + 1$. Consequently, $d = n - k + 1$, and the code C_k is an $[n, k, n-(k-1)]$ code over $\mathbb{F}_q$. This demonstrates that Reed-Solomon codes are MDS.

Because $\{1, x, x^2, \ldots, x^{k-1}\}$ is a basis for L_k and ev is injective, $\{ev(1), ev(x), ev(x^2), \ldots, ev(x^{k-1})\}$ is a basis for C_k. These basis elements can be written explicitly as $ev(x^i) = (\alpha_1^i, \alpha_2^i, \ldots, \alpha_n^i)$. It follows that a generator matrix for C_k is given by

$$G = \begin{array}{c} \\ 1 \\ x \\ x^2 \\ \vdots \\ x^{k-1} \end{array} \begin{array}{c} \alpha_1 \quad\; \alpha_2 \quad \cdots \quad \alpha_n \\ \left[\begin{array}{cccc} 1 & 1 & \cdots & 1 \\ \alpha_1 & \alpha_2 & \cdots & \alpha_n \\ \alpha_1^2 & \alpha_2^2 & \cdots & \alpha_n^2 \\ \vdots & \vdots & & \vdots \\ \alpha_1^{k-1} & \alpha_2^{k-1} & \cdots & \alpha_n^{k-1} \end{array}\right] \end{array} \in \mathbb{F}_q^{k \times n}$$

where we leave the rows and columns indexed as shown to emphasize the origin of the entries. Notice that a message $a = (a_0, \ldots, a_{k-1}) \in \mathbb{F}_q^k$ is encoded by the Reed-Solomon code C_k as

$$aG = \left(\sum_{i=0}^{k-1} a_i \alpha_1^i, \sum_{i=0}^{k-1} a_i \alpha_2^i, \ldots, \sum_{i=0}^{k-1} a_i \alpha_{k-1}^i \right) = ev(f)$$

where $f(x) = \sum_{i=0}^{k-1} a_i x^i$ is the polynomial whose coefficients are prescribed by the message a.

A generalized Reed-Solomon code is obtained by scaling the coordinates of a Reed-Solomon code as follows. Given a vector $v \in \mathbb{F}^n$ with all coordinates nonzero, set

$$C_{k,v} = \{(v_1 f(\alpha_1), v_2 f(\alpha_2), \ldots, v_n f(\alpha_n)) : f \in L_k\} \subseteq \mathbb{F}_q^n.$$

This code has as a generator matrix

$$\begin{array}{c c} & \begin{array}{cccc} v_1, \alpha_1 & v_2, \alpha_2 & \cdots & v_n, \alpha_n \end{array} \\ \begin{array}{c} 1 \\ x \\ x^2 \\ \vdots \\ x^{k-1} \end{array} & \left[\begin{array}{cccc} v_1 & v_2 & \cdots & v_n \\ v_1 \alpha_1 & v_2 \alpha_2 & \cdots & v_n \alpha_n \\ v_1 \alpha_1^2 & v_2 \alpha_2^2 & \cdots & v_n \alpha_n^2 \\ \vdots & \vdots & & \vdots \\ v_1 \alpha_1^{k-1} & v_2 \alpha_2^{k-1} & \cdots & v_n \alpha_n^{k-1} \end{array} \right] \end{array}.$$

Using this notation, $C_k = C_{k,1}$. The code $C_{k,v}$ can be considered as the image of the evaluation map

$$\begin{aligned} ev_v : L_k &\to \mathbb{F}_q^n \\ f &\mapsto (v_1 f(\alpha_1), v_2 f(\alpha_2), \ldots, v_n f(\alpha_n)). \end{aligned}$$

Because ev_v is injective, a codeword $c \in C_{k,v}$ has a unique preimage f_c; that is, given $c \in C_{k,v}$, there exists a unique $f_c \in L_k$ such that

$$ev_v(f_c) = c. \tag{2.5}$$

At times, when we wish to emphasize the dependence of the code on the field elements $\alpha_1, \ldots, \alpha_n$, we may write $C_{k,v,\alpha}$ instead of $C_{k,v}$.

The dual of a Reed-Solomon code C_k is the generalized Reed-Solomon code

$$C_k^\perp = \{(v_1 f(\alpha_1), v_2 f(\alpha_2), \ldots, v_n f(\alpha_n)) : f \in L_{n-k}\} \subseteq \mathbb{F}_q^n$$

where $v_i := \left(\prod_{j \in [n] \setminus \{i\}} \alpha_i - \alpha_j \right)^{-1}$. In the case where $n = q$, it can be shown that $v_i = \prod_{\alpha \in \mathbb{F}_q \setminus \{0\}} \alpha = 1$ and the dual of C_k is a Reed-Solomon code: $C_k^\perp = C_{n-k}$.

There are numerous generalizations of Reed-Solomon codes, including algebraic geometry codes, toric codes, and Cartesian codes.

2.2.2.2 Goppa Codes

Goppa codes were defined by V. D. Goppa (1970) and may be specified over any finite field. The McEliece cryptosystem is based on binary Goppa codes. Goppa codes may be explained in a variety of ways. Here, we take the subfield subcode point of view.

To define a code over $\mathbb{F}_{q^l}$ of length n, let $g(x) \in \mathbb{F}_{q^l}[x]$ be a polynomial of degree $t < n$, which is irreducible over $\mathbb{F}_{q^l}$. Choose a set of n distinct field elements, $\{\alpha_1, \ldots, \alpha_n\} \subseteq \mathbb{F}_{q^l}$. Because $g(x)$ is irreducible over $\mathbb{F}_{q^l}$, $g(x)$ has no roots in $\mathbb{F}_{q^l}$. Thus, $g(\alpha_i) \neq 0$ for all $i, 1 \leq i \leq n$, and $g(\alpha_i)$ has a multiplicative inverse $g(\alpha_i)^{-1} \in \mathbb{F}_{q^l}$. Due to this, $g(x)$ may be used to define a parity-check matrix by setting

$$H = \begin{bmatrix} g(\alpha_1)^{-1} & g(\alpha_2)^{-1} & \cdots & g(\alpha_n)^{-1} \\ \alpha_1 g(\alpha_1)^{-1} & \alpha_2 g(\alpha_2)^{-1} & \cdots & \alpha_n g(\alpha_n)^{-1} \\ \alpha_1^2 g(\alpha_1)^{-1} & \alpha_2^2 g(\alpha_2)^{-1} & \cdots & \alpha_n^2 g(\alpha_n)^{-1} \\ \vdots & \vdots & & \vdots \\ \alpha_1^{t-1} g(\alpha_1)^{-1} & \alpha_2^{t-1} g(\alpha_2)^{-1} & \cdots & \alpha_n^{t-1} g(\alpha_n)^{-1} \end{bmatrix} \in \mathbb{F}_{q^l}^{t \times n}.$$

Then

$$C = NS(H) \subseteq \mathbb{F}_{q^l}^n$$

is a code over $\mathbb{F}_{q^l}$. Using the subfield subcode construction, the Goppa code determined by $g(x)$ is

$$\Gamma := C|_{\mathbb{F}_q} \subseteq \mathbb{F}_q^n.$$

It can be shown that Γ is an $[n, \geq k - lt, \geq 2t + 1]$ code over $\mathbb{F}_q$.

Moreover, Goppa codes are subfield subcodes of generalized Reed-Solomon codes:

$$\Gamma = C_{t,\lambda}^{\perp}|_{\mathbb{F}_q}$$

where $\lambda = \left(g(\alpha_1)^{-1}, \ldots, g(\alpha_n)^{-1}\right)$. Notice that there are nontrivial binary Goppa codes, whereas there are no such Reed-Solomon codes. Indeed, a Reed-Solomon code has length $n \leq q$ where q is the cardinality of the alphabet. Thus, a Reed-Solomon code over $\mathbb{F}_2$ has as its length either 1 or 2 and has no error-correcting capability.

2.2.2.3 Parity-Check Codes

With few exceptions, the first 50 years of coding theory revolved around classical code constructions such as those in Examples 2.2 and 2.3 and the previous subsections along with associated decoding algorithms. The early 1990s ushered in a new era coined modern coding theory. Modern coding theory relies on coupling code and decoding algorithm design along with a probabilistic approach to achieve superior performance in terms of error correction and decoding efficiency.

A parity-check code is a code given as the nullspace of a specified matrix, $C = NS(H)$ where $H \in \mathbb{F}_q^{(n-k) \times n}$. Since every linear code has a parity-check matrix, every linear code can be described as a parity-check code. However, this point of view is typically taken with codes whose description is specified via a sparse parity-check matrix, meaning one with few nonzero entries, which lacks apparent structure (unlike, for instance, the Reed-Solomon or Goppa codes described in the previous subsections). Iterative decoding algorithms are typically associated with such codes, and the particular matrix H impacts the decoder performance (as opposed to classical decoding algorithms which depend solely on the code space C). Loosely speaking, we say that a code C is a low-density parity-check (or LDPC) code if and only if $C = NS(H)$ where H has columns of weight $\mathcal{O}(\log n)$; alternatively, C is a moderate-density parity-check (or MDPC) code if and only if H has columns of weight $\mathcal{O}(\sqrt{n})$. LDPC codes are described by parity-check matrices with very few nonzero entries, while MDPC codes have more (but still not too many) nonzero entries in their associated parity-check matrices. LDPC codes have very fast (meaning linear) decoding algorithms based on belief propagation and iterative decoding. While the same algorithms may be applied to more general parity-check codes, such as MDPC codes, they typically yield inferior performance in terms of error correction.

LDPC codes were first defined by Gallagher (1962). The topic lay mostly dormant until the 1990s, with the important exception of the introduction of the graphical representation of an LDPC code introduced by Tanner in 1981 (Tanner, 1981). We pause to recall that a graph G is described in terms of its vertex set $V(G)$ and edge set $E(G)$, whose elements are subsets of pairs of elements of $V(G)$. Elements of $V(G)$ are called vertices, and elements of $E(G)$ are called edges. Two vertices $u, v \in V(G)$ are adjacent if and only if $\{u, v\} \in E(G)$. The graph G is bipartite if and only if its vertex set can be expressed as a disjoint union of two sets so that no edge contains two elements from one of the two sets; that is, G is bipartite if and only if $V(G) = V' \dot\cup V''$, meaning that $V(G) = V' \cup V''$ and $V' \cap V'' = \emptyset$, so that each edge contains one element of V' and one element of V''.

Example 2.6. Consider the graph C_6 given by $V(C_6) = \{v_1, v_2, v_3, v_4, v_5, v_6\}$ and $E(C_6) = \{\{v_1, v_2\}, \{v_2, v_3\}, \{v_3, v_4\}, \{v_4, v_5\}, \{v_5, v_6\}, \{v_6, v_1\}\}$, which can be depicted as shown in Figure 2.3. Setting $V' = \{v_1, v_3, v_5\}$ and $V'' = \{v_2, v_4, v_6\}$, it is immediate that each edge has precisely one element in each of the two sets. This proves that C_6 is bipartite.

Consider the graph C_5 given by $V(C_5) = \{v_1, v_2, v_3, v_4, v_5\}$ and $E(C_5) = \{\{v_1, v_2\}, \{v_2, v_3\}, \{v_3, v_4\}, \{v_4, v_5\}, \{v_5, v_1\},\}$, which can be depicted as shown in Figure 2.3. Suppose that C_5 is bipartite. Then, without loss of generality (meaning up to relabeling), $v_1, v_3 \in V'$ and $v_2, v_4 \in V''$. However, since $v_5 \in V(C_5)$, either $v_5 \in V'$ or $v_5 \in V''$. The former implies the edge $\{v_5, v_1\}$ has both its elements in V' whereas the latter implies the edge $\{v_4, v_5\}$ has both its elements in V''. This contradiction proves that C_5 is not bipartite.

The arguments above may be generalized to show that that the n-cycle C_n, defined by $V(C_n) = \{v_1, v_2, v_3, \ldots, v_n\}$ and $E(C_n) = \{\{v_1, v_2\}, \{v_2, v_3\}, \ldots, \{v_n, v_1\}\}$ is bipartite if and only if n is even.

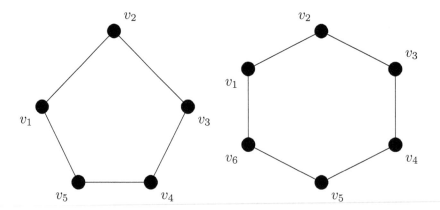

FIGURE 2.3 5-cycle and 6-cycle.

It is more common to represent a bipartite graph so that the vertices in V' are on the left and the vertices in V'' are on the right, as in Figure 2.4.

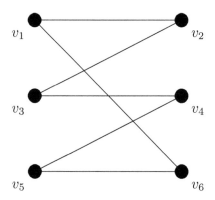

FIGURE 2.4 The 6-cycle is illustrated as a bipartite graph.

Additional examples of bipartite graphs appear in Example 2.7.

Next, we see how a code gives rise to a bipartite graph, based on how the code is represented in terms of a parity-check matrix. The Tanner graph of a binary code C with parity-check matrix $H \in \mathbb{F}_2^{r \times n}$ is a bipartite graph $T(H)$ with vertex set $X \dot{\cup} F$ where $X = \{v_i : i \in [n]\}$ and $F = \{f_i : i \in [r]\}$ and edge set E such that

- Each vertex in X, called a bit node, represents a coordinate position of the code; and each vertex in F, called a check node, represents a parity-check condition given by a row of H.

- $\{v_i, f_j\} \in E$ if and only if $h_{ji} \neq 0$.

Notice that the vertices in X are in one-to-one correspondence with the columns of H (equivalently, with the elements of $[n]$), and the vertices in F correspond to the rows of H. There is an edge containing vertices corresponding to $j \in [n]$ and $i \in F$

if and only if $Row_i H$ has a nonzero entry in position j, meaning $H_{ij} \neq 0$. This is demonstrated in Example 2.7 below.

Example 2.7. Consider the Hamming code C with parity-check matrix

$$H = \begin{bmatrix} 1 & 1 & 0 & 1 & 1 & 0 & 0 \\ 1 & 0 & 1 & 1 & 0 & 1 & 0 \\ 0 & 1 & 1 & 1 & 0 & 0 & 1 \end{bmatrix} \in \mathbb{F}_q^{3 \times 7}$$

from Example 2.3. Here, $X = \{v_1, v_2, v_3, v_4, v_5, v_6, v_7\}$ and $F = \{f_1, f_2, f_3\}$. Hence, $V(T(H)) = \{v_1, v_2, v_3, v_4, v_5, v_6, v_7, f_1, f_2, f_3\}$. Examining the first row of H, we see that $h_{11} = h_{12} = h_{14} = h_{15} = 1$ while $h_{13} = h_{16} = h_{17} = 0$. As a result, we see that f_1 is adjacent to vertices v_1, v_2, v_4, and v_5 and no others; said differently, $\{v_1, f_1\}, \{v_2, f_1\}, \{v_4, f_1\}, \{v_5, f_1\} \in E(T(H))$. The nonzero entries of $Row_2(H)$ are $h_{21}, h_{23}, h_{24}, h_{26}$ which implies that f_2 is adjacent to vertices v_1, v_3, v_4, and v_6 and no others. Considering $Row_3(H)$, we observe that f_3 is adjacent to vertices v_2, v_3, v_4, and v_7 and no others, since $h_{32} = h_{33} = h_{34} = h_{37} = 1$ and there are no other nonzero entries in the third row of H. Therefore, the edge set of the Tanner graph $T(H)$ is

$$E(T(H)) = \left\{ \begin{array}{l} \{v_1, f_1\}, \{v_2, f_1\}, \{v_4, f_1\}, \{v_5, f_1\}, \\ \{v_1, f_2\}, \{v_3, f_2\}, \{v_4, f_2\}, \{v_6, f_2\}, \\ \{v_2, f_3\}, \{v_3, f_3\}, \{v_4, f_3\}, \{v_7, f_3\} \end{array} \right\}.$$

This is depicted in Figure 2.5.

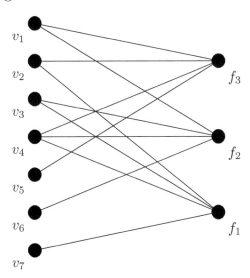

FIGURE 2.5 The Tanner graph of the Hamming code $C(H)$ with H as specified in Example 2.7.

The Tanner graph of a parity-check code over a field $\mathbb{F}_q$, where $q > 2$ is a weighted bipartite graph, with vertices and edges as described above and edge weights given by the entries of H, meaning edge (i, j) is assigned a weight $h_{ji} \in \mathbb{F}_q$. Please see the next example for an illustration.

Example 2.8. Consider the code C over $\mathbb{F}_3$ given by the parity-check matrix

$$H = \begin{bmatrix} 1 & 2 & 2 & 1 \\ 2 & 0 & 1 & 2 \end{bmatrix} \in \mathbb{F}_3^{2 \times 4}.$$

Its associated Tanner graph is displayed in Figure 2.6. It is interesting to note that the matrix

$$H' = \begin{bmatrix} 1 & 2 & 2 & 1 \\ 2 & 0 & 1 & 2 \end{bmatrix} \in \mathbb{F}_3^{2 \times 4}$$

defines the same code C; indeed, $NS(H) = NS(H')$ as the matrices H and H' have identical first rows and the second row of H' is a linear combination of the two rows of H. However, the associated Tanner graph (given in Figure 2.7) is different. This illustrates how the same code can be represented with graphs with different numbers of edges.

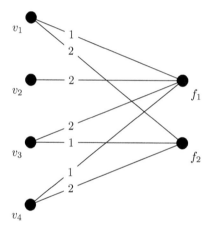

FIGURE 2.6 The Tanner graph of the Hamming code $C(H)$ with H as specified in Example 2.8.

Notice that a Tanner graph of a code C depends not just on the code itself but also on the particular choice of parity-check matrix H. The Tanner graph of an LDPC code is very sparse, meaning it has few edges, while that of an MDPC code is less sparse.

It is important to note that LDPC codes have no algebraic structure, unlike the Reed-Solomon and Goppa codes described above. However, their sparse graphical representations allow for fast decoding algorithms (MacKay, 1996) (MacKay, 1999).

2.2.3 Decoding

Decoding can be accomplished in a number of ways, depending on the particular code. In general, given an $[n, k, d = 2t + 1]$ code C over $\mathbb{F}_q$, a minimum distance

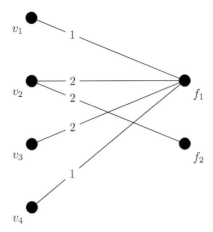

FIGURE 2.7 The Tanner graph of the Hamming code $C(H')$ with H' as specified in Example 2.8.

decoder $\mathcal{D}$ is a map
$$\begin{aligned} \mathcal{D}: \mathbb{F}_q^n &\to C \\ x &\mapsto c \end{aligned}$$
where c has the property that $d(x,c) = \min\{d(x,w) : w \in C\}$.

Specific families of codes may have specialized decoders, such as those for LDPC codes which operate on an associated Tanner graph. However, generic decoders exist. For instance, the syndrome decoding method outlined here applies to any linear code. In addition, it is a fundamental to the definition of the Niederreiter cryptosystem.

Definition 2.7. Consider a $[n,k,d]$ linear code over $\mathbb{F}_q$ given by $C = NS(H)$, meaning H is a parity-check matrix for C. The syndrome of $x \in \mathbb{F}_q^n$ is Hx^T.

Each codeword $c \in C$ has syndrome $Hc^T = 0$, because $c \in NS(H)$. Notice that if $x = c + e$ where $c \in C$ and $e \in \mathbb{F}_q^n$, then
$$Hx^T = H(c+e)^T = Hc^T + He^T = He^T;$$
that is, a received word has the same syndrome as its error vector. Moreover, if $x, x' \in \mathbb{F}^n$ and there exists a codeword $c \in C$ with $x - x' = c$, then
$$0 = Hc^T = H(x-x')^T = Hx^T - Hx'^T,$$
which implies that x and x' have the same syndrome. This demonstrates that all elements of the set $x + C := \{x + c : c \in C\}$ have the same syndrome. The set $x + C$ is called the coset of C in $\mathbb{F}_q^n$ determined by x. A set of this form is called a coset of C in $\mathbb{F}_q^n$; cosets may be reviewed more fully in an algebra course or text. Here, we note the important facts relative to decoding.

The set of all cosets of an $[n,k,d]$ code C over $\mathbb{F}_q$ is denoted $\mathbb{F}_q^n/C$, meaning

$$\mathbb{F}_q^n/C := \{x + C : x \in \mathbb{F}_q^n\}.$$

Observe that given two cosets $x + C$ and $y + C$ of C in $\mathbb{F}_q^n$,

$$x + C = y + C \Leftrightarrow x - y \in C. \tag{2.6}$$

Indeed, $x+C = y+C$ implies $x+0 = y+c$ for some codeword $c \in C$, which in turn implies that $x-y = c \in C$. On other hand, $x-y = c \in C$ forces $x = y+c$ and in turn $x+c' = y+c+c' \in y+C$ for any $c' \in C$ (since $c+c' \in C$); similarly, $x-y = c \in C$ forces $y = x - c$ and in turn $y + c' = x + (c' - c) \in y + C$ for any $c' \in C$ (since $c' - c \in C$). The cardinality of each coset $x+C$ is $|x+C| = |C|$, and every element of $x \in \mathbb{F}_q^n$ is in the coset it determines; in particular, $x = x+0 \in x+C$ as $0 \in C$. The total number of cosets of an $[n,k,d]$ code C over $\mathbb{F}_q$ can be found by dividing the total number of elements of $\mathbb{F}_q^n$ by the number of elements of $\mathbb{F}_q^n$ that determine the same coset. According to (2.6), each coset of C in $\mathbb{F}_q^n$ is determined by $|C|$ different elements; in particular, the coset $x+C$ is determined by any vector of the form $x+c$ where $c \in C$. Because $|C| = q^k$, there are q^k vectors of the form $x + c$. Thus, the total number of cosets of an $[n,k,d]$ code C is $|\mathbb{F}_q^n/C| = \frac{q^n}{q^k} = q^{n-k}$. For each of the q^{n-k} cosets $w_i + C$, $i = 1, \ldots, q^{n-k}$, take $s_i \in w_i + C$ such that $wt(s_i) \le wt(w_i + c)$ for all $c \in C$; the vector s_i is called a coset leader and has the smallest weight possible among those vectors in the same coset. In syndrome decoding, a list of coset leaders $s_1, \ldots, s_{n-k}$ and their corresponding syndromes $Hs_1^T, \ldots, Hs_{n-k}^T$ is kept for table lookup. The syndrome decoding process is summarized below.

Syndrome Decoding

Consider an $[n,k,d]$ code C with parity-check matrix $H \in \mathbb{F}_q^{k \times n}$. Suppose that $x \in \mathbb{F}_q^n$ is received.

1. Calculate the syndrome of x: Hx^T.

2. Using table lookup, determine a coset leader, say s. This is done by reviewing the list of syndromes $Hs_1^T, \ldots, Hs_{n-k}^T$ to find the one equal to Hx^T and noting its associated coset leader.

3. Decode x as $x - s$

To confirm that syndrome decoding enables the correction of up to t errors, suppose $x \in \mathbb{F}_q^n$ is received and $x = c + e$ with $c \in C$ and vector $e \in \mathbb{F}_q^n$ with $wt(e) \le t$. At this point, the decomposition of x into the vectors c and e is not known, and it is the goal of the decoding process to determine these vectors. Step 2. gives a coset leader s of $x + C$. Recall that $s \in x + C$, and observe that $e = x - c = x + (-c) \in x + C$ also. Hence, s and e have the same syndrome, as they are both elements of the coset $x + C$. By (2.6), there exists $c' \in C$ so that $s - e = c'$.

However, $wt(c') = wt(s-e) \leq wt(s) + wt(-e) = wt(s) + wt(e) \leq 2t$ which forces $c' = 0$. Thus, $s = e$, and $x - s = x - e = c$. As a result, taking $x - s$ in Step 3. is equivalent to taking $x - e$, and x is correctly decoded to the nearest codeword $x - e = c$.

2.3 CODE-BASED CRYPTOSYSTEMS

2.3.1 Public-Key Cryptography

Error-correcting codes were developed to mitigate the impact of noise in the channel or otherwise uncontrollable errors, and they are still designed with this in mind. However, they can also be used to intentionally insert errors that can be eliminated by those with knowledge of the code. This is the basis of code-based cryptography, which is one type of public-key cryptosystem. Before examining the details of code-based systems, we first provide a sketch of some key ideas of public-key cryptography. For a more complete introduction, there are many informative textbooks on the subject, such as (Hoffstein, Pipher, and Silverman, 2008), (Smart, 2015), and (Stinson and Paterson, 2018), which may be consulted.

Public-key cryptography is sometimes called asymmetric cryptography to emphasize that it utilizes two types of keys (public and private) which are not easily derivable from one another. In this scenario, the sender uses a public key to encipher the message, whereas the receiver uses a private key to decipher it. This is in contrast to symmetric key cryptography, in which the sender and receiver possess a key that is shared between them (which is used for encryption and decryption) and must be kept secret so as to not compromise the system. Symmetric key cryptography requires establishment of a secret key between users, which is a major drawback. How can such a key be agreed upon securely, prior to the formation of a secure channel? Even though symmetric key cryptography had been in use for thousands of years, it was not until 1976 that an answer to this question was provided by the method now known as Diffie-Hellman key exchange (Diffie and Hellman, 1976) (Merkle, 1978). This idea laid the groundwork for public-key cryptosystems such as RSA.

In a public-key cryptosystem, each user A has both a public key, denoted K_A and a private key P_A. Public keys are meant for wide-spread distribution, whereas private keys are not shared; that is, only user A has access to their private key P_A while the public keys are available to all. Loosely speaking, to send a message m to user A, their public key is used for encryption: $K_A(m)$. User A is able to decipher the message by applying P_A, so that $P_A(K_A(m))$ reveals the original message m. We typically think of K_A as a function, say f, that is easy to apply while P_A is its inverse and is difficult to find. Such a function K_A is known as a trapdoor function.

The most commonly implemented public-key cryptosystems are RSA and elliptic curve cryptography. The security of RSA relies on the complexity of factoring products of large primes, and that of elliptic curve cryptography relies on the complexity of solving the discrete log problem. While both remain intractable via classical algorithms, Shor's Algorithm solves both in polynomial time. It is precisely this that leads to the current interest in code-based cryptography.

2.3.2 The McEliece Public-Key Cryptosystem

Code-based cryptography was first defined by McEliece in 1978 (McEliece, 1978). We abstract the key components here for use with an arbitrary code C, calling this a McEliece-based public-key cryptosystem. The original cryptosystem defined by McEliece is as described below, implemented with binary Goppa codes.

McEliece-based Cryptosystem

Let $\mathbb{F}$ be a finite field and

- $G \in \mathbb{F}^{k \times n}$ be a generator matrix for an $[n, k, \geq 2t+1]$ code C.
- D_C be an efficient decoding algorithm for C.
- $S \in \mathbb{F}^{k \times k}$ be a nonsingular matrix.
- $P \in \mathbb{F}^{n \times n}$ be a permutation matrix.

Set
$$G^{pub} = SGP \in \mathbb{F}^{k \times n}.$$

Public key:
$$(G^{pub}, t)$$

Private key:
$$(S, D_C, P)$$

Encryption

Message $m = (m_1, m_2, ..., m_k) \in \mathbb{F}^k$ is enciphered as

$$w = mG^{pub} + e = mSGP + e \in \mathbb{F}^n$$

where $e \in \mathbb{F}^n$ is chosen uniformly at random from vectors of weight t.

Suppose $w \in \mathbb{F}^n$ is intercepted without knowledge of the private key. In general, knowledge of a generator matrix of a code C' is not enough to provide an efficient decoding algorithm for C'. Consequently, if G cannot be determined, then decoding w to determine m is equivalent to decoding a general linear code. In 1978, this was proven by Berlekamp, McEliece, and van Tilborg to be an NP-complete problem Berlekamp (McEliece). More precisely, they obtained the following result.

Theorem 2.2. *Given a code $C \subseteq \mathbb{F}_q^n$, vector $w \in \mathbb{F}_q^n$, and nonnegative integer $t \leq n$, the problem of determining if there exists $c \in C$ such that $d(w, c) \leq t$ is NP-complete.*

This theorem provides the security guarantee for the McEliece cryptosystem, as it suggests that there is not a polynomial-time algorithm that solves the general decoding problem for linear codes. It should be noted, however, that the NP-hardness in Theorem 2.2 is in the context of worst-case scenario; the average-case hardness may be relevant here and is a topic of current research.

Of course, deciphering a message is immediate for the intended receiver who has access to the private key, as outlined below.

Decryption

Suppose
$$w = mG^{pub} + e = mSGP + e \in \mathbb{F}^n$$
is received. Received w is deciphered as follows:

1. Multiply w by P^{-1} to obtain
$$wP^{-1} = mSG + eP^{-1} = mSG + e' \in C + e'$$
where $e' = eP^{-1}$ and $wt(e') = wt(e) \leq t$.

2. Apply $\mathcal{D}_C$ to wP^{-1} to get
$$\mathcal{D}_C\left(mSG + e'\right) = mS.$$

3. Multiply mS by S^{-1} to obtain
$$mSS^{-1} = m.$$

McEliece proposed using binary Goppa codes, and the primary drawback is the large public key sizes. For nearly 40 years, researchers have attempted to reduce the key size by proposing McEliece-based cryptosystems with different families of underlying codes. More on this may be found in Section 2.4.4. It is worth noting that security proofs which consist of reducing to the bounded distance decoding problem operate under the following assumption: The uniform distribution on the public code family, meaning the family of codes from which C is drawn, is computationally indistinguishable from the uniform distribution on the whole family of $[n, k]$ codes over the same alphabet.

2.3.3 The Niederreiter Public-Key Cryptosystem

The Niederreiter cryptosystem (Niederreiter, 1986) is based on a parity-check matrix for a code rather than a generator matrix (as in a McEliece-based system) and includes a preprocessing step so that enciphered messages are syndromes. It has some advantages in terms of key size and speed. We abstract the key components below for use with an arbitrary linear code, calling such a system Niederreiter-based; the original Niederreiter cryptosystem is based on Reed-Solomon codes. Before doing so, it is worth pointing out a key difference in Niederreiter and McEliece.

In a McEliece-based system, messages $m \in \mathbb{F}_q^k$ are enciphered directly, whereas in Niederreiter there is a preprocessing step in which messages $m \in \mathbb{F}_q^k$ are written as strings of restricted weight in $\mathbb{F}_q^n$. This is because, in practice, the Niederreiter cryptosystem utilizes a Reed-Solomon code which corrects t errors. Because an arbitrary message vector $m \in \mathbb{F}_q^k$ may not have weight t, it needs to be altered so that this encryption scheme can be used.

Next, we describe the Niederreiter-based cryptosystem.

Niederreiter-based Cryptosystem

Let $\mathbb{F}$ be a finite field and

- $H \in \mathbb{F}^{(n-k) \times n}$ be a parity-check matrix for an $[n, k, \geq 2t+1]$ code C.
- $S \in \mathbb{F}^{(n-k) \times (n-k)}$ be a nonsingular matrix.
- $P \in \mathbb{F}^{n \times n}$ be a permutation matrix.

Set
$$H^{pub} = SHP \in \mathbb{F}^{(n-k) \times n}.$$

Public key:
$$(H^{pub}, t)$$

Private key:
$$(S, H, P)$$

Encryption

Message $m = (m_1, m_2, ..., m_k) \in \mathbb{F}^k$ is converted to a string $m' \in \mathbb{F}^n$ such that $wt(m') \leq t$, which is then enciphered as
$$w = H^{pub} m'^T \in \mathbb{F}^{n-k},$$
the syndrome of m'.

The Niederreiter scheme has the same level of security as the McEliece cryptosystem; this can be seen as follows. Suppose
$$w = H^{pub} m'^T \in \mathbb{F}^{n-k}$$
is received via the Niederreiter scheme. Let C' denote the code defined by H^{pub}; that is, set $C' = NS(H^{pub})$. Because H^{pub} is public, it may be used to find a generator matrix for C'. Call this generator matrix G^{pub}. Then, with linear algebra, a vector $c \in \mathbb{F}^n$ of weight at least t can be found such that
$$w = H^{pub} c^T$$

and
$$c = mG^{pub} + m'.$$

Notice that this is the form of the McEliece encryption, and so if McEliece can be broken, so can Niederreiter.

For the other relation, suppose we use McEliece to encode a message as
$$w = mG^{pub} + e$$

If we multiply both sides of this equation by $(H^{pub})^T$, where H^{pub} is the parity-check matrix for G^{pub}, we get

$$w(H^{pub})^T = (mG^{pub} + e)(H^{pub})^T = mG^{pub}(H^{pub})^T + e(H^{pub})^T = e(H^{pub})^T$$

since $G^{pub}(H^{pub})^T = 0$ (according to (2.3)). Noting that H^{pub} and w are both public, we then have the form of encryption of Niederreiter. So, if Niederreiter can be broken, then McEliece can be broken.

Next, we state how access to the private key allows for decryption.

Decryption

Received $w = H^{pub}m'^T \in \mathbb{F}^{n-k}$ is deciphered as follows:

1. Multiply w by S^{-1} to obtain
$$S^{-1}w = S^{-1}SHPm'^T = HPm'^T.$$

2. Apply syndrome decoding to HPm'^T to get
$$Pm'^T.$$

3. Multiply Pm'^T by P^{-1} to obtain
$$P^{-1}Pm'^T = m'^T.$$

The original Niederreiter cryptosystem employed Reed-Solomon codes which was later rendered insecure by Sidelnikov and Shestakov (Sidelnikov and Shestakov, 1992); their attack is outlined in Subsection 2.4.1. However, a Niederreiter-based system remains secure when implemented with binary Goppa codes.

2.4 ATTACKS AND POTENTIAL REPAIRS

Attacks on code-based cryptosystems fall into two primary categories. To understand these broad groupings, suppose that a message m (or m' if the system is Niederreiter-based) is enciphered as w using a code-based cryptosystem as described in Section 2.3.

1. Structural: Structural attacks try to determine G (or H) from G^{pub} (or H^{pub}).

2. Decoding: Decoding attacks attempt to determine m (or m' directly from w without access to the matrix G or decoding algorithm $\mathcal{D}$ (or H).

In this part, we describe some of these attacks and pivots made to avoid them. We will see that some of these variants created in response to a particular attack can also be rendered insecure through the use of additional mathematics.

2.4.1 Sidelnikov-Shestakov Attack

As an example of a structural attack, we consider the basic idea underlying the Sidelnikov-Shestakov attack (Sidelnikov and Shestakov, 1992). This attack, presented in 1992, reveals the insecurity of a Niederreiter-based system which incorporates Reed-Solomon codes. Here, we present it as it would apply to a McEliece-based system with Reed-Solomon codes. It addresses the following problem: Given G^{pub} where $G^{pub} = SGP$, determine G.

Suppose that G is a generator matrix for a generalized Reed-Solomon code $C_{k,v,\alpha}$. To determine G, we must determine α and v.

We may assume that $P = I_k$. Indeed, multiplication by P on the right simply permutes the columns of SG which corresponds to a relabeling of the field elements $\alpha_1, \ldots, \alpha_n$. Hence, we must determine α and v from SG. Below, we outline an approach which first identifies the α_i, $k+1 \leq i \leq n$, then the α_i, $1 \leq i \leq k$, and finally the v_i, for $1 \leq i \leq n$.

1. Compute the echelon form E of G^{pub}. Then $E = [I_k \mid A]$ where $A \in \mathbb{F}^{k \times (n-k)}$.

2. Next, we determine $\alpha_{k+1}, \ldots, \alpha_n$. Let $f_i := f_{Row_i E}$, so that $f_i = ev^{-1}(Row_i E)$ in keeping with (2.5). Then $f_i(\alpha_j) = 0$ for all $j \in [k] \setminus \{i\}$, which means f has roots $\alpha_1, \ldots, \alpha_{i-1}, \alpha_{i+1}, \ldots, \alpha_k$. Consequently, since $\deg f_i \leq k-1$, there exists $a_i \in \mathbb{F}_q^*$ such that

$$f_i(x) = a_i \prod_{j \in [k] \setminus \{i\}} (x - \alpha_j).$$

It follows that

$$\frac{f_i(x)}{f_l(x)} = \frac{a_i \prod_{j \in [k] \setminus \{i\}} (x - \alpha_j)}{a_l \prod_{j \in [k] \setminus \{i\}} (x - \alpha_j)} = \frac{a_i}{a_l} \frac{(x - \alpha_l)}{(x - \alpha_i)}. \quad (2.7)$$

We may assume that $\alpha_1 = 0$ and $\alpha_2 = 1$. Then

$$\frac{f_1(x)}{f_2(x)} = \frac{a_1}{a_2} \frac{(x - \alpha_2)}{(x - \alpha_1)} = \frac{a_1}{a_2} \frac{(x - 1)}{x}.$$

Hence, for $k+1 \leq j \leq n$,

$$\frac{E_{1j}}{E_{2j}} = \frac{a_1}{a_2} \frac{(\alpha_j - 1)}{\alpha_j}.$$

It follows that
$$\alpha_j = \frac{-\frac{a_1}{a_2}}{\frac{E_{1j}}{E_{2j}} - \frac{a_1}{a_2}}.$$

Note that $\frac{a_1}{a_2} \in \mathbb{F}_q$, so there are $q-1$ possibilities for $\frac{a_1}{a_2}$. Once the correct one is determined, $\alpha_{k+1}, \ldots, \alpha_n$ can be determined.

3. Next, $\alpha_3, \ldots, \alpha_k$ are found as follows. Notice that (2.7) implies
$$\frac{E_{1j}}{E_{lj}}(\alpha_j - 0) = \frac{a_1}{a_l}(\alpha_j - \alpha_l) \tag{2.8}$$

for all $j \in [n]$. In particular, for $l = 3$, we have the following two equations given by taking $j = k+1, k+2$ in (2.8):
$$\begin{aligned} \frac{E_{1k+1}}{E_{3k+1}}\alpha_{k+1} &= \frac{a_1}{a_3}(\alpha_{k+1} - \alpha_3) \\ \frac{E_{1k+2}}{E_{3k+2}}\alpha_{k+2} &= \frac{a_1}{a_3}(\alpha_{k+2} - \alpha_3) \end{aligned} \tag{2.9}$$

Observe that (2.9) is two equations in two unknowns, α_3 and $\frac{a_1}{a_3}$, which allows both to be determined. Similarly, for each $l = 4, \ldots, k$, we may determine α_l using
$$\begin{aligned} \frac{E_{1k+1}}{E_{lk+1}}\alpha_{k+1} &= \frac{a_1}{a_l}(\alpha_{k+1} - \alpha_l) \\ \frac{E_{1k+2}}{E_{lk+2}}\alpha_{k+2} &= \frac{a_1}{a_l}(\alpha_{k+2} - \alpha_l) \end{aligned} \tag{2.10}$$

At this point, the evaluation points $\alpha_1, \ldots, \alpha_n$ are determined, subject to knowledge of the field element $\frac{a_1}{a_2}$.

4. Next, the vector v is recovered. Recall that $G^{pub} \in \mathbb{F}_q^{k \times n}$. Write G^{pub} as
$$G^{pub} = [(G^{pub})' \mid M]$$

where $(G^{pub})' \in \mathbb{F}_q^{k \times (k+1)}$ and $M \in \mathbb{F}_q^{k \times (n-(k-1))}$. Consider
$$(G^{pub})'y = 0. \tag{2.11}$$

Since (2.11) yields a system of k equations in $k+1$ unknowns, it has a non-trivial solution y. Notice that G has a decomposition into
$$G = [G' \mid N]$$

where $G' \in F_q^{k \times (k+1)}$ and $N \in \mathbb{F}_q^{k \times (n-(k-1))}$. It follows that $(G^{pub})' = SG'$, since $G^{pub} = SG$. Because S is nonsingular and $SG'y = 0$, $G'y = 0$. Consequently,
$$\begin{bmatrix} v_1 & v_2 & \cdots & v_{k+1} \\ v_1\alpha_1 & v_2\alpha_2 & \cdots & v_{k+1}\alpha_{k+1} \\ v_1\alpha_1^2 & v_2\alpha_2^2 & \cdots & v_{k+1}\alpha_{k+1}^2 \\ \vdots & \vdots & & \vdots \\ v_1\alpha_1^{k-1} & v_2\alpha_2^{k-1} & \cdots & v_{k+1}\alpha_{k+1}^{k-1} \end{bmatrix} \begin{bmatrix} y_1 \\ y_2 \\ \vdots \\ y_{k+1} \end{bmatrix} = 0$$

which can be expressed as

$$\begin{bmatrix} y_1 & y_2 & \cdots & y_{k+1} \\ y_1\alpha_1 & y_2\alpha_2 & \cdots & y_{k+1}\alpha_{k+1} \\ y_1\alpha_1^2 & y_2\alpha_2^2 & \cdots & y_{k+1}\alpha_{k+1}^2 \\ \vdots & \vdots & & \vdots \\ y_1\alpha_1^{k-1} & y_2\alpha_2^{k-1} & \cdots & y_{k+1}\alpha_{k+1}^{k-1} \end{bmatrix} \begin{bmatrix} v_1 \\ v_2 \\ \vdots \\ v_{k+1} \end{bmatrix} = 0, \qquad (2.12)$$

a system of k equations in $k+1$ unknowns $v_1, \ldots, v_{k+1}$. For a fixed value $v_1 \in \mathbb{F}_q$, we may view (2.12) as k equations in k unknowns.

5. Finally, denoting the first k columns of G' by $G'' \in \mathbb{F}_q^{k \times k}$, and the first k columns of $(G^{pub})'$ by $(G^{pub})'' \in \mathbb{F}_q^{k \times k}$, calculate $S = (G^{pub})''(G'')^{-1}$. Then we may calculate $G = S^{-1}G^{pub}$, which means the private key has been recovered.

2.4.2 Distinguisher Attacks

As made clear above, the security of a code-based cryptosystem depends on the inability of the attacker to determine the underlying code (equivalently, generator matrix, decoding algorithm, or parity-check matrix). To guard against this, it is desirable that the code utilized behave or appear as a random linear code; that is, it should be difficult to distinguish it from a random code or one with no underlying structure. One measure of this is given by the $*$-product. Such products (and related squares) of codes have arisen in the contexts of decoding, secret sharing, and multiparty computation. Starting with Wieschebrink's work in 2010 (Wischebrink, 2010), there has been work to apply this approach to give a distinguisher attack for certain families of codes; for instance, Couvreur, Márquez-Corbella, and Pellikaan (2017) and Couvreur, Gaborit, Gauthier-Umaña, Otmani, and Tillich (2014).

Definition 2.8. The $*$-product of vectors $a = (a_1, \ldots, a_n), b = (b_1, \ldots, b_n) \in \mathbb{F}^n$ is the vector

$$a * b = (a_1 b_1, a_2 b_2, \ldots, a_n b_n) \in \mathbb{F}^n.$$

The square of a is $a^2 := a * a$.

It is an exercise to verify that the $*$-product is distributive and commutative. Moreover, for $a, b \in \mathbb{F}_2^n$,

$$a * b = \sum_{i \in \mathrm{Supp}(a) \cap \mathrm{Supp}(b)} e_i. \qquad (2.13)$$

Definition 2.9. The square of a code $C \subseteq \mathbb{F}^n$ is

$$C^2 = \langle a * b : a, b \in C \rangle,$$

the vector space over $\mathbb{F}$ generated by all $*$-products of codewords of C.

Notice that if C is an $[n,k]$ code, then C^2 is a code of length n. There are folklore results suggesting that the $[n,k,d]$ code C in a McEliece-based cryptosystem should have dim (C^2) roughly $\binom{k+1}{2}$; loosely speaking, the dimension of C^2 should be comparable (meaning not too small when compared) to that of a random code of the same dimension. One can find precise statements to support this in Cascudo and Cramer (2015). Because their description goes beyond the scope of the chapter at hand, we omit them here and instead consider how this notion applies to some codes considered in this chapter.

Example 2.9. Consider the generalized Reed-Solomon code $C_{k,\alpha,v}$. Then

$$\begin{aligned}
C^2_{k,v,\alpha} &= \langle (v_1 f(\alpha_1), \ldots, v_n f(\alpha_n)) * (v_1 h(\alpha_1), \ldots, v_n h(\alpha_n)) : f, h \in L_k \rangle \\
&= \langle (v_1 f(\alpha_1) v_1 h(\alpha_1), \ldots, v_n f(\alpha_n) v_n h(\alpha_n)) : f, h \in L_k \rangle \\
&= \langle (v_1^2 f(\alpha_1) h(\alpha_1), \ldots, v_n^2 f(\alpha_n) h(\alpha_n)) : f, h \in L_k \rangle \\
&= \langle (v_1^2 (fh)(\alpha_1), \ldots, v_n^2 (fh)(\alpha_n)) : f, h \in L_k \rangle \\
&= \langle (v_1^2 g(\alpha_1), \ldots, v_n^2 g(\alpha_n)) : g \in L_{2k-1} \rangle \\
&= C_{2k-1, v^2}.
\end{aligned}$$

Hence, the square of a generalized Reed-Solomon code is also a Reed-Solomon code. Moreover,

$$\dim\left(C^2_{k,v,\alpha}\right) = \dim\left(C_{2k-1, v^2, \alpha}\right) = 2k - 1.$$

In particular, the dimension of the square of a Reed-Solomon code of dimension k is linear in k rather than quadratic in k (as $\binom{k+1}{2}$ is).

We can see that Example 2.9 suggests that Reed-Solomon codes do not behave as general linear codes with respect to the square. This is another reason that Reed-Solomon codes typically render code-based systems insecure.

2.4.3 Information Set Decoding

Information set decoding was introduced by Prange in 1962 (Prange, 1962) as a procedure for decoding cyclic codes. To date, the best known generic decoding algorithms all derive from this approach which extends well beyond the original setting considered by Prange. We refer the reader to recent papers detailing these advancements and restrict ourselves to the fundamental ideas. The basis for this subsection is work by Lee and Brickell (1988), which was originally detailed for binary codes, later improved by Stern (1988), and ultimately generalized to arbitrary fields by Peters (2010).

Recall that if G is in systematic form, then $G|_{[k]} = I_k$ and decoding is immediate as $c = mG$ implies $c_i = m_i$ for $i \in [k]$. More generally, we say that a set of indices $\mathcal{I} = \{j_1, \ldots, j_k\}$ is an information set if and only if $G|_\mathcal{I}$ is nonsingular. In this setting, decoding is also straightforward. Indeed, if $c = mG$ for some $m \in \mathbb{F}^k$, then

$$\begin{aligned}
c|_\mathcal{I} &= (mG)|_\mathcal{I} \\
&= [Col_{j_1} mG \mid \cdots \mid Col_{j_k} mG] \\
&= (m_1, \ldots, m_k) [Col_{j_1} G \mid \cdots \mid Col_{j_k} G] \\
&= [m \cdot Col_{j_1} G \mid \cdots \mid m \cdot Col_{j_k} G] \\
&= mG|_\mathcal{I},
\end{aligned} \quad (2.14)$$

and
$$m = c|_{\mathcal{I}} \left(G|_{\mathcal{I}}\right)^{-1}.$$

Information set decoding essentially seeks to determine an information set which can then be utilized to reveal the plaintext m as demonstrated above. This all relies on determining an information set initially. To do so,

1. Select a set S of k columns of G.

2. Determine if $G|_S$ is nonsingular (via Gaussian elimination).

3. If $G|_S$ is nonsingular, then S is an information set. Otherwise, repeat with a different set S.

Clearly, at most $\binom{n}{k}$ steps are required. Alternatively, an information set may be built column by column where at each stage a column is added if and only if it is not a linear combination of those already in the set.

Of course, the scenario considered in (2.14) above assumes that no errors have been introduced. If instead $w \in \mathbb{F}^n$ is received such that $d(w,c) \leq r$ for some specified $r > 0$, then more work is to be done. For this, we establish a bit of notation. Given an information set $\mathcal{I}$, the columns of $G|_{\mathcal{I}}^{-1} G$ consist of the columns of the identity matrix I_k together with the columns of $G|_{\mathcal{I}}^{-1} G|_{[n]\setminus \mathcal{I}}$ in some order. For $i \in I$, $Col_i G|_{\mathcal{I}}^{-1} G = e_{j_i}$ for some $j_i \in [k]$. This scenario can be addressed as follows.

1. Determine an information set $\mathcal{I}$.

2. Compute
$$w_y := wt\left(w - \sum_{l=1}^{r} y_{i_l} Row_{j_i} G|_I^{-1} G\right)$$

for $y \in (\mathbb{F}^*)^r$. If $w_y \leq r$, then stop. Otherwise, repeat 1. until such a vector y is found.

While some computation (or trial and error) is required to determine an information set, it is clear that this approach does indeed reveal the desired information. This may be seen as the quintessential decoding attack.

2.4.4 Variants

To demonstrate how innovation and desire to decrease the public key size can lead to vulnerabilities in a cryptosystem, we consider some variants of the McEliece and Niederreiter cryptosystems.

2.4.4.1 Berger-Loidreau

Given the Sidelnikov-Shestakov attack (as well as the poor performance under the *-product), it is generally accepted that Reed-Solomon codes are not a good choice for use with a code-based cryptosystem. However, their nice structure prompts researchers to explore variants of them that may be better suited to the McEliece-based and Niederreiter-based cryptosystems.

One such instance is the approach by Berger and Loidreau (2005), which makes use of a subcode of a generalized Reed-Solomon code. Starting with an $[n, k]$ generalized Reed-Solomon code $C_{k,v,\alpha}$, they construct a subcode C with parameters $[n, n-\ell, d']$, where ℓ is small and $d' \geq d$, by adding ℓ rows to a parity-check matrix of $C_{k,v,\alpha}$.

Berger-Loidreau-based Cryptosystem

Let G be the generator matrix for an $[n, k, d]$ generalized Reed-Solomon code $C_{k,v,\alpha}$. Observe that $d = n - k + 1$, using the same argument as for the case $v = 1$. Construct a subcode $C \subseteq C_{k,v,\alpha}$ as follows. Let H be a generator matrix for $C_{k,v,\alpha}^{\perp}$; in other words H is a $(d-1) \times n$ parity-check matrix for $C_{k,v,\alpha}$.

From this, let

- $A \in \mathbb{F}_q^{\ell \times n}$ be a full rank matrix such that $Row_i A \notin \langle Row_j : j \in [d-1] \rangle$ for all $i \in [l]$ and

- $S \in \mathbb{F}_q^{(d-1+\ell) \times (d-1+\ell)}$ be a nonsingular matrix.

Set
$$H^{pub} = S \begin{bmatrix} H \\ A \end{bmatrix} \in \mathbb{F}_q^{(d-1+\ell) \times n}.$$

Public key:
$$(H^{pub}, t)$$

Private key:
$$(H, A, S)$$

Notice that part of the public key is a parity-check matrix of the subcode C, and the system can encrypt a word of weight at most $t = \lfloor \frac{n-k}{2} \rfloor$ as described below.

Encryption

Message $m \in \mathbb{F}_q^n$ of weight at most $t = \lfloor \frac{n-k}{2} \rfloor$ is encrypted by computing
$$w = m(H^{pub})^T \in \mathbb{F}_q^{d-1+\ell}.$$

The above encryption scheme is nearly identical to that in the Niederreiter cryptosystem, with the key difference being the underlying code being more complicated than a Reed-Solomon code. Decryption parallels that in the Niederreiter scheme as shown below.

> ### Decryption
>
> The received word
> $$w = m(H^{pub})^T = \begin{bmatrix} mH_d^T | mA^T \end{bmatrix} S^T$$
> is decrypted in the same way as in the Niederreiter cryptosystem, using syndrome decoding.

Here, the remarkable feature is that the Sidelnikov-Shestakov attack does not apply directly since C is a subcode of a generalized Reed-Solomon code rather than a generalized Reed-Solomon code itself. Even so, in 2010 an attack (Wischebrink, 2010) was made on the Berger-Loidreau cryptosystem. The main steps are outlined below to demonstrate how techniques may be layered.

1. Compute C^2. Wieschebrink proved that with high probability
$$C^2 = C_{k,v,\alpha}^2.$$
As a result, with high probability,
$$C^2 = C_{2k-1,v^2,\alpha}.$$

2. Apply the Sidelnikov-Shestakov attack to $C_{2k-1,v^2,\alpha}$ to ascertain α and v^2.

3. Calculate v so that $C_{k,\alpha,v}$ may be determined.

2.4.4.2 Random Linear Code Encryption

Next, we consider another way in which Reed-Solomon codes may be modified to give rise to a potentially more robust cryptosystem when combined with the McEliece structure. RLCE, or Random Linear Code Encryption, was introduced by Yongge Wang in 2016 (Wang, 2016).

> ### RLCE (Random Linear Code Encryption) Cryptosystem
> Let $w, n, k \in \mathbb{Z}^+$ with $w \le n - k$ and
>
> -
> $$G_1 = \begin{bmatrix} | & & | & | & | & & | & | \\ g_1 & \cdots & g_{n-w} & g_{n-w+1} & r_1 & \cdots & g_n & r_w \\ | & & | & | & | & & | & | \end{bmatrix} \in \mathbb{F}_q^{k \times (n+w)}$$
>
> where $r_1, \ldots, r_w \in \mathbb{F}_q^k$ are vectors chosen uniformly at random and
>
> $$\begin{bmatrix} | & & | \\ g_1 & \cdots & g_n \\ | & & | \end{bmatrix} \in \mathbb{F}_q^{k \times n}$$

is a generator matrix for an $[n, k, d]$ Reed-Solomon code.

- $$A = \begin{bmatrix} I_{n-w} & & & \\ & A_1 & & \\ & & \ddots & \\ & & & A_w \end{bmatrix} \in \mathbb{F}_q^{(n+w) \times (n+w)}$$

where $A_1, \ldots, A_w \in \mathbb{F}_q^{2 \times 2}$ are nonsingular and chosen uniformly at random.

Set
$$G^{pub} := G_1 A \in \mathbb{F}_q^{k \times (n+w)}.$$

Let D_C be a decoding algorithm for the code with generator matrix G_1.

Public key:
$$(G^{pub}, t)$$

Private key:
$$(A, D_C)$$

Encryption and decryption are as in the standard McEliece-based system, so we omit the description here. The reader is referred to (Wang, 2016) for a specialized decoding algorithm that can serve as D_C above.

Again, since G_1 is not a generator matrix for a Reed-Solomon code, no attack (such as Sidelnikov-Shestakov) seems to apply directly. Even so, in 2018, Couvreur, Lequesne, and Tillich gave a polynomial time key recovery attack on RLCE for $w < n - k$ (Couvreur, Lequesne, and Tillich, 2019). The key steps are outlined below to demonstrate the use of shortening and puncturing in preparation for a structural attack. We only need one more concept specific to this scenario. By abuse of notation, we may write

$$G^{pub} = \begin{bmatrix} | & & | & \| & \| & & \| \\ g_1 & \cdots & g_{n-w} & B_1 & B_2 & \cdots & B_w \\ | & & | & \| & \| & & \| \end{bmatrix}$$

where
$$B_i := \begin{bmatrix} | & | \\ g_i & r_i \\ | & | \end{bmatrix}$$

for $i \in [w]$.

1. Choose a positive integer $l \in [n]$.

2. Determine all pairs of twin positions, meaning those corresponding to the B_i column pairs, by calculating and comparing
$$\dim S_\mathcal{I}(C)^2$$

and
$$\dim P_{\{i\}}(C)^2$$
for $\mathcal{I} \subseteq [n+2]$ with $|\mathcal{I}| = l$ and $i \in [n+w] \setminus \mathcal{I}$.

3. Puncture at the twin positions, which yields a generalized Reed-Solomon code $C_{k,\alpha,v}$.

4. Apply the Sidelnikov Shestakov attack to determine α and v.

5. For each pair of twin positions, recover the corresponding 2×2 non-singular matrix A_i; a method for doing so may be found in the work of Couvreur, Lequesne, and Tillich.

6. Finish to recover the structure of the underlying GRS code.

This attack applies when
$$\dim\left(S_{\mathcal{I}}(C)^2\right) \leq \min\left\{\frac{k+1-|\mathcal{I}|}{2}, n+w-|\mathcal{I}|\right\},$$
meaning whenever
$$w + \frac{3\sqrt{16w+1}}{2} < n - k.$$
Consequently, taking $w = n - k$ avoids this approach. Of course, this means a greater key size than originally proposed by Wang.

2.4.4.3 LDPC and MDPC

The use of LDPC codes in a cryptosystem was first considered by Monico, Rosenthal, and Shokrollahi in 2000 (Monico, Rosenthal, and Shokrollahi, 2000), where they outline the system below.

LDPC Code-based Cryptosystem

Let

- $H \in \mathbb{F}_2^{(n-k) \times n}$ be a parity-check matrix for an $[n, k, \geq 2t+1]$ code C.

- $S \in \mathbb{F}_2^{k \times k}$ and $T \in \mathbb{F}_2^{(n-k) \times (n-k)}$ be sparse, nonsingular matrices.

Set
$$H^{pub} = TH \in \mathbb{F}_2^{(n-k) \times n}.$$

Public key:
$$(H^{pub}, S, t)$$

Private key:
$$(H, T)$$

Encryption

Message $m = (m_1, m_2, ..., m_k) \in \mathbb{F}_2^k$ is enciphered as

$$w = m\tilde{G} + e = mS^{-1}G + e \in \mathbb{F}_2^n$$

where $e \in \mathbb{F}_2^n$ is chosen uniformly at random from vectors of weight t, $G \in \mathbb{F}_2^{k \times n}$ is a generator matrix for C in systematic form, and $\tilde{G} := S^{-1}G$.

Because H is an LDPC code, it is accompanied by an efficient belief-propagation decoding algorithm. Hence, access to the private key yields fast decoding. Notice that the matrices H and H^{pub} define the same code, since T is nonsingular and $NS(H^{pub}) = NS(TH) = NS(H)$. The same holds true for G and $\tilde{G}$. However, the public matrix H^{pub} is not as sparse as H; hence, an efficient decoding algorithm based on H^{pub} is out of reach. This is one of the key ideas of the method. The other is the choice to provide keys based on parity-check matrices; generator matrices for LDPC codes tend to be very dense. One can note that the sparsity of the public and private key matrices leads to a more compact representation, meaning smaller key sizes. However, as we will see shortly, this sparsity also presents a vulnerability.

Before doing so, we outline the deciphering procedure for the intended receiver who has access to the private key.

Decryption

Suppose

$$w = m\tilde{G} + e = mS^{-1}G + e \in \mathbb{F}^n$$

is received. Received w is deciphered as follows:

1. Apply $\mathcal{D}_C$ to w to get $mS^{-1}G$.
2. Extract the first k coordinates of $mS^{-1}G$ to obtain mS^{-1}.
3. Multiply mS^{-1} by S on the right to obtain $(mS^{-1})S = m$.

Here, no algebraic structure underlies the codes used. Instead, if the matrix T is too sparse, the product TH may reveal H or another matrix H' with the same nullspace, where H' is suitable for belief-propagation. This is detailed in the same paper by Monico, Rosenthal, and Shokrollahi (2000) where the cryptosystem is introduced. We outline the basic idea below.

Recall that H and H^{pub} have the same rowspace, since $H^{pub} = TH$ and T is nonsingular. Moreover, for each $j \in [n-k]$, by writing out the product, we verify that

$$Row_j H^{pub} = \sum_{i=1}^{n-k} T_{ji} Row_i H. \tag{2.15}$$

Since T is very sparse, $T_{ji} = 0$ for most $(j,i) \in [n-k] \times [n]$; hence, the right-hand side of (2.15) has very few terms. This may be captured by writing $Row_j H^{pub} = \sum_{i \in \text{Supp}(Row_j T)} Row_i H$; writing $\text{Supp}(Row_j T) = \{i_1, \ldots, i_{n_j}\}$, this becomes $Row_j H^{pub} = \sum_{m=1}^{n_j} Row_{i_m} H$. It follows that for any $l \in [n-k]$,

$$\begin{aligned} Row_j H^{pub} * Row_l H &= \left(\sum_{m=1}^{n_j} Row_{i_m} H\right) * Row_l H \\ &= \sum_{m=1}^{n_j} \left(Row_{i_m} H * Row_l H\right). \end{aligned}$$

Given that the rows of H are sparse, n_j is small, and in light of (2.13), most of these terms will be 0; one exception is $Row_l H * Row_l H = Row_l H$. Hence, in this case, we expect that $Row_j H^{pub} * Row_l H = Row_l H$ for many $(j,l) \in [n-k]^2$. Notice that this implies $Row_l H \leq Row_j H^{pub}$ where we write $a \leq b$ for $a, b \in \mathbb{F}_2^n$ to mean $\text{Supp}(a) \subseteq \text{Supp}(b)$. Then we can see that

$$Row_i H^{pub} * Row_j H^{pub} \geq Row_i H^{pub} * Row_l H \geq Row_l H.$$

This motivates the following structural attack.

1. Compute $*$-products of a pair of rows of H^{pub}: $Row_i H^{pub} * Row_j H^{pub}$.

2. For each product $Row_i H^{pub} * Row_j H^{pub}$, check $Row_i H^{pub} * Row_j H^{pub} \in RS(H^{pub})$.

3. Repeat to find some rows of H.

To avoid the attack described above, several new ways of using parity-check codes in McEliece-based and Niederreiter-based systems have been introduced. To increase the density of the matrices H and H^{pub}, MDPC codes (Misoczki, Tillich, Sendrier, and Barreto, 2013) have been considered as have quasi-cyclic LDPC codes (Baldi, Bianchi, Chiaraluce, Rosenthal, and Schipani, 2016). Each attempts to balance the sparsity, which gives smaller key sizes with density to guard against attack.

2.5 CONCLUSION

Code-based cryptography is an exciting topic that continues to develop to meet the challenges of securing information in a post-quantum setting. In this chapter, we have outlined the basics of the McEliece and Niederreiter cryptosystems, along with the necessary background in the theory of error-correcting codes while assuming just linear algebra as a prerequisite. The promise of these cryptosystems is the basis for code-based cryptography. Researchers are especially interested in decreasing the associated key sizes through the use of a variety of codes. However, as we have seen, this may introduce vulnerabilities into the system. Some attacks are outlined in this chapter to provide a sense of the resilience required. Potential fixes of these attacks are also presented, some of which are ultimately shown to be insecure via the clever application of mathematics. This field of study has blossomed, with a whole range of improvements, breaks, and possible fixes of attacks and defenses, and continues to be a fluid and vibrant area of research. While this chapter is meant to capture that, it is by no means comprehensive. For further reading, more on post-quantum cryptography may be found in the very useful reference (Bernstein, 2009), and more on code-based cryptography may be found in (Overbeck and Sendrier, 2009).

ACKNOWLEDGMENTS

The authors would like to thank the reviewers who provided suggestions which significantly improved this contribution. They wish to acknowledge the generous support of Virginia Tech's Integrated Security Destination Area. The first author's work is supported in part by NSF DMS-2037833, NSF DMS-1802345, and the Commonwealth Cyber Initiative.

REFERENCES

Baldi, M., Bianchi, M., Chiaraluce, F., Rosenthal, J., & Schipani, D. (2016). Enhanced public key security for the McEliece cryptosystem. *Journal of Cryptology*, 29(1), 1–27.

Berger, T.P., & Loidreau, P. (2005). How to mask the structure of codes for a cryptographic use. *Designs, Codes and Cryptography*, 35(1), 63–79.

Berlekamp, E., McEliece, R., & van Tilborg, H. (1978). On the inherent intractability of certain coding problems. *IEEE Transactions on Information Theory*, 24(3), 384–386.

Bernstein, D.J. (2009). *Introduction to post-quantum cryptography*, in *Post-quantum cryptography*, pages 1–14. Berlin, Germany: Springer Publishing Company, Incorporated.

Cascudo, I., Cramer, R., Mirandola, D., & Zémor, G. (2015). Squares of random linear codes. *IEEE Transactions on Information Theory*, 61(3), 1159–1173.

Couvreur, A., Márquez-Corbella, I., & Pellikaan, R. (2017). Cryptanalysis of McEliece cryptosystem based on algebraic geometry codes and their subcodes. *IEEE Transactions on Information Theory*, 63(8), 5404–5418.

Couvreur, A., Gaborit, P., Gauthier-Umaña, V., Otmani, A., & Tillich, J.P. (2014). Distinguisher-based attacks on public-key cryptosystems using Reed–Solomon codes. *Designs, Codes and Cryptography*, 73(2), 641–666.

Couvreur, A., Lequesne, M., & Tillich, J.P. (2019). *Recovering short secret keys of RLCE in polynomial time*, in *International Conference on Post-Quantum Cryptography*, pages 133–152. Berlin, Germany: Springer Publishing Company, Incorporated.

Diffie, W., & Hellman, M.E. (1976). New directions in cryptography. *IEEE Transactions on Information Theory*, 22(6), 644–654.

Fraleigh, J., & Brand, N. (2020) *A First Course in Abstract Algebra* (8th ed.). Pearson.

Gallagher, R. (1962). Low-density parity-check codes. *IRE Transactions on Information Theory*, 8(1), 21–28.

Gallian, J. (2016). *Contemporary Abstract Algebra* (9th ed.). Cengage Learning.

Goppa, V.D. (1970). A new class of linear error-correcting codes. *Probl. Peredachi Inf.*, 6(3), 24–30.

Hamming, R.W. (1950). Error detecting and error correcting codes. *The Bell System Technical Journal*, 29(2), 147–160.

Hoffstein, J., Pipher, J., & Silverman, J.H. (2008). *An Introduction to Mathematical Cryptography* (1st ed.). Berlin, Germany: Springer Publishing Company, Incorporated.

Huffman, W.C., & Pless, V. (2003). *Fundamentals of Error-Correcting Codes*. Cambridge, England: Cambridge University Press.

Koblitz, N. (1987). Elliptic curve cryptosystems. *Mathematics of Computation*, 48(177), 203–209.

Lee, P.J., & Brickell, E.F. (1988). *An observation on the security of McElieces public-key cryptosystem*, in *Workshop on the Theory and Application of Cryptographic Techniques*, pages 275–280. Berlin, Germany: Springer Publishing Company, Incorporated.

MacKay, D.J.C. (1999). Good error-correcting codes based on very sparse matrices. *IEEE Transactions on Information Theory*, 45(2), 399–431.

MacKay, D.J.C. (1996). Near Shannon limit performance of low density parity check codes. *Electronic Letters*, 32, 1645–1646.

MacWilliams, F.J., & Sloane, N.J.A. (1977). *The Theory of Error-Correcting Codes*. Amsterdam, Netherlands: North-Holland Publishing Co.

McEliece, R.J. (1978). A public-key cryptosystem based on algebraic coding theory. *Coding Thv*, 4244, 114–116.

Merkle, R.C. (1978). Secure communications over insecure channels. *Commun. ACM*, 21(4), 294–299.

Misoczki, R., Tillich, J., Sendrier, N, & Barreto, P.S.L.M. (2013). *MDPC-McEliece: New McEliece variants from moderate density parity-check codes*, in *2013 IEEE International Symposium on Information Theory*, pages 2069–2073. Piscataway, New Jersey: IEEE.

Monico, C., Rosenthal, J., & Shokrollahi, A. (2000). *Using low density parity check codes in the McEliece cryptosystem*, in *2000 IEEE International Symposium on Information Theory (Cat. No.00CH37060)*, page 215. Piscataway, New Jersey: IEEE.

Mullen, G., & Mummert, C. (2007) *Finite Fields and Applications*. Student Mathematical Library. American Mathematical Society.

Niederreiter, H. (1986). Knapsack-type cryptosystems and algebraic coding theory. *Problems of Control and Information Theory*, 15, 159–166.

Overbeck, R., & Sendrier, N. (2009). *Code-based cryptography*, in *Post-quantum cryptography*, pages 95–145. Berlin, Germany: Springer Publishing Company, Incorporated.

Peters, C. (2010). *Information-set decoding for linear codes over F q*, in *International Workshop on Post-Quantum Cryptography*, pages 81–94. Berlin, Germany: Springer Publishing Company, Incorporated.

Prange, E. (1962). The use of information sets in decoding cyclic codes. *IRE Transactions on Information Theory*, 8(5), 5–9.

Reed, I.S., & Solomon, G. (1960). Polynomial codes over certain finite fields. *Journal of the Society for Industrial and Applied Mathematics*, 8(2), 300–304.

Rivest, R.L., Shamir, A., & Adelman, L. (1978). A method for obtaining digital signatures and public-key cryptosystems. *Communications of the ACM*, 21(2), 120–126.

Shannon, C.E. (2001). A mathematical theory of communication. *ACM SIGMOBILE mobile computing and communications review*, 5(1), 3–55.

Shor, P.W. (1994). *Algorithms for quantum computation: discrete logarithms and factoring*, in *Proceedings 35th Annual Symposium on Foundations of Computer Science*, pages 124–134. Piscataway, New Jersey: IEEE.

Sidelnikov, V.M., & Shestakov, S.O. (1992). On insecurity of cryptosystems based on generalized Reed-Solomon codes. *Discrete Mathematics and Applications*, 2(4), 439–444.

Smart, N.P. (2015). *Cryptography Made Simple* (1st ed.). Berlin, Germany: Springer Publishing Company, Incorporated.

Stern, J. (1988). *A method for finding codewords of small weight*, in *International Colloquium on Coding Theory and Applications*, pages 106–113. Berlin, Germany: Springer Publishing Company, Incorporated.

Stinson, D. & Paterson, M. (2018) *Cryptography: Theory and Practice* (4th ed.) Chapman and Hall/CRC.

Tanner, R. (1981). A recursive approach to low complexity codes. *IEEE Transactions on information theory*, 27(5), 533–547.

Wang, Y. (2016). *Quantum resistant random linear code based public key encryption scheme RLCE*, in *2016 IEEE International Symposium on Information Theory (ISIT)*, pages 2519–2523. Piscataway, New Jersey: IEEE.

Wieschebrink, C. (2010). *Cryptanalysis of the Niederreiter public key scheme based on GRS subcodes*, in *International Workshop on Post-Quantum Cryptography*, pages 61–72. Berlin, Germany: Springer Publishing Company, Incorporated.

CHAPTER 3

Algebraic Geometry

Lubjana Beshaj

CONTENTS

3.1	Introduction	97
3.2	Preliminaries	99
	3.2.1 Finite Fields	99
	3.2.2 Algebraic Curves	101
	3.2.3 Affine and Projective Varieties	102
	3.2.4 Plane Curves	103
	3.2.5 Coordinate Ring of an Affine Variety	105
	3.2.6 Graph Theory	107
3.3	Elliptic Curves and Cryptography	109
	3.3.1 Elliptic Curves	109
	3.3.2 Isogenies of Elliptic Curves	112
	3.3.3 Velu's Formula	113
	3.3.4 Modular Polynomials	114
	3.3.5 Division Polynomials	115
	3.3.6 Elliptic Curve Cryptography	117
	3.3.7 Elliptic Curve Diffie-Hellman Key Exchange	118
	3.3.8 Post-Quantum Cryptography with Elliptic Curves	119
	3.3.9 Supersingular Isogeny Diffie-Hellman Key Exchange	120
3.4	Genus 2 Curves and Cryptography	122
	3.4.1 Hyperelliptic Curves	122
	3.4.2 Divisors	123
	3.4.3 Computing Principal Divisors on a Curve	126
	3.4.4 The Group Law of Divisors	126
	3.4.4.1 Geometrically	126
	3.4.4.2 Algebraically	127
	3.4.5 Cryptography on Hyperelliptic Curves	128
	3.4.6 Post-Quantum Hyperelliptic Curve Cryptography	129
3.5	Final Remarks and Future Work	130

3.1 INTRODUCTION

Solving algebraic equations has been the central focus of mathematics throughout history. Algebraic curves, as a special kind of equations, have played an important

DOI: 10.1201/9780429354649-3

role in modern mathematics. The development of new computational tools and a lot more computer power have made possible applications of algebraic curves in areas such as computer vision, coding theory, computer security and cryptography, physics, and other areas. Hence, algebraic curves play a central role in theoretical mathematics and practical applications.

Algebraic curves over finite fields are extensively used in the design of public-key cryptographic schemes. Ever since their introduction to public-key cryptography by Miller (1986) and Koblitz (1987), elliptic curves have been of interest to the cryptographic community. But elliptic curves are not the only type of algebraic curves that can be used efficiently in cryptography, so can hyperelliptic curves of genus two and three. By using the group of points on an appropriately chosen algebraic curve where the discrete logarithm problem (DLP) is assumed to be hard, many standard protocols can be instantiated.

However, with the advent of quantum computing, an adversary can efficiently break universally adopted public-key cryptographic schemes (e.g. RSA, DSA, and algebraic curve cryptography). In his paper, Shor (1994) gives an algorithm showing how a quantum computer can be used to solve the discrete log problem. The threat of large-scale quantum computers has initiated the search for alternative algorithms that also resist quantum adversaries. In order to mitigate this imminent threat, cryptographic schemes that are resistant against quantum computers have drawn great attention from academia and industry. These schemes are collectively referred to as post-quantum cryptography (PQC), see Chen et al. (2016) for more details. In Jao and De Feo (2011) the authors proposed supersingular isogeny Diffie-Hellman (SIDH) as a key exchange protocol that leverages elliptic curves and offers post-quantum security. Moreover, Costello (2018) explores new possibilities for efficient supersingular isogeny-based cryptography leveraging hyperelliptic curves. Isogeny-based algorithms rely on the structure of large isogeny graphs, and the cryptographically interesting properties of these graphs are tied to their expansion properties.

The main focus of this chapter is the study of elliptic and hyperelliptic curves of genus two and their applications in cryptography. We start with a brief introduction on the mathematical background of algebraic curves, focusing on elliptic and hyperelliptic curves, and then focus on their applications on cryptography, i.e. elliptic and hyperelliptic curve cryptography (ECC and HCC) as well as isogeny-based cryptography.

This chapter is organized as follows. As we have stated above, elliptic and hyperelliptic curves are types of algebraic curves. In order to better understand them we will start this chapter with covering some basic concepts in algebraic geometry about plane curves. In the preliminaries, we start with some basic concepts from abstract algebra, such as finite fields and field extensions. Then we move our focus towards affine and projective varieties, plane curves, and the coordinate ring of varieties. Lastly we will cover some concepts from graph theory, such as the definition of a graph, Ramanujan graph, and expander graphs. These concepts from graph theory play a crucial role in algebraic curve cryptography in a post-quantum world.

Then, in Section 3.3, we focus on elliptic curves and cryptography. We start this section with defining elliptic curves, and then we define all the tools that we will need to describe their application in classical cryptography, which in most literature is referred to as elliptic curve cryptography (ECC), and in a post-quantum world, which we will refer to as isogeny-based cryptography.

Section 3.4 is on hyperelliptic curves of genus two and their applications to cryptography. The structure of this section is similar to the previous one. We will cover the definition of hyperelliptic curves and then the necessary tools to describe their application in classical and post-quantum cryptography. It is important to point out here that there are some fundamental differences between elliptic curves and hyperelliptic curves as we will see at the beginning of this section.

Lastly, we give some final remarks about what the status quo is in this area of research and what are some important questions that researchers are considering.

The graphics in this chapter were done in SageMath and Mathematica.

3.2 PRELIMINARIES

In this section, we aim to give a brief introduction of some basic concepts about algebraic geometry. We will start with giving a brief introduction to algebraic curves, which will provide us the foundation to then discuss elliptic and hyperelliptic curves, which are special cases of algebraic curves with applications in cryptography. Moreover, we will cover some concepts in graph theory which we will use to discuss the use of elliptic and hyperelliptic curves in cryptography in a post-quantum world.

3.2.1 Finite Fields

We will start with an overview of some concepts from abstract algebra, which will be used extensively in this chapter. A *field* K is a commutative ring in which every nonzero element has a multiplicative inverse. Some examples of fields are the real numbers $\mathbb{R}$, the field of rational numbers $\mathbb{Q}$, and the field of complex numbers $\mathbb{C}$. The characteristic of K, denoted by $char(K)$, is defined to be the smallest positive integer such that

$$nx = 0, \ \ for\ all\ x \in K.$$

If no such integer exists then we say that the field has characteristic zero. The characteristic of a field is either zero or a prime p. If we let p be a prime, then $(\mathbb{Z}/p\mathbb{Z}, +, \times)$, the set of integers modulo p with the corresponding addition and multiplication modulo p is a field. Furthermore, this set has a finite number of elements. Thus $\mathbb{Z}/p\mathbb{Z}$ is called a *finite field* and is often denoted by $\mathbb{F}_p$. In this chapter, we will only use the $\mathbb{F}_p$ notation when discussing finite fields. Finite fields are critical in modern cryptography.

If K and L are fields such that $K \subset L$ then L is called a *field extension* of K, denoted by L/K. It is also very useful to use diagrams to denote such extensions, as in Figure 3.1.

It is an elementary exercise in linear algebra to show that L is a vector space over K. The dimension of this vector space is called the *degree* of the extension and

FIGURE 3.1 A field extension.

denoted by $[L : K]$. If $[L : K]$ is finite (resp., infinite) then the field extension is a *finite* (resp., *infinite*) *extension*.

Now let us consider polynomials in x with coefficients from K, given as

$$p(x) = a_n x^n + \cdots + a_1 x + a_0$$

for $n \geq 0$ and $a_i \in K$. The set of all polynomials with coefficients from K is called the polynomial ring, and it is denoted with $K[x]$. Similarly, we can define the polynomial ring of several variables $x_1, \ldots, x_n$ and denote it with $K[x_1, \ldots, x_n]$. Let $f(x) \in K[x]$ be given. We say that $f(x)$ *splits over a field L* if $f(x)$ factors completely into linear factors in $L[x]$.

Let L/K be a field extension and $\alpha \in L$. If there is a polynomial $f(x) \in K[x]$ such that $f(\alpha) = 0$, then α is called an *algebraic element* over K. An *algebraic extension* is a field extension L/K such that every element of L is algebraic over K.

A field F is called *algebraically closed* if every non-constant polynomial in $K[x]$ has a root (splits) in K. The *algebraic closure* of a field K will be called a field $\overline{K}$ if it is an algebraic extension of K and it is algebraically closed. For example, $\mathbb{C}$ is the algebraic closure of $\mathbb{R}$. A polynomial $p(x) \in K[x]$ that cannot be factored into the product of two non-constant polynomials over $K[x]$ is called irreducible. We call K a *perfect* field if every irreducible polynomial over K has distinct roots. Let us now illustrate these concepts with the following examples.

Example 3.1. *The splitting field of $x^2 - 2$ over $\mathbb{Q}$ is $\mathbb{Q}(\sqrt{2})$. Indeed, the two roots are $\pm\sqrt{2}$ and $-\sqrt{2} \in \mathbb{Q}(\sqrt{2})$.*

Example 3.2. *The splitting field of $f(x) = x^4 + 4$ over $\mathbb{Q}$ is $\mathbb{Q}(i)$. We can factor $f(x)$ into irreducible factors over $\mathbb{Q}$ as follows*

$$f(x) = (x^2 - 2x + 2)(x^2 + 2x + 2.)$$

The roots are $\pm 1 \pm i$. Hence the splitting field is $\mathbb{Q}(i)$.

Example 3.3 (Splitting field of cubics)**.** *Let $f(x) \in \mathbb{Q}[x]$ be irreducible and be given as*

$$f(x) = x^3 + ax^2 + bx + c$$

Let $\alpha_1, \alpha_2, \alpha_3$ be the roots of the cubic. Then, $\mathbb{Q}(\alpha_1, \alpha_2, \alpha_3) = \mathbb{Q}(\alpha_1, \alpha_2)$ since $\alpha_3 = -a - (\alpha_1 + \alpha_2)$. Since $f(x)$ is irreducible, then $[\mathbb{Q}(\alpha_1) : \mathbb{Q}] = 3$. The polynomial $\frac{f(x)}{x - \alpha_1}$ has degree 1 or 2 and so $[\mathbb{Q}(\alpha_1, \alpha_2) : \mathbb{Q}(\alpha_1)] = $ 1 or 2. Hence, $[\mathbb{Q}(\alpha_1, \alpha_2) : \mathbb{Q}] = 3$ or 6.

It is an easy exercise to show that the splitting field of a cubic has degree 3 if and only if the discriminant of $f(x)$ is a square in $\mathbb{Q}$.

Extending a finite field F_q with degree n extensions will form the field F_{q^n}. To construct F_{q^n}, we will take $F_{q^n} = F_q(\alpha)$ where $f(\alpha) = 0$ and $f(x)$ is an irreducible polynomial of degree n in $F_q[x]$.

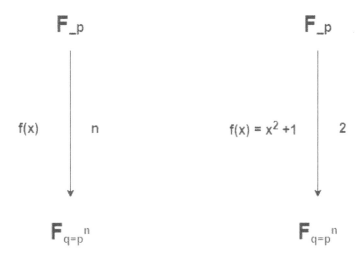

FIGURE 3.2   A degree n and $n = 2$ extension of finite field F_p.

Figure 3.2 visually depicts an n-degree extension of F_p to F_q where $q = p^n$. By setting $n = 2$, we may set the extension to be F_{p^2} and the function to be $f(x) = x^2 + 1$, an irreducible polynomial of degree 2.

3.2.2 Algebraic Curves

Riemann surfaces can be thought of as 'deformed copies' of the complex plane. Locally near every point, they look like patches of the complex plane. Every algebraic curve with coefficients in $\mathbb{C}$ is a **compact Riemann surface**. Moreover, every compact Riemann surface is a sphere with some handles attached (see Figure 3.3).

The number of handles is an important topological invariant called the topological **genus** of the surface. Hence, the genus of an algebraic curve is the number of handles on the surface. Let us consider next some examples of some very recognizable families of curves. We will get into more details about some of these families in the sections that follow.

Example 3.4. *An **elliptic curve** is a curve with equation*

$$y^2 = f(x), \text{ where } \deg f = 3 \text{ or } 4 \text{ and } \mathcal{D}_f \neq 0$$

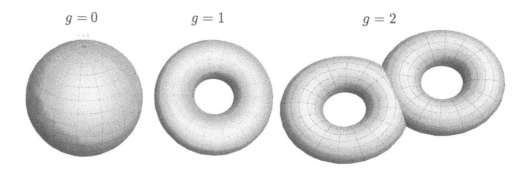

FIGURE 3.3 Algebraic curves.

A **hyperelliptic curve** is a curve with equation

$$y^2 = f(x), \text{ where } \deg f \geq 5 \text{ and } \mathcal{D}_f \neq 0$$

A **superelliptic curve** is a curve with equation

$$y^n = f(x), \text{ where } n \geq 2, \deg f \geq 3, \text{ and } \mathcal{D}_f \neq 0$$

Next we will formally define algebraic curves.

3.2.3 Affine and Projective Varieties

Before defining algebraic curves we need to introduce the space where they are defined. Let K be a perfect field and $\overline{K}$ it's algebraic closure. The n-dimensional projective space $\mathbb{P}_K^n$ over K is defined as follows

$$\mathbb{P}^n(\overline{K}) = \{(x_0, \ldots, x_n) \mid x_i \in \overline{K}, \text{ such that at least one } x_i \neq 0\}/\sim$$

where $\sim$ is the equivalence relation

$$(x_0, \ldots, x_n) \sim (y_0, \ldots, y_n) \iff \exists \lambda \in \overline{K}, \forall i : x_i = \lambda y_i \in \overline{K}.$$

The equivalence classes are called *projective points*. The set of L-rational points $\mathbb{P}^n(L)$ is defined to be equal to the subset of $\mathbb{P}^n$ such that

$$\mathbb{P}^n(\overline{K}) = \{(x_0, \ldots, x_n) \mid \exists \lambda \in \overline{K}, \forall i : \lambda x_i \in L\}.$$

Recall that a polynomial $f(x_0, \ldots, x_n) \in K[x_0, \ldots, x_n]$ is called *homogenous of degree d* if it is the sum of monomials of the same degree d, i.e. this is equivalent to requiring that

$$f(\lambda x_0, \ldots, \lambda x_n) = \lambda^d f(x_0, \ldots, x_n), \text{ for all } \lambda \in \overline{K}.$$

This implies that the set

$$D_f(L) = \{P \in \mathbb{P}^n(L) \mid f(P) \neq 0\}$$

is well defined. An ideal $I \subseteq K[x_0, \ldots, x_n]$ is homogenous if it is generated by a homogenous polynomial. For $I \neq \langle x_0, \ldots, x_n \rangle$, define

$$V(I) = \{P \in \mathbb{P}^n(\overline{K}) \mid f(P) = 0, \forall f \in I\}.$$

It is easy to see that $V(I)$ is well defined and it is called a *projective variety*.

The *affine space of dimension n over K* is defined to be the set of n-tuples

$$\mathbb{A}^n = \{(x_1, \ldots, x_n) \mid x_i \in \overline{K}\}$$

and the set of L-rational points is given by

$$\mathbb{A}^n = \{(x_1, \ldots, x_n) \mid x_i \in L\}.$$

For $f \in K[x_1, \ldots, x_n]$ let

$$D_f(L) = \{P \in \mathbb{A}^n(L) \mid f(P) \neq 0\}$$

and for an ideal $I \subseteq K[x_1, \ldots, x_n]$ let

$$V(I) = \{P \in \mathbb{A}^n(L) \mid f(P) = 0, \forall f \in I\}$$

be an *affine variety*.

3.2.4 Plane Curves

In this section, we will give a brief overview of algebraic curves. Most of the material is standard and can be found at Silverman (1986) amongst many others.

Definition 3.1. *Let $F \in k[x, y]$ be a polynomial in two variables with coefficients in k. We define the following subset of the affine plane over k*

$$\mathcal{X}_F(k) = \{(a, b) \in \mathbb{A}^2 \mid F(a, b) = 0\}.$$

*Consider the set of pairs $\{(\mathcal{X}_F, F) \mid \mathcal{X} \subset \mathbb{A}^2(K), F \in k[x, y] \text{ is nonzero}\}$. We will identify two pairs $(\mathcal{X}_1, F_1)$ and $(\mathcal{X}_2, F_2)$ if $\mathcal{X}_1 = \mathcal{X}_2$ and $F_1 = \lambda F_2$ for some $\lambda \in k$. An **affine plane curve** over k is a pair $(\mathcal{X}_F(k), F)$ for some nonzero $F \in k[x, y]$.*

Let K be a field containing k (possibly $K = k$). The K-rational points of the curve $\mathcal{X}$ given by the function above are the solutions of $F(x, y) = 0$ in $x, y \in K$ and we will denote this set with $\mathcal{X}(K)$. We call $\mathcal{X}$ *irreducible* over k if F is an irreducible polynomial in the ring $k[x, y]$. We call $\mathcal{X}$ *absolutely irreducible* if F is irreducible in $\bar{k}[x, y]$, where $\bar{k}$ is the algebraic closure of k.

Example 3.5. *It is easy to check that the polynomial $y^2 = (x^3 - x)$ is absolutely irreducible, whereas the real points of $y^2 = x^3 - x$ look like the curve displayed in Figure 3.4.*

As we can see from the graph $C(\mathbb{R})$ (Figure 3.4) consists of two topological components, even though we pointed out that the curve is absolutely irreducible. This will prompt us to look at the complex points of an algebraic curve rather than the real ones.

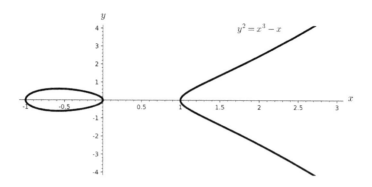

FIGURE 3.4 Curve with equation $y^2 = x^3 - x$.

Any affine curve $\mathcal{X}$ can be extended to a projective curve. Given $\mathcal{X}$ with equation $F(x,y) = 0$ such that F has degree n we homogenize by taking the projective equation $z^n F(x/z, y/z) = 0$. This process is called homogenization of the affine equation. Let us illustrate this with the following example.

Example 3.6. *It is easy to check that the polynomial $y^2 = (x^3 - x)$ can be extended as a projective curve by writing it as*

$$\left(\frac{y}{z}\right)^2 = \left(\frac{x}{z}\right)^3 - 2\frac{x}{z}$$

and then multiplying by z^3 we get $y^2 z = x^3 - 2xz^2$

Let us now consider the intersection of two algebraic curves. Given two algebraic curves (with affine equations)

$$\mathcal{X} : F(x,y) = 0 \text{ and } \mathcal{Y} : G(x,y) = 0$$

we say that they have a **common component** if F and G have non-constant common divisor $H \in K[x,y]$. The common component is the curve given by the equation $H(x,y) = 0$. If F and G do not have a non-constant common factor we say that the curves do not have a common component. Bezout's inequality says that given $\mathcal{X}$ and $\mathcal{Y}$ two algebraic curves of degree m and n, respectively, such that the curves have no common component, then the number of intersection points of $\mathcal{X}$, $\mathcal{Y}$ is at most mn. Hence, we have that two algebraic curves without common component intersect in finitely many points.

Let $\mathcal{X}$ and $\mathcal{Y}$ be two plane projective curves given respectively by the equations $F(x,y,z) = 0$ and $G(x,y,z) = 0$ defined over a field k. Using the intersection multiplicity we can state the following theorem.

Theorem 3.1 (Bezout). *Let $\mathcal{X}$ and $\mathcal{Y}$ be two plane projective curves defined over a field k of degrees m, n. Suppose that they have no common components. Let $\mathcal{P}$ be the set of intersection points $\mathcal{X}(\overline{k}) \cap \mathcal{Y}(\overline{k})$. Then,*

$$\sum_{P \in \mathcal{P}} \nu_p(\mathcal{X}, \mathcal{Y}) = mn.$$

Next, we will briefly describe how we can find the common solutions. Let $\mathcal{X}$, $\mathcal{Y}$ be two curves defined over the reals $\mathbb{R}$. Suppose the curves are given by $F(x,y) = 0$ and $G(x,y) = 0$, then we are interested in their common solutions. In order to solve this system of equations we need to use elimination theory for polynomial equations. In general this is done using Gröbner basis techniques (see Cox and O'Shea (2007) for more details on this topic). Another method of finding the intersection of curves would be using the resultant.

3.2.5 Coordinate Ring of an Affine Variety

In mathematics, we often understand an object by studying the functions on that object. For example:

1. To understand groups, we study homomorphisms
2. To understand topological spaces, we study continuous functions
3. To understand manifolds in differential geometry we study smooth functions
4. Since varieties are defined by polynomials it is appropriate to consider polynomial functions on them in algebraic geometry.

Recall, any polynomial in n-variables, say $f \in k[x_1, \ldots, x_n]$ is a function mapping $k^n \to k$, for example:
$$F : k^n \to k$$
$$(x_1, \ldots, x_n) \to f(x_1, \ldots, x_n)$$

Given an affine algebraic variety $V \subset K^n$, the restriction of f to V defines a map from V to k, that is:
$$f|_V : V \to k$$

Under the usual pointwise operations of addition and multiplication, these functions form a k-algebra (an associative ring A with nonzero unit that has the structure of a k-vector space)
$$k[x_1, x_2, \ldots, x_n]\,|_V$$
which we call the coordinate ring of V and denote it by $k[V]$.

Definition 3.2. *Given an affine variety $V \subseteq K^n$, the coordinate ring of V is denoted by $k[V]$ and defined to be the set of polynomials*
$$f|_V : V \to k$$
where $f \in k[x_1, \ldots, x_n]$. The ring $k[V]$ is often described as 'the ring of polynomial functions on V.'

Example 3.7 (The coordinate ring of lines). *Consider*

$$L : f(x, y) = 0$$

for

$$f(x, y) = y - mx + b.$$

We have $k[L] = k[x, y] = k[x, y]/(f)$. The representatives of the cosets $g + (f)$ are polynomials in $k[x, y]$; we can replace every y in g by $mx + b$. Thus every element in $k[L]$ can be written as $g(x) + (f)$ for some polynomial in x; and these elements are all pointwise distinct:

$$g(x) + f = h(x) + (f),$$

where $g(x) - h(x)$ is divisible by $f(x, y) = y - mx - b$, which is only possible if $g = h$. The map: $k[L] \to k[x]$, $g(x) + (f) \to g(x)$ is a ring homomorphism. Hence $k[L] \simeq k[x]$.

Example 3.8 (The coordinate ring of a parabola).

$$C : y - x^2 = 0 \text{ is given by } k[c] = k[x, y]/(y - x^2)$$

Any element $g(x, y) + (y - x^2)$ can be represented by a polynomial in x alone since we may replace each y by x^2 without changing the coset; in particular we have $g(x, y) + (y - x^2) = g(x, x^2) + (y - x^2)$. Again, the map: $g(x, y) + (y - x^2) \to g(x, x^2)$ is a ring homomorphism; thus $k[c] \simeq k[x]$.

Example 3.9 (The coordinate ring of the unit circle).

$$C : f(x, y) = x^2 + y^2 - 1$$

and

$$k[c] = k[x, y]/(f).$$

The polynomial $g(x, y) = x^4 + x^2y + x(1 - x^2) + (f) = x^4 - x^3 + x^2y + x + (f)$ has image $g + (f) \in k[c]$. Note that:

$$g + (f) = x^4 + x^2y + x(1 - x^2) + (f) = x^4 - x^3 + x^2y + x + (f).$$

In general, every element $g + (f)$ can be written in the form

$$g(x, y) + (f) = h_1(x) + yh_2(x) + (f)$$

since we may have replaced y^2 by $1 - x^2$. $K[c] \not\simeq k[x]$, this is because $k[x]$ is a unique factorization domain, but $k[c]$ is not; in fact we have $y^2 + (f) = (1 - x)(1 + x) + (f)$ and the elements $y + f$, $1 + x + (f)$ and $1 - x + (f)$ are irreducible.

Example 3.10 (Coordinate ring of an algebraic curve). *Let us now look at the function field of an algebraic curve given by the equation $C : y^2 = h(x)y + f(x)$ defined over a field k.*

- *The coordinate ring of C over k is the quotient ring:*

$$k[C] = k[x,y]/(y^2 + h(x)y - f(x))$$

- *Similarly, the coordinate ring of C over $\bar{k}$ is the quotient ring:*

$$\bar{k}[C] = \bar{k}[x,y]/(y^2 + h(x)y - f(x))$$

  An element of $\bar{k}[C]$ is called a polynomial function on C.

- *The function field $\bar{k}[C]$ of C over $\bar{k}$ is the field of fractions of $\bar{k}[C]$. The elements of $k(\bar{C})$ are called rational functions on C.*

In the next subsection, we will provide some basic concepts in graph theory as well as define expander graphs which we will need to define elliptic curve cryptography in a post-quantum world. We will restrict the discussion to only undirected graphs.

3.2.6 Graph Theory

An undirected graph G is a tuple or a pair (V, E) where V is a finite set of vertices and $E : E \subset V \times V$ is a set of unordered vertex pairs called edges. A series of definitions follows in order to better understand graphs and their properties. Additionally, some key definitions are visually depicted in Figure 3.5. Two vertices v and v' are said to be connected by an edge if $\{v, v'\}$ exists and $\{v, v'\} \in E$. A vertex v's neighbors is the set of all vertices in V connected to v by an edge. A path between two vertices v and v' is the sequence of vertices such that each vertex is connected to the next vertex by an edge. We will denote a path by $v \to v_1 \to ... \to v'$. The distance between two vertices is the length of the shortest path between the vertices. If a path between the two vertices does not exist, then we say the vertices are at an infinite distance. A graph G is called connected if any two vertices are connected by a path and otherwise is called disconnected. The degree of a vertex v is the number of edges emerging from v or the number of v's neighbors. If every vertex in a graph has degree k, then it is said to be k-regular.

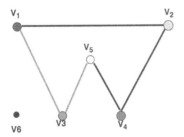

- Vertex V_1 has degree 2 since has two neighbors V_2 and V_3
- The shortest path from V_1 to V_5 is the path V_1-> V_3 -> V_5
- The distance between V_1 to V_5 is 2
- Vertex V_6 is at infinity distance from all other vertices

FIGURE 3.5 Example of a graph with visual and written depictions of common graph terminology.

There are a few methods of keeping track of the graph's vertices and edges efficiently. One such method is an adjacency matrix of G, denoted $A(G)$, with vertices $V = \{v_1, ..., v_n\}$ and edges E. The adjacency matrix is an $n \times n$ matrix with each entry (i, j) set to 1 if an edge exists between v_i and v_j; otherwise, the (i, j) entry is set to 0. Notice that a loop is represented by a 2.

Example 3.11. *The adjacency matrix for the above graph is given as follows:*

$$A(G) = \begin{bmatrix} 0 & 1 & 1 & 0 & 0 & 0 \\ 1 & 0 & 0 & 1 & 0 & 0 \\ 1 & 0 & 0 & 0 & 1 & 0 \\ 0 & 1 & 0 & 0 & 1 & 0 \\ 0 & 0 & 1 & 1 & 0 & 0 \\ 0 & 0 & 0 & 0 & 0 & 0 \end{bmatrix}$$

Since we are restricted to undirected graphs, the adjacency matrix is symmetric and will have zeros along the diagonal if there are no loops. Being a real symmetric matrix $A(G)$ has n-real eigenvalues, say $\lambda_1, \ldots, \lambda_n$ such that

$$\lambda_1 \geq \ldots \geq \lambda_n.$$

If a graph G is a k-regular graph, then its largest and smallest eigenvalues, respectively λ_1 and λ_n are bound such that

$$k = \lambda_1 \geq \lambda_n \geq -k.$$

A Ramanujan graph is a graph such that for any v_i with $i > 1$, $|\lambda_i| \leq 2\sqrt{k-1}$. This property will be useful in identifying supersingular graphs. An *expander graph* has a slightly nuanced definition as follows. Let $\epsilon > 0$ and $k \geq 1$. A k-regular graph is called a one-sided ϵ-expander if

$$\lambda_2 \leq (1 - \epsilon)k$$

and a two-sided ϵ-expander if

$$\lambda_n \geq -(1 - \epsilon)k.$$

A sequence $G_i = (V_i, E_i)$ of k-regular graphs with an infinite number of vertices in G_i is called a one-sided expander family if there exists an $\epsilon > 0$ such that G_i is a one-sided ϵ-expander for all sufficiently large i.

Edge expansion is a ratio that quantifies how well subsets of vertices are connected to the whole graph. From another perspective, this ratio also indicates how far the graph is from being disconnected. Let $F \subset V$ and the boundary of F, denoted $\partial F \subset E$, be the subset of the edges in G that connect F to V/F. The edge expansion ratio of G, $h(G)$, is

$$h(G) = \min_{\#F \leq \#V/2} \frac{\#\partial F}{\#F}.$$

If G is disconnected, then $h(G) = 0$. For every connected graph, the diameter is the longest distance between any two of its vertices. The diameter of a k-regular one-sided ϵ-expander graph is bounded by $O(\log n)$, depending only on k and ϵ. For a graph, a random walk of length i is a path from v_1 to v_i that is selected with a random process. This process selects v_i uniformly at random among the other neighbors of v_{i-1}. Thus the entire path is selected randomly. As a result of this uniformly random selection process, random walks on expander graphs of lengths close to the graph's diameter terminate on any vertex with a probability nearly uniform.

These definitions are critical in understanding isogeny graphs and how they are used in cryptography. Moreover, expander families have pseudo-randomness properties that are very useful in computer science. They can be used to create pseudo-random number generators, error-correcting codes, probabilistic checkable proofs, and cryptographic primitives, and are the backbone of isogeny graphs and provide properties that enable randomness for isogeny-based cryptography.

3.3 ELLIPTIC CURVES AND CRYPTOGRAPHY

A relatively new yet well understood field of cryptography, elliptic curves allow for a reasonable level of security with much less bits (160-250 compared to 1024-3072 bits for RSA). With a sufficiently short key size, RSA can be faster than elliptic curve cryptography (ECC), but ECC still results in fewer computations and takes up less bandwidth. Elliptic curve cryptography (ECC) is used in public-key cryptography and is primarily used in key agreement and digital signature verification but may be used in many other applications. As we look towards the future, we can assume ECC will be more and more common as we develop smaller devices that need to be secured (see Washington (2008) for more details).

In this section, we will give a brief overview of elliptic curves and the group operation on them. Next, since for cryptographic purposes we are interested in finite fields, we will discuss elliptic curves over finite fields. Then, we will define some of the tools that we need in order to describe isogeny-based cryptography. We will start with defining what an isogeny is and then describe methods of computing isogenies on elliptic curves. Lastly, we will cover elliptic curve cryptography as well as isogeny-based cryptography. For more details about the material covered in this section see Silverman (1986), Silverman and Tate (2015), Hoffstein et al. (2010), Galbraith (2017), Lange (2016), and Feo (2017) amongst many others.

3.3.1 Elliptic Curves

An elliptic curve is a type of algebraic curve given as follows

$$E = \left\{ (x,y) \in k^2 \mid y^2 = x^3 + ax + b,\ 4a^3 + 27b^2 \neq 0 \right\} \cup \{\mathbf{O}\},$$

where k is a field and $\mathbf{O}$ the point at infinity. An elliptic curve over the reals is given in Figure 3.6.

If we let the point at infinity play the role of the identity and define addition on the points of the curve as we will describe below, then it is easy to prove that the points on the curve form a group. Note that any point on the curve can be assigned

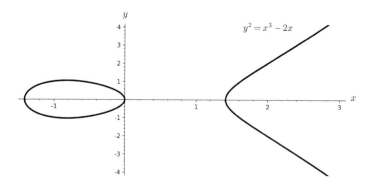

FIGURE 3.6 Elliptic curve.

to be the identity of the group, but then addition also will have to be modified accordingly.

Let us first describe geometrically how the addition of points on elliptic curves works. Given $P = (x_p, y_p)$, and $Q = (x_q, y_q)$ points on an elliptic curve E, let $\mathcal{L}$ be a line passing between P and Q. Then from Bezout's theorem, we know that this line will intersect the curve E at a third point R'. Let $R = P \oplus Q$ be symmetric to the point R' with respect to the x-axis. This is graphically represented in Figure 3.7.

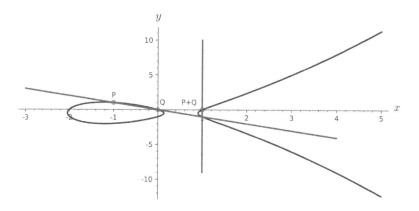

FIGURE 3.7 Addition of points on an elliptic curve.

To add points on an elliptic curve algebraically, we do the following. Let

$$E : y^2 = x^3 + Ax + B$$

be the equation for an elliptic curve and P_1 and P_2 points on E.

1. If $P_1 = \mathbf{O}$, then $P_1 \oplus P_2 = P_2$
2. Otherwise we write $P_1 = (x_1, y_1)$, and $P_2 = (x_2, y_2)$

3. If we denote with λ the slope of the line passing through P_1 and P_2, then we have:
$$\lambda = \begin{cases} \frac{y_2-y_1}{x_2-x_1} & \text{if } P_1 \neq P_2 \\ \frac{3x^2+A}{2y_1} & \text{if } P_1 = P_2 \end{cases}$$

Let $y = \lambda x + b$ be the equation of the line passing through the points P_1 and P_2, where λ can be computed as above and b is the y-intercept. To compute $P_1 \oplus P_2$, we need to find the intersection of this line with the equation of the elliptic curve, i.e. solve the following system of equations:

$$y = \lambda x + b$$
$$y^2 = x^3 + Ax + B$$

and then take the reflection with respect to the x-axis. If we denote by $P_3 = P_1 \oplus P_2 = (x_3, y_3)$, then we have:

$$x_3 = \lambda^2 - x_1 - x_2 \text{ and } y_3 = \lambda(x_1 - x_3) - y_1$$

Given an elliptic curve E, the j-invariant for elliptic curves, j, of E is defined to be

$$j = 1728 \frac{4a^3}{4a^3 + 27b^2}$$

Two elliptic curves E and E' over the finite field k are isomorphic if and only if $j(E) = j(E')$.

In order to use elliptic curves for cryptography, we must consider elliptic curves whose points have coordinates in a finite field $\mathbb{F}_p$. Let E be an elliptic curve over $\mathbb{F}_p$ such that $E : Y^2 = X^3 + AX + B$ where $A, B \in \mathbb{F}_p$ and $4A^3 + 27B^2 \neq 0 \mod p$. Additionally, we will denote the points on E with coordinates in $\mathbb{F}_p$ by

$$E(\mathbb{F}_p) = \{(x, y) : x, y \in \mathbb{F}_p | y^2 = x^3 + Ax + B\} \cup \{\mathbf{O}\}.$$

Example 3.12. *Find all the points on the elliptic curve $E : Y^2 = X^3 + 3X + 8$ over the field $\mathbb{F}_{13}$. The set of all possible X-coordinates of points on this curve is $X = \{ 0, 1, 2, 3, ..., 12\}$. In order to find all the points of $E(\mathbb{F}_{13})$, we must check each possible X value and check whether $X^3 + 3X + 8$ is a square, modulo 13. First we check $X = 0$*

$$(0)^3 + 3(0) + 8 = 8.$$

Since 8 is not a square modulo 13, there are no points with the X-coordinate of 0. Next we check $X = 1$

$$(1)^3 + 3(1) + 8 = 12.$$

Now, 12 is a square modulo 13 and has two corresponding Y-coordinate values.

$$5^2 \equiv 12 \mod 13 \text{ and } 8^2 \equiv 12 \mod 13.$$

Iterating though each option will yield the following result:

$$E(\mathbb{F}_{13}) = \Big\{\, \boldsymbol{O}, (1,5), (1,8), (2,3), (2,10), (9,6), (9,7), (12,2), (12,11) \Big\}.$$

These nine points are all the points on this elliptic curve over $\mathbb{F}_{13}$.

Theorem 3.2. *The points on $E(\mathbb{F}_p)$ form an Abelian group.*

For the proof of this, see Silverman (1986). The following theorem is called Hasse's Theorem and gives a bound on the number of points $\#E(\mathbb{F}_p)$ of an elliptic curve E over $\mathbb{F}_p$.

Theorem 3.3 (Hasse's Theorem). *Let E be an elliptic curve over the finite field $\mathbb{F}_p$, then*

$$\mid \#E(\mathbb{F}_p) - (p+1) \mid \leq 2\sqrt{p}$$

For large values of $q = p^n$, the number of points on the curve is in a narrow range of width $4\sqrt{q}$ around $(q+1)$. This provides some insight into how to pick elements of $E(\mathbb{F}_q)$ with an almost uniform distribution.

3.3.2 Isogenies of Elliptic Curves

In the upcoming subsections, we will define some of the tools we need in order to describe isogeny-based cryptography. Let E and E' be elliptic curves defined over a field K. Before giving the definition of an isogeny, let us recall some concepts from abstract algebra. A group homomorphism is a map $f : G \to G$ between two groups such that the group operation is preserved:

$$f(g_1 g_2) = f(g_1) f(g_2)$$

for all $g_1, g_2 \in G$, where the product on the left-hand side is in G and on the right-hand side in H. As a result, a group homomorphism maps the identity element in G to the identity element in H: $f(e_G) = e_H$. Note that a homomorphism must preserve the inverse map because

$$f(g)f(g^{-1}) = f(gg^{-1}) = f(e_G) = e_H, \text{ so } f(g)^{-1} = f(g^{-1}).$$

In particular, the image of G is a subgroup of H and the group kernel, i.e., $f^{-1}(e_H)$ is a subgroup of G. The kernel is actually a normal subgroup, as is the pre-image of any normal subgroup of H. Hence, any (nontrivial) homomorphism from a simple group must be injective.

An *isogeny* $\phi : E \to E'$ is an algebraic morphism satisfying $\phi(\infty) = \infty$. The degree of the isogeny is its degree as an algebraic map. The endomorphism ring End(E) is the set of isogenies from E to itself, together with the constant morphism. This set forms a ring under pointwise addition and composition.

When K is a finite field, the rank of End(E) as a $\mathbb{Z}$-module is either 2 or 4. We say E is *supersingular* if the rank is 4, and ordinary otherwise. A supersingular curve cannot be isogenous to an ordinary curve. Supersingular curves are all defined over $\mathbb{F}_{p^2}$, and for every prime l that does not divide p, there exist $l+1$ isogenies (counting multiplicities) of degree l originating from any given such supersingular curve.

Two elliptic curves are called *isogenous* if there exists an isogeny between them. A theorem of Tate states that over a finite field $\mathbb{F}_q$, two elliptic curves E and E' are isogenous if and only if $\#E_1(\mathbb{F}_q) = \#E_2(\mathbb{F}_q)$, see Tate (1966).

One common type of endomorphism is the multiplication-by-m map defined by

$$[m] : P \to [m]P = P + ... + P$$

where P is a point on an elliptic curve E and $[m]P$ is the result of adding P to itself m times. A point P on elliptic curve E is called a *torsion point* if there exists an integer $m > 0$ such that

$$[m]P = P + ... + P = \mathbf{O}$$

Then, the m-torsion subgroup of E, $E[m]$, is the set of order m in E of torsion points. We may write this as

$$E[m] = \{P \in E : [m]P = \mathbf{O}\}.$$

Thus the kernel of the multiplication-by-m endomorphism is the m-th torsion subgroup, $E[m]$.

Given an elliptic curve E and a finite group G of E, there is up to an isomorphism a unique isogeny $E \to E' = E/G$ having kernel G, (Silverman, 1986, Section II.4.12). Hence we can identify an isogeny by specifying its kernel, and conversely given a kernel subgroup the corresponding isogeny can be found using Vélu's formulas, as defined in the next subsection.

3.3.3 Velu's Formula

Let $E : y^2 = x^3 + ax + b$ be an elliptic curve defined over field k, and let $G \subset E(k)$ be a finite subgroup. For every finite subgroup of E called G, we can construct a quotient isogeny as $\phi : E \to E/G$. We let $E' = E/G$. Thus, $\phi : E \to E'$ is an isogeny between the elliptic curves E and E', and the two elliptic curves are isogenous with each other. To make this map between the elliptic curves precise we use Velu's formulas (see Velú (1971) for more details). See Washington (2008, Thm. 12.16) for the proof of the following.

Theorem 3.4 (Vélu)**.** *Let* $E : y^2 = x^3 + ax + b$ *be an elliptic curve over* k *and let* G *be a finite subgroup of* $E(\bar{k})$ *of odd order. For each nonzero* $Q = (x_Q, y_Q)$ *in* G *define*

$$t_Q := 3x_Q^2 + a, \quad u_Q := 2y_Q^2, \quad w_Q := u_Q + t_Q x_Q$$

and let

$$t := \sum_{Q \in G \setminus \{O\}} t_Q, \quad w := \sum_{Q \in G \setminus \{O\}} w_Q,$$

$$r(x) := x + \sum_{Q \in G \setminus \{O\}} \left(\frac{t_Q}{x - x_Q} + \frac{u_Q}{(x - x_Q)^2} \right).$$

The rational map

$$\phi(x, y) := \Big(r(x), r'(x)y \Big)$$

is a separable isogeny from E to $E' : y^2 = x^3 + a'x + b'$, where $a' := a - 5t$ and $b' := b - 7w$.

Using Velu's Formula, we can now find the equation of the quotient isogeny for any finite subgroup of E, calculate points across the map, and find the mapping itself.

3.3.4 Modular Polynomials

Another important question to consider here is the following. Given two elliptic curves E and E', how can we check whether or not they are isogenous to each other. In order to verify that two elliptic curves are isogenous, we use modular polynomials, $\phi(x, y)$. Modular polynomials are polynomials in two variables x and y, where for any two elliptic curves E and E', $x = j(E)$ and $y = j(E')$. The elliptic curves E and E' are isogenous of degree l if and only if $\phi_l(x, y) = \phi(j(E), j(E')) = 0$.

Here we will examine some sample modular polynomials for selected common degrees. The modular polynomial for degree $l = 1$ is

$$\phi_1 = x + y$$

The modular polynomial for degree $l = 2$ is

$$\phi_2 = x^3 + y^3 - x^2 y^2 - 162000(x^2 + y^2) + 1488xy(x + y) \\ + 40773375xy + 8748000000(x + y) - 157464000000000.$$

The modular polynomial for degree $l = 3$ is

$$\phi_3 = x^4 + 7y^4 - x^3 y^3 + 2232x^2 y^2(x + y) - 1069956xy(x^2 + y^2) \\ + 36864000(x^3 + y^3) + 2587918086x^2 y^2 + 8900222976000xy(x + y) \\ + 452984832000000(x^2 + y^2) - 770845966336000000xy \\ + 1855425871872000000000(x + y).$$

The modular polynomial for degree $l = 5$ is

$$\begin{aligned}\phi_5 =\ & x^6 + y^6 - x^5 y^5 + 3720 x^4 y^4 (x+y) - 4550940 x^3 y^3 (x^2 + y^2) \\ & + 2028551200 x^2 y^2 (x^3 + y^3) - 246683410950 xy (x^4 + y^4) \\ & + 1963211489280 (x^5 + y^5) + 107878928185336800 x^3 y^3 (x+y) \\ & + 1665999364600 x^4 y^4 + 383083609779811215375 x^2 y^2 (x^2 + y^2) \\ & + 128541798906828816384000 xy (x^3 + y^3) \\ & + 1284733132841424456253440 (x^4 + y^4) \\ & - 441206965512914835246100 x^3 y^3 \\ & + 26898488858380731577417728000 x^2 y^2 (x+y) \\ \\ & - 19245793461892829965510823116800 0 xy (x^2 + y^2) \\ & + 28024477782843952780432156529786880 0 (x^3 + y^3) \\ & + 511094177755241808311076519936000 0 x^2 y^2 \\ & + 3655473658394962929570647233265664000 0 xy (x+y) \\ & + 669250004262799770848714941501506846720 0 (x^2 + y^2) \\ & - 2640734570766205962597157902479787829493 76 xy \\ & + 53274330803424425450420160273356509151232000 (x+y) \\ & + 141359947154721358697753474691071362751004672000.\end{aligned}$$

The modular polynomials increase in complexity as the degree increases and can be computed up to degree 300. Using the appropriate l-degree modular polynomial on the j-invariants of the two elliptic curves will indicate if the elliptic curves are l-isogenous. This is especially useful in verifying the elliptic curve produced by Velu's Formula or in relating two given elliptic curves.

3.3.5 Division Polynomials

Let $E : y^2 = x^3 + ax + b$ be an elliptic curve and ψ_m be the m-th division polynomial of E. The division polynomials are defined recursively with the initial values of

$$\psi_1 = 1,$$
$$\psi_2 = 2y^2,$$
$$\psi_3 = 3x^4 + 6ax^2 + 12bx - a,$$
$$\psi_4 = 2y^2 (2x^6 + 10ax^4 + 40bx^3 - 10a^2 x^2 - 8abx - 2a^3 - 16b^2),$$

and the recursive relation of

$$\psi_{2m+1} = \psi_{m+2} \psi_m^3 - \psi_{m-1} \psi_{m+1}^3 \quad \text{for } m \geq 2,$$
$$\psi_2 \psi_{2m} = \psi_m (\psi_{m+2} \psi_{m-1}^2 - \psi_{m-2} \psi_{m+1}^2) \quad \text{for } m \geq 3.$$

If we let $P(x,y)$ be a point on the curve E, then we can explicitly find the formulas for $[m]P$, the multiplication-by-m mapping of point P, as

$$[m]P \neq \left(\frac{\phi_m(P)}{\psi_m(P)^2}, \frac{\omega_m(P)}{\psi_m(P)^3} \right)$$

for any point $P \neq \mathbf{O}$, where ϕ_m and ω_m are defined as

$$\phi_m = x\psi_m^2 - \psi_{m+1}\psi_{m-1},$$

$$\omega_m = \psi_{m-1}^2 \psi_{m+2} + \psi_{m-2}\psi_{m+1}^2.$$

Note that the m-th division polynomial ψ_m vanishes on $E[m]$. To have $[m]P = \mathbf{O}$, we need to set up the denominators equal to 0.

Now we will present a specific example to find the isogenies of the elliptic curve over a finite field.

Example 3.13. *Consider the elliptic curve $E/\mathbb{F}_{11} : y^2 = x^3 + 4$. We want to find all the possible 3-isogeny maps that are mapped out of F. For this finite field, $\mathbb{F}_{11}$, the possible values for x are $x = \{0, 1, 2, 3, ...10\}$. These values correspond to the y-values of $y = \{0, 1, ..., 10\}$ and the points on this elliptic curve are*

$$E(\mathbb{F}_{11}) = \{(0,2), (0,9), (1,4), (1,7), (2,1), (2,10), (3,3), (3,8), (6,0), \mathbf{O}\}$$

Since we want to compute 3-isogenies of this curve we have to find $E[3]$, hence we use the 3-division polynomial. The division polynomial is

$$\psi_3 = 3x^4 + 6ax^2 + 12bx - a$$

Since for this elliptic curve, $a = 0$ and $b = 4$, we obtain the division polynomial

$$\psi_3(x) = 3x^4 + 48x$$

and we can partially split it into the factors

$$\psi_3(x) = x(x+3)(x^2 + 8x + 9),$$

which solves into $x = 0$ and $x = -3$. These x-values result in 3-torsion points. Thus the points $(0,2)$ and $(0,9)$ are in $E(\mathbb{F}_{11})$ but the rest of the points are in the extension field, $E(\mathbb{F}_{11^2})$. Write

$$\mathbb{F}_{11^2} = \mathbb{F}_{11}(i), \text{ where } i^2 = -1$$

Then, ψ_3 splits over $\mathbb{F}_{11^2}$ as

$$\psi_3(x) = x(x+3)(x+9i+4)(x+2i+4)$$

We want to compute all possible degree 3-isogenies. Consider all possible cyclic subgroups of order 3 of $E_{\mathbb{F}_{11}^2}$. We have,

$$G_1 = \left\{ (0,2), (0,9), \mathbf{O} \right\}$$
$$G_2 = \left\{ (8,i), (8,10i), \mathbf{O} \right\}$$
$$G_3 = \left\{ (2i+7,i), (2i+7,10i), \mathbf{O} \right\}$$
$$G_4 = \left\{ (9i+7,i), (9i+7,10i), \mathbf{O} \right\}$$

Using Thm. 3.4 for the above groups we get the following four isogenies

$$\phi_1 : E \to E_1/\mathbb{F}_{11} : y^2 = x^3 + 2$$
$$\phi_2 : E \to E_2/\mathbb{F}_{11} : y^2 = x^3 + 5x$$
$$\phi_3 : E \to E_3/\mathbb{F}_{11^2} : y^2 = x^3 + (7i+3)x$$
$$\phi_4 : E \to E_4/\mathbb{F}_{11}^2 : y^2 = x^3 + (4i+3)x$$

The four curves E_1, E_2, E_3, E_4 are all 3-isogenous to E. Moreover, each of these curves will be 3-isogenous to four more elliptic curves and so on.

3.3.6 Elliptic Curve Cryptography

In this subsection we would like to give a simplified idea of how one uses elliptic curves to construct a cryptosystem. The reason why elliptic curves are used for cryptography stems from how difficult it is to solve the discrete log problem on points on the elliptic curve. Let G be a finite group. Given $a, b \in G$, find an integer x such that
$$b^x = a.$$

This is known as the **discrete log problem**. The elliptic curve discrete log problem (ECDLP) relies on the fact that the points on the curve over a finite field form a cyclic group and is formulated as follows. For two points P and Q on $E(\mathbb{F}_p)$, it is difficult to find n such that the following holds

$$Q = \underbrace{P + P + \ldots + P}_{\text{'}n\text{' many additions of } P} = nP$$

Trying to solve for n is translated as solving the DLP for elliptic curves. On a classical computer with classical computer adversaries, the elliptic curve discrete log problem is sufficiently difficult. The fastest algorithms solve it in $\Omega(\sqrt{p})$. However, with the threat of quantum computers, the discrete log problem is no longer secure.

3.3.7 Elliptic Curve Diffie-Hellman Key Exchange

In this subsection, we will discuss the Diffie-Hellman key exchange algorithm using elliptic curves. The Elliptic Curve Diffie-Hellman (ECDH) key exchange follows this algorithm: Alice and Bob choose a prime p and then they agree on a particular elliptic curve $E(\mathbb{F}_p)$ and a specific point on the curve $P \in E(\mathbb{F}_p)$ and publicize these values. Then, Alice and Bob each must select secret integers n_A and n_B respectively, and keep them secret, and compute

$$Q_A = n_A P \quad \text{and} \quad Q_B = n_B P$$

where $n_A P$ is the n_A-th multiplication of point P, i.e. a point on the elliptic curve as well as Q_B. Alice and Bob exchange the values Q_A and Q_B, which are points on the elliptic curve E, on the public channel. Then, Alice uses the point that Bob sent and her private number n_A to compute $n_A Q_B$, and Bob does the same and computes $n_B Q_A$. They end up computing the same shared secret key since

$$n_A Q_B = (n_A n_B) P = n_B Q_A.$$

The only way that Eve can find Alice's and Bob's secret key is she uses the elliptic curve E and the three points on it P, Q_A, and Q_B to solve the ECDLP. If Eve solved this problem, then she would be able to uncover both secrets. We sum up this protocol in Table 3.1.

TABLE 3.1 Elliptic curve Diffie-Hellman key exchange

Public Parameters	
The parameters p (a large prime number), E over $\mathbb{F}_p$, and point P in $E(\mathbb{F}_p)$ are made publicly available	
Private Computations	
Alice: $Q_A = n_A P$	Bob: $Q_B = n_B P$
Public Exchange of Values	
Alice sends Q_A	Bob sends Q_B
Private Computations	
Alice: Computes $n_A Q_B$	Bob: Computes $n_B Q_A$
Now both parties have $n_A Q_B = n_A(n_B P) = n_B(n_A P) = n_B Q_A$	

Let us illustrate this with the following example.

Example 3.14. *Let $E : y^2 = x^3 + 324x + 1287$, be an elliptic curve and $P = (920, 303) \in E(\mathbb{F}_{3851})$. Alice and Bob choose arbitrary integers $n_A = 1194$ and $n_B = 1759$. They then calculate:*

$$\text{Alice: } Q_A = 1194P = (2607, 2178) \in E(\mathbb{F}_{3851}),$$
$$\text{Bob: } Q_B = 1759P = (3684, 3125) \in E(\mathbb{F}_{3851}).$$

Once they exchange their respective Q values, they compute:

$$\text{Alice: } nQ_B = 1194(3684, 3125) = (3347, 1242) \in E(\mathbb{F}_{3851}),$$
$$\text{Bob: } nQ_A = 1759(2607, 2178) = (3347, 1242) \in E(\mathbb{F}_{3851}).$$

So now both Alice and Bob have the secret shared point $(3347, 1242)$, *which is the private shared key, and in order for a third party to obtain this value they would have to solve the ECDLP for* $nP = Q_{A,B}$.

In the following example we are considering an elliptic curve taken from the NIST database of elliptic curves used for cryptography.

Example 3.15. *Consider P–192 NIST curve* $y^2 = x^3 - 3x + b$ *over a finite field* $\mathbb{F}_q$ *where* $q, n, G,$ *and* b *are as follows.*

$b = 2455155546008943817740293915197451784769108058161191238065;$
$q = 6277101735386680763835789423207666416083908700390324961279$
$n = 6277101735386680763835789423176059013767194773182842284081$
$G = \{602046282375688656758213480587526111916698976636884684818,$
$\phantom{G = \{}174050332293622031404857552280219410364023488927386650641\};$

It is clear that these cryptosystems deal with some fairly large numbers, however, one of the advantages of ECC is that the key size is smaller than RSA. This allows for it to be used in smaller devices.

In the next section, we will explore how elliptic curves can be used in cryptography in a post-quantum world.

3.3.8 Post-Quantum Cryptography with Elliptic Curves

As we saw in the previous sections, classical elliptic curve cryptography is based on the addition on the set of rational points on an elliptic curve over a finite field. Furthermore, classical ECC relies on the difficulty of the elliptic curve discrete log problem. However, since Shor's algorithm breaks the elliptic curve discrete log problem, it is necessary to study how we can leverage elliptic curves for new types of cryptosystems in order to make them quantum resistant.

In this subsection, we will describe how we can use elliptic curves to come up with a new type of cryptosystem, isogeny-based cryptography, believed to be quantum-resistant. In contrast to classical elliptic curve cryptography, isogeny-based cryptography does not involve point addition on the curve at all. Rather it is based on supersingular isogeny graphs on which multiple elliptic curves are depicted. The relationships between isogenies of elliptic curves is described in detail in the following sections and is based on the j-invariant of the curve.

For the rest of this section we will focus on isogeny-based cryptography and explain in detail the quantum-resistant supersingular Diffie-Hellman key exchange scheme. Most of the material presented in this section can be found in Alwen (2018), Costello and Hisil (2017), Feo (2017), and De Feo et al. (2014).

3.3.9 Supersingular Isogeny Diffie-Hellman Key Exchange

In this section, we present a key exchange protocol using supersingular elliptic curves, see Feo (2017); De Feo et al. (2014) for a more complete description of this protocol as well as zero-knowledge proof of identity and a public-key encryption based on supersingular isogenies.

This protocol requires supersingular curves of smooth order. If we let p and l be distinct primes then the following statements are true about supersingular curves. All supersingular j-invariants of curves in $\mathbb{F}_p$ are also defined in $\mathbb{F}_{p^2}$. And, the graph of supersingular curves in $\mathbb{F}_p$ with l-isogenies is connected, $l+1$-regular, and has the Ramanujan property.

Fix $\mathbb{F}_q = \mathbb{F}_{p^2}$, where $p = l_A^{e_A} l_B^{e_B} \cdot f \pm 1$ and l_A, l_B are small primes, and f is a cofactor such that p is prime. Construct a supersingular elliptic curve E defined over $\mathbb{F}_q$ of cardinality $(l_A^{e_A} l_B^{e_B} \cdot f)^2$. By construction, $E[l_A^{e_A}]$ is $\mathbb{F}_q$-rational and contains $l_A^{e_A - 1}(l_A + 1)$ cyclic subgroups of order $l_A^{e_A}$, each defining a different isogeny; the analogous statement holds for $E[l_B^{e_B}]$.

More precisely, the supersingular Isogeny Diffie-Hellman (SIDH) key exchange follows this algorithm. Pick as the public parameters a supersingular elliptic curve E over $\mathbb{F}_{p^2}$, and bases $\{P_A, Q_A\}$ and $\{P_B, Q_B\}$ which generate respectively $E[\lambda_A^{e_A}] = \langle P_A, Q_A \rangle$, and $E[\lambda_B^{e_B}] = \langle P_B, Q_B \rangle$. Then Alice chooses two random numbers $m_A, n_A \in \mathbb{Z}$ not both divisible by l_A, and computes an isogeny $\alpha : E \to E/\langle A \rangle$ with kernel $\langle A \rangle = \langle [m_A]P_A + [n_A]Q_A \rangle$. Alice computes also $\alpha(P_B)$ and $\alpha(Q_B)$ and then sends them to Bob together with E_A.

Bob on the other side chooses two random numbers $m_B, n_B \in \mathbb{Z}$ not both divisible by l_B, and computes an isogeny $\beta : E \to E/\langle B \rangle$ with kernel $\langle B \rangle = \langle [m_B]P_AB + [n_B]Q_B \rangle$ as well as $\beta(P_A)$ and $\beta(Q_A)$ and then sends them to Alice. Upon receipt of the respective information, both of the parties can go ahead and compute the secret shared key. Alice computes

$$E/\langle A, B \rangle = E_B/\langle \beta(A) \rangle$$

where

$$\langle \beta(A) \rangle = \langle [m_A]\beta(P_A) + [n_A]\beta(Q_A) \rangle$$

and Bob similarly computes

$$E/\langle A, B \rangle = E_A/\langle \alpha(B) \rangle$$

where

$$\langle \alpha(A) \rangle = \langle [m_B]\alpha(P_B) + [n_B]\alpha(Q_B) \rangle$$

so that they have the shared secret key $E/\langle A, B \rangle$.

Given two elliptic curves E, E' over a finite field, isogenous of known degree d, to find an isogeny $\phi : E \to E'$ of degree d is a notoriously difficult problem for which only algorithms exponential in $\log \#E$ are known in general.

The security of this scheme is based on the difficulty of finding a path connecting two given vertices in a graph, called the isogeny graph of supersingular isogenies. An *isogeny graph* is a multi-graph in which the nodes are the j-invariants of isogenous elliptic curves, defined over $\mathbb{F}_q$, and edges are isogenies between them, up to $\overline{\mathbb{F}}_q$ isomorphism.

It was shown in Pizer (1990) and Pizer (1998) that the isogeny graphs for supersingular elliptic curves have the Ramanujan property. The general concept for the key exchange is that Alice and Bob both agree to start at the same curve E on the agreed-upon isogeny graph G. Both Alice and Bob take random walks from E of a secret number of steps to some curves E_A and E_B, respectively. They exchange the curves E_A and E_B but keep the number of steps it took them to get there privately. Then, Alice begins at curve E_B and repeats the 'same' secret steps she took while Bob begins at curve E_A and repeats the 'same' secret steps he took. They will both arrive at the same secret shared curve E_S.

If ϕ and ψ are random walks in the graph of isogenies of degree l_A and l_B then the following diagram is commutative.

$$\begin{array}{ccc} E & \xrightarrow{\phi} & E/\langle P \rangle \\ \downarrow \psi & & \downarrow \\ E/\langle Q \rangle & \longrightarrow & E/\langle P, Q \rangle \end{array}$$

Next, we will briefly discuss a method of exploiting isogeny paths. If two elliptic curves E and E' are provided over a finite field k, such that the curves have the same cardinality, then finding an isogeny $\phi : E \to E'$ is a common but relatively difficult problem. The only algorithms that are known to solve it compute in exponential time, $O(\log \#E)$. This approach relies on a meet in the middle random walk. This kind of exploitation involves selecting an expander graph G that contains the elliptic curves E and E'. Then, beginning a random walk from each curve. The two random walks are expected to collide, or meet, after $O(\sqrt{\#G})$ by the birthday paradox. Once detecting that collision, we can simply compose the walks, resulting in the isogeny. This solution technique was actually applied in a wide-spread attack on elliptic curves over extension fields, but it only affects a portion of an isogeny class. As a result, an attacker must find an isogeny from an attack-immune elliptic curve to a weak elliptic curve and then map the discrete log problem from one curve to the other. Since the average size of an isogeny class of ordinary elliptic curves is $O(\sqrt{\#E})$, this meet in the middle random walk strategy yields an $O(\#E^{1/4})$ computing time on any single elliptic curve in the class. This time is actually more efficient than the general attack on the discrete log problem. This example demonstrates the need for an improved cryptosystem that relies on supersingular isogenies and is quantum resistant.

In De Feo et al. (2014), they give a precise formulation of the necessary computational assumptions (of supersingular isogeny Diffie-Hellman key exchange, zero-knowledge proof of identity, and a public-key encryption based on supersingular isogenies) along with a discussion of their validity, and prove the security of these protocols under those assumptions.

3.4 GENUS 2 CURVES AND CRYPTOGRAPHY

The field of elliptic curve cryptography is a well-established field, so as we look to push the boundary of knowledge on this subject, we look to a field of mathematics that is very similar, hyperelliptic curve cryptography. Elliptic curves are known as genus one curves. Curves of higher order are higher genus curves. Hyperelliptic curves of genus two and three are known to be good for cryptography purposes, while there are known methods of attacks for curves of higher genus. Hence, for cryptographic purposes, we are interested in studying only curves of genus 2 and 3.

In this chapter, we will focus on hyperelliptic curves of genus 2. They have been an object of much mathematical interest since the eighteenth century and there exists continued interest to date. They have become an important tool in many algorithms in cryptographic applications, such as factoring large numbers, hyperelliptic curve cryptography, etc. The study of hyperelliptic curves is articulated well in Cohen and Frey (2006), Frey and Shaska (2019), Simpson (2019), and Silverman (1986), and that is where much of this material is derived from.

As we look to extend our study to hyperelliptic curves, we have to note a few differences from elliptic curves. Most notably the points on a hyperelliptic curve do not inherently form a group, as it was the case for elliptic curves. Instead we must define divisors and what is known as the Jacobian variety of the curve. In the upcoming section, we define hyperelliptic curves of genus G, the divisors and then show how we can define an addition operation on them which makes the Jacobian of a hyperelliptic curve into a group. This enables us to create analogous methods such as the hyperelliptic curve discrete logarithm problem. Then, we will focus our attention on hyperelliptic curves of genus $g = 2$ and describe how they are used in classical cryptography as well as some potential applications in post-quantum cryptography.

3.4.1 Hyperelliptic Curves

Let k be a field and $\bar{k}$ be an algebraic closure of k. A hyperelliptic curve C of genus g over k are algebraic curves given by an equation of the form:

$$C = \left\{ (x,y) \in k \,|\, Y^2 = X^{2g+1} + A_1 X^{2g} + \ldots + A_{2g} X + A_{2g+1} \right\} \cup \left\{ \mathcal{O} \right\},$$

where g is the genus of the curve and the discriminant $\Delta \neq 0$. Hence, a hyperelliptic curve of $g = 2$ is given by the equation:

$$y^2 = a_6 x^6 + \cdots + a_1 x + a_0$$

and the discriminant $\Delta \neq 0$. For a geometric description of this type of curve over the reals see the next example.

Example 3.16. *Let*

$$C : y^2 = \frac{1}{9}\left(x^5 - 5x^3 + 4x\right) = \frac{1}{9}x(x-1)(x-2)(x+1)(x+2)$$

be a hyperelliptic curve over the reals. Then, the set of points that satisfies the above equation looks like Figure 3.8.

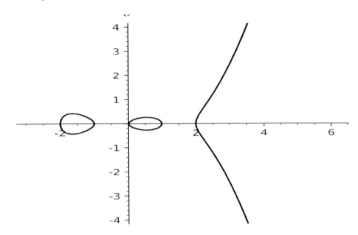

FIGURE 3.8 Hyperelliptic curve.

Next we define divisors of hyperelliptic curves.

3.4.2 Divisors

Let C be a smooth algebraic curve of genus 2. A divisor D is a finite formal sum of points P_i on C, such that:

$$D = \sum_i m_i P_i,$$

where $m_i \in \mathbb{Z}$. The set of points $P \in C$ where $D(P) \neq 0$ is called the **support** of D and the set of all divisors under the addition

$$D_1 + D_2 = \sum_i m_i P_i + \sum_i n_i P_i = \sum_i (m_i + n_i) P_i$$

form a group which is denoted by $Div(C)$. The finiteness of the support of genus 2 curves allows us to account for the degree of the divisor, which is the sum:

$$n = \deg D = \sum_i m_i.$$

All divisors with degree n form a group denoted $Div^n(C)$ and the group of divisors of degree zero is denoted $Div^0(C)$. Now, for each function $\varphi(x,y)$ on the curve, we can associate a divisor $Div(\varphi) \subset Div(C)$.

Definition 3.3. *Let P be a point on the curve and $\varphi(x,y)$ a function on the curve, then:*

$$Div(\varphi) = \sum_{P \in C} ord_P \varphi \cdot P$$

is called a principal divisor.

Let $PDiv(C)$ be the set of all principal divisors. We call a divisor of zeros the sum:
$$Div_0 \varphi = \sum_{P, ord_P \varphi > 0} ord_P \varphi \cdot P$$
and we call a divisor of poles the sum:
$$Div_\infty \varphi = \sum_{P, ord_P \varphi < 0} ord_P \varphi \cdot P.$$

It is a basic fact that D is a principal divisor if and only if $\deg D = 0$. The quotient group $Div(C)/PDiv(C)$ is called the Picard group and is denoted Pic (C). All the elements of degree 0 of Pic (C) form a subgroup which we will denote Pic $^0(C)$, and this is called the Jacobian of the curve Jac (C). So,
$$\text{Jac } (C) \cong \text{Pic } ^0(C) \cong Div(C)/PDiv(C).$$

A semi-reduced divisor D_1 is of the form $D_1 = \sum_i m_i(P_i) - \sum_i m_i(P_\infty)$ such that:

1. All $m_i \geq 0$, and the P_i are finite points
2. If $P_i \neq -P_i$, then only one of them occurs in the sum, with $m_i \neq 0$
3. If $P_i = -P_i$, then $m_i \leq 1$

A convenient representation of reduced (and semi-reduced) divisors is due to Mumford.

Theorem 3.5 (Mumford representation). *Let $\mathcal{X}$ be a genus g hyperelliptic curve with affine part given by $y^2 + h(x)y - f(x)$, where $h, f \in K[x]$, $\deg f = 2g+1$, $\deg h \leq g$. Each nontrivial group element $\overline{D} \in \text{Pic }^0_{\mathcal{X}}$ can be represented via a unique pair of polynomials $[\mathfrak{u}(x), \mathfrak{v}(x)] \in K[x]$, where*

1. *$\mathfrak{u}$ is a monic,*
2. *$\deg \mathfrak{v} < \deg \mathfrak{u} \leq g$*
3. *$\mathfrak{u} | \mathfrak{v}^2 + \mathfrak{v}h - f$.*

Let $\overline{D}$ be uniquely represented by $D = \sum_{i=1}^r P_i - rP_\infty$, where $P_i \neq P_\infty$, $P_i \neq -P_j$ for $i \neq j$ and $r \leq g$. Put $P_i(x_i, y_i)$. Then the corresponding polynomials are defined by
$$\mathfrak{u} = \prod_{i=1}^r (x - x_i)$$
and the property that if P_i occurs n_i times then
$$\left(\frac{d}{dx}\right)^j [\mathfrak{v}(x)^2 + \mathfrak{v}(x)h(x) - f(x)]|_{x=x_i} = 0, \text{ for } 0 \leq j \leq n_i - 1.$$

A divisor with at most g points in the support satisfying $P_i \neq P_\infty$, $P_i \neq -P_j$ for $i \neq j$ is called a *reduced divisor*. The first part states that each class can be represented by a reduced divisor. The second part of the theorem means that for all points $P_i = (x_i, y_i)$ occurring in D we have $\mathfrak{u}(x_i) = 0$ and the third condition guarantees that $\mathfrak{v}(x_i) = y_i$ with appropriate multiplicity.

The Jacobian, denoted Jac (C) is the set of reduced divisors, and we can define an addition operation on the reduced divisor, which makes Jac (C) into a group. The unique divisor of weight 0 is the neutral element of the addition law. A convenient representation of reduced (and semi-reduced) divisors, due to Mumford, uses a pair of polynomials $[\mathfrak{u}(x), \mathfrak{v}(x)]$, which for genus 2 curves are given

$$\mathfrak{u}(x) := x^2 + \mathfrak{u}_1 x + \mathfrak{u}_0 \text{ and } \mathfrak{v}(x) := \mathfrak{v}_1 x + \mathfrak{v}_0$$

In every class of divisors in Jac (C), there exists a unique divisor

$$D = P_1 + P_2 + \ldots + P_n - nP_\infty,$$

such that for all $i \neq j$, P_i and P_j are not symmetric points, and there is a unique representation of D by two polynomials $[\mathfrak{u}, \mathfrak{v}]$, such that $\deg \mathfrak{v} < \deg \mathfrak{u} \leq 2$, and $\mathfrak{u}$ divides $\mathfrak{v}^2 - f(x)$.

Conversely, given any polynomial $\mathfrak{u}, \mathfrak{v}$, such that $\mathfrak{u}$ divides $\mathfrak{v}^2 - f(x)$, where $\deg \mathfrak{v} < \deg \mathfrak{u} \leq 2$, then we get the divisor D of a point on the x-line; over each zero of $\mathfrak{u}$ the corresponding values of $\mathfrak{v}$ gives a square root of $f(x)$, thus the divisor of the point on the curve so obtained is in the group $Div(\mathcal{C})$.

Note that we are focusing on hyperelliptic curves C of genus $g = 2$ and note that in every class of divisors in Jac (C), there exists a unique divisor

$$D = P_1 + P_2 - 2P_\infty. \tag{3.1}$$

We illustrate this fact with the following example.

Example 3.17. *Consider the hyperelliptic curve C over $\mathbb{F}_5$ with equation*

$$C : y^2 = x^5 + 3x^3 + 2x^2 + 3$$

The following are points on the curve:

$$P_1 = (3, 0), P_2 = (1, 2), P_3 = (4, 1), P_4 = (1, 3)$$

We define the divisors:

$$D_1 = P_1 + P_2 - 2P_\infty, \qquad D_2 = P_3 + P_4 - 2P_\infty$$

3.4.3 Computing Principal Divisors on a Curve

Consider the smooth curve $C \subset \mathbb{R}^2$. Take $g \in k(C)^*$. Then the principal divisor is

$$(g) = \sum v_p(g) p \in Div_C^0$$

where $v_p(g) \in \mathbb{Z}$ is the order of the zero, if p is a zero, or minus the order of the pole, if p is a pole. Let us illustrate with the following example.

Example 3.18. *Let $C : y^2 = x^3 + 1$ over $\mathbb{F}_{13}$. Compute the Principal Divisor of the rational function $g = \frac{x^2}{y}$. Let us first homogenize:*

$$g = \frac{x^2}{yz} \text{ and } C : y^2 z = x^3 + z^3$$

Next we have to find zeroes and poles. To find the poles we have that

$$x^2 = 0 \text{ and } y^2 z = x^3 + z^3,$$

so $y^2 z = z^3$ and therefore $z(y^2 - z^2) = 0$. From there we have that $y^2 = z^2$. This gives us the zeroes: $(0, 0, 0)$; $(0, 1, 1)$; and $(0, -1, 1)$ as our zeroes. For poles, we know that $yz = 0$, and $y^2 z = x^3 + z^3$. We need to find where $y = 0$ or $z = 0$. For $y = 0$,

$$x^3 = -z^3$$

Because this is over $\mathbb{F}_{13}$, we have not only the points $(-1, 0, 1)$ and $(1, 0, 1)$ but also $(4, 0, -1)$ and $(-3, 0, 1)$. When $z = 0$ we get $x^3 = 0$ and so we get the pole $(0, 1, 0)$. We then write the divisor as

$$D = 2P_0 + 1 \cdot P_1 + 1 \cdot P_2 - 1 \cdot P_3.$$

In the next section, we will make the addition of divisors on a hyperelliptic curve precise geometrically as well as algebraically.

3.4.4 The Group Law of Divisors

There are two different types of approaches to describe addition in the Jacobians of hyperelliptic curves. One approach is by using Cantor's algorithm (see Cohen and Frey (2006) for more details). The other approach is addition by interpolation. Below we will describe addition by interpolation.

3.4.4.1 Geometrically

In Figure 3.9, we show geometrically how one can add divisors of genus 2 curves. The idea is similar to the addition of points on elliptic curves but with the right adjustments to divisors.

Note that, from Eq. (3.1), we have that for every class of divisors in Jac (C), there exists a unique divisor $D = P_1 + P_2 - 2P_\infty$ (here we define $2P_\infty$ as being two

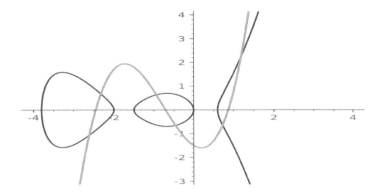

FIGURE 3.9 Addition of divisors on a hyperelliptic curve of genus $g = 2$.

points at infinity). We want to add two divisors D_1 and D_2. Let $D = P_1 + P_2 - 2P_\infty$ and $D = R_1 + R_2 - 2P_\infty$, as shown in Figure 3.9. The divisor

$$D_3 = D_1 + D_2 = Q_1 + Q_2 - 2P_\infty$$

can be obtained from the intersection of the degree three interpolating polynomial passing through the points, P_1, P_2, R_1, R_2, and the given hyperelliptic curve.

3.4.4.2 Algebraically

Let C be a genus 2 curve and $D_1 = P_1 + P_2 - 2P_\infty$ and $D_2 = P_3 + P_4 - 2P_\infty$ be two divisors on the hyperelliptic curve. We want to compute $D_3 = P_5 + P_6 - 2P_\infty = D_1 + D_2$. Let

$$b(x) = b_0 x^3 + b_1 x^2 + b_2 x + b_3 = \prod_{j \neq i} \frac{(x - x_i)}{(x_i - x_j)} \tag{3.2}$$

We can find the coordinates $b_0, \ldots, b_3$ using Cramer's rule and the fact that the points $P_1, \ldots, P_4$ satisfy Eq. (3.2). Then, our next step is to find the points where this degree three interpolating curve intersects with our given hyperelliptic curve. For the x-coordinates of the intersections with our curve $y^2 = a_0 x^5 + a_1 x^4 + \ldots + a_5$ we have:

$$(b_0 x^3 + b_1 x^2 + b_2 x + b_3)^2 - a_0 x^5 - a_1 x^4 - \ldots - a_5 = 0.$$

Expanding the above equation we get a degree six polynomial and $x_1, \ldots, x_6$ satisfy it. From Vieta's formulas we have that

$$\sum_{1 \leq i \leq 6} x_i = \frac{a_0 - 2b_0 b_1}{b_0^2}, \quad \sum_{1 \leq i < j \leq 6} x_i x_j = \frac{b_1^2 + 2b_0 b_2 - a_1}{b_0^2}$$

$$\sum_{1 \leq i < j < k \leq 6} x_i x_j x_k = \frac{a_2 - 2b_0 b_3 - 2b_1 b_2}{b_0^2} \quad \sum_{1 \leq i < j < k < l \leq 6} x_i x_j x_k x_l = \frac{b_2^2 + 2b_1 b_3 - a_3}{b_0^2}$$

$$\sum_{1\leq i<j<k<l<m\leq 6} x_i x_j x_k x_l x_m = \frac{a_4 - 2b_2 b_3}{b_0^2}, \quad \prod_{1\leq i\leq 6} x_i = \frac{b_3^2 - a_5}{b_0^2}$$

Since we want to solve for x_5 and x_6 then we consider only

$$x_1 + x_2 + x_3 + x_4 + x_5 + x_6 = \frac{a_0 - 2b_0 b_1}{b_0^2}, \quad x_1 x_2 x_3 x_4 x_5 x_6 = \frac{b_3^2 - a_5}{b_0^2},$$

and note that x_5 and x_6 are solutions to the following quadratic equation:

$$x^2 + \left(x_1 + x_2 + x_3 + x_4 - \frac{a_0 - 2b_0 b_1}{b_0^2} \right) x + \frac{b_3^2 - a_5}{b_0^2 x_1 x_2 x_3 x_4} = 0.$$

So we have that:

$$\bar{P}_5 = (x_5, -b_0 x_5^3 - b_1 x_5^2 - b_2 x_5 - b_3), \quad \bar{P}_6 = (x_6, -b_0 x_6^3 - b_1 x_6^2 - b_2 x_6 - b_3),$$

and $D_3 = D_1 + D_2 = P_5 + P_6 - 2P_\infty$. Hence, we defined algebraically the addition of two divisors.

Efficient group law for Jacobian varieties has many applications in cryptography. The level of security for elliptic curves has been widely studied, but the literature for higher dimensional cases is less developed. It is known that, in the case of $g = 2$ curves, the addition law can be computed by Cantor's algorithm (see Cantor (1994) for more details). But unfortunately these formulas do not provide the same level of efficiency as elliptic curves for a similar level of security. To improve arithmetic for higher genus curves, an idea is to consider the Kummer variety associated with the algebraic curve (see Costello (2018) for more details).

3.4.5 Cryptography on Hyperelliptic Curves

Now that we have made precise the group law on the Jacobian variety associated to a hyperelliptic curve, the Discrete Logarithm Problem for divisors on the HC can be defined the same way as for EC. For two given divisors D_1 and D_2 on $Jac(C)$, it is difficult to find n such that the following holds

$$D_1 = \underbrace{D_2 + D_2 + ... + D_2}_{\text{'n' many additions}} = nD_2.$$

The Diffie Hellman key exchange for hyperelliptic curves can be defined exactly the same way as for elliptic curves. Table 3.2 can be used to model the DHKE for HC. Much as is the same with EC, the two parties have shared values, but this time they are a predetermined prime number p, a hyperelliptic curve H, and a divisor D. They then calculate their values D_A and D_B respectively, and exchange. After calculating in the same way they found the previous values, they now each have the private key.

TABLE 3.2 Hyperelliptic curve Diffie-Hellman key exchange

Public Parameters	
The parameters p (a large prime number), H over $\mathbb{F}_p$, and divisor D in $H(\mathbb{F}_p)$ are made publicly available	
Private Computations	
Alice: $\quad D_A = n_A D$	Bob: $\quad D_B = n_B D$
Public Exchange of Values	
Alice sends D_A	Bob sends D_B
Private Computations	
Alice: Computes $n_A D_B$	Bob: Computes $n_B D_A$
Now both parties have $n_A D_B = n_A(n_B D) = n_B(n_A D) = n_B D_A$	

3.4.6 Post-Quantum Hyperelliptic Curve Cryptography

Note that even though genus 2 crypto is an aggressive alternative for elliptic curve crypto, both of them are based on the hardness of the discrete logarithm problem (DLP) which Shor's quantum algorithm solves in polynomial time. In the previous section, we briefly introduced supersingular elliptic curve isogeny-based cryptography as a post-quantum cryptosystem. There is a lot of research being done in exploring new possibilities for efficient supersingular isogeny-based cryptography with hyperelliptic curves. The generalization of supersingular isogeny-based cryptography to isogeny-based hyperelliptic curve cryptography gets rather complicated. There are several reasons why this is the case such as, for example, computing isogenies for hyperelliptic curves is much more complicated than for the case of elliptic curves.

The latest developments in supersingular isogeny-based cryptography have increased the interest in the problem of decomposition of Jacobian varieties and reducible Jacobians. Reducible Jacobian varieties have been studied extensively since the nineteenth century, most notably by Friecke, Clebch, and Bolza. In the late twentieth century they became the focus of many mathematicians through the work of Frey (1995), Frey and Kani (2009), Shaska and Völklein (2004), Shaska (2004), Magaard et al. (2009), Kumar (2015) and many others.

We say that a Jacobian variety is decomposed if the following holds. Let $\mathcal{X}$ be a genus 2 curve and

$$\psi_1 : \mathcal{X} \to E_1$$

be a degree n map to an elliptic curve. There is always another map

$$\psi_2 : \mathcal{X} \to E_2$$

to another elliptic curve. Th Jacobian Jac $(\mathcal{X})$ is n^2 isogenous to $E_1 \times E_2$, where E_i, $i = 1, 2$ are 1-dimensional.

$$\text{Jac } \mathcal{X} \mapsto E_1 \times E_2$$

Hence, the Jacobian of the given hyperelliptic curve $\mathcal{X}$, Jac $(\mathcal{X})$, is decomposed into a product of elliptic curves, E_1 and E_2.

In Costello (2018), the author focuses on the $(2, 2)$-reducible Jacobians, i.e. $n = 2$, where the addition is done via the Kummer surface. More importantly, it seems as a most interesting case when E_1 is isogenous to E_2. In this case, since the decomposition of the Abelian varieties is determined up to the isogeny, the 2-dimensional Jacobian is isogenous to E^2. We will not get into more detail about this area of research since we have not covered all the necessary tools. Hence, for a more complete description of this area of research and all the tools necessary, see Costello (2018), Beshaj and Hall (2020), and Beshaj et al. (2019) amongst many others.

3.5 FINAL REMARKS AND FUTURE WORK

The use of cryptography is ubiquitous in the modern world, often times without the user knowing of its presence. From browsing the web to using a credit card and even in medical devices, cryptography grants the security needed to live in this modern world. Looking to the future, there will only be more need for this security as we transition to the Internet of Things (IoT) and self-driving cars. But this transition will arguably disrupt the status quo less than the invention of the computer. Before then, cryptography was generally reserved for large institutions, governments, and militaries.

Elliptic curves allow for a reasonable level of security with much less bits (160-250 compared to 1024-3072 bits for RSA). With a sufficiently short key size, RSA can be faster than ECC, but ECC still results in fewer computations and takes up less bandwidth. ECC also generally exhibits really good one-way characteristics as it makes it such that the only known attacks are generic Discrete Logarithm algorithms. Much of this depends on the curve that is chosen, and this is a large part of the research currently being conducted. Looking at practicality purposes, ECC is very common in RFID (Radio Frequency Identification) tags because of their shorter key lengths. As we look towards the future, we can assume ECC will be more and more common as we develop smaller devices that need to be secured (see Washington, 2008).

In general, the number of operations for hyperelliptic curves is so large that the performance is worse than for elliptic curves. However, if one is willing to trade off generality for higher speed, there exist genus two curves offering fast arithmetic. Comparable choices can not be made for elliptic curves. These curves constitute families, i.e. they are not considered special curves. Utilizing these curves could provide for better security, with smaller hardware and even faster performance. However, performing computations on hyperelliptic curves is more expensive than for ordinary elliptic curves. But on the other hand, hyperelliptic curves provide superior bit strength security with regard to the size of the base field that they are defined over.

Moreover, it seems that elliptic and hyperelliptic curves will be good candidates for post-quantum cryptography, which makes them even more interesting to study.

REFERENCES

Alwen, J. (2018). What is lattice-based cryptography and why should you care. *Medium, Wickr Cryptography*.

Beshaj, L., Elezi, A., and Shaska, T. (2019). Isogenous components of jacobian surfaces. *European Journal of Mathematics*, doi: 10.1007/s40879-019-00375-y.

Beshaj, L. and Hall, A. (2020). Recent developments in cryptography. *12th International Conference on Cyber Conflict. 20/20 Vision: The Next Decade. Proceedings 2020.* 351–368.

Cantor, D. G. (1994). On the analogue of the division polynomials for hyperelliptic curves. *J. Reine Angew. Math.*, 447:91–145.

Chen, L., Jordan, S., Liu, Y., Moody, D., Peralta, R., and Smith-Tone, D. (2016). Report on post-quantum cryptography. *National Institute of Standards and Technology NISTIR 8105*, Gaithersburg, MD.

Cohen, H. and Frey, G. (2006). Handbook of elliptic and hyperelliptic curve cryptography. *Discrete Mathematics and its Applications*. Chapman & Hall/CRC, Boca Raton, FL.

Costello, C. (2018). Computing supersingular isogenies on kummer surfaces. In: Peyrin T., Galbraith, S. (eds), Advances in Cryptology, *Proceedings of the ASIACRYPT Conference*, 11274.

Costello, C. and Hisil, H. (2017). A simple and compact algorithm for sidh with arbitrary degree isogenies. *Proceedings of the ASIACRYPT Conference*.

Cox, D., Little, J., and O'Shea, D. (2007). *Ideals, varieties, and algorithms. An introduction to computational algebraic geometry and commutative algebra.* Springer International Publishing, Switzerland.

De Feo, L., Jao, D., and Plút, J. (2014). Towards quantum resistant cryptosystems from supersingular elliptic curve isogenies. *J. Math. Cryptology*, 8:209–247.

Feo, L. (2017). Mathematics of isogeny based cryptography. arXiv: 1711.04062.

Frey, G. (1995). On elliptic curves with isomorphic torsion structures and corresponding curves of genus 2. In *Elliptic curves, modular forms, & Fermat's last theorem (Hong Kong, 1993)*, International Press, Cambridge, MA, 79–98.

Frey, G. and Kani, E. (2009). Curves of genus 2 with elliptic differentials and associated Hurwitz spaces. In *Arithmetic, geometry, cryptography and coding theory*, Contemp. Math., 487:33–81.

Frey, G. and Shaska, T. (2019). Curves, jacobians, and cryptography. In Beshaj, L., editor, *Algebraic curves and their applications*, Contemporary Math., 724:280–350.

Galbraith, S. (2017). Isogeny-based post-quantum crypto. *EllipticNews*. Available at: https://ellipticnews.wordpress.com.

Hoffstein, J., Pipher, J., and Silverman, J. (2010). *An introduction to mathematical cryptography*. Springer Science and Business Media, LLC, New York, NY.

Jao, D. and De Feo, L. (2011). *Towards quantum-resistant cryptosystems from supersingular elliptic curve isogenies*, In: Yang, B.Y. (eds) Post-Quantum Cryptography. PQCrypto, Berlin, Germany.

Koblitz, N. (1987). Elliptic curve cryptosystems. *Math. Comp.*, 48(177):203–209.

Kumar, A. (2015). Hilbert modular surfaces for square discriminants and elliptic subfields of genus 2 function fields. *Res. Math. Sci.*, 2:Art. 24, 46.

Lange, T. (2016). Code-based cryptography. *Post-Quantum Cryptography Winter School*. PQCrypto, Eindhoven, Netherlands.

Magaard, K., Shaska, T., and Völklein, H. (2009). Genus 2 curves that admit a degree 5 map to an elliptic curve. *Forum Math.*, 21(3):547–566.

Miller, V. S. (1986). *Use of elliptic curves in cryptography*, In: Williams, H.C. (eds), Advances in Cryptology, CRYPTO '85, 218: 417-426.

Pizer, A. (1990). Ramanujan graphs and hecke operators. *Bull. Amer. Math. Soc.*, 23(1):127–137.

Pizer, A. (1998). Ramanujan graphs in computational perspectives on number theory. *AMS/IP Stud. Adv. Math.*, Providence, RI, 7:159–178.

Shaska, T. (2004). Genus 2 fields with degree 3 elliptic subfields. *Forum Math.*, 16(2):263–280.

Shaska, T. and Völklein, H. (2004). Elliptic subfields and automorphisms of genus 2 function fields. In *Algebra, arithmetic and geometry with applications*, West Lafayette, IN, 703–723.

Shor, P. W. (1994). Algorithms for quantum computation: discrete logarithms and factoring. *Proceedings of the 35th Annual Symposium on Foundations of Computer Science*, Santa Fe, NM.

Silverman, J. (1986). *The arithmetic of elliptic curves*. Springer Science and Business Media, New York, NY.

Silverman, J. and Tate, J. (2015). *Rational points on elliptic curves*. Springer International Publishing, Switzerland.

Simpson, A. Hyperelliptic curve. *MathWorld*, Wolfram Web Resources, created by Weisstein, E.W.

Tate, J. (1966). Endomorphisms of abelian varieties over finite fields. *Inventiones Math.*, 2:134–144.

Velú, J. (1971). Isogénies entre courbe elliptiques. *C. R. Acad. Sci. Paris Séries A* 273:238–241.

Washington, L. C. (2008). *Elliptic curves: Number theory and cryptography, second edition*. Chapman & Hall/CRC, Boca Raton, FL.

CHAPTER 4

Topology

Steve Huntsman

Jimmy Palladino

Michael Robinson

CONTENTS

4.1	Introduction	133
4.2	Dowker Homology to Analyze Complexity of Source and Binary Code	135
	4.2.1 Simplicial Complexes and Their Homology	136
	4.2.2 Dowker Homology	141
4.3	Path Homology to Analyze Graphical Structures	147
4.4	Topological Data Analysis and Unsupervised Learning in One Dimension	152
4.5	Critical Node Detection in Wireless Networks Using Sheaves	154
	4.5.1 Historical Context and Contributions	154
	4.5.2 Interference From a Transmission	155
	4.5.3 An Algebraic Interlude: Relative Homology	159
	4.5.4 Using Activation Patterns	160
	4.5.5 Cohomological Analysis	163
4.6	Conclusion	164

4.1 INTRODUCTION

Basic topological notions of connectivity are at the center of the cyber domain. For example, graph theory addresses many problems relating to connectivity and global or qualitative structure in computer science and cybersecurity using techniques that trace their lineage to Euler. Euler asked (and answered) the foundational problem in both graph theory and topology: is there a round trip that crosses every bridge in Königsberg (Figure 4.1) exactly once? The answer is *no*, because no orientation of the edges in the corresponding graph can give the same in- and out-degree to vertices with odd degree. The topological insight of Euler was that a combinatorial structure can faithfully represent connectivity properties of continuous bodies.

DOI: 10.1201/9780429354649-4

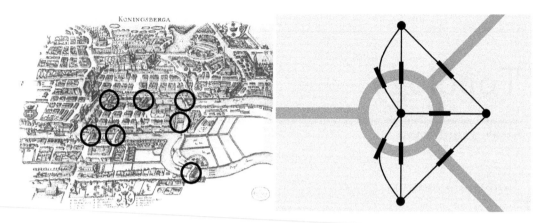

FIGURE 4.1 (Left) The seven bridges of Königsberg, indicated by circles. (Right) A graphical representation of the bridges.

More distinctly modern topological approaches can provide richer information. Several such approaches are sketched in this chapter. Taking connectivity as the base case, modern topological methods provide fine invariants that are useful for addressing complex cyber problems. This chapter reviews various relevant constructions, focusing on discrete structures that are naturally suited for addressing cyber-oriented problems and that require a minimum of background.

The chief advantage that topological techniques provide is a sort of invariance with respect to well-behaved maps on an underlying space. The more general manifestation of this idea is called functoriality: it is at the root of category theory, algebraic topology, modern algebraic geometry, and many other major disciplines within mathematics. This sort of invariance yields qualitative and/or robust measures of structures that can be particularly useful in the face of uncertain or missing data, a lack of canonical coordinates or parameters, etc.

Because so many aspects of computer and network architecture involve specific details that can propagate in unintended ways through layers of abstraction, such as platform-specific implementation bugs, the topological perspective carries a special power that can inform cyber-oriented problems. This chapter aims to give a sense of how to exploit topological constructions in various aspects of the cyber domain. Because the data and structures involved are inevitably discrete and finite, the flavor of our treatment is idiosyncratic, and the novice may want to complement it with, e.g. Sato (1999); Hatcher (2002); Ghrist (2014).

The chapter is organized in four parts that respectively treat simplicial homology (Section 4.2), the recent and related theory of path homology (Section 4.3), topological data analysis (Section 4.4), and sheaf theory (Section 4.5). Overall, homology measures deficits in structure via the construction of certain well-behaved quotients using tools of linear algebra. More specifically, simplicial homology is the most conceptually and computationally ubiquitous algebraic invariant of a reasonably generic space. We apply simplicial homology to the analysis of computer code by considering special simplicial complexes that encode relations between program assignments and variables, and that do not even have to be explicitly formed in order to obtain useful invariants. Meanwhile, path homology is an important and quite new theory that defines high-dimensional topological invariants of directed graphs and, as such, is very promising for cyber-oriented applications such as the analysis of control flow. Our treatment of topological data analysis is very brief and restricted to one dimension, where it is possible to introduce and exploit the morals of topological persistence to the useful end of statistical mixture estimation without invoking the algebraic machinery of persistent homology. Finally, our treatment of sheaves is largely self-contained and developed in the service of detecting critical nodes in wireless networks.

Throughout this chapter, our focus on intrinsically discrete structures, realistic applications, and space constraints entail a somewhat idiosyncratic treatment. For example, the word 'functor' and its variants do not occur outside this section, though we point out the functoriality of simplicial homology without invoking the formalism of category theory.

4.2 DOWKER HOMOLOGY TO ANALYZE COMPLEXITY OF SOURCE AND BINARY CODE

In this section, we introduce a class of data structures called *abstract simplicial complexes* that model interactions of arbitrary order, and generalizing graphs, which model interactions of order two. We illustrate how these data structures can model well-behaved shapes and compute the basic topological invariant of homology by transporting these structures into the realm of linear algebra. Finally, we demonstrate how these ideas can characterize source and binary code. The same ideas could be applied to bipartite structures such as interactions between processes and files, clients, servers, etc.

The basic idea of simplicial homology is that structural deficits in abstract simplicial complexes can be straightforwardly and efficiently encoded in quotient vector spaces (or modules) indexed by the dimension of the deficit. In dimensions zero, one, and two, this means that connected components, holes, and 'bubbles' are identified by suitable representatives of the corresponding homology vector spaces, and the dimensions of these quotients give the numbers of the corresponding deficits.

Dowker homology rests upon two observations. The first observation is that a relation (i.e., a subset of a Cartesian product) corresponds to two abstract simplicial complexes that turn out to be topologically equivalent. The second observation is that as a computational matter, it is possible to bypass this construction entirely when over the binary field $\mathbb{F}_2$, computing homology directly in terms of the underlying relation. As suggested above, the ubiquity of relational data in the cyber domain implies fertile ground for applications, though we focus on a demonstration that Dowker homology provides features that are well-suited for clustering algorithms. Along the way, we point out how the notion of an algorithm itself is mathematically ill-posed except in carefully constrained circumstances.

4.2.1 Simplicial Complexes and Their Homology

Although topology is generally thought of as the study of spaces under continuous transformations, its intellectual roots are in combinatorial models of spaces. While these combinatorial models are typically discarded once the theory is developed, they are ideally suited for describing cyber applications. Abstract simplicial complexes are among the easiest of these combinatorial models to define and apply. This section introduces these structures and the tools to compute topological invariants of them as a warmup to actual applications in the sequel.

Definition 4.1. *An* abstract simplicial complex *is a family Δ of finite subsets (called* simplices*) of a set $V = \{v_0, \ldots, v_p\}$ of* vertices *such that if $X \in \Delta$ and $Y \subseteq X$ is nonempty, then $Y \in \Delta$. In other words, an abstract simplicial complex is a hypergraph with all sub-hyperedges. Usually, we write simplices with square brackets $[v_0, \ldots, v_p] \equiv 2^{\{v_0, \ldots, v_p\}} \setminus \varnothing$, where the set difference of a power set and the empty set are indicated on the right. The* dimension *of a simplex $[v_0, \ldots, v_p]$ is p, which is one less than its cardinality as a set. A simplex that is the subset of no other simplex is called a* facet.

When describing the local structure of a simplicial complex, it is often useful to delineate which simplices are subsets of each other. If a and b are simplices of a simplicial complex X and $a \subseteq b$, we say that 'a is a *face* of b' or equivalently that 'b is a *coface* of a.' These relationships determine the topology of an abstract simplicial complex, in terms of its open and closed subsets. A *closed set* A of a simplicial complex contains every possible subset of every element of A. The *star* of a subset A of a simplicial complex consists of the set of all simplices containing an element of A. An *open set* of an abstract simplicial complex is one that can be written as a union of stars.

For example, let Δ be given by all nonempty subsets of sets in $\{\{1,2\}, \{1,3\}, \{2,3,4\}, \{5\}\}$. Then Δ is an abstract simplicial complex of dimension $2 = |\{2,3,4\}| - 1$; Figure 4.2 shows a geometric realization of Δ. The 2-simplex $[2,3,4]$ is indicated by shading, and simplices in dimensions 1 and 0 are respectively depicted using line segments or dots. The expression of an abstract simplicial complex as a nondegenerate union of power sets manifestly reflects its facets.

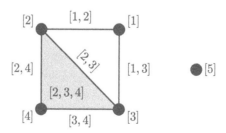

FIGURE 4.2 A geometric realization of the abstract simplicial complex $\Delta = (2^{\{1,2\}} \cup 2^{\{1,3\}} \cup 2^{\{2,3,4\}} \cup 2^{\{5\}}) \setminus \varnothing \subseteq 2^V$ with $V = \{1, 2, 3, 4, 5\}$.

In the abstract simplicial complex Δ, the set $A = \{[2,4],[2],[4]\}$ is a closed set but $B = \{[1,2],[1,3],[1],[5]\}$ is not, because $[3]$ is a face of $[1,3]$ that is not contained in B. On the other hand, B is the union of the star over $[1]$ and the star over $[5]$, so B is an open set. A set can be both open and closed; $\{[5]\}$ is such as set.

Functions that preserve the simplices of abstract simplicial complexes are afforded special status and are called *simplicial maps*. These help characterize salient features of abstract simplicial complexes.

Definition 4.2. *A simplicial map $f : \Delta \to \Gamma$ from one abstract simplicial complex Δ to another Γ is a function of vertices such that each simplex $\sigma = [v_0, \ldots, v_p]$ of Δ is taken to a simplex $f(\sigma) = [f(v_0), \ldots, f(v_p)]$ of Γ.*

In the image $f(\sigma)$, repeated vertices count as one vertex. This means that simplicial maps may decrease the dimension of a simplex but not increase it.

Simplicial maps immediately give rise to the notion of isomorphic abstract simplicial complexes: Δ and Γ are *isomorphic* if there are simplicial maps $f : \Delta \to \Gamma$ and $g : \Gamma \to \Delta$ such that $f = g^{-1}$ and $g = f^{-1}$. Isomorphisms are a natural equivalence relation on abstract simplicial complexes and generalize the idea of relabeling vertices in a simplicial complex.

It is rather computationally difficult to study abstract simplicial complexes and simplicial maps directly. It is much easier to work by analogy: transform abstract simplicial complexes into vector spaces and simplicial maps into linear maps. The way we will do this is by way of a construction called *simplicial homology*. The construction is a two-step process, in which we first transform each abstract simplicial complex into an algebraic construction called a *chain complex* (schematically depicted in Figure 4.3), and each simplicial map transforms into a *chain map*. From there, chain complexes and chain maps allow us to compute topological invariants via linear algebra.

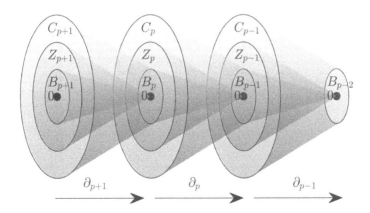

FIGURE 4.3 Schematic picture of a chain complex.

Definition 4.3. *A chain complex $(C_\bullet, \partial_\bullet)$ over a field $\mathbb{F}$ is a pair of sequences (indexed by $p \in \mathbb{N}$ or $\mathbb{Z}$ depending on context) of $\mathbb{F}$-vector spaces $C_\bullet = \{\ldots, C_{p-1}, C_p, \ldots\}$ and linear boundary operators $\partial_p : C_p \to C_{p-1}$ such that $\partial_{p-1} \circ \partial_p \equiv 0$. This can be schematically depicted as in Figure 4.3, and written as (for $p \in \mathbb{N}$)*

$$\cdots \xrightarrow{\partial_{p+1}} C_{p+1} \xrightarrow{\partial_p} C_p \xrightarrow{\partial_{p-1}} C_{p-1} \xrightarrow{} \cdots \xrightarrow{\partial_1} C_0 \xrightarrow{\partial_0} 0.$$

Given an abstract simplicial complex Δ, let $C_p(\Delta)$ be the $\mathbb{F}$-vector space generated by basis elements $e_{(v_0,\ldots,v_p)}$ corresponding to *oriented simplices* of *dimension p* in Δ. This essentially means that if σ is a permutation acting on $(v_0, \ldots, v_p)$, then $e_{(v_0,\ldots,v_p)} = (-1)^\sigma e_{(v_{\sigma(0)},\ldots,v_{\sigma(p)})}$, where $(-1)^\sigma$ indicates the sign of the permutation σ. Thus, for example, $e_{(v_0,v_1,v_2)} = -e_{(v_0,v_2,v_1)}$. Note that an order on V induces orders (and hence orientations) on simplices in Δ.

The simplicial boundary operator ∂_p is now defined to be the linear map acting on basis elements as

$$\partial_p e_{(v_0,\ldots,v_p)} := \sum_{j=0}^{p} (-1)^j e_{\nabla_j(v_0,\ldots,v_p)} \qquad (4.1)$$

where ∇_j deletes the jth entry of a tuple. It turns out that this construction yields a chain complex, called the *simplicial chain complex* for $C_\bullet(\Delta)$. To compute $\partial_{p-1} \circ \partial_p$, we delete two entries from $e_{(v_0,\ldots,v_p)}$ according to Equation 4.1. There are two different ways we can do this: first i and then j, or first j and then i. These two ways yield opposite signs, which cancel all of the terms in the sum.

Like the structure-preserving nature of simplicial maps for abstract simplicial complexes, there are structure preserving *chain maps* for chain complexes. They are defined by way of diagrams

$$\begin{array}{ccccccccc}
\cdots & \longrightarrow & C_{p+1} & \xrightarrow{\partial_{p+1}} & C_p & \xrightarrow{\partial_p} & C_{p-1} & \xrightarrow{\partial_{p-1}} & \cdots \\
& & \downarrow m_{p+1} & & \downarrow m_p & & \downarrow m_{p-1} & & \\
\cdots & \longrightarrow & C'_{p+1} & \xrightarrow{\partial'_{p+1}} & C'_p & \xrightarrow{\partial'_p} & C'_{p-1} & \xrightarrow{\partial'_{p-1}} & \cdots
\end{array}$$

in which the composition of consecutive maps is path-independent. Because in such a diagram

$$m_{p-1} \circ \partial_p = \partial'_p \circ m_p, \qquad (4.2)$$

it is said to *commute*.

A somewhat involved but straightforward calculation establishes the following key result about the simplicial chain complex.

Proposition 4.1. *Every simplicial map $f : \Delta \to \Gamma$ between abstract simplicial complexes induces a chain map $f_\bullet : C_\bullet(\Delta) \to C_\bullet(\Gamma)$ between their simplicial chain complexes.*

While chain complexes distill abstract simplicial complexes into the realm of algebra, they are still rather complicated. Moreover, the simplicial chain complex contains combinatorial, non-topological information. Homology is a convenient, linear algebraic summary for a chain complex that still preserves the structure of chain maps. Additionally, the homology of the simplicial chain complex is a topological invariant. Although we restrict our considerations to fields for practical purposes, homology is readily defined over rings, with the integers $\mathbb{Z}$ serving as the case through which all others factor via the universal coefficient theorem (which, incidentally, gave rise to the topics of category theory and homological algebra).

Definition 4.4. *Writing $Z_p := \ker \partial_p$ and $B_p := \operatorname{im} \partial_{p+1}$, the homology of the chain complex $(C_\bullet, \partial_\bullet)$ is the sequence of quotient spaces*
$$H_p := Z_p/B_p.$$
The Betti numbers are $\beta_p := \dim H_p = \dim Z_p - \dim B_p$.

The essential point of this construction is that homology transforms chain complexes into vector spaces and chain maps into linear maps.

Proposition 4.2. *Every chain map $m_\bullet : C_\bullet \to D_\bullet$ induces a family of linear maps $(m_*)_p : H_p(C_\bullet) \to H_p(D_\bullet)$ between homology spaces, one for each p.*

As an immediate consequence, a simplicial map $f : \Delta \to \Gamma$ induces a family of linear maps $H_p(C_\bullet(\Delta)) \to H_p(C_\bullet(\Gamma))$ between the homologies of the corresponding simplicial chain complexes. We will call
$$H_p(\Delta) := H_p(C_\bullet(\Delta)) \tag{4.3}$$
the *p-simplicial homology* of the abstract simplicial complex Δ. What this means is that if two simplicial complexes are isomorphic, then their simplicial homologies will also be isomorphic vector spaces for every index. Conversely, if two abstract simplicial complexes have different simplicial homologies, we know that they cannot be isomorphic as simplicial complexes.

For our purposes here, simplicial homology is practically valuable because it underlies *cyclomatic complexity* (McCabe, 1976), which is essentially the first (and only nontrivial) Betti number of a control flow graph treated as an abstract simplicial complex (i.e., edges correspond to 1-simplices as in Figure 4.1). Cyclomatic complexity is an archetypal and widely used (Ebert & Cain, 2016) software metric that can guide fuzzing (Duran et al., 2011; Iozzo, 2010) and identification of fault-prone or vulnerable code (Alves et al., 2016; Du et al., 2019; Medeiros et al., 2017). In Section 4.3, we briefly discuss *path homology*, which has promise for generalizing cyclomatic complexity to higher dimensions.

Definition 4.5. *In the event that all but finitely many β_p are zero, the Euler characteristic $\chi := \sum_p (-1)^p \beta_p$ is well-defined. For abstract simplicial complexes, we get the familiar formula $\chi = V - E + F - \ldots$, where the terms on the right-hand side respectively indicate the numbers of vertices/0-simplices, edges/1-simplices, faces/2-simplices, etc.*

Moreover, the simplicial Betti numbers β_p count the number of voids of dimension p in a geometric realization of an abstract simplicial complex. Here, 0-dimensional voids amount to connected components.

For example, consider $\Delta = 2^{\{1,2,3\}} \setminus \emptyset$: now $C_2(\Delta) = \langle e_{(1,2,3)} \rangle$, where $\langle \cdot \rangle$ indicates the vector span (say, over $\mathbb{R}$); $C_1(\Delta) = \langle e_{(1,2)}, e_{(1,3)}, e_{(2,3)} \rangle$; $C_0(\Delta) = \langle e_{(1)}, e_{(2)}, e_{(3)} \rangle$; and all other $C_p(\Delta)$ are 0. Using lexicographic indexing for basis elements, we have the matrix representations

$$\partial_2 = \begin{pmatrix} 1 \\ -1 \\ 1 \end{pmatrix}; \quad \partial_1 = \begin{pmatrix} -1 & -1 & 0 \\ 1 & 0 & -1 \\ 0 & 1 & 1 \end{pmatrix}; \quad (4.4)$$

all other boundary operators are zero. For example, the boundary of the 2-simplex or 'triangle' is

$$\partial_2 e_{(1,2,3)} = e_{(1,2)} - e_{(1,3)} + e_{(2,3)}, \quad (4.5)$$

or in matrix form

$$\begin{pmatrix} 1 \\ -1 \\ 1 \end{pmatrix} (1) = \begin{pmatrix} 1 \\ -1 \\ 1 \end{pmatrix}. \quad (4.6)$$

Its boundary, in turn, is

$$\partial_1 \left(e_{(1,2)} - e_{(1,3)} + e_{(2,3)} \right) = 0, \quad (4.7)$$

or in matrix form

$$\begin{pmatrix} -1 & -1 & 0 \\ 1 & 0 & -1 \\ 0 & 1 & 1 \end{pmatrix} \begin{pmatrix} 1 \\ -1 \\ 1 \end{pmatrix} = \begin{pmatrix} 0 \\ 0 \\ 0 \end{pmatrix}. \quad (4.8)$$

Thus the homology of the boundary of a triangle has $\beta_p = \delta_{p1}$: there is a single void in dimension one, and none in other dimensions.

As a slightly more detailed example, take $V = \{1, \ldots, 5\}$ and Δ to be all nonempty subsets of sets in $\{\{1,2\}, \{1,3\}, \{2,3,4\}, \{5\}\}$, as in Figure 4.2. We have the chain complex (over $\mathbb{R}$)

$$0 \xrightarrow{\partial_3} C_2 \xrightarrow{\partial_2} C_1 \xrightarrow{\partial_1} C_0 \xrightarrow{\partial_0} 0 \quad (4.9)$$

where $C_2 = \langle e_{(2,3,4)} \rangle$, $C_1 = \langle e_{(1,2)}, e_{(1,3)}, e_{(2,3)}, e_{(2,4)}, e_{(3,4)} \rangle$, $C_0 = \langle e_{(1)}, e_{(2)}, e_{(3)}, e_{(4)}, e_{(5)} \rangle$, and the nontrivial boundary operators are (again, lexicographically ordering basis elements)

$$\partial_2 = \begin{pmatrix} 0 \\ 0 \\ 1 \\ -1 \\ 1 \end{pmatrix}; \quad \partial_1 = \begin{pmatrix} -1 & -1 & 0 & 0 & 0 \\ 1 & 0 & -1 & -1 & 0 \\ 0 & 1 & 1 & 0 & -1 \\ 0 & 0 & 0 & 1 & 1 \\ 0 & 0 & 0 & 0 & 0 \end{pmatrix}. \quad (4.10)$$

TABLE 4.1 Betti numbers
for the chain complex (4.9)

p	$\dim Z_p$	$\dim B_p$	β_p
0	5	3	2
1	2	1	1
2	0	0	0

A few row reductions yield that $\mathrm{rank}(\partial_1) = 1$ and $\mathrm{rank}(\partial_2) = 3$, which gives the hard part of Table 4.1. It follows that $\beta_\bullet = (2, 1, 0, \dots)$. Indeed, a geometric realization of Δ has two connected components and one hole.

As a more intricate example, take $V = \{1, \dots, 18\}$ and Δ to be all nonempty subsets of sets in

$$\{\{1,2,3\}, \{1,4\}, \{5\}, \{6,7,8,9\}, \{9,10,12\}, \{10,11,12\},$$
$$\{12,13,16\}, \{13,14\}, \{13,15\}, \{14,15\}, \{16,17\}, \{17,18\}\}.$$

Now ∂_p acts on the span of all vectors of the form $e_{(v_0,\dots,v_p)}$ where $\{v_0, \dots, v_p\} \in \Delta$. Meanwhile, a brief calculation shows that $\ker \partial_2 = \mathrm{im}\,\partial_3 \oplus 0$, $\ker \partial_1 = \mathrm{im}\,\partial_2 \oplus \langle e_{(13,14)} - e_{(13,15)} + e_{(14,15)} \rangle$, and $\ker \partial_0 = \mathrm{im}\,\partial_1 \oplus \langle e_{(1)}, e_{(5)}, e_{(6)} \rangle = \langle e_{(1)}, \dots e_{(18)} \rangle$. It follows that $H_p = 0$ for $p \geq 2$, $H_1 = \langle e_{(13,14)} - e_{(13,15)} + e_{(14,15)} \rangle$, and $H_0 = \langle e_{(1)}, e_{(5)}, e_{(6)} \rangle$. It follows that $\beta_\bullet = (3, 1, 0, \dots)$. A geometric realization of Δ has three connected components and one hole, as shown in Figure 4.4.

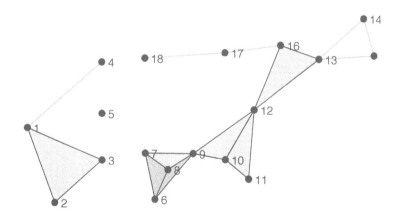

FIGURE 4.4 A geometric realization of an abstract simplicial complex with Betti numbers $\beta_\bullet = (3, 1, 0, \dots)$.

4.2.2 Dowker Homology

In this section, we show how relational data gives rise to abstract simplicial complexes whose topology can be computed with a shortcut that avoids actually forming the abstract simplicial complexes themselves. Although relational data is commonplace in cyber applications, with databases providing a meta-example, we focus on a specific use case: characterizing algorithms at the level of low-level source and/or binary code.

For finite sets X and Y, recall that a relation between X and Y is just a subset $R \subseteq X \times Y$. From this relation, we can form two abstract simplicial complexes. The first has vertex set X and simplices generated by finite subsets of $R(\cdot, y) := \{x \in X : (x,y) \in R\}$ for $y \in Y$; the second has vertex set Y and simplices generated by finite subsets of $R(x, \cdot)$ for $x \in X$. Remarkably, these two abstract simplicial complexes are topologically equivalent under the very strong notion of homotopy (Dowker, 1952) and either is referred to as a *Dowker complex* of the relation R. Almost as remarkably, the $\mathbb{F}_2$ homology of Dowker complexes can be computed directly from the relation R in about 50 lines of straightforward MATLAB® code, and the only computationally intensive part is computing the rank of the boundary matrices.

For example, consider the relation specified by the 0-1 matrix (adapted from (Ghrist, 2014) and also informing a previous example):

$$R = \begin{pmatrix} 0 & 1 & 0 & 0 & 0 & 1 & 0 \\ 0 & 0 & 0 & 1 & 0 & 1 & 1 \\ 1 & 1 & 0 & 1 & 0 & 0 & 1 \\ 1 & 0 & 0 & 1 & 0 & 0 & 0 \\ 0 & 0 & 1 & 0 & 1 & 0 & 0 \end{pmatrix}. \tag{4.11}$$

Taking the choice of vertex set $X = \{1, \ldots, 5\}$, we get the same chain complex as in Equation 4.9 and Equation 4.10, but over $\mathbb{F}_2$, i.e., all signs in Equation 4.10 are ignored. The matrix element $(\partial_p)_{jk}$ indicates whether or not the set corresponding to the jth basis element in C_{p-1} is contained in the set corresponding to the kth basis element in C_p. Computer calculations yield the same results as in Table 4.1.

Dowker complexes have a long history of applications to social disciplines under the aegis of 'Q-analysis' (Atkin, 1974); however, only recently have applications gained any wider traction, e.g. to navigation and mapping (Ghrist et al., 2012), lower bounds in privacy analyses (Erdmann, 2017), and analyses of weighted digraphs (Chowdhury & Mémoli, 2018). The preceding example highlights a circle of ideas that is very interesting for cyber applications. In the following, we detail another application (in many ways mirroring (Robinson, 2017)) of Dowker complexes to the analysis and characterization of 'straight-line' source code and/or basic blocks in binary code (i.e., sequences of instructions without control flow).

Programs are fairly simple to define and have a simple decision procedure for determining equality: a program is a string in some language, and two programs are equal if and only if they are equal as strings. Meanwhile, functions are also fairly simple to define (even in the context of computers, via the theory of denotational semantics (Nielson & Nielson, 1992)), though the problem of determining equality of functions within even simple classes is undecidable (Richardson, 1969). However, *algorithms* are notoriously hard to define, and though there is a sort of order structure on reasonable definitions (Yanofsky, 2017), all of the definitions that are substantively different from programs also lead to undecidable equality problems.

For the sorts of heuristics used in practice for the equality decision problem, see Taherkhani et al. (2010); Mesnard et al. (2016); Shalaby et al. (2017).

To illustrate this notion, consider the sets of 'algorithms' in Figure 4.5 and 4.6: each set has the same inputs (a, b, c, and d) and outputs (q and x), albeit computed differently. Absent notions of control flow (e.g., conditional branches or loops), it is easy to define a relation between variables and assignments and construct the corresponding Dowker complex. In these examples, the homology classes associated to 'primitive' algorithms on the left are preserved under compilation-like rewrites, though additional homology classes can be introduced by 'tearing apart' high-arity assignments into low-arity ones. More formally and suggestively, the primitive notion of 'decompilation' indicated here is an injective simplicial map, and thus induces a homomorphism on homology.

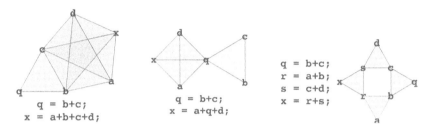

FIGURE 4.5 From left to right: Dowker complexes for a toy algorithm, a similar algorithm, and a 'compiled' version.

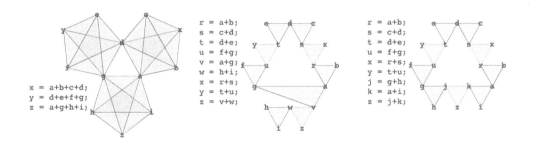

FIGURE 4.6 From left to right: Dowker complexes for a toy algorithm and two 'compiled' versions.

Notwithstanding the problems of defining algorithms and the undecidability of gauging equality (much less computing a principled similarity) of algorithms in general, Dowker homology can capture salient information about straight-line or basic block algorithms. In this restricted setting, it is not hard to identify various narrow classes of algorithms that admit reasonable definitions.

(To handle control flow nicely, F. R. Genovese [private communication] has suggested considering a so-called *étalé space* building on the sheaf implied by considering subsets of instructions/assignments. However, given *any* construct capable of dealing with basic control flow in the present context, it should be possible to 'desugar' more complex language semantics to deal with correspondingly more complex control flow and data structures.)

For example, sorting networks are fixed compositions of pairwise compare/swap operations that guarantee to sort an input tuple of a given size (Knuth, 1997), and fixed-size matrix multiplication algorithms are essentially rank-1 decompositions of a particular tensor (Landsberg, 2017). In both cases, the formulation of optimal algorithms is a nontrivial problem. For matrix multiplication, the 'naïve' algorithm was originally improved upon by Strassen (1969); Winograd (1971), which showed how to multiply two 2×2 matrices with only seven scalar multiplications (versus eight for the naïve approach). Although these instances are known to be optimal, all that is known for the 3×3 case is that somewhere between 19 and 23 scalar multiplications are required (versus 27 for the naïve approach), and over noncommutative (respectively, commutative) rings, 23 (respectively, 22) scalar multiplications are the best-known result, achieved by many inequivalent algorithms (Laderman, 1976; Johnson & McLoughlin, 1986; Makarov, 1986; Courtois et al., 2011) which we analyze below along with the naïve algorithms and some 'compiled' variants where all assignments have two inputs. There is also recent work producing still more 3×3 algorithms (see Chokaev & Shumkin (2018); Ballard et al. (2019); Heule et al. (2019)) and notions of matrix multiplication algorithm equivalence for more general sizes (Berger et al., 2019).

Figures 4.7 and 4.8 illustrate how Dowker homology can distinguish between optimal sorting networks: using the negative Euler characteristic as a measure of topological complexity highlights networks that exhibit more comparator reuse and symmetry. The graphical representations are shaded by $-\chi$ (lower values are paler) of Dowker complexes formed from code (by treating the statements `if k < j` as vertices i_{jk}). For the case $n = 4$, the graphical representations are topologically equivalent (specifically, both are homotopic to a figure eight), but the Dowker complexes are respectively homotopic to a figure eight and a circle.

```
if b < a, ab = a; a = b; b = ab; end
if d < c, cd = c; c = d; d = cd; end
if c < a, ac = a; a = c; c = ac; end
if d < b, bd = b; b = d; d = bd; end
if c < b, bc = b; b = c; c = bc; end
```

```
if d < a, ad = a; a = d; d = ad; end
if c < b, bc = b; b = c; c = bc; end
if b < a, ab = a; a = b; b = ab; end
if d < c, cd = c; c = d; d = cd; end
if c < b, bc = b; b = c; c = bc; end
```

FIGURE 4.7 From left to right: code representation of an optimal sorting network for $n = 4$; graphical representation of the same network (with inputs on left labeled **a** through **d** from top down and outputs on right); graphical representation of the other optimal network; code representation of the other optimal network.

Meanwhile, Figures 4.9 and 4.10 give a sense of how matrix multiplication algorithms cluster in meaningful ways when the Betti numbers for Dowker homology are used as features.

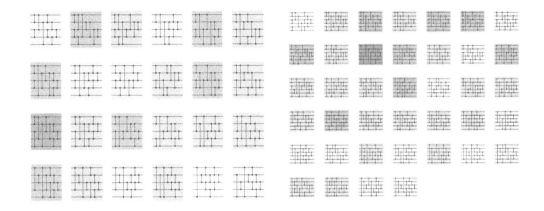

FIGURE 4.8 Representative sorting networks for $n = 5$ (left) and $n = 6$ (right) shaded by $-\chi$. Reuse of comparators and symmetry turn out to be signaled by lower (= paler) values.

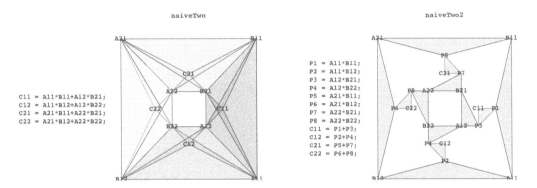

FIGURE 4.9 Dowker complexes for the naïve 2×2 matrix multiplication algorithm (left) and for a 'compiled' version (right).

By computing homologies over local windows of instructions, assignments, or line numbers (Figures 4.11-4.13), we obtain detailed structurally aware features evocative of spectrograms. For example, Figure 4.13 shows that lines 13-15 of the `Makarov` algorithm correspond to a local maximum in complexity. These lines embody a nontrivial homology class in dimension one corresponding to `A12`, `A22`, and `B23` that is isolated from the impact of lines 12 and 16 (i.e., there are no shared variables). Lines 8-19 of the Makarov algorithm turn out to correspond to a local extremum in χ, which is apparent in a thresholded version of Figure 4.13. The corresponding simplicial complex has six holes and one void (or bubble) that are not practical to visualize directly.

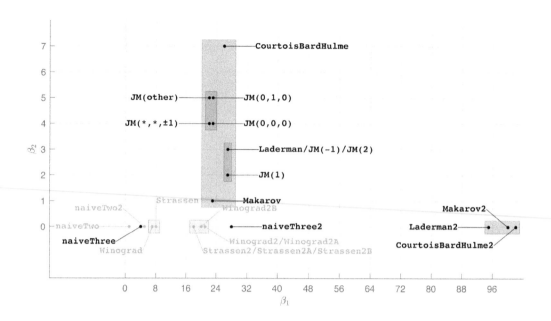

FIGURE 4.10 The Betti numbers for the Dowker homology of matrix multiplication algorithms are useful features for clustering. There are respectively 18 and seven inequivalent but similar algorithms from the three-parameter Johnson-McLoughlin family (Johnson & McLoughlin, 1986) that respectively correspond to the labels JM(*,*,±1) and JM(other).

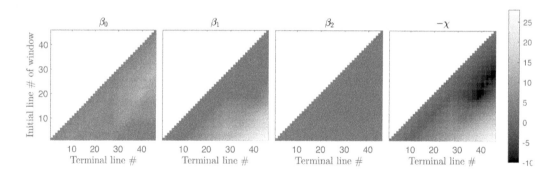

FIGURE 4.11 Windowed Dowker homology for the naiveThree2 algorithm, i.e., the 'compiled' version of naïve 3×3 matrix multiplication. Structural features of lines 1-9, 2-10, ..., 19-27 versus lines 28-45 are apparent.

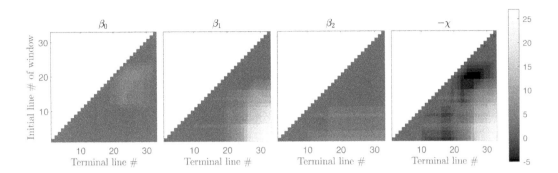

FIGURE 4.12 Windowed Dowker homology for the `Laderman` 3×3 matrix multiplication algorithm.

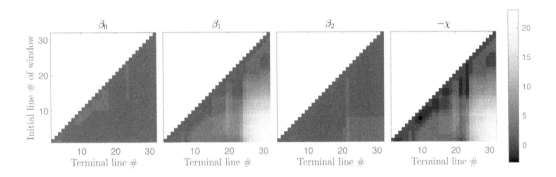

FIGURE 4.13 Windowed Dowker homology for the `Makarov` 3×3 matrix multiplication algorithm.

Finally, we can apply this same sort of construction at the binary level. In Figure 4.14, we show a snippet of Reverse Engineering Intermediate Language (REIL) (Dullien & Porst, 2009) code, the corresponding abstract simplicial complex (accounting for memory locations in a natural way), and the corresponding 'spectrograms.' By limiting the size of windows considered, this sort of feature construction can be performed in linear time (albeit with a possibly large overhead constant) and used to analyze basic blocks in disassembled binaries or their rough equivalents.

4.3 PATH HOMOLOGY TO ANALYZE GRAPHICAL STRUCTURES

In this section, we introduce what turns out to be a generalization of many of the ideas in the preceding one, though we treat it on its own. Instead of computing topological invariants of shape-like data structures, we compute topological invariants of oriented path-like data structures. Because these are ubiquitous in cyber applications, we do not attempt an exhaustive treatment but instead, limit ourselves to sketching an application to the control flow of computer programs.

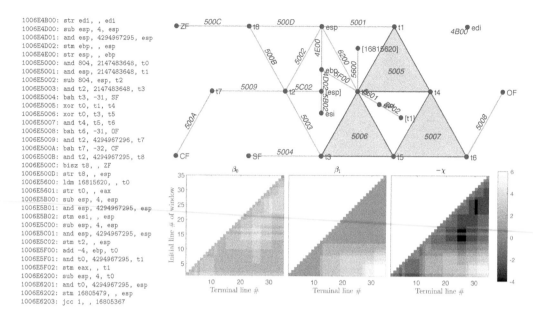

FIGURE 4.14 (Left) Some REIL code. (Right; upper) The corresponding 2-complex. (Right; lower) A 'spectrogram' of Betti numbers and Euler characteristic χ as a function of windowed code. Lines 7-13 (memory addresses ending in 5001-5007) exhibit clearly visible extrema in β_1 and χ (corner of a light rectangular region in center and right panels). Indeed, the registers `esp`, `t0`, ..., `t6` are all involved in multiple instructions in this range, leading to a single connected component with two holes. A secondary locus of topological complexity is lines 11-13 (memory addresses ending in 5005-5007).

Although, as we have already pointed out, undirected graphs can be regarded as one-dimensional abstract simplicial complexes with nontrivial homology in dimensions zero and one, digraphs have much subtler structure. For example, although simplicial homology deals with oriented simplices, the theory is fundamentally unequipped to treat digraphs. Instead, a generalization called *path homology* subsumes the ideas of simplicial homology into a framework that also gives meaning to topological deficits in arbitrary dimension. For example, it turns out that a directed 2-cycle corresponds to a 'hole' and that repeated 'suspensions' correspond to deficits in higher dimensions. (Here, the suspension of a digraph is basically formed by adding arcs from every vertex to both of two new 'polar' vertices.)

Although it is only beginning to be applied in many circumstances, path homology holds particular promise for cyber applications because of its ability to capture 'high-dimensional' phenomena as well as the prevalence of digraphs in the cyber realm (e.g., attack graphs (Lallie et al., 2020), flow graphs, call and dependency graphs, etc.) Our outline of path homology is based on Grigor'yan et al. (2012); Chowdhury & Mémoli (2018). For additional background on path homology, see the series of papers Grigor'yan, Muranov, & Yau (2014); Grigor'yan, Yong, et al. (2014); Grigor'yan et al. (2015, 2017); Grigor'yan, Jimenez, et al. (2018); Grigor'yan, Muranov, et al. (2018).

For convenience, we replace the chain complex with its *reduction*

$$\ldots C_{p+1} \xrightarrow{\partial_{p+1}} C_p \xrightarrow{\partial_p} C_{p-1} \xrightarrow{\partial_{p-1}} \ldots \xrightarrow{\partial_1} C_0 \xrightarrow{\tilde{\partial}_0} \mathbb{F} \longrightarrow 0 \qquad (4.12)$$

which (using an obvious notational device and assuming the original chain complex is nondegenerate) has the minor effect $\tilde{H}_0 \oplus \mathbb{F} \cong H_0$, while $\tilde{H}_p \cong H_p$ for $p > 0$. Similarly, $\tilde{\beta}_p = \beta_p - \delta_{p0}$, where $\delta_{jk} = 1$ if and only if $j = k$ and $\delta_{jk} = 0$ otherwise.

For a loopless digraph $D = (V, A)$, the set $\mathcal{A}_p(D)$ of *allowed p-paths* is

$$\{(v_0, \ldots, v_p) \in V^{p+1} : (v_{j-1}, v_j) \in A, 1 \leq j \leq p\}. \qquad (4.13)$$

As a convention, we set $\mathcal{A}_0 := V$, $V^0 \equiv \mathcal{A}_{-1} := \{0\}$, and $V^{-1} \equiv \mathcal{A}_{-2} := \emptyset$. For a field $\mathbb{F}$ and a finite set X, let $\mathbb{F}^X \cong \mathbb{F}^{|X|}$ be the free $\mathbb{F}$-vector space on X, with the convention $\mathbb{F}^\emptyset := \{0\}$. The *non-regular boundary operator* $\partial_{[p]} : \mathbb{F}^{V^{p+1}} \to \mathbb{F}^{V^p}$ is the linear map acting on the standard basis as

$$\partial_{[p]} e_{(v_0, \ldots, v_p)} = \sum_{j=0}^{p} (-1)^j e_{\nabla_j(v_0, \ldots, v_p)}. \qquad (4.14)$$

It is not hard to verify that $\partial_{[p-1]} \circ \partial_{[p]} \equiv 0$, so $(\mathbb{F}^{V^{p+1}}, \partial_{[p]})$ is a chain complex.

(Path homology can be defined over rings as well. This definition gives additional power: M. Yutin has exhibited digraphs on as few as six vertices that have *torsion*, i.e., a finite abelian summand of homology complementing the free abelian summand that a Betti number characterizes (Chowdhury et al., 2020).)

Path homology is obtained from a different chain complex derived from the immediate preceding one. Set

$$\Omega_p := \{\omega \in \mathbb{F}^{\mathcal{A}_p} : \partial_{[p]} \omega \in \mathbb{F}^{\mathcal{A}_{p-1}}\}, \qquad (4.15)$$

$\Omega_{-1} := \mathbb{F}^{\{0\}} \cong \mathbb{F}$, and $\Omega_{-2} := \mathbb{F}^\emptyset = \{0\}$. We have that $\partial_{[p]} \Omega_p \subseteq \mathbb{F}^{\mathcal{A}_{p-1}}$, so $\partial_{[p-1]} \partial_{[p]} \Omega_p = 0 \in \mathbb{F}^{\mathcal{A}_{p-2}}$ and $\partial_{[p]} \Omega_p \subseteq \Omega_{p-1}$. We can, therefore, define the *(non-regular) path complex* of D as the chain complex (Ω_p, ∂_p), where $\partial_p := \partial_{[p]}|_{\Omega_p}$. The homology of this path complex is the *(non-regular) path homology* of D. (The implied *regular path complex* prevents a directed 2-cycle from having nontrivial 1-homology. While Grigor'yan et al. (2012) advocate regular path homology, in our view, non-regular path homology is simpler, richer, and more likely useful in applications.)

For example, consider the digraphs D_1 and D_2 in Figure 4.15. $\mathcal{A}_1(D_1)$ and $\mathcal{A}_1(D_2)$ are given by the directed edges, $\mathcal{A}_2(D_2) = \emptyset$, and $\mathcal{A}_2(D_2) = \{(w,x,z),(w,y,z)\}$. Now $\partial_{[2]} e_{(w,x,z)} = e_{(x,z)} - e_{(w,z)} + e_{(w,x)} \notin \mathbb{F}^{\mathcal{A}_1(D_2)}$ and $\partial_{[2]} e_{(w,y,z)} = e_{(y,z)} - e_{(w,z)} + e_{(w,y)} \notin \mathbb{F}^{\mathcal{A}_1(D_2)}$ (because the edge $w \to z$ is missing), so

$$\partial_{[2]}(e_{(w,x,z)} - e_{(w,y,z)}) = e_{(x,z)} - e_{(w,z)} + e_{(w,x)} - e_{(y,z)} + e_{(w,z)} - e_{(w,y)}$$
$$= e_{(x,z)} + e_{(w,x)} - e_{(y,z)} - e_{(w,y)} \in \mathbb{F}^{\mathcal{A}_1(D_2)}. \qquad (4.16)$$

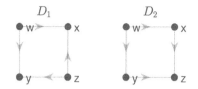

FIGURE 4.15 The digraph D_1 has trivial path homology but the digraph D_2 does not.

Consequently the dimensions of the path homology vector spaces (denoted by the Betti numbers β) are different: $\beta_1(D_1) = 1$ and $\beta_1(D_2) = 0$.

The ubiquity of digraphs in the cyber domain suggests that path homology can find a multitude of applications, and we briefly mention a few, though research in this area is only beginning. A generic application is the structural analysis of complex and/or temporal networks, which reveals common email traffic motifs (Chowdhury et al., 2020). Another generic application is feature extraction for cyber data in machine learning pipelines. A more specific and interesting possibility in line with Chowdhury et al. (2019) may be to characterize *attack graphs* that encode sequences of events that can produce a successful cyber attack. Because there are many different representations for attack graphs (Lallie et al., 2020), topological techniques may be useful or even necessary to identify commonalities between graphs and/or particularly salient structures within graphs.

In a similar vein, path homology provides a rich software metric applicable to source and binary code. Figure 4.16 shows a control flow graph with nontrivial path homology in dimension two. The binary is directly compiled from C source (albeit with `gotos` and inline assembly). The common instruction motif in most of the basic blocks clearly indicates how to construct binaries with essentially arbitrary control flow. Note that inserting operations without control flow (e.g., arithmetic operations in the instruction set) and reindexing memory addresses at various points would leave the control flow unaffected.

It turns out that it is possible to construct control flow graphs (at the assembly level) with arbitrary path homology, and experiments suggest that path homology generalizes cyclomatic complexity in a way that can detect unstructured control flow (Huntsman, 2020). The proof that control flow graphs can exhibit arbitrary path homology follows from a result of Chowdhury et al. (2019), which itself has more direct applications to the characterization of neural networks.

Meanwhile, the first author's analyses of UK and global air transportation networks (to be reported in detail elsewhere) suggest that changes in the path homology of 'backbone' digraphs (obtained by retaining only arcs corresponding to passenger volume above a threshold) as a function of the backbone threshold are strongly correlated with measures such as betweenness centrality due largely to the presence of 'bow-tie' structures in which two triangles share a vertex. That is, path homology may provide network metrics that simultaneously complement and correlate with existing metrics.

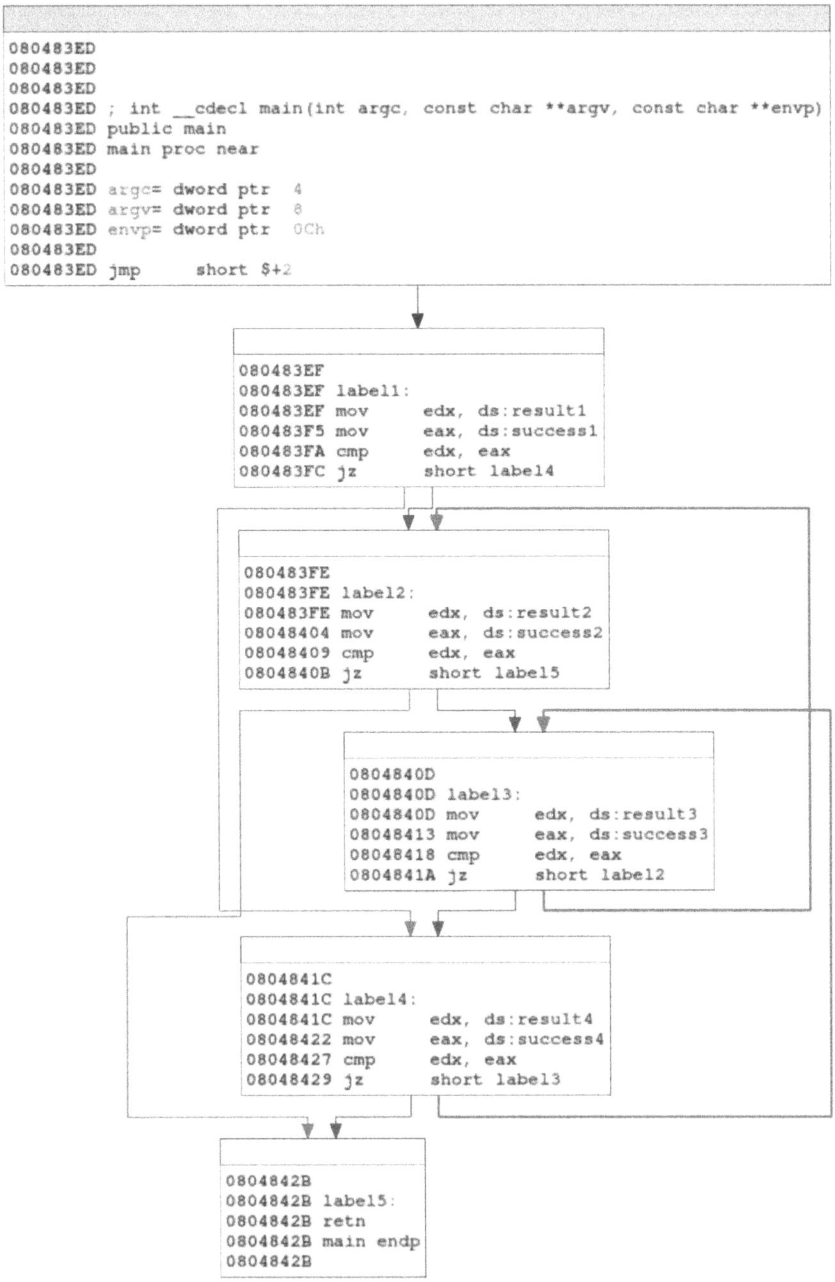

```
080483ED
080483ED
080483ED
080483ED ; int __cdecl main(int argc, const char **argv, const char **envp)
080483ED public main
080483ED main proc near
080483ED
080483ED argc= dword ptr  4
080483ED argv= dword ptr  8
080483ED envp= dword ptr  0Ch
080483ED
080483ED jmp     short $+2
```

```
080483EF
080483EF label1:
080483EF mov     edx, ds:result1
080483F5 mov     eax, ds:success1
080483FA cmp     edx, eax
080483FC jz      short label4
```

```
080483FE
080483FE label2:
080483FE mov     edx, ds:result2
08048404 mov     eax, ds:success2
08048409 cmp     edx, eax
0804840B jz      short label5
```

```
0804840D
0804840D label3:
0804840D mov     edx, ds:result3
08048413 mov     eax, ds:success3
08048418 cmp     edx, eax
0804841A jz      short label2
```

```
0804841C
0804841C label4:
0804841C mov     edx, ds:result4
08048422 mov     eax, ds:success4
08048427 cmp     edx, eax
08048429 jz      short label3
```

```
0804842B
0804842B label5:
0804842B retn
0804842B main endp
0804842B
```

FIGURE 4.16 A control flow graph with $\tilde{\beta}_\bullet = (0, 1, 1, 0, \dots)$, obtained by disassembly in IDA Pro (Eagle, 2011).

4.4 TOPOLOGICAL DATA ANALYSIS AND UNSUPERVISED LEARNING IN ONE DIMENSION

In this section, we sketch the basic ideas of the rapidly expanding field of topological data analysis by considering a simple application in one dimension that simultaneously advances the state of the art in the fundamental area of nonparametric statistical estimation and avoids much of the technical baggage of persistent homology. The basic idea is to describe a data set at multiple scales, then identify the best scale as having the most common value of a topological invariant, and finally to use that same invariant to initiate decomposition of the data into a topologically and information-theoretically optimal way.

Topological data analysis (TDA) has had a profound effect on data science and statistics over the last 15 years. Perhaps the most widely recognized and utilized tool in TDA is *persistent homology* (Zomorodian, 2005; Ghrist, 2008; Carlsson, 2009; Edelsbrunner & Harer, 2010; Ghrist, 2014; Oudot, 2015). The basic idea is to associate an inclusion-oriented family (i.e., a *filtration*) of simplicial complexes to a point set in a metric space. Each simplicial complex in the filtration is formed by considering the intersections of balls of a fixed radius about each data point. As the radius varies, different simplicial complexes are produced, and their homologies are computed. In the example of Figure 4.17, we consider a sample of 100 uniformly distributed points in a thin annulus about the unit circle. From left to right, we place disks of radius 0.1, 0.15, and 0.95 around each point. The topology of the data set is morally that of a circle, and the (persistent) homology of simplicial complexes formed from the intersections of disks reveals this: a 1-homology class 'persists' over an interval slightly bigger than [.15, .95].

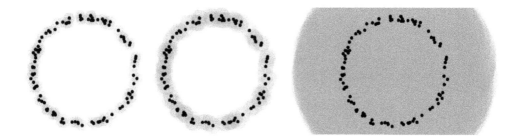

FIGURE 4.17 The topology of a data set can be probed at different scales.

Although the theory of topological persistence involves a considerable amount of algebra for bookkeeping associated to the 'births' and 'deaths' of homology classes as a function of the radius/filtration parameter, in practice, simply treating the Betti numbers as functions of that parameter gives considerable information. Along similar lines, we can consider how other topological invariants behave as a function of scale. For example, a kernel density estimate is a sort of 'smooth histogram' that represents sample data by an average of copies of a 'kernel' probability distribution centered around the data points. The bandwidth of a kernel density estimate is a scaling factor for the kernel: small bandwidths scale the kernel to be narrow and tall, and large bandwidths scale the kernel to be wide and short. A kernel density estimate has a very simple topological invariant: i.e., the minimal number of unimodal functions that it can be decomposed into.

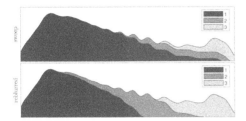 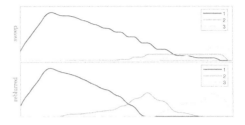

FIGURE 4.18 Topological mixture estimation. Left panels: area plots of (top) initial and (bottom) information-theoretically optimized unimodal decompositions of an estimated probability distribution. Right panels: line plots of the same decompositions.

Call $\phi : \mathbb{R}^n \to [0, \infty)$ *unimodal* if ϕ is continuous and the excursion set $\phi^{-1}([y, \infty))$ is contractible (i.e., homotopy equivalent to a point) for all $0 < y \le \max \phi$. For $n = 1$, contractibility means that these excursion sets are all intervals, which coincides with the intuitive notion of unimodality. For $f : \mathbb{R}^n \to [0, \infty)$ sufficiently nice, define the *unimodal category* of f to be the smallest number M of functions such that f admits a *unimodal decomposition* of the form $f = \sum_{m=1}^{M} \pi_m \phi_m$ for some $\pi > 0$, $\sum_m \pi_m = 1$, and ϕ_m unimodal (Ghrist, 2014).

The unimodal category is a topological (homeomorphism) invariant and a 'sweep' algorithm efficiently produces a unimodal decomposition in $n = 1$. As Figure 4.18 demonstrates, the unimodal category can be much less than the number of extrema. (The case $n = 2$ is still beyond the reach of current techniques, and only partial results are known. Moreover, for n sufficiently large, there is provably no algorithm for computing the unimodal category!)

The unimodal category of a kernel density estimate for a probability distribution can be used to select an appropriate bandwidth for sample data and, as shown in Figure 4.18, to decompose the resulting estimated distribution into well-behaved unimodal components *using no externally supplied parameters whatsoever* (Huntsman, 2018). The key ideas behind *topological mixture estimation* are to identify the most common unimodal category as a function of bandwidth and to exploit convexity properties of the mutual information between the mixture weights and the distribution itself. The result is an extremely general (though also computationally expensive) unsupervised learning technique in one dimension that can automatically set thresholds for anomaly detectors or determine the number of clusters in data (by taking random projections). The bandwidth for the kernel density estimate for the distribution and the number of unimodal mixture components are both determined using the same topological considerations.

4.5 CRITICAL NODE DETECTION IN WIRELESS NETWORKS USING SHEAVES

The abstract simplicial complex tools developed in the previous sections of this chapter can also be applied to understand the structure of wireless communication networks. The special feature of wireless networks that distinguishes them from wired networks in this context is that wireless networks can support point-to-multipoint transmissions or multicasting at the physical layer of the network stack, whereas wired networks can only support this behavior in logical layers of the network stack. As before, the combinatorial nature of such a network aligns neatly with the combinatorial structure of an abstract simplicial complex. Qualitative intuition about how the network responds to stress can be transformed into quantitative analytic tools using the topology of these simplicial complexes.

When a carrier sense multiple access/collision detection (CSMA/CD) media access model is used in a wireless network, only one node in a given vicinity can transmit while the others must wait. Although the physical layer protocols of wireless networks can be quite complex, the basic topology of the network plays an important role in determining network performance. This section addresses the problem of identifying critical nodes and links within a network by using local invariants derived from the local topology of the network. Recognizing that although protocol plays an important role, we are specifically concerned with those effects that are *protocol independent*.

This section provides theoretical justification for the 'right' local neighborhood in a wireless network with a CSMA/CD media access model using the structure of network activation patterns, and then validates the resulting topological invariants using simulated network traffic generated with **ns2**.

4.5.1 Historical Context and Contributions

Graph theory methods have been used extensively (for instance, Nandagopal et al. (2002); Yang & Vaidya (2002); Jain et al. (2003); Lee et al. (2007)) for identifying critical nodes in a network that carry a disproportionate amount of traffic. However, direct application of graph theory to locate these nodes is computationally expensive (Di Summa et al., 2011; Dinh et al., 2012). Furthermore, graphs are better suited to *wired* networks and don't necessarily address the multi-way interactions inherent in wireless networks (Chiang et al., 2007).

We can extend the ideas discussed earlier in this chapter to wireless networks by using higher-dimensional abstract simplicial complexes instead of graph connectivity as a measure of network health. Although connectivity can be a useful measure of health (Noubir, 2004; Gueye et al., 2010), it is rather coarse. We remedy this with a more systematic study of an 802.11b wireless network using the **ns2** network simulator (*The NS-2 Network simulator*, n.d.).

4.5.2 Interference From a Transmission

One of the main differences between a wired and a wireless communication network is the prevalence of interference on shared channels. Channels that are shared by more users or nodes are more likely to be congested. An abstract simplicial complex called the interference complex can model the shared channel usage within a wireless network and forms the basis of its topological analysis.

Let a wireless network consist of a single channel, with nodes $N = \{n_1, n_2, \ldots, n_i, \ldots\}$ in a region R. Associate an open set $U_i \subset R$ to each node n_i that represents its *transmitter coverage region*. For each node n_i, a continuous function $s_i : U_i \to \mathbb{R}$ represents its *signal level* at each point in U_i. Without loss of generality, we assume that there is a global threshold T for accurately decoding the transmission from any node. In Robinson (2014a), two abstract simplicial complex models were developed: the *interference* and *link* complexes.

Definition 4.6. *The* interference complex *is the abstract simplicial complex* $I = I(N, U, s, T)$ *consisting of all subsets of N of the form $[n_{i_1}, \ldots, n_{i_m}]$ for which $U_{i_1} \cap \cdots \cap U_{i_m}$ contains a point $x \in R$ for which $s_{i_k}(x) > T$ for all $k = 1, \cdots m$.*

The vertices of the interference complex are the nodes N of the network. There is a simplex for each list of transmitters that when transmitting will result in at least one mobile receiver location receiving multiple signals simultaneously. (The interference complex is a Čech complex (Ghrist, 2014; Hatcher, 2002).)

Proposition 4.3. *Each facet of the interference complex corresponds to a maximal collection of nodes that mutually interfere.*

Proof: Let c be a simplex of the interference complex. Then c is a collection of nodes whose coverages have a nontrivial intersection. The decoding threshold is exceeded for all nodes at some point x in this intersection. If any two nodes in c transmit simultaneously, they will interfere at x. If c is a facet, it is contained in no larger simplex, so it is clearly maximal.

Definition 4.7. *The* link graph *is a one-dimensional simplicial complex defined by the following collection of subsets of N:*

(i) $[n_i] \in N$ for each node n_i, and

(ii) $[n_i, n_j] \in N$ if $s_i(n_j) > T$ and $s_j(n_i) > T$.

The link complex $L = L(N, U, s, T)$ *is the clique complex of the link graph, which means that it contains all elements of the form $[n_{i_1}, \ldots, n_{i_m}]$ whenever this set is a clique in the link graph.*

Figure 4.19 shows three transmitters, labeled 1, 2, and 3, with their coverage regions U_1, U_2, and U_3 for a particular threshold T. Assuming that all points within U_i can receive the signal from transmitter i, the link complex for each configuration is shown in the middle row of Figure 4.19. Notice that in the second column,

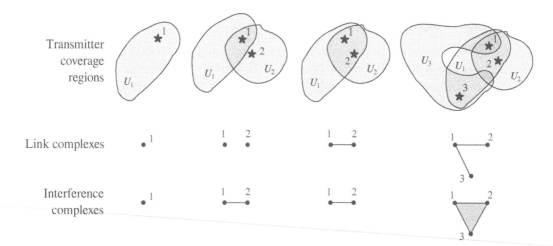

FIGURE 4.19 Several transmitters marked by stars, their coverage regions (top row), their link complexes (middle row), and their interference complexes (bottom row).

transmitter 1 can receive transmitter 2's signal but not conversely. This explains the absence of an edge in the link complex. However, since there are points in the intersection between their two coverage regions, the interference complex contains an edge. This also happens in the rightmost column, in which neither of transmitter 2 or 3 can receive each other's signal, but there are points where all three transmitters can be received.

Proposition 4.4. *Each facet in the link complex is a maximal set of nodes that can communicate directly with one another (with only one transmitting at a time).*

Proof: Let c be a simplex of the link complex. By definition, for each pair of nodes, $i, j \in c$ implies that $s_i(n_j) > T$ and $s_j(n_i) > T$. Therefore, i and j can communicate with one another.

Corollary 4.1. *Facets of the link complexes represent common broadcast resources.*

Since the CSMA/CD protocol is implemented locally, it can be modeled as follows:

Definition 4.8. *Suppose that X is a simplicial complex (such as an interference or link complex) whose set of vertices is N. Consider the following assignment $\mathcal{A}$ of additional information to capture which nodes are transmitting and decodable:*

(i) To each simplex $c \in X$, assign the set

$$\mathcal{A}(c) = \{n \in N : \text{there exists a simplex } d \in X \text{ with } c \subset d \text{ and } n \in d\} \cup \{\bot\}$$

of nodes that have a coface in common with c, along with the symbol $\bot$. We call $\mathcal{A}(c)$ the stalk *of $\mathcal{A}$ at c.*

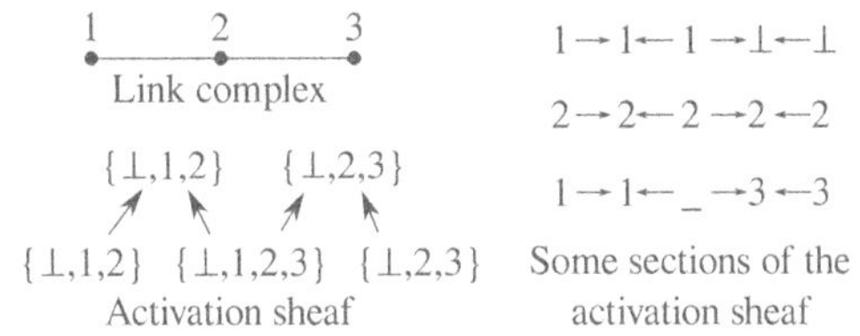

FIGURE 4.20 A link complex (left top), sheaf $\mathcal{A}$ (left bottom), and three sections (right). The restrictions are shown with arrows. There is a global section when node 1 transmits (right top), a global section when node 2 transmits (right middle), and a local section with nodes 1 and 3 attempting to transmit, interfering at node 2 (right bottom). An underscore in the right bottom frame indicates where an element is outside the support of the section.

(ii) To each pair $c \subset d$ of simplices, assign the restriction function

$$\mathcal{A}(c \subset d)(n) = \begin{cases} n & \text{if } n \in \mathcal{A}(d) \\ \bot & \text{otherwise} \end{cases}$$

For instance, if $c \in X$ is a simplex of a link complex, $\mathcal{A}(c)$ specifies which nearby nodes are transmitting and decodable, or $\bot$ if none are. The restriction functions relate the decodable transmitting nodes at the nodes to which nodes are decodable along an attached wireless link. Similarly, if $c \in X$ is a simplex of an interference complex, $\mathcal{A}(c)$ also specifies which nearby nodes are transmitting and effectively locks out any interfering transmissions from other nodes.

Definition 4.9. *The assignment $\mathcal{A}$ is called the* activation sheaf *and is a sheaf on an abstract simplicial complex.*

The theory of sheaves explains how to extract consistent information called *sections*, which in the present context consists of nodes whose transmissions do not interfere with one another.

Definition 4.10. *A section of $\mathcal{A}$ supported on a subset $Y \subseteq X$ is a function $s : Y \to N$ so that for each $c \subset d$ in Y, $s(c) \in \mathcal{A}(c)$ and $\mathcal{A}(c \subset d)(s(c)) = s(d)$. We call the subset Y the* support *of the section. A section supported on X is called a* global section.

Specifically, global sections are complete lists of nodes that can be transmitting without interference.

Figure 4.20 shows a network with three nodes labeled 1, 2, and 3. When node 1 transmits, node 2 receives. Because node 2 is busy, its link to node 3 must remain inactive (right top). When node 2 transmits, both nodes 1 and 3 receive (right middle). The right bottom diagram shows a local section that cannot be extended to the simplex marked with a blank. This corresponds to the situation where nodes 1 and 3 attempt to transmit but instead cause interference at node 2.

Definition 4.11. *Suppose that s is a global section of $\mathcal{A}$. The* active region *associated to a node $n \in X$ in s is the set*

$$active(s, n) = \{a \in X : s(a) = n\},$$

which is the set of all nodes that are currently waiting on n to finish transmitting.

Lemma 4.1. *The active region of a node is a connected, closed subset of X that contains n.*

Proof: Consider a simplex $c \in \text{active}(s, n)$. If c is not a vertex, then there exists a $b \subset c$; we must show that $b \in \text{active}(s, n)$. Since s is a global section $\mathcal{A}(b \subset c)s(b) = s(c) = n$. Because $s(c) \neq \bot$, the definition of the restriction function $\mathcal{A}(b \subset c)$ implies that $s(b) = n$. Thus $b \in \text{active}(s, n)$, so $\text{active}(s, n)$ is closed.

If $c \in \text{active}(s, n)$, then c and n have a coface d in common. Since s is a global section $s(d) = \mathcal{A}(c \subset d)s(c) = \mathcal{A}(c \subset d)n = n$. Thus, $n \in \text{active}(s, n)$, because n is a face of d and $\text{active}(s, n)$ is closed. This also shows that every simplex in $\text{active}(s, n)$ is connected to n.

Lemma 4.2. *The star over the active region of a node does not intersect the active region of any other node.*

Proof: Let $c \in \text{star active}(s, n)$. Without loss of generality, assume that $c \notin \text{active}(s, n)$. Therefore, there is a $b \in \text{active}(s, n)$ with $b \subset c$. By the definition of the restriction function $\mathcal{A}(b \subset c)$, the assumption that $c \notin \text{active}(s, n)$, and the fact that s is a global section $s(c)$ must be $\bot$.

Corollary 4.2. *If s is a global section of an activation sheaf $\mathcal{A}$, then the set of simplices c where $s(c) \neq \bot$ consists of a disjoint union of active regions of nodes.*

Lemma 4.3. *The active region of a node is independent of the global section. More precisely, if r and s are global sections of $\mathcal{A}$ and the active regions associated to $n \in X$ are nonempty in both sections, then $\text{active}(s, n) = \text{active}(r, n)$.*

Notice that if either of r or s has an empty active region, then Lemma 4.3 makes no assertions.

Proof: Without loss of generality, we need only show that $\text{active}(s, n) \subseteq \text{active}(r, n)$. If $c \in \text{active}(s, n)$, there must be a simplex $d \in X$ that has both n and c as faces. Now $s(n) = r(n) = n$ by Lemma 4.1, which means that $r(d) = \mathcal{A}(n \subset d)r(n) = n$. Therefore, since $\text{active}(r, n)$ is closed, this implies that $c \in \text{active}(r, n)$.

Figure 4.21 shows an example of a link complex in which two transmitters, labeled 1 and 2, are indicated. Their active regions are shown in the top row of Figure 4.21. Because of Lemma 4.1, each of these active regions is a closed set. The stars over their active regions are shown in the bottom row of Figure 4.21. Notice

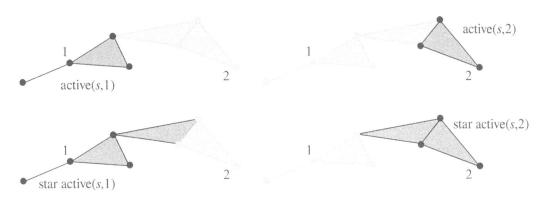

FIGURE 4.21 The active regions of two transmitters within a link complex (top row) and the stars over their active regions (bottom row).

that because of Lemma 4.2, the star over the active region of transmitter 1 does not intersect the active region of transmitter 2 or vice versa. Additionally, according to Lemma 4.3, it is unnecessary to specify the global section of the activation sheaf used to construct these regions.

Corollary 4.3. *The space of global sections of an activation sheaf consists of all sets of nodes that can be transmitted simultaneously without interference.*

4.5.3 An Algebraic Interlude: Relative Homology

Homology is a global topological invariant, which is to say that it applies to the entirety of a topological space. Since we wish to identify portions of the network that are more critical, it is useful to construct a local version of homology. This can be achieved by an algebraic construction that temporarily removes a portion of the space from consideration, called *relative homology*.

Suppose that $Y \subseteq X$ is a subcomplex of an abstract simplicial complex.

Definition 4.12. *The* relative k-chain space $C_k(X,Y)$ *is the vector space whose basis consists of the k-dimensional simplices of X that are not in Y. We can define the* relative boundary map $\partial_k : C_k(X,Y) \to C_{k-1}(X,Y)$ *using*

$$\partial_k([v_0,\ldots,v_k]) = \sum_{j=0}^{k}(-1)^j \begin{cases} \nabla_j[v_0,\ldots,v_k] & \text{if } \nabla_j[v_0,\ldots,v_k] \notin Y, \\ 0 & \text{otherwise.} \end{cases}$$

This is really a more elaborate form of the simplicial chain complex defined in Equation 4.1, and the same proof as before establishes that $(C_\bullet(X,Y), \partial_\bullet)$ is a chain complex. Naturally enough, there is a notion of *relative simplicial homology*.

Definition 4.13. *For a subcomplex $Y \subseteq X$ of an abstract simplicial complex X,*

$$H_k(X,Y) := H_k(C_\bullet(X,Y), \partial_\bullet)$$

is called the relative homology *of the pair (X,Y).*

As before, there is a notion of simplicial maps inducing maps on the relative homology. However, not every simplicial map works: it needs to respect the subcomplexes!

Proposition 4.5. *(Hatcher, 2002, Props. 2.9, 2.19) Every simplicial map $f : X \to Z$ from one abstract simplicial complex to another which restricts to a simplicial map $Y \to W$ induces a linear map $H_k(X, Y) \to H_k(Z, W)$ for each k. We call (X, Y) and (Z, W)* simplicial pairs *and f a* pair map $(X, Y) \to (Z, W)$.

4.5.4 Using Activation Patterns

The structure of the global sections of an activation sheaf leads to a model in which an active node silences all other nodes in its vicinity. In this section, we develop the concept of the local homology dimension and show how it can identify topological 'pinch points' within the network.

Definition 4.14. *Because of the Lemmas, we call the star over an active region associated to a node n the* region of influence. *The region of influence of a facet is the star over the closure of that facet. The region of influence for a collection of facets F can be written as a union*

$$\text{roi } F = \bigcup_{f \in F} \text{star cl } f.$$

One can, therefore, interpret the bottom row of Figure 4.21 as showing the regions of influence of transmitters 1 and 2.

In our previous work (Robinson, 2014a), the region of influence was used without detailed justification; the following Corollary provides this needed justification.

Corollary 4.4. *The complement of the region of influence of a facet is a closed subcomplex.*

Given this justification, Robinson (2014a) shows that critical nodes or links are those simplices c for whom the *local homology dimension* (see also Joslyn et al. (2016))

$$LH_k(c) = \dim H_k(X, X \setminus \text{roi } c) \qquad (4.17)$$

is larger than the average.

This implies the following experimental hypothesis: *If a node is critical, it will have a large local homology dimension.* Since the `ns2` network simulator provides complete transcripts of all packets, we can define a critical node to be one that *forwards* a large number of packets compared to other nodes in the network (Arulselvan et al., 2009).

We constructed a small simulation with 50 nodes as shown in Figure 4.22. Packets were randomly assigned source and destination nodes within the network, and all packet histories were recorded for analysis.

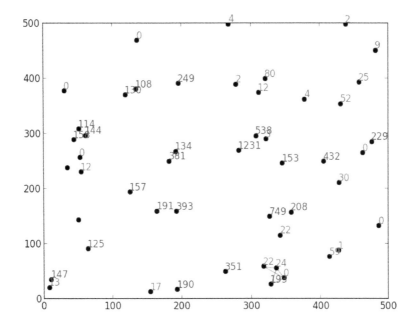

FIGURE 4.22 Locations of nodes and forwarded packet counts (axes in meters).

Figure 4.23 shows the probability that a node will forward a random packet. (The node numbers have been sorted from greatest to least probability.) The figure shows that most nodes forward only a small number of packets, while a few nodes carry considerably more traffic.

Figure 4.24 shows the dimension of local homology over all nodes and links in the network. In this particular network, the local homology dimension is only zero, one, or two. It is clear that nodes with high LH_1 occupy certain 'pinch points' in the network.

Figure 4.25 shows the probability that a node forwarding a certain number of packets will have the given value of LH_1. (We did not find a strong correspondence between forwarded packets and LH_2.) It is immediately clear that all nodes forwarding a large number of packets are assigned a high local homology, but the converse is not necessarily true. Local homology dimension is an indication that a node may be critical but does not guarantee criticality.

The ability to topologically identify bottlenecks without actually monitoring traffic has significance to the wireless network planner and/or to an attacker. Conventional centralities designed for graphs have deficiencies in the wireless context, though the idea of using centralities to identify and/or attack critical nodes has by now a long history (Lalou et al., 2018).

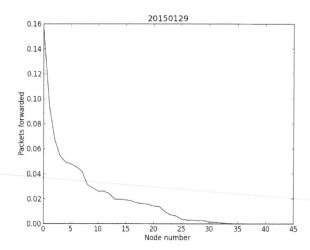

FIGURE 4.23 Probability that a given packet will be forwarded by a specific node.

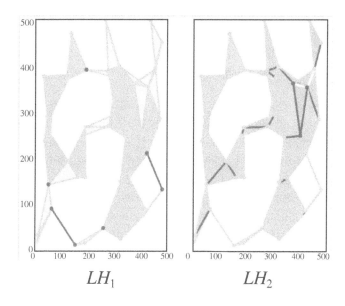

FIGURE 4.24 Dimension of local homology LH_1 (left) and LH_2 (right). Axes in meters: gray, black, and white respectively indicate dimensions zero, one, and two.

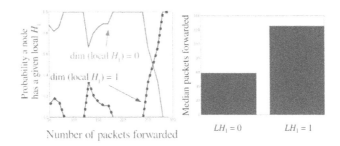

FIGURE 4.25 Probability a node has a certain local homology dimension given the number of packets it forwards.

4.5.5 Cohomological Analysis

(The basic idea of sheaf cohomology is to algebraically characterize obstructions to gluing local sections together into global sections. For background on and other practical applications of sheaf cohomology, see Robinson (2014b); Ghrist (2014).)

Although the space of global sections for an activation sheaf is a useful invariant, its sheaf cohomology is rather uninteresting. We need to enrich their structure somewhat to see this, though.

Definition 4.15. *If $\mathcal{A}$ is an activation sheaf on an abstract simplicial complex X, the vector activation sheaf $\widehat{\mathcal{A}}$ is given by specifying its stalks and restrictions:*

(i) *To each simplex $c \in X$, let $\widehat{\mathcal{A}}(c)$ be the vector space whose basis is $\mathcal{A}\backslash\{\bot\}$ (so the dimension of this vector space is the cardinality of $\mathcal{A}$ without counting $\bot$)*

(ii) *The restriction map $\widehat{\mathcal{A}}(c \subset d)(n)$ is the basis projection, which is well-defined since $\mathcal{A}(d) \subseteq \mathcal{A}(c)$.*

Proposition 4.6. *The dimension of the cohomology spaces of a vector activation sheaf $\widehat{\mathcal{A}}$ on a link complex X are*

$$dim\ H^k(\widehat{\mathcal{A}}) = \begin{cases} \text{the total number of nodes} & \text{if } k = 0 \\ 0 & \text{otherwise} \end{cases}$$

Proof: Every global section of $\mathcal{A}$ corresponds to a global section of $\widehat{\mathcal{A}}$, but formal linear combinations of global sections of $\mathcal{A}$ are also global sections of $\widehat{\mathcal{A}}$. Therefore, a global section of $\widehat{\mathcal{A}}$ merely consists of a list of those nodes that are transmitting, without regard for whether they interfere.

The fact that the other cohomology spaces are trivial is considerably more subtle. Consider the decomposition

$$X = \bigcup_i F_i$$

of the link complex into the set of its facets. Suppose that F_i is a facet of dimension k, and define $\mathcal{F}_i$ to be the direct sum of $k+1$ copies of the constant sheaf supported on F_i. (Each copy corresponds one of the vertices of F_i.) Then there is an exact sequence of sheaves

$$0 \to \widehat{\mathcal{A}} \xrightarrow{\Delta} \bigoplus_i \mathcal{F}_i \xrightarrow{m} \mathcal{S} \to 0$$

where Δ is a map that takes a basis vector corresponding to a given node to the linear combination of all corresponding basis vectors in each copy of the constant sheaves, and m is, therefore, a kind of difference map. This exact sequence leads to a long exact sequence

$$\cdots H^{k-1}(\mathcal{S}) \to H^k(\widehat{\mathcal{A}}) \to \bigoplus_i H^k(\mathcal{F}_i) \to H^k(\mathcal{S}) \cdots$$

Since each $\mathcal{F}_i$ is a direct sum of constant sheaves supported on a closed subcomplex, it only has nontrivial cohomology in degree 0.

Observe that $\mathcal{S}$ is a sheaf supported on sets of simplices lying in the intersections of facets. By Corollary 4.4, $\mathcal{S}$ must be a direct sum of copies of constant sheaves supported on closed subcomplexes, like each $\mathcal{F}_i$. Thus $\mathcal{S}$ only has nontrivial cohomology in degree 0, which means that for $k > 1$, $H^k(\widehat{\mathcal{A}}) = 0$.

It, therefore, remains to address the $k = 1$ case, which comes about from the exact sequence

$$\bigoplus_i H^0(\mathcal{F}_i) \to H^0(\mathcal{S}) \to H^1(\widehat{\mathcal{S}}) \to 0.$$

The leftmost map is surjective since every global section of $\mathcal{S}$ is given by specifying a single transmitting node. By picking exactly one facet containing that node, a global section of the corresponding $\mathcal{F}_i$ may be selected in the preimage. Thus the map $H^0(\mathcal{S}) \to H^1(\widehat{\mathcal{S}})$ must be the zero map and yet also surjective. This completes the proof.

4.6 CONCLUSION

We have only scratched the surface of topological techniques that can be fruitfully applied to problems in the cyber domain. Discrete Morse theory (Scoville, 2019), the algebraic topology of finite topological spaces (Barmak, 2011), and connections between simplicial complexes and partially ordered sets (Wachs, 2007) provide just a few opportunities for applications that we have not discussed at all here. For example, a notion of a weighted Dowker complex and an associated partial order can be used for *topological differential testing* to discover files that similar programs handle inconsistently (Ambrose et al., 2020). Another emerging topological tool for the analysis of enriched categories (such as furnished by sub-flow graphs in the sense of Huntsman (2019) or by generalized metric spaces including digraphs) is the theory of magnitude (co)homology (Leinster & Shulman, 2017; Hepworth, 2018).

More generally, both discrete and continuous topological methods can provide unique capabilities for problems in the cyber domain. The analysis of concurrent protocols and programs highlights this. While simplicial complexes have been used to solve problems in concurrency (Herlihy et al., 2014), the entire (recently developed) theory of directed topology traces its origin to static analysis of concurrent programs (Fajstrup et al., 2016).

In short, while there are many cyber-oriented problems that present a large attack surface for mainstream topological data analysis, the space of applicable techniques is much larger. Besides specifically tailored applications, topological approaches can provide robust and qualitatively interpretable features for machine learning pipelines. They can also address cyber problems for which other tools are fundamentally inadequate. The fundamentally discrete nature of cyber data entails opportunities for novel applications and capabilities, which in turn provide a substrate for future theoretical developments.

ACKNOWLEDGMENTS

The authors thank Samir Chowdhury, Fabrizio Romano Genovese, Jelle Herold, and Matvey Yutin for helpful discussions; Greg Sadosuk for producing Figure 4.16 and its attendant code; and Richard Latimer for providing code and data relating to sorting networks. The authors also thank the editors for their patience and attention to detail.

This research was partially supported with funding from the Defense Advanced Research Projects Agency (DARPA) via Federal contracts HR001115C0050 and HR001119C0072. The views, opinions, and/or findings expressed are those of the authors and should not be interpreted as representing the official views or policies of the Department of Defense or the U.S. Government.

REFERENCES

Alves, H., Fonseca, B., & Antunes, N. (2016). Software metrics and security vulnerabilities: dataset and exploratory study. In *European Dependable Computing Conference (EDCC)*.

Ambrose, K., Huntsman, S., Robinson, M., & Yutin, M. (2020). Topological differential testing. arXiv preprint arXiv:2003.00976.

Arulselvan, A., Commander, C. W., Elefteriadou, L., & Pardalos, P. M. (2009). Detecting critical nodes in sparse graphs. *Computers and Operations Research*, 36(7), 2193–2200.

Atkin, R. (1974). *Mathematical structure in human affairs*. Heinemann.

Ballard, G., Ikenmeyer, C., Landsberg, J., & Ryder, N. (2019). The geometry of rank decompositions of matrix multiplication II: 3×3 matrices. *Journal of Pure and Applied Algebra*, 223(8), 3205–3224.

Barmak, J. A. (2011). *Algebraic topology of finite topological spaces and applications*. Springer.

Berger, G. O., Absil, P.-A., De Lathauwer, L., Jungers, R. M., & Van Barel, M. (2019). Equivalent polyadic decompositions of matrix multiplication tensors. arXiv preprint arXiv:1902.03950.

Carlsson, G. (2009). Topology and data. *Bulletin of the American Mathematical Society*, 46, 255–308.

Chiang, M., Low, S., Calderbank, R., & Doyle, J. (2007, January). Layering as optimization decomposition: a mathematical theory of network architectures. *Proceedings of the IEEE*, 95(1).

Chokaev, B. V., & Shumkin, G. N. (2018). Two bilinear (3 × 3)-matrix multiplication algorithms of complexity 25. *Moscow University Computational Mathematics and Cybernetics*, 42(1), 23–30.

Chowdhury, S., Gebhart, T., Huntsman, S., & Yutin, M. (2019). Path homologies of deep feedforward networks. In *IEEE International Conference on Machine Learning and Applications (ICMLA)*.

Chowdhury, S., Huntsman, S., & Yutin, M. (2020). Path homology and temporal networks. In *International Conference on Complex Networks and their Applications*.

Chowdhury, S., & Mémoli, F. (2018). A functorial Dowker theorem and persistent homology of asymmetric networks. *Journal of Applied and Computational Topology*, 2(1), 115.

Courtois, N. T., Bard, G. V., & Hulme, D. (2011). A new general-purpose method to multiply 3 × 3 matrices using only 23 multiplications. arXiv preprint arXiv:1108.2830.

Di Summa, M., Grosso, A., & Locatelli, M. (2011). Complexity of the critical node problem over trees. *Computers and Operations Research*, 38(12), 1766–1774.

Dinh, T. N., Xuan, Y., Thai, M. T., Pardalos, P. M., & Znati, T. (2012). On new approaches of assessing network vulnerability: hardness and approximation. *IEEE/ACM Transactions on Networking*, 20(2), 609–619.

Dowker, C. H. (1952). Homology groups of relations. *Annals of Mathematics*, 56:84.

Du, X., Chen, B., Li, Y., Guo, J., Zhou, Y., Liu, Y., & Jiang, Y. (2019). LEOPARD: identifying vulnerable code for vulnerability assessment through program metrics. In *International Conference on Software Engineering (ICSE)*.

Dullien, T., & Porst, S. (2009). REIL: a platform-independent intermediate representation of disassembled code for static code analysis. In *CanSecWest*.

Duran, D., Weston, D., & Miller, M. (2011). Targeted taint driven fuzzing using software metrics. In *CanSecWest*.

Eagle, C. (2011). *The IDA Pro book: The unofficial guide to the world's most popular disassembler.* No Starch Press.

Ebert, C., & Cain, J. (2016). Cyclomatic complexity. *IEEE Software*, 33:27.

Edelsbrunner, H., & Harer, J. L. (2010). *Computational topology: An introduction.* AMS.

Erdmann, M. (2017). Topology of privacy: lattice structures and information bubbles for inference and obfuscation. arXiv preprint arXiv:1712.04130.

Fajstrup, L., Goubault, E., Haucourt, E., Mimram, S., & Raussen, M. (2016). *Directed algebraic topology and concurrency.* Springer.

Ghrist, R. (2008). Barcodes: the persistent topology of data. *Bulletin of the American Mathematical Society*, 45(1), 61.

Ghrist, R. (2014). *Elementary applied topology.* Createspace.

Ghrist, R., Lipsky, D., Derenick, J., & Speranzon, A. (2012). Topological landmark-based navigation and mapping. *preprint*.

Grigor'yan, A., Jimenez, R., Muranov, Y., & Yau, S.-T. (2018). On the path homology theory of digraphs and Eilenberg-Steenrod axioms. *Homology, Homotopy and Applications*, 20, 179.

Grigor'yan, A., Muranov, Y., Vershinin, V., & Yau, S.-T. (2018). Path homology theory of multigraphs and quivers. *Forum Mathematicum*, 30, 1319.

Grigor'yan, A., Muranov, Y., & Yau, S.-T. (2014). Graphs associated with simplicial complexes. *Homology, Homotopy and Applications*, 16, 295.

Grigor'yan, A., Muranov, Y., & Yau, S.-T. (2017). Homologies of graphs and Künneth formulas. *Communications in Analysis and Geometry*, 25, 969.

Grigor'yan, A., Yong, L., Muranov, Y., & Yau, S.-T. (2012). Homologies of path complexes and digraphs. arXiv preprint arXiv:1207.2834.

Grigor'yan, A., Yong, L., Muranov, Y., & Yau, S.-T. (2014). Homotopy theory for digraphs. *Pure and Applied Mathematics Quarterly*, 10, 619.

Grigor'yan, A., Yong, L., Muranov, Y., & Yau, S.-T. (2015). Cohomology of digraphs and (undirected) graphs. *Asian Journal of Mathematics*, 19, 887.

Gueye, A., Walrand, J. C., & Anantharam, V. (2010). Design of network topology in an adversarial environment. In *International Conference on Decision and Game Theory for Security (GameSec)*.

Hatcher, A. (2002). *Algebraic topology.* Cambridge.

Hepworth, R. (2018). Magnitude cohomology. arXiv preprint arXiv:1807.06832.

Herlihy, M., Kozlov, D., & Rajsbaum, S. (2014). *Distributed computing through combinatorial topology.* Morgan Kaufmann.

Heule, M. J. H., Kauers, M., & Seidl, M. (2019). New ways to multiply 3 × 3-matrices. arXiv preprint arXiv:1905.10192.

Huntsman, S. (2018). Topological mixture estimation. In *International conference on machine learning (ICML).*

Huntsman, S. (2019). The multiresolution analysis of flow graphs. In *International workshop on logic, language, information, and computation (WoLLIC).*

Huntsman, S. (2020). Generalizing cyclomatic complexity via path homology. arXiv preprint arXiv:2003.00944.

Iozzo, V. (2010). 0-knowledge fuzzing. In *Black Hat DC.*

Jain, K., Padhye, J., Padmanabhan, V., & Qiu, L. (2003). Impact of interference on multi-hop wireless network performance. In *International Symposium on Mobile Ad Hoc Networking and Computing (MobiHoc).*

Johnson, R. W., & McLoughlin, A. M. (1986). Noncommutative bilinear algorithms for 3 × 3 matrix multiplication. *SIAM J. Comp.*, 15(2), 595–603.

Joslyn, C., Praggastis, B., Purvine, E., Sathanur, A., Robinson, M., & Ranshous, S. (2016). Local homology dimension as a network science measure. In *SIAM Workshop on Network Science.*

Knuth, D. E. (1997). *The art of computer programming* (Vol. 3). Pearson.

Laderman, J. D. (1976). A noncommutative algorithm for multiplying 3 × 3 matrices using 23 multiplications. *Bulletin of the American Mathematical Society*, 82(1), 126–128.

Lallie, H. S., Debattista, K., & Bal, J. (2020). A review of attack graph and attack tree visual syntax in cyber security. *Computer Science Review*, 35, 100219.

Lalou, M., Tahraoui, M. A., & Kheddouci, H. (2018). The critical node detection problem in networks: A survey. *Computer Science Review*, 28, 92–117.

Landsberg, J. M. (2017). *Geometry and complexity theory.* Cambridge.

Lee, J.-W., Chiang, M., & Calderbank, R. (2007). Utility-optimal random-access control. *IEEE Transactions on Wireless Communications*, 6(7), 2741–2751.

Leinster, T., & Shulman, M. (2017). Magnitude homology of enriched categories and metric spaces. arXiv preprint arXiv:1711.00802.

Makarov, O. M. (1986). An algorithm for multiplying 3 × 3 matrices. *USSR Computational Mathematics and Mathematical Physics*, 26(1), 179–180.

McCabe, T. J. (1976). A complexity measure. *IEEE Transactions on Software Engineering*, SE-2, 308.

Medeiros, N., Ivaki, N., Costa, P., & Vieira, M. (2017). Software metrics as indicators of security vulnerabilities. In *International Symposium on Software Reliability Engineering (ISSRE)*.

Mesnard, F., Payet, E., & Vanhoof, W. (2016). Towards a framework for algorithm recognition in binary code. In *International Symposium on Principles and Practice of Declarative Programming (PPDP)*.

Nandagopal, T., Kim, T.-E., Gao, X., & Bharghavan, V. (2002). Achieving MAC layer fairness in wireless packet networks. In *International Symposium on Mobile Ad Hoc Networking and Computing (MobiHoc)*.

Nielson, H. R., & Nielson, F. (1992). *Semantics with applications*. Springer.

Noubir, G. (2004). On connectivity in *ad hoc* networks under jamming using directional antennas and mobility. In *IFIP International Conference on Wired/Wireless Internet Communications (WWIC)* (pp. 186–200).

The NS-2 network simulator. (n.d.). http://www.nsnam.org/. (Accessed: 2016-05-23)

Oudot, S. Y. (2015). *Persistence theory: From quiver representations to data analysis*. AMS.

Richardson, D. (1969). Some undecidable problems involving elementary functions of a real variable. *Journal of Symbolic Logic*, 33(4), 514–520.

Robinson, M. (2014a). Analyzing wireless communication network vulnerability with homological invariants. In *IEEE Global Conference on Signal and Information Processing (GlobalSIP)*.

Robinson, M. (2014b). *Topological signal processing*. Springer.

Robinson, M. (2017). Sheaf and duality methods for analyzing multi-model systems. In I. Pesenson, Q. T. L. Gia, A. Mayeli, H. Mhaskar, & D.-X. Zhou (Eds.), *Recent applications of harmonic analysis to function spaces, differential equations, and data science*. Springer.

Sato, H. (1999). *Algebraic topology: An intuitive approach*. AMS.

Scoville, N. A. (2019). *Discrete morse theory*. American Mathematical Society.

Shalaby, M., Mehrez, T., El-Mougy, A., Abdulnasser, K., & Al-Safty, A. (2017). Automatic algorithm recognition of source-code using machine learning. In *IEEE International Conference on Machine Learning and Applications (ICMLA)*.

Strassen, V. (1969). Gaussian elimination is not optimal. *Numerische Mathematik*, 13(4), 354–356.

Taherkhani, A., Korhonen, A., & Malmi, L. (2010). Recognizing algorithms using language constructs, software metrics, and roles of variables: an experiment with sorting algorithms. *The Computer Journal*, 54, 1049-1066.

Wachs, M. L. (2007). Poset topology: tools and applications. In E. Miller, V. Reiner, & B. Sturmfels (Eds.), *Geometric combinatorics*. American Mathematical Society.

Winograd, S. (1971). On multiplication of 2×2 matrices. *Linear Algebra and Applications*, 4(4), 381–388.

Yang, X., & Vaidya, N. (2002). Priority scheduling in wireless *ad hoc* networks. In *International Symposium on Mobile Ad Hoc Networking and Computing (MobiHoc)*.

Yanofsky, N. S. (2017). Galois theory of algorithms. In C. Başkent, L. S. Moss, & R. Ramanujan (Eds.), *Rohit Parikh on logic, language and society*. Springer.

Zomorodian, A. J. (2005). *Topology for computing*. Cambridge.

CHAPTER 5

Differential Equations

Parisa Fatheddin

CONTENTS

5.1	Introduction	171
5.2	Epidemiological Models	173
5.3	Applications to Computer Networks	178
5.4	Cyber Attack Models	188
5.5	Conclusion	196

5.1 INTRODUCTION

As this monograph intends to demonstrate, there are many ways to study cyber networks based on mathematical concepts. Here we concentrate on results in the literature that are based on models involving differential equations and aim to give motivation and self-contained explanation. We assume that the reader has some background on differential equations and linear algebra. We try to keep the exposition at a uniform level and provide an Appendix to recall most of the concepts needed, which is also intended to help the reader find connections to the material he/she learned in these courses. For standard introductory textbooks on differential equations, we recommend Boyce and DiPrima (2009) and Polking, Boggess, and Arnold (2018) and for linear algebra, we refer the reader to Anton and Rorres (2005) and Larson (2013).

In the literature, there are two main ideas on how to model problems faced in cyber security by differential equations. The first type interprets the spread of a computer virus amongst computer networks as the spread of a disease in a population and uses models in epidemiology. The most classical model of the spread of a disease in a population is the Susceptible-Infected-Recovered (SIR) model, where individuals either have the disease or have recovered or are susceptible to the disease. The main task in epidemiology is to determine under what conditions the disease will die out. This is given by a threshold called the basic reproductive number, denoted as R_0. It is shown that if $R_0 > 1$, then the system is unstable and the disease will become epidemic and if $R_0 \leq 1$, then the system will be locally asymptotically stable and the disease will eventually die out. We will explain the SIR model along with its properties and how to obtain the basic reproductive number,

DOI: 10.1201/9780429354649-5

then the results from recent research articles that have used these techniques to investigate the spread of a computer virus will be explained. Using concepts from differential equations and linear algebra, the stability of the equilibrium points is determined for each system. To explain the background required in epidemiology, we have chosen material from Castillo-Chavez, Feng, and Huang (2002), Li (2018), and Weiss (2013) and refer to Martcheva (2015) for more information. Afterwards, results in Gan, Yang, Liu, Zhu, and Zhang (2012), Ren, Yang, Zhu, Yang, and Zhang (2012), Toutonji, Yoo, and Park (2012) and Xiao, Fu, Dou, Li, Hu, and Xia (2017) in the area of cyber security are described in which concepts and models from epidemiology have been applied.

Another type of model based on differential equations that we have observed in the literature is the model of cyber attacks. Many major infrastructures such as water distribution systems, gas lines, communication networks, and electric power lines are now operated by smart grids, which incorporate modern technologies offered by computers and are monitored online. Thus, they are vulnerable to cyber attacks, especially to false data injection attack. The Supervisory Control and Data Acquisition (SCADA) system is used to monitor and control these power systems by reporting measurements every few seconds or minutes. For examples of incidents and analysis of cyber attacks on water systems see Amin, Litrico, Sastry, and Bayen (2012) Part I and Part II, on electric power grids see Liu, Ning, and Reiter (2011), Sidhar, Hahn, and Govindarasu (2011), and for a general smart grid, we refer to Farraj, Hammad, and Kundur (2018) and Manandhar, Cao, Hu, and Liu (2014). Using geometric control theory, the model of attacks on cyber physical systems is given as a system of differential equations, with one equation providing information about the state of the process and the input and another equation modeling the effects of the attack on the output. Here, we will discuss results in Pasqualetti, Dörfler, and Bullo (2011) and Pasqualetti et al. (2013) on the detection and identification of an attack on the system. These articles use algorithmic monitors and prove when an attack is undetectable or unidentifiable and apply these results to an IEEE 14 bus electric circuit system. Also, we will present findings on input-to-state stability of a cyber system under attack presented in Persis and Tesi (2015).

This chapter is organized as follows. We begin in Section 5.2 by giving background and results relevant to models in epidemiology. Then in Section 5.3 we apply these concepts and explain results in articles in cyber security that have used epidemiological models to investigate the spread of a computer virus in computer networks. Afterwards, Section 5.4 is devoted to models of cyber attacks and their stability and when they cannot be detected or identified. We then conclude the chapter and provide definitions, notations, and concepts needed in the Appendix. To avoid diverting the attention of the reader from the concepts discussed in this chapter, we refer him/her to the Appendix for the complete information on the notations used in each example.

5.2 EPIDEMIOLOGICAL MODELS

The most classical and well-known model used to study the spread of a disease in a population is the Susceptible-Infected-Recovered (SIR) model. It was developed by Sir Ronald Ross in his study of malaria in a mosquito population, for which he received the Nobel prize in 1902. The SIR model can be described as follows. It is assumed that no birth or death occurs in the population, making the size of the population at any time t, denoted as N, to be fixed. The individuals in the population are categorized as either, Susceptible: those who never had the disease, Infected: those who have the disease, or Recovered: those who had the disease and are now immune and in this model, it is assumed that they will never again get the disease after becoming immune. In the setting of cyber networks, this might be a particular virus code that the computer recognizes and avoids.

The number of individuals in each category at time t are denoted as $S(t), I(t)$, and $R(t)$, respectively. In the SIR model, it is also assumed that there is no latent period, which is the time period from being in contact with an infected individual to being infected. The rate of encounter between susceptible and infected individuals is considered to be directly proportional to the product of their sizes, referred to as mass action mixing. Hence,
$$S'(t) = -\beta SI,$$
where $\beta > 0$ is the transmission rate. This is because the number of susceptible individuals decreases as the number of interactions (found by $S(t)I(t)$) between susceptible and infected individuals increases.

As for the rate of change of the number of infected individuals, as the number of interactions between susceptible and infected individuals increases, the number of infected individuals increases. Letting the recovery rate be proportional to the infected rate, we have the rate of recovery to be νI for a positive constant ν leading to,
$$I'(t) = \beta SI - \nu I.$$

Along the same lines of reasoning, $R'(t) = \nu I$. Therefore, the SIR model is given by the following system of differential equations,

$$S'(t) = -\beta SI, \tag{5.1}$$
$$I'(t) = \beta SI - \nu I, \tag{5.2}$$
$$R'(t) = \nu I, \tag{5.3}$$
$$N = S(t) + I(t) + R(t), \tag{5.4}$$

where $\beta > 0$ is the rate at which susceptible individuals become infected and $\nu > 0$ is the rate at which infected individuals recover. Note that if we add the right-hand side of equations (5.1) - (5.3) we obtain, $S'(t) + I'(t) + R'(t) = 0$ and this agrees with equation (5.4) since the derivative of the constant N is zero.

We now study the SIR model given by equations (5.1)-(5.4) to derive more information about the population. It can be concluded that after a sufficiently long period of time, the disease will eventually die out in the population. If not then $\lim_{t\to\infty} I(t) \neq 0$ which implies that $I(t) > k$, for some constant k. Then,

$$\lim_{t\to\infty} R'(t) = \lim_{t\to\infty} \nu I(t) > k\nu,$$

however, this expression only holds if $\lim_{t\to\infty} R(t) = \infty$, which is a contradiction since $R(t) \leq N$ for all time $t > 0$. We may also find information on the number of susceptible individuals as time is set to go to infinity by using equations in the system directly. Namely, by equation (5.1) and (5.3) we have,

$$\frac{\frac{dS}{dt}}{\frac{dR}{dt}} = \frac{dS}{dR} = \frac{-\beta S}{\nu},$$

which is a separable differential equation leading to,

$$\int \frac{1}{S} dS = \int -\frac{\beta}{\nu} dR,$$

giving $\ln|S(t)| = -\frac{\beta}{\nu} R(t) + C$ for an arbitrary constant C and finally we obtain, $S(t) = S(0) e^{-\frac{\beta}{\nu} R(t)}$ and noting that $R(t) < N$, we arrive at $\lim_{t\to\infty} S(t) > S(0) e^{-\frac{\beta}{\nu} N}$.

In addition, finding the maximum number of infected individuals at a particular time can also be valuable. We use equations (5.1) and (5.2) to obtain,

$$\frac{\frac{dS}{dt}}{\frac{dI}{dt}} = \frac{dS}{dI} = \frac{-\beta S}{\beta S - \nu} \implies \int \frac{-\beta S + \nu}{\beta S} dS = \int dI$$

$$\implies -S(t) + \frac{\nu}{\beta} \ln|S(t)| = I(t) + C,$$

which yields,

$$-I(t) - S(t) + \frac{\nu}{\beta} \ln|S(t)| = -I(0) - S(0) + \frac{\nu}{\beta} \ln|S(0)|.$$

Recall that the maximum value of $I(t)$ occurs when $I'(t) = 0$. Thus, using equation (5.2), we have that $S(t)$ must be $\frac{\nu}{\beta}$ and so the maximum value for $I(t)$ is given by,

$$I_{max}(t) = -\frac{\nu}{\beta} + \frac{\nu}{\beta} \ln \frac{\nu}{\beta} + I(0) + S(0) - \frac{\nu}{\beta} \ln|S(0)|.$$

Furthermore, observe that for the SIR model at equations (5.1)-(5.4), since I and S are always positive, then based on equation (5.1), $S'(t) \leq 0$ making $S(t)$ decreasing as t increases. If $S(0) < \frac{\nu}{\beta}$, then $I'(t)|_{t=0} < 0$ and since $S(t)$ is decreasing then $S(t) \leq S(0) < \frac{\nu}{\beta}$ and based on equation (5.2), $I'(t) < 0$ for all $t \geq 0$. Hence, $I(t)$ (number of infected) strictly decreases, implying that no epidemic will occur. On the other hand, if $S(0) > \frac{\nu}{\beta}$, then there exists a $t_0 > 0$ such that $S(t) > \frac{\nu}{\beta}$ for all $t \in [0, t_0]$. Therefore, using equation (5.2), $I(t)$ is strictly increasing for $t \in [0, t_0]$, and an epidemic will occur. The threshold $\frac{\nu}{\beta}$ is called the basic reproductive number,

R_0. Thus, determining this number is crucial in epidemiology, since it offers a condition that ensures that the disease will eventually die out.

The basic reproductive number has been investigated by many authors for different population models. The convention is that if $R_0 \leq 1$, then the model reaches the disease-free equilibrium, that is the disease dies out and if $R_0 > 1$, then the population experiences an outbreak of the disease. R_0 is the expected number of new infectious individuals produced by a typical infected individual when exposed to the susceptible group. If $R_0 \leq 1$, then the disease-free equilibrium is locally asymptotically stable and if $R_0 > 1$, then it is unstable. In many applications, it is important to determine the behavior of the solution after a long period of time in the context of solution's stability. For a review of stability of equilibrium points see 5.1.1 in the Appendix. We now discuss a way to compute R_0 and offer some examples.

For the SIR model, the equilibrium point is where,

$$-\beta SI = \beta SI - \nu I = \nu I = 0,$$

giving, $(S_0, I_0, R_0) = (0, 0, k)$ for an arbitrary real number k, since $R'(t) = 0$. The next-generation operator approach, introduced by Diekmann, Heesterbeek, and Metz (1990), is commonly used to determine R_0 and can be explained as follows. Consider the system,

$$X'(t) = f(X, E, I),$$
$$E'(t) = g(X, E, I),$$
$$I'(t) = h(X, E, I),$$

where $X = (S, R)$ and E represents the number of individuals who are exposed to the disease and do not transmit it. The disease-free equilibrium, denoted as $U_0 = (X^*, 0, 0)$, is the equilibrium point assuming there is no disease at time zero. Solving $g(X^*, E, I) = 0$ for E, we obtain, $E = \tilde{g}(X^*, I)$. Let,

$$A = D_I h\left(X^*, \tilde{g}(X^*, I), I\right)\bigg|_{I=0},$$

where D_I is the derivative with respect to I. Afterwards, we write A as $A = M - D$, where M is a positive definite matrix ($M_{ij} \geq 0$ for all i, j) and D is a diagonal matrix. The basic reproductive number, R_0, is then $R_0 = \rho(MD^{-1})$, where $\rho(B)$ is the spectral radius of matrix B, i.e. the largest eigenvalue of B. We now explain some examples given in Castillo-Chavez et al. (2002) and Li (2018) in some detail and refer to the Appendix for more information on the notation used.

Example 1(b) in Castillo-Chavez et al. (2002): Consider the system,

$$S'(t) = \Lambda - \beta S \frac{I}{N} - \mu S,$$
$$I'(t) = \beta S \frac{I}{N} - (\mu + \gamma)I,$$
$$R'(t) = \gamma I - \mu R,$$
$$N = S + I + R.$$

See 5.1.8 in the Appendix for a description on parameters used in the above system. Note that $h(x, E, I) = \beta S \frac{I}{N} - (\mu + \gamma)I$ and if we set $R(0) = I(0) = 0$, we obtain $S^* = S(0) = \frac{\Lambda}{\mu}$, giving the disease-free equilibrium, $U_0 = \left(\frac{\Lambda}{\mu}, 0, 0\right)$. Then,

$$A = D_I h\left(X^*, \tilde{g}(X^*, I), I\right)\bigg|_{I=0} = D_I \left(\beta S^* \frac{I}{N} - (\mu + \gamma)I\right)\bigg|_{I=0} = \beta - (\mu + \gamma),$$

where we have used the fact that $S^* = \frac{\Lambda}{\mu} = N$, since $R^* = I^* = 0$. Hence, $R_0 = \frac{\beta}{\mu+\gamma}$, which implies that if $\frac{\beta}{\mu+\gamma} > 1$, there is an outbreak of the disease and if $\frac{\beta}{\mu+\gamma} \leq 1$, there will be no outbreak and the disease will die out.

Example 2 in Castillo-Chavez et al. (2002): Similar to the previous example, consider the Susceptible, Exposed, Infected, Recovered (SEIR) model given by,

$$S'(t) = \Lambda - \beta S \frac{I}{N} - \mu S,$$
$$E'(t) = \beta S \frac{I}{N} - (\mu + k)E,$$
$$I'(t) = kE - (\gamma + \mu)I,$$
$$R'(t) = \gamma I - \mu R.$$

Here E is the number of individuals who have been exposed to the disease, but don't yet have the disease. See 5.1.8 in the Appendix for information on the rest of the notation. Setting $I^* = E^* = R^* = 0$, and solving for S^*, we obtain, $S^* = \frac{\Lambda}{\mu}$ leading to $U_0 = \left(\frac{\Lambda}{\mu}, 0, 0, 0\right)$. In the above system, $h(X, E, I) = kE - (\gamma + \mu)I$ and $g(X, E, I) = \beta S \frac{I}{N} - (\mu+k)E$. Recall we may find $\tilde{g}(X^*, I)$ by setting $g(X^*, E, I) = 0$ and solving for E. That is, $\tilde{g}(X^*, I) = \frac{\beta I}{\mu+k}$. Now,

$$A = D_I h\left(X^*, \tilde{g}(X^*, I), I\right)\bigg|_{I=0} = D_I \left(k \frac{\beta I}{\mu + k} - (\gamma + \mu)I\right)\bigg|_{I=0}$$
$$= \frac{k\beta}{\mu + k} - (\gamma + \mu),$$

leading to $R_0 = MD^{-1} = \frac{k\beta}{(\mu+k)(\mu+\gamma)}$.

Example 3 in Castillo-Chavez et al. (2002): The following system is used to model the transmission dynamics of Tuberculosis (TB), where after individuals become exposed to the disease, they are treated with a drug and each individual may be sensitive or resistant to the drug. Letting $i = 1$ denote the sensitive strain, $i = 2$ denote the resistant strain and T represent the number of treated individuals, we have,

$$S'(t) = \Lambda - \beta_1 S \frac{I_1}{N} - \beta_2 S \frac{I_2}{N} - \mu S,$$

$$E'_1(t) = \beta_1 S \frac{I_1}{N} - (\mu + k_1 + r) E_1 + p\tilde{r} I_1 + \sigma \beta_1 T \frac{I_1}{N} - \beta_2 E_1 \frac{I_2}{N},$$

$$I'_1(t) = k_1 E_1 - (\mu + d_1 + \tilde{r}) I_1,$$

$$T'(t) = rE_1 + (1 - p - q)\tilde{r} I_1 - \sigma \beta_1 T \frac{I_1}{N} - \beta_2 T \frac{I_2}{N} - \mu T,$$

$$E'_2(t) = q\tilde{r} I_1 - (\mu + k_2) E_2 + \beta_2 (S + E_1 + T) \frac{I_2}{N},$$

$$I'_2(t) = k_2 E_2 - (\mu + d_2) I_2,$$

$$N = S + E_1 + I_1 + T + E_2 + I_2,$$

where we refer the reader to 5.1.8 in the Appendix for more details. Then $X = (S, T)$, $E = (E_1, E_2)$, $I = (I_1, I_2)$ and $U_0 = \left(\frac{\Lambda}{\mu}, 0, 0, 0\right)$. Now setting $E'_1(t)$ and $E'_2(t)$ equal to zero and solving for E_1 and E_2, respectively, we obtain,

$$\tilde{g}(X^*, I) = (\tilde{g}_1(X^*, I), \tilde{g}_2(X^*, I))$$
$$= \left(\frac{(\beta_1 + p\tilde{r}) I_1}{\mu \beta_2 \frac{I_2}{\Lambda} + \mu + k_1 + r}, \frac{q\tilde{r} I_1 + \beta_2 I_2}{\mu + k_2} + O(2)\right),$$

where $O(2) \leq 2K$ for some constant $K > 0$. Denoting $I'_1(t)$ and $I'_2(t)$ as $h_1(X, E, I)$ and $h_2(X, E, I)$, respectively, we obtain,

$$D_I (h_1(X, \tilde{g}_1(X^*, I), I)) = D_I \left(k_1 \frac{(\beta_1 + p\tilde{r}) I_1}{\mu \beta_2 \frac{I_2}{\Lambda} + \mu + k_1 + r} - (\mu + d_1 + \tilde{r}) I_1 \right),$$

$$D_I (h_2(X, \tilde{g}_2(X^*, I), I)) = D_I \left(k_2 \frac{q\tilde{r} I_1 + \beta_2 I_2}{\mu + k_2} + O(2) - (\mu + d_2) I_2 \right).$$

Hence,

$$A = \begin{bmatrix} \frac{k_1(\beta_1 + p\tilde{r})}{\mu + k_1 + r} & 0 \\ \frac{k_2 q\tilde{r}}{\mu + k_2} & \frac{k_2 \beta_2}{\mu + k_2} \end{bmatrix} - \begin{bmatrix} (\mu + d_1 + \tilde{r}) & 0 \\ 0 & (\mu + d_2) \end{bmatrix},$$

giving,

$$MD^{-1} = \begin{bmatrix} \frac{k_1(\beta_1 + p\tilde{r})}{\mu + k_1 + r} & 0 \\ \frac{k_2 q\tilde{r}}{\mu + k_2} & \frac{k_2 \beta_2}{\mu + k_2} \end{bmatrix} \frac{1}{(\mu + d_1 + \tilde{r})(\mu + d_2)} \begin{bmatrix} (\mu + d_2) & 0 \\ 0 & (\mu + d_1 + \tilde{r}) \end{bmatrix}$$

$$= \begin{bmatrix} \frac{k_1(\beta_1 + p\tilde{r})}{(\mu + d_1 + \tilde{r})(\mu + k_1 + r)} & 0 \\ \frac{k_2 q\tilde{r}}{(\mu + k_2)(\mu + d_1 + \tilde{r})} & \frac{k_2 \beta_2}{(\mu + d_2)(\mu + k_2)} \end{bmatrix},$$

and since the above is a triangular matrix, then we may find the eigenvalues to be the entries of the diagonal and R_0 becomes $\max\{\lambda_1, \lambda_2\}$, where,
$\lambda_1 = \frac{k_1(\beta_1 + p\tilde{r})}{(\mu + d_1 + \tilde{r})(\mu + k_1 + r)}$ and $\lambda_2 = \frac{k_2 \beta_2}{(\mu + d_2)(\mu + k_2)}$.

We now investigate the stability of the equilibrium points to find information on the behavior of the solutions. In the following we explain some results given in Section 2.3 in Li (2018) and for the notations used, we refer to 5.1.9 in the Appendix.

The system,

$$S'(t) = b - \beta IS - bS,$$
$$I'(t) = \beta IS - \gamma I - bI,$$

gives the disease-free equilibrium point $U_0 = (1, 0)$ and another equilibrium point $U^* = (S^*, I^*) = \left(\frac{\gamma+b}{\beta}, \frac{b(\beta-(b+\gamma))}{\beta(b+\gamma)}\right)$, called the endemic equilibrium for which the number of infected is not assumed to be zero. To determine the stability of each of these points, we form the Jacobian matrix,

$$J = \begin{bmatrix} -\beta I - b & -\beta S \\ \beta I & \beta S - \gamma - b \end{bmatrix}.$$

Following the method described in A.2 to A.4 of the Appendix, we have,

$$J(U_0) = \begin{bmatrix} -b & -\beta \\ 0 & \beta - (b+\gamma) \end{bmatrix}.$$

Then the eigenvalues will be the entries on the diagonal: $\lambda_1 = -b$ and $\lambda_2 = \beta - (b+\gamma)$. Therefore, for the equilibrium point, U_0 to be asymptotically stable, we need $\beta < b + \gamma$; otherwise, $\lambda_2 > 0$, making the equilibrium point to be a saddle and unstable. Furthermore, for the endemic equilibrium, we obtain,

$$J(U^*) = \begin{bmatrix} -\beta I^* - b & -\beta S^* \\ \beta I^* & 0 \end{bmatrix}.$$

Here, if $\beta > b + \gamma$, then $I^* > 0$ arriving at,

$$\text{Tr}(J(U^*)) = -\beta I^* - b < 0,$$
$$\det(J(U^*)) = \beta^2 I^* S^* > 0.$$

Hence, by the method of a Trace-Determinant (TD) diagram, U^* is asymptotically stable if $\beta > b + \gamma$ (see 5.1.3 in the Appendix for a review on this technique).

5.3 APPLICATIONS TO COMPUTER NETWORKS

The techniques described in the previous section have been used to determine the basic reproductive number, R_0 and the stability of equilibrium points of the systems modeling computer viruses as a disease spreading in computer networks. By connecting to the internet, computers become vulnerable to viruses. The common malicious computer attacks are viruses, worms, and attacks called a Trojan Horse. To be precise, a computer virus attacks the file system and a computer worm uses the system to search and attack the computer. The Trojan Horse attacks convince the users to download a particular file to enter the system and then damage it.

We describe the results of some of the research in the literature that have applied epidemiological models and concepts from Section 5.2 to cyber networks. In all examples, N is used to denote the total number of computers in the network.

Gan et al. (2012) use the following system to model the spread of a computer virus in computers connected to the internet,

$$S'(t) = \delta - \alpha_1 SI - \delta S + \gamma_2 I - \beta SI + \alpha_2 R, \quad (5.5)$$
$$I'(t) = \beta SI - \gamma_2 I - \delta I - \gamma_1 I, \quad (5.6)$$
$$R'(t) = \gamma_1 I + \alpha_1 SI - \delta R - \alpha_2 R, \quad (5.7)$$

where as in SIR model, the categories are S: Susceptible, I: Infected, R: Recovered computers and here $S(t)$, $I(t)$ and $R(t)$ are percentages making $S(t)+I(t)+R(t) = 1$ (see 5.1.10 in the Appendix). In this example, it is assumed that all computers are virus-free at the start of their connection with the network and the recovered computers have immunity until they once again become Susceptible. Following the method presented in Section 5.2, by setting $I = R = 0$, we obtain, $S^* = 1$, making $U_0 = (1, 0)$ be the disease-free equilibrium point. Then, denoting the right-hand side of equation (5.6) as $h(X, I)$, we obtain,

$$A = D_I\left(\beta I - \gamma_2 I - \delta I - \gamma_1 I\right)\Big|_{I=0} = \beta - (\gamma_2 + \gamma_1 + \delta).$$

Thus, $R_0 = \frac{\beta}{\gamma_1+\gamma_2+\delta}$. Notice that here there is no exposed category and we do not need to consider the function $\tilde{g}(X^*, 0)$. Now we use $S + I + R = 1$ to write the system as,

$$S'(t) = \delta - \alpha_1 SI - \delta S + \gamma_2 I - \beta SI + \alpha_2(1 - S - I), \quad (5.8)$$
$$I'(t) = \beta SI - \gamma_2 I - \delta I - \gamma_1 I. \quad (5.9)$$

To prove the stability of U_0, we use the La Salle's Invariance Principle (see 5.1.5 in the Appendix) as follows. Let $V(t) = I(t)$ then,

$$V'(t) = \beta SI - \gamma_2 I - \delta I - \gamma_1 I = \beta I\left(S - \frac{1}{R_0}\right).$$

We know that $S + I \leq 1$ and if we assume $R_0 \leq 1$, we have $V'(t) \leq 0$ for all t. Also $V'(t) = 0$ if and only if $(S, I) = (1, 0)$. Therefore, we obtain that U_0 is globally asymptotically stable if $R_0 \leq 1$.

As for the endemic equilibrium, we set each equation in the system of equations (5.8)-(5.9) to zero and solve to obtain,

$$U^* = (S^*, I^*) = \left(\frac{1}{R_0}, \frac{(\delta + \alpha_2)(R_0 - 1)}{\alpha_1 + (\delta + \gamma_1 + \alpha_2)R_0}\right).$$

Moreover, the Jacobian matrix is,

$$J = \begin{bmatrix} -\alpha_1 I - \delta - \beta I - \alpha_2 & -\alpha_1 S + \gamma_2 - \beta S - \alpha_2 \\ \beta I & \beta S - \gamma_2 - \delta - \gamma_1 \end{bmatrix},$$

which at U^* becomes,

$$J(U^*) = \begin{bmatrix} -(\alpha_1 + \beta)I^* - \delta - \alpha_2 & -(\alpha_2 + \gamma_1 + \delta + \alpha_1 S^*) \\ \beta I^* & 0 \end{bmatrix},$$

where we used $S^* = \frac{1}{R_0}$. Now if we assume $R_0 > 1$, then $I^* > 0$ and,

$$\text{Tr}(J(U^*)) = -(\alpha_1 + \beta)I^* - \delta - \alpha_2 < 0,$$
$$\det(J(U^*)) = \beta I^*(\alpha_2 + \gamma_1 + \delta + \alpha_1 S^*) > 0,$$

implying that the real part of both eigenvalues of the Jacobian matrix must be negative, making the endemic equilibrium point locally asymptotically stable if $R_0 > 1$.

Note that in the context of their cyber model, the requirement of $R_0 = \beta/(\gamma_1 + \gamma_2 + \delta) \leq 1$ complies with what one expects would end the virus infection. Namely, if the number of computers that transfer from Infected to Recovered (γ_1) or to Susceptible (γ_2) increases and is more than the number of computers that transfer from Susceptible to Infected (β), then the infection due to the virus will eventually be eliminated. Also an increase in the number of Susceptible computers joining or internal computers leaving the network (denoted as δ) helps eradicate the infection.

Ren et al. (2012) consider the SIR model for computer virus, where anti-virus software is used to recover the infected computers. They denote the recovery rate as ε and consider the recovery function,

$$T(I) = \begin{cases} \varepsilon I & 0 \leq I \leq I_0; \\ m & I > I_0, \end{cases} \quad (5.10)$$

where $m = \varepsilon I_0$ if the anti-virus has not yet been fully run. Then since $S'(t)$ and $I'(t)$ do not depend on $R(t)$, their system of equations is given as follows,

$$S'(t) = rS\left(1 - \frac{S}{k}\right) - \lambda SI - dS, \quad (5.11)$$
$$I'(t) = \lambda SI - T(I) - dI,$$

which by equation (5.10), may be given by the two systems below depending on the value of I:

$$S'(t) = rS\left(1 - \frac{S}{k}\right) - \lambda SI - dS, \quad (5.12)$$
$$I'(t) = \lambda SI - I(d + \varepsilon),$$

if $0 \leq I \leq I_0$ and,

$$S'(t) = rS\left(1 - \frac{S}{k}\right) - \lambda SI - dS, \quad (5.13)$$
$$I'(t) = \lambda SI - m - dI,$$

if $I_0 < I$ (see 5.1.12 in the Appendix). For system (5.12), note that the virus-free equilibrium is $U_0 = \left(\frac{k(r-d)}{r}, 0\right)$ leading to,

$$A = D_I\left(\lambda\left(\frac{k(r-d)}{r}\right)I - I(d+\varepsilon)\right)\bigg|_{I=0} = \frac{k\lambda(r-d)}{r} - (d+\varepsilon),$$

which gives, $R_0 = \frac{k\lambda(r-d)}{r(d+\varepsilon)}$. Now for the endemic equilibrium we have, from the second equation in system (5.12), $S^* = \frac{d+\varepsilon}{\lambda}$ and hence we obtain,

$$U^* = (S^*, I^*) = \left(\frac{d+\varepsilon}{\lambda}, \frac{r\lambda k - r(d+\varepsilon) - d\lambda k}{\lambda^2 k}\right) \tag{5.14}$$

$$= \left(\frac{k(r-d)}{rR_0}, \frac{(R_0-1)(r-d)}{\lambda R_0}\right).$$

To determine the stability of the two equilibrium points of system (5.12), we find the Jacobian matrix,

$$J = \begin{bmatrix} r - \frac{2rS}{k} - \lambda I - d & -\lambda S \\ \lambda I & \lambda S - (d+\varepsilon) \end{bmatrix}.$$

Then

$$J(U_0) = \begin{bmatrix} -(r-d) & -\frac{\lambda k(r-d)}{r} \\ 0 & \frac{\lambda k(r-d)}{r} - (d+\varepsilon) \end{bmatrix},$$

giving,

$$\text{Tr}(J(U_0)) = -(r-d) + \left(\frac{\lambda k(r-d)}{r} - (d+\varepsilon)\right),$$

$$\det(J(U_0)) = -(r-d)\left(\frac{\lambda k(r-d)}{r} - (d+\varepsilon)\right).$$

Therefore, for U_0 to be locally asymptotically stable we need $\frac{\lambda k(r-d)}{r} - (d+\varepsilon) < 0$. That is $\frac{\lambda k(r-d)}{r(d+\varepsilon)} < 1$ but that is exactly the condition $R_0 < 1$.

For the endemic equilibrium,

$$J(U^*) = \begin{bmatrix} -\frac{2(r-d)}{R_0} + r - d - \frac{(R_0-1)(r-d)}{R_0} & -\frac{\lambda k(r-d)}{rR_0} \\ \frac{(R_0-1)(r-d)}{R_0} & \frac{\lambda k(r-d)}{rR_0} - (d+\varepsilon) \end{bmatrix}.$$

Now from the second equation of system (5.12), $\frac{\lambda k(r-d)}{rR_0} - (d+\varepsilon) = 0$, thus, we have, assuming $R_0 > 1$,

$$\text{Tr}(J(U^*)) = -(r-d)\left(\frac{2}{R_0} - 1 + \frac{R_0 - 1}{R_0}\right) = -(r-d)\frac{1}{R_0} < 0,$$

$$\det(J(U^*)) = \frac{\lambda k(r-d)}{rR_0}\frac{(R_0-1)(r-d)}{R_0} = \frac{\lambda k(R_0-1)(r-d)^2}{rR_0^2} > 0,$$

where for the sign of $\text{Tr}(J(U^*))$, we have observed that $r - d > 0$ based on the condition, $R_0 > 1$. Hence, the real part of the eigenvalues must be negative, and the endemic equilibrium point is locally asymptotically stable. Similarly, the authors in Ren et al. (2012) find the equilibrium points and their stability for system (5.13).

We remark that in this example,

$$R_0 = \frac{k\lambda(r-d)}{r(d+\varepsilon)}, \tag{5.15}$$

and the condition, $R_0 \leq 1$ predicted to end the spread of the virus is intuitively clear, since increasing the intrinsic growth rate of Susceptible computers (r) and the Recovery Rate (ε) and slowing the rate at which Susceptible computers become infected (λ), would help eliminate the virus. Here d is the rate at which new Susceptible computers join the network and as indicated by (5.15), its increase is preferred as it helps lower the value of R_0.

In Ren et al. (2012), the authors support their results by numerical simulations. They use the notation E_0^* to denote the virus-free equilibrium point, U_0 and the endemic equilibrium, U^* is denoted by E^*. In Figure 2 in their paper, they provide the phase portrait of I versus S with parameters,

$$k = 1000, \lambda = 0.01, r = 1, d = 0.5, \varepsilon = 5.2,$$

which lead to $R_0 \approx 0.877$ and

$$U_0 = \left(\frac{k(r-d)}{r}, 0\right) = (500, 0).$$

We found that if $R_0 < 1$, then the virus-free equilibrium is asymptotically stable, and this is confirmed by their Figure 2, where the point, E_0^* is a sink.

For the endemic equilibrium, E^*, they use the following parameters,

$$k = 1000, \lambda = 0.01, r = 1, d = 0.5, \varepsilon = 4.2,$$

and obtain the phase portrait given in the paper's Figure 3. The parameters give $R_0 \approx 1.064$ and $E^* = (470, 3)$, where (5.14) was used. Thus, from the previous computations, the endemic equilibrium is asymptotically stable verified by the spiral sink shown in their Figure 3.

The authors continue to find information on the stability of system (5.13). They prove in Theorem 3.3(c) that the equilibrium point, E_2 is asymptotically stable if $R_0 > p_0, 2r = \lambda k, m < m_0$ are satisfied and in addition, either $p_2 < R_0$ or $p_1 < R_0 < p_2$ holds true. In Figures 4 and 5 of the article they provide phase portraits with different parameters that verify the result of Theorem 3.3(c), and in the figures, they obtain a spiral sink and a center, respectively.

Toutonji et al. (2012) study the spread of computer worms by Vulnerable, Exposed, Infectious, Secured, Vulnerable (VEISV) model, where the term vulnerable has the same interpretation as susceptible as in previous examples, and Secured state is when a computer has been given some security measure to obtain temporary immunity. Using $S(t) = N - V(t) - E(t) - I(t)$, they give the system as follows,

$$V'(t) = \phi N - fEV - (\psi_1 + \phi)V - \phi E - \phi I, \tag{5.16}$$
$$E'(t) = fEV - (\alpha + \psi_2)E, \tag{5.17}$$
$$I'(t) = \alpha E - (\gamma + \theta)I. \tag{5.18}$$

with $f := (\alpha\beta)/N$ (see A.11). Setting $E = I = 0$, we find the worm-free equilibrium to be $U_0 = EQ_{wf} = (V_1^*, E_1^*, I_1^*) = \left(\frac{\phi}{\psi_1+\phi}N, 0, 0\right)$. To determine the stability of the equilibrium point, we find the Jacobian matrix,

$$J = \begin{bmatrix} -fE - (\psi_1 + \phi) & -fV - \phi & -\phi \\ fE & fV - (\alpha + \psi_2) & 0 \\ 0 & \alpha & -(\gamma + \theta) \end{bmatrix},$$

which for the worm-free equilibrium is,

$$J(EQ_{wf}) = \begin{bmatrix} -(\psi_1 + \phi) & -fV_1^* - \phi & -\phi \\ 0 & fV_1^* - (\alpha + \psi_2) & 0 \\ 0 & \alpha & -(\gamma + \theta) \end{bmatrix}.$$

We obtain,

$$\det(J(EQ_{wf}) - \lambda I) = \bigl(-(\psi_1 + \phi) - \lambda\bigr)\bigl(fV_1^* - (\alpha + \psi_2) - \lambda\bigr)\bigl(-(\gamma + \theta) - \lambda\bigr),$$

giving the following eigenvalues,

$$\lambda_1 = -(\psi_1 + \phi),$$
$$\lambda_2 = fV_1^* - (\alpha + \psi_2),$$
$$\lambda_3 = -(\gamma + \theta).$$

Moreover, for the worm-free equilibrium to be locally asymptotically stable we need, $fV_1^* - (\alpha + \psi_2) < 0$ that is with $f = \frac{\alpha\beta}{N}$,

$$\frac{\phi N}{\psi_1 + \phi} < (\alpha + \psi_2)\left(\frac{N}{\alpha\beta}\right),$$

giving

$$\frac{\phi\alpha\beta}{(\psi_1 + \phi)(\alpha + \psi_2)} < 1.$$

Thus,

$$R_0 = \frac{\phi\alpha\beta}{(\psi_1 + \phi)(\alpha + \psi_2)}.$$

We can also confirm that with $R_0 < 1$, we have the required condition, $E^* > 0$, since $R_0 < 1$ ensures that the numerator of E^* is positive.

To prove that the worm-free equilibrium point is globally asymptotically stable if $R_0 \leq 1$, we consider the Lyapunov function, $L(t) = E(t)$. Then using the fact that the maximum number of vulnerable computers is $V_1^* = \frac{\phi N}{\psi_1 + \phi}$, we arrive at,

$$L'(t) = fE(t)V(t) - (\alpha + \psi_2)E(t)$$
$$\leq fE(t)\left(\frac{\phi N}{\psi_1 + \phi}\right) - (\alpha + \psi_2)E(t)$$
$$= E(t)\left(\frac{\phi N}{\psi_1 + \phi}\left(\frac{\alpha\beta}{N}\right) - (\alpha + \psi_2)\right)$$
$$= (\alpha + \psi_2)E(t)\left(\frac{\alpha\beta\phi}{(\psi_1 + \phi)(\alpha + \psi_2)} - 1\right)$$
$$= (\alpha + \psi_2)E(t)(R_0 - 1).$$

Now $R_0 < 1$ implies that $L'(t) < 0$. Also $L'(t) = 0$ for all $t \geq 0$ if and only if $(V, E, I) = (\frac{\phi N}{\psi_1 + \phi}, 0, 0)$, since $E(t) = 0$ for all $t \geq 0$, implies that $I(t) = 0$ for every $t \geq 0$. By La Salle's Principle, we have that the worm-free equilibrium is globally asymptotically stable.

As for the endemic equilibrium, observe that from (5.17), $E^*(fV^* - (\alpha + \psi_2)) = 0$ and to ensure that E^* is not zero, we must have $V^* = \frac{\alpha + \psi_2}{f}$. Then solving (5.18) for E^* and plugging both E^* and V^* into (5.16), we are able to find I^* and the endemic equilibrium becomes,

$$U^* = EQ_{we} = (V^*, E^*, I^*)$$
$$= \left(\frac{\alpha + \psi_2}{\beta\alpha}N, \frac{\phi - \frac{\alpha + \psi_2}{\beta\alpha}(\psi_1 - \phi)}{\alpha + \psi_2 + \phi\left(1 + \frac{\alpha}{\gamma + \theta}\right)}N, \frac{\alpha}{\gamma + \theta}E^*\right).$$

For this case,

$$J(U^*) = \begin{bmatrix} -(fE_2^* + \psi_1 + \phi) & -\alpha - \psi_2 - \phi & -\phi \\ fE_2^* & 0 & 0 \\ 0 & \alpha & -(\gamma + \theta) \end{bmatrix}.$$

Due to the technical steps involved in determining the stability of the endemic equilibrium, we omit the proof and refer the interested reader to the proof of Lemma 2 in Toutonji et al. (2012).

In regards to cyber security, note that the condition, $R_0 \leq 1$ is equivalent to requiring $\phi\alpha\beta \leq (\psi_1 + \phi)(\alpha + \psi_2)$. That is, the transition rates from Exposed to Infected (α) and from Secured to Vulnerable (ϕ), need to be less than the transition rates from Vulnerable to Secured (ψ_1) and from Exposed to Secured (ψ_2). Control over the contact rate (β) also helps weaken the spread of the worm in the network. Note that the effect of the increase in the parameters α and ϕ in R_0 is greater in the numerator than in the denominator.

The authors confirm the results by Figures 4 and 5 in their paper. In Figure 4, the size of population in exposed (E) and infected (I) are graphed versus time, and afterwards in Figure 5, the graph of the number of vulnerable (V) and secured (S) versus time is depicted. Both figures confirm the fact that as R_0 decreases, the number of exposed (E), infected (I) and vulnerable (V) go to zero faster and the number of secured (S) increases at a faster rate.

Xiao et al. (2017) examine the spread of worms on the internet, accessed by mobile devices through Wi-Fi. Susceptible, Exposed, Infectious, Quarantined, Recovered (SEIQR) model is considered, in which the infected devices are quarantined by their Wi-Fi base station, where the connection to the internet may be turned off. Mathematically, their model is given by the following system,

$$S'(t) = -\beta SI + \mu N - \mu S,$$
$$E'(t) = \beta SI - \eta E - \varepsilon E - \mu E,$$
$$I'(t) = \eta E - \mu I - \xi I - \gamma I,$$
$$Q'(t) = \xi I - \varphi Q - \mu Q,$$
$$R'(t) = \varepsilon E + \gamma I + \varphi Q - \mu R,$$
$$S(t) + E(t) + I(t) + Q(t) + R(t) = N,$$

where more information on the system is given in 5.1.13 of the Appendix. They find the worm-free and endemic equilibrium points as follows,

$$P_0 = (S_0, E_0, I_0, Q_0, R_0) = (N, 0, 0, 0, 0),$$
$$P^* = (S^*, E^*, I^*, Q^*, R^*) =$$
$$\left(\frac{(\eta + \varepsilon + \mu)(\xi + \gamma + \mu)}{\eta \beta}, \frac{\mu(N - S^*)}{\eta + \varepsilon + \mu}, \frac{\mu(N - S^*)}{\beta S^*}, \frac{\xi I^*}{\varphi + \mu}, \frac{\varepsilon E^* + \gamma I^* + \varphi Q^*}{\mu} \right),$$

and use the next-generation operator method to determine the basic reproductive number. Namely, setting, $E'(t)\big|_{S=S_0} = 0$, we obtain, $E = \frac{\beta S_0 I}{\eta + \varepsilon + \mu}$. Thus,

$$A = D_I \left(\frac{\eta \beta S_0}{\eta + \varepsilon + \mu} I - (\xi + \gamma + \mu)I \right) \bigg|_{I=0},$$

giving,

$$R_0 = \frac{\eta \beta N}{(\eta + \varepsilon + \mu)(\xi + \gamma + \mu)}.$$

To determine the stability of the equilibrium points, they find the Jacobian matrix,

$$J = \begin{bmatrix} -\beta I - \mu & 0 & -\beta S & 0 & 0 \\ \beta I & -\eta - \varepsilon - \mu & \beta S & 0 & 0 \\ 0 & \eta & -\mu - \xi - \gamma & 0 & 0 \\ 0 & 0 & \xi & -\varphi - \mu & 0 \\ 0 & \varepsilon & \gamma & \varphi & -\mu \end{bmatrix}.$$

At the worm-free equilibrium, the above matrix becomes,

$$J(P_0) = \begin{bmatrix} -\mu & 0 & -\beta N & 0 & 0 \\ 0 & -\eta - \varepsilon - \mu & \beta N & 0 & 0 \\ 0 & \eta & -\mu - \xi - \gamma & 0 & 0 \\ 0 & 0 & \xi & -\varphi - \mu & 0 \\ 0 & \varepsilon & \gamma & \varphi & -\mu \end{bmatrix},$$

leading to,

$$\det(J(P_0) - \lambda I)$$
$$= (-\mu - \lambda)\Big[(-\eta - \varepsilon - \mu - \lambda)(-\mu - \xi - \gamma - \lambda)(-\varphi - \mu - \lambda)(-\mu - \lambda)$$
$$- (\beta N)(\eta)(-\varphi - \mu - \lambda)(-\mu - \lambda)\Big]$$
$$= (\mu + \lambda)^2 (\varphi + \mu + \lambda)\Big[(\eta + \varepsilon + \mu + \lambda)(\mu + \xi + \gamma + \lambda) - \eta\beta N\Big].$$

Hence, $\lambda_1 = \lambda_2 = -\mu$, $\lambda_3 = -(\varphi + \mu)$ and for the third multiple, we let $a := \eta + \varepsilon + \mu$, $b := \mu + \xi + \gamma$ to obtain,

$$(\eta + \varepsilon + \mu + \lambda)(\mu + \xi + \gamma + \lambda) - \eta\beta N = (a + \lambda)(b + \lambda) - \eta\beta N$$
$$= \lambda^2 + (a + b)\lambda + ab - \eta\beta N.$$

Now, adopting the notation given in A.6, we notice that $a_1 = a + b > 0$ and

$$\Delta_2 = \det \begin{bmatrix} a_1 & a_3 \\ a_0 & a_2 \end{bmatrix} = \det \begin{bmatrix} a+b & 0 \\ 1 & ab - \eta\beta N \end{bmatrix} = (a+b)(ab - \eta\beta N) > 0.$$

Since $\eta\beta N < ab$ under the assumption $R_0 < 1$, then by the Hurwitz condition the real part of all roots are negative and thus, the worm-free equilibrium, P_0, is locally asymptotically stable.

To show that P_0 is globally asymptotically stable, we consider the Lyapunov function,

$$L(t) = \eta E(t) + (\eta + \varepsilon + \mu)I(t),$$

and find,

$$L'(t) = \eta\Big(\beta SI - (\eta + \varepsilon + \mu)E\Big) + (\eta + \varepsilon + \mu)\Big(\eta E - (\mu + \xi + \gamma)I\Big)$$
$$= \Big(\eta\beta S - (\mu + \varepsilon + \mu)(\mu + \xi + \gamma)\Big)I$$
$$\leq \Big(\eta\beta N - (\mu + \varepsilon + \mu)(\mu + \xi + \gamma)\Big)I < 0,$$

using the condition, $R_0 < 1$. Furthermore, $L'(t) = 0$ only when $(S, E, I, Q, R) = (N, 0, 0, 0, 0)$, since if $I(t) = 0$ for all $t \geq 0$, then $Q(t) = R(t) = 0$ for all $t \geq 0$.

Therefore, the La Salle Invariance Principle may be applied to obtain the global asymptotically stable condition of P_0.

For endemic equilibrium we have,

$$J(P^*) = \begin{bmatrix} -\beta I^* - \mu & 0 & -\beta S^* & 0 & 0 \\ \beta I^* & -a & \beta S^* & 0 & 0 \\ 0 & \eta & -b & 0 & 0 \\ 0 & 0 & \xi & -\varphi - \mu & 0 \\ 0 & \varepsilon & \gamma & \varphi & -\mu, \end{bmatrix}$$

giving,

$$\det(J(P^*) - \lambda I)$$
$$= (-\beta I^* - \mu - \lambda)\Big[(-a-\lambda)(-b-\lambda)(-\varphi-\mu-\lambda)(-\mu-\lambda)$$
$$- \beta S^*\big(\eta(-\varphi-\mu-\lambda)(-\mu-\lambda)\big)\Big] - \beta S^*\Big(\beta I^*\big(\eta(-\varphi-\mu-\lambda)(-\mu-\lambda)\big)\Big)$$
$$= (\varphi + \mu + \lambda)(\mu + \lambda)\Big(-(\beta I^* + \mu + \lambda)(ab + (a+b)\lambda + \lambda^2 - \beta S^*\eta)$$
$$- \beta^2 S^* I^* \eta\Big)$$
$$= -(\varphi + \mu + \lambda)(\mu + \lambda)\Big(\lambda^3 + (a + b + \mu + \beta I^*)\lambda^2$$
$$+ \big(\beta I^*(a+b) + \mu(a+b) + ab - \beta S^*\eta\big)\lambda + \big(\beta I^* ab + \mu ab - \mu\beta S^*\eta\big)\Big).$$

Thus, $\lambda_1 = -\varphi - \mu$, $\lambda_2 = -\mu$ and for the third multiple, note that applying the notation in A.6, we have a_0 and a_1 being positive. In addition,

$$\Delta_2 = \det \begin{bmatrix} a_1 & a_3 \\ a_0 & a_2 \end{bmatrix} = (a + b + \mu + \beta I^*)\Big(\beta I^*(a+b) + \mu(a+b) + ab - \beta S^*\eta\Big)$$
$$- \beta I^* ab - \mu ab + \mu\beta S^*\eta.$$

Since $S^* = ab/(\eta\beta)$ we further obtain,

$$\Delta_2 = (a + b + \mu + \beta I^*)\big(\beta I^*(a+b) + \mu(a+b)\big) - \beta I^* ab > 0,$$

and

$$\Delta_3 = \det \begin{bmatrix} a_1 & a_3 & 0 \\ 1 & a_2 & 0 \\ 0 & a_1 & a_3 \end{bmatrix} = a_3(a_1 a_2 - a_3) = a_3 \Delta_2$$
$$= \Big(\beta I^* ab + \mu ab - \mu\beta S^*\eta\Big)\Delta_2 = \beta I^* ab > 0,$$

where again the definition of S^* was used and by the Hurwitz condition we obtain the local asymptotically stability of P^*.

As for numerical simulations, the authors graph the population size in each category versus time in Figure 2 in their paper for which they use the parameters,

$S(0) = 40,000, \ E(0) = 30,000, \ I(0) = 30,000, \ Q(0) = 0, \ R(0) = 0,$
$N = 100,000, \ \beta = 5.0 \times 10^{-7}, \ \mu = 3.0 \times 10^{-7}, \ \varepsilon = 3.0 \times 10^{-7},$
$\eta = 0.05, \ \gamma = 0.01, \ \xi = 0.05, \ \varphi = 0.05,$

from the above we may find that $R_0 \approx 0.833$ and thus the worm-free equilibrium point, R_0 is asymptotically stable and the worm affecting the computers will eventually disappear. This can be observed in their Figure 2, since the number of exposed (E), infected (I), and quarantined (Q) decrease to zero over time and the number of recovered (R) increases as time increases. They also consider the parameters below for the endemic equilibrium, P^*,

$S(0) = 99,000, \ E(0) = 500, \ I(0) = 500, \ Q(0) = 0, \ R(0) = 0,$
$N = 100,000, \ \beta = 4.0 \times 10^{-6}, \ \mu = 3.0 \times 10^{-7}, \ \varepsilon = 3.0 \times 10^{-7},$
$\eta = 0.05, \ \gamma = 0.01, \ \xi = 0.05, \ \varphi = 0.05,$

giving, $R_0 \approx 6.667$. Thus, the endemic equilibrium is asymptotically stable. These parameters are used for their Figure 3, in which similar to the previous case, as time increases, the number of individuals that are susceptible (S), exposed (E), infected (I), and quarantined (Q) decrease to zero and the number of recovered (R) increases. In the paper, they apply the fourth-fifth order Runge-Kutta method for their simulations. Background on this discretization technique see Section 6.2 in Polking et al. (2018) or Section 8.3 of Boyce and DiPrima (2009).

For other similar results on the use of epidemiological models to determine the threshold R_0 for the spread of malicious objects in cyber networks, we recommend Han and Tan (2010), Mishra and Jha (2010), Mishra and Pandey (2011), and Roberto, Piqueira, and Araujo (2009).

5.4 CYBER ATTACK MODELS

In the literature, attacks on cyber physical systems are typically categorized as either deception attacks or denial of service (DoS) attacks. Deception attacks manipulate the data and deceive the receiver. On the other hand, the DoS attacks prevent the receiver from obtaining the information by jamming and delaying the connection or causing loss of data, referred to as packet drop. The common types of deception attacks are false data injection attacks, stealthy attacks, replay attacks, and covert attacks. Stealthy attacks secretly change the data without being detected; whereas, covert attacks secretly change the behavior of the cyber physical system. Replay attacks record the readings, then change it while replaying the reading to the receiver. For more background and results we refer the reader to Foroush and Martínez (2013), and Lu and Yang (2018) for DoS attacks, Mo, Garone, Casavola, and Sinopoli (2010), and Mo and Sinopoli (2010) for false data injection attacks, to Teixeira, Shames, Sandberg, and Johansson (2012) for stealthy attacks, to Mo and Sinopoli (2009), and Teixeira, Pérez, Sandberg and Johansson (2012) for replay attacks and to Hoehn and Zhang (2016), and Sá, Carmo, and Machado (2017) for covert attacks. For a general survey on these attacks we recommend Din, Han, Xiang, Ge, and Zhang (2018).

We first discuss some results regarding the stability of cyber systems under attacks. Persis and Tesi (2015) study the cyber system under Denial of Service (DoS) attack given by,

$$x'(t) = Ax(t) + Bu(t) + w(t), \qquad (5.19)$$
$$u(t) = Kx(t_{k(t)}),$$

where for $m, n \in \mathbb{N}$, $x \in \mathbb{R}^n$ is the state of the system, $u \in \mathbb{R}^m$ is the control input, and $w \in \mathbb{R}^n$ represents the effects that cannot be controlled in the system. A, B, K are matrices with constant entries, where matrix K is formed such that all eigenvalues of $\Phi := A + BK$ have negative real part. The times at which the control action is set to be updated are denoted by $\{t_k\}_{k \in \mathbb{N}_0}$, where $t_0 = 0$. When the system is working properly, the data is sent and received at any time $t \in \mathbb{R}^+$. Namely, for any $t \in [t_k, t_{k+1})$,

$$u_{\text{ideal}}(t) = Kx(t_k). \qquad (5.20)$$

The DoS attack jams the communication or drops the data packets, so that (5.20) does not hold true for every time t and instead we consider the control input $u(t)$, given in system (5.19), where for each $t \in \mathbb{R}^+$,

$$k(t) = \begin{cases} -1 & \Theta(0, t) = \emptyset, \\ \sup\{k \in \mathbb{N}_0 : t_k \in \Theta(0, t)\} & \text{otherwise,} \end{cases}$$

denotes the last control update that went through successfully and the following notation is applied,

$$H_n := \{h_n\} \cup [h_n, h_n + \tau_n),$$
$$\Xi(\tau, t) := \bigcup_{n \in \mathbb{N}_0} H_n \cap [\tau, t],$$
$$\Theta(\tau, t) := [\tau, t] \setminus \Xi(\tau, t).$$

The set $\{h_n\}_{n \in \mathbb{N}_0}$ with $h_0 \geq 0$ represents the times when the DoS attack switches from zero to one, where zero means no attack has occurred and one indicates that the DoS attack has blocked the communication. This implies that H_n is the nth DoS time interval of length $\tau_n \in \mathbb{R}^+$, in which there can be no communication. In the case that $\tau_n = 0$, the DoS attack will be at the instant pulse time, h_n. Furthermore, for the given $\tau, t \in \mathbb{R}^+$, with $t \geq \tau$, since H_n gives the time interval where communication is not possible, in the time interval, $[\tau, t]$, $\Xi(\tau, t)$ is the set of times in which communication cannot take place and $\Theta(\tau, t)$ gives the set of times in which there is no interruption or damage made to the communication.

For the results, the authors first write the system in terms of the error between values of the state of the process at the last control update and at the current time. That is,
$$e(t) := x(t_{k(t)}) - x(t), \qquad (5.21)$$
for $t \in \mathbb{R}^+$. Note that to ensure the validity of the results, $e(t)$ needs to be as small as possible. Then, using (5.19),

$$\begin{aligned} x'(t) &= Ax(t) + Bu(t) + w(t) \\ &= Ax(t) + B(Kx(t_{k(t)})) + BKx(t) - BKx(t) + w(t) \\ &= Ax(t) + BKx(t) + BK(x(t_{k(t)}) - x(t)) + w(t), \end{aligned}$$

hence system (5.19) becomes,
$$\begin{aligned} x'(t) &= \Phi x(t) + BKe(t) + w(t), \\ u(t) &= Kx(t_{k(t)}), \end{aligned} \qquad (5.22)$$

where all the eigenvalues of $\Phi := A + BK$ are again assumed to have negative real part. Next in the theorem below, they determine conditions under which system (5.19) is input-to-state stable (see 5.1.7 in the Appendix for a definition). In the theorem and its proof, $\|\cdot\|$ indicates the Euclidean norm.

Theorem 5.1 (Theorem 1 in Persis and Tesi (2015)). *Suppose,*
$$\|e(t)\| \leq \sigma \|x(t)\| + \sigma \sup_{s \in [0,t]} \|w(s)\|, \qquad (5.23)$$

holds for $\sigma \in \mathbb{R}^+$, being the design parameter satisfying, $\gamma_1 - \sigma\gamma_2 > 0$, where γ_1 is the smallest eigenvalue of Q and $\gamma_2 := \|2PBK\|$. Also assume that there exists $\underline{\Delta} \in \mathbb{R}^+$, such that for all $k \in \mathbb{N}_0$, $t_{k+1} - t_k \geq \underline{\Delta}$. Then, system (5.19) is input-to-state stable.

Proof: Let $Q \in \mathbb{R}^{n \times n}$ be a positive definite matrix, P be the unique solution to the Lyapunov equation,
$$\Phi^T P + P\Phi + Q = 0,$$
and $V(x) = x^T P x$ be a Lyapunov function. Using system (5.22), the authors find that for all $t \in \mathbb{R}^+$,

$$\alpha_1 \|x(t)\|^2 \leq V(x(t)) \leq \alpha_2 \|x(t)\|^2, \qquad (5.24)$$
$$\frac{d}{dt}V(x(t)) \leq -\gamma_1 \|x(t)\|^2 + \gamma_2 \|x(t)\| \|e(t)\| + \gamma_3 \|x(t)\| \|w(t)\|, \qquad (5.25)$$

where α_1 and α_2 are the smallest and largest eigenvalues of P, respectively and $\gamma_3 := \|2P\|$. Note that by (5.23) and (5.25) we may write,

$$\frac{d}{dt}V(x(t)) \leq -\gamma_1 \|x(t)\|^2 + \gamma_2 \|x(t)\| \left(\sigma\|x(t)\| + \sigma \sup_{s \in [0,t)} \|w(s)\|\right)$$
$$+ \gamma_3 \|x(t)\| \|w(t)\|$$
$$\leq -\gamma_4 \|x(t)\|^2 + \gamma_5 \|x(t)\| v(t)$$
$$\leq -\gamma_4 \|x(t)\|^2 + \frac{\gamma_4}{2}\|x(t)\|^2 + \frac{\gamma_5^2}{2\gamma_4}v(t)^2$$
$$= -\frac{\gamma_4}{2}\|x(t)\|^2 + \gamma_6 v(t)^2,$$

where, the Young's inequality was applied to obtain the third inequality and $v(t) := \sup\{\|w(t)\|, \sup_{s \in [0,t)} \|w(s)\|\}$, $\gamma_4 := (\gamma_1 - \sigma\gamma_2)$, $\gamma_5 := (\gamma_3 + \sigma\gamma_2)$, $\gamma_6 := \gamma_5^2/(2\gamma_4)$. Now using (5.24), we arrive at,

$$\frac{d}{dt}V(x(t)) \leq -\frac{\gamma_4}{2\sigma_2}V(x(t)) + \gamma_6 v(t)^2.$$

Noting that $\sup_{s \in [0,t)} \|v(s)\| = \sup_{s \in [0,t)} \|w(s)\|$ and letting $w_1 = \gamma_4/(2\sigma_2)$, we have,

$$\frac{d}{dt}V(x(t)) \leq -w_1 V(x(t)) + \gamma_6 \sup_{s \in [0,t)} \|w(s)\|^2,$$

which is a linear differential equation and can be solved to obtain,

$$V(x(t)) \leq e^{-w_1 t}V(x(0)) + \frac{\gamma_6}{w_1} \sup_{s \in [0,t)} \|w(s)\|^2.$$

Therefore, by (5.24),

$$\alpha_1 \|x(t)\|^2 \leq e^{-w_1 t}\alpha_2 \|x(0)\|^2 + \frac{\gamma_6}{w_1} \sup_{s \in [0,t)} \|w(s)\|^2.$$

We now use the fact that $a^2 + b^2 \leq (a+b)^2$ with $a = \sqrt{\alpha_2/\alpha_1}e^{\frac{-w_1 t}{2}}\|x(0)\|$ and $b = \sqrt{\gamma_6/w_1}\sup_{s \in [0,t)}\|w(s)\|$, to arrive at,

$$\|x(t)\| \leq \left(\sqrt{\frac{\alpha_2}{\alpha_1}}e^{-\frac{w_1}{2}t}\|x(0)\| + \sqrt{\frac{\gamma_6}{w_1 \alpha_1}} \sup_{s \in [0,t)} \|w(s)\|\right).$$

Since for each fixed $t \in \mathbb{R}^+$, $\beta(\|x(0)\|, t) := \sqrt{\alpha_2/\alpha_1}e^{-(w_1 t)/2}\|x(0)\|$ and $\gamma(\sup_{s \in [0,t)} \|w(s)\|) := \sqrt{\gamma_6/(w_1 \alpha_1)}\sup_{s \in [0,t)}\|w(s)\|$ are continuous, strictly increasing and unbounded as functions of $\|x(0)\|$ and $\sup_{s \in [0,t)} \|w(s)\|$, respectively, and $\lim_{t \to \infty} \beta(\|x(0)\|, t) = 0$, then we have the input-to-state stability of the system.

□

For similar results on input-to-state stability of a cyber-physical model under DoS attacks, see for instance, Lu and Yang (2018). We now present some results in Pasqualetti et al. (2011) and Pasqualetti et al. (2013) on the detectability and identifiability of cyber attacks and conditions under which cyber attacks are undetected.

In Pasqualetti et al. (2013), the authors model the cyber physical system under attack as follows,

$$Ex'(t) = Ax(t) + Bu(t), \qquad (5.26)$$
$$y(t) = Cx(t) + Du(t),$$

where for some $n, m, p \in \mathbb{N}$ with $m = n + p$, $x(t) \in \mathbb{R}^n$ is the state of the system, and $y(t) \in \mathbb{R}^p$. A, B, C, D, E are matrices with constant entries such that $(Ax(0) + Bu(0)) \in Im(E)$, $det(sE - A)$ is not zero for any $s \in \mathcal{C}$, and E may be singular. Also we assume, $A \in \mathbb{R}^{n \times n}, B \in \mathbb{R}^{n \times m}, C \in \mathbb{R}^{p \times n}, D \in \mathbb{R}^{p \times m}, E \in \mathbb{R}^{n \times n}$. The attack signal is given by $t \mapsto u(t) \in \mathbb{R}^m$ and terms Bu and Du represent signals that cause disturbance in the system and are unknown. We suppose that there is a fixed attack set, $K \subset \{1, ..., m\}$ with $|K| = k$, where $|\cdot|$ denotes the cardinality of the set. Notation u_K is used to denote u with attack set K. Namely, for every $i \in K$, there is a $t \in \mathbb{R}^+$ such that $u_i(t) \neq 0$ and for every $j \notin K$ and $u_j(t) = 0$ for all time t. Moreover, let B_K and D_K be submatrices of B and D, respectively with columns in K, that is, $Bu(t) = B_K u_K(t)$ and $Du(t) = D_K u_K(t)$, with the pair, (B_K, D_K) being referred to as the attack signature.

To detect and identify an attack, a monitor in the form of an algorithm is placed to oversee the system. With the given initial state, x_0 and attack input, u_K, we let $y(x_0, y_k, t)$ represent the output signal of system (5.26). Matrices A, C, E are assumed to be given and the input of the monitor is $\Lambda = \{E, A, C, y(x_0, u_K, t) \, \forall t \in \mathbb{R}^+\}$. The output of the monitor is $\Psi(\Lambda) = \{\Psi_1(\Lambda), \Psi_2(\Lambda)\}$, where $\Psi_1(\Lambda) \in \{\text{True}, \text{False}\}$, indicates if an attack has occurred (referred to as detection) and $\Psi_2(\Lambda) \subseteq \{1, ..., m\}$, gives what kinds of attacks have occurred (referred to as identification). Hence, the attack $(B_K u_K, D_K u_K)$ is detected by the monitor Φ, if $\Psi_1(\Lambda) = \text{True}$ and is identified by the monitor, if $\Psi_2(\Lambda) = K$. Based on these properties of the monitor, we have the following lemma.

Lemma 5.1 (Lemma 3.1 in Pasqualetti et al. (2013)). *For system (5.26), the nonzero attack, $(B_K u_K, D_K u_K)$ is undetectable if and only if $y(x_1, u_K, t) = y(x_2, 0, t)$ for some initial states $x_1, x_2 \in \mathbb{R}^n$ and for all $t \in \mathbb{R}^+$.*

Proof: $\Rightarrow$ Suppose to the contrary that $(B_K u_K, D_K u_K)$ is undetectable and $y(x_1, u_K, t) \neq y(x_2, 0, t)$ for all initial states x_1, x_2. Then the outputs $\Phi(E, A, C, y(x_1, u_K, t)) \neq \Phi(E, A, C, y(x_2, 0, t))$. Since the output of $\Phi(E, A, C, y(x_2, 0, t))$, which has $u_K = 0$, is $\Psi_1 = \text{False}$, then the output of $\Phi(E, A, C, y(x_1, u_K, t))$ is $\Psi_1 = \text{True}$ indicating that the attack is detectable and leading to a contradiction.

$\Leftarrow$ Suppose for some initial states, $x_1, x_2 \in \mathbb{R}^n$ and for all $t \in \mathbb{R}^+$, $y(x_1, u_K, t) = y(x_2, 0, t)$. Then, $\Phi(E, A, C, y(x_1, u_K, t)) = \Phi(E, A, C, y(x_2, 0, t))$, and thus, $\Phi(E, A, C, y(x_1, u_K, t))$ will give Ψ_1 = False and the attack will be undetectable.

□

Analogous to the above proof, the following lemma offers a condition that ensures that the attack is unidentifiable.

Lemma 5.2 (Lemma 3.2 in Pasqualetti et al. (2013)). *For system (5.26), the nonzero attack, $(B_K u_K, D_K u_K)$ is unidentifiable if and only if $y(x_1, u_K, t) = y(x_2, u_R, t)$ for some initial states, $x_1, x_2 \in \mathbb{R}^n$ and all time $t \in \mathbb{R}^+$ and some attack $(B_R u_R, D_R u_R)$ with $|R| \leq |K|$ and $R \neq K$.*

In Pasqualetti et al. (2011), the same authors as above, consider a power network model similar to Liu et al. (2011), that is used to model a IEEE 14 bus system, where here they use a static or dynamic detector. In power networks, a bus is referred to the place where power can be injected or transmitted from the electric circuit. There are three types of buses: Load bus, Slack bus and Generator bus. A Load bus is where active and reactive powers are injected to the network, a Slack bus is where active or reactive powers are emitted or absorbed from the system, and the Generator bus is where a power generator is connected to the network. For some $n, m \in \mathbb{N}$, the generator's rotor angle and frequency are given by $\delta \in \mathbb{R}^n$ and $\omega \in \mathbb{R}^n$, respectively, and the bus voltage angle is denoted by $\theta \in \mathbb{R}^m$. It is assumed that there are n generators, $\{g_1, ..., g_n\}$, with n associated generator terminal buses, $\{b_1, ..., b_n\}$ and m Load buses, $\{b_{n+1}, ..., b_{n+m}\}$. Then, the network in the presence of a cyber attack is given by (5.26) with,

$$E = \begin{bmatrix} I & 0 & 0 \\ 0 & M & 0 \\ 0 & 0 & 0 \end{bmatrix}, \quad A = -\begin{bmatrix} 0 & -I & 0 \\ \mathcal{L}_{gg} & D_g & \mathcal{L}_{g1} \\ \mathcal{L}_{1g} & 0 & \mathcal{L}_{11} \end{bmatrix}, \quad B = \begin{bmatrix} F & 0 \end{bmatrix}, \quad D = \begin{bmatrix} 0 & L \end{bmatrix},$$

$$u(t) = \begin{bmatrix} f(t) \\ \ell(t) \end{bmatrix}, \quad x = \begin{bmatrix} \delta \\ \omega \\ \theta \end{bmatrix} \in \mathbb{R}^{2n+m},$$

$$C = \begin{bmatrix} C_\delta \\ C_\omega \\ C_\theta \end{bmatrix} \in \mathbb{R}^{p \times n}, \quad L = \begin{bmatrix} L_\delta \\ L_\omega \\ L_\theta \end{bmatrix} \in \mathbb{R}^{p \times p}, \quad F = \begin{bmatrix} F_\delta \\ F_\omega \\ F_\theta \end{bmatrix} \in \mathbb{R}^{(2n+m) \times (2n+m)}.$$

Letting the generators and buses present vertices of a graph and the transmission lines given by, $\{b_i, b_j\}$ to be the edges, we obtain the weighted graph given by the symmetric matrix,

$$\begin{bmatrix} \mathcal{L}_{gg} & \mathcal{L}_{g1} \\ \mathcal{L}_{1g} & \mathcal{L}_{11} \end{bmatrix} \in \mathbb{R}^{(2n+m) \times (2n+m)},$$

where the first n entries correspond to the generators and the last $n + m$ entries correspond to the buses. Matrix M in E and matrix D_g in A are diagonal matrices representing the generators' inertia and damping constants, respectively and matrix

C is the output matrix. The attack set is $K \subset \{1, ..., 2n+m+p\}$ and $t \mapsto u(t) \in \mathbb{R}^{2n+m+p}$ is the attack mode with $f(t) \in \mathbb{R}^{2n+m}$ being the state attack mode and $\ell(t) \in \mathbb{R}^p$ being the output attack mode, where both $f(t)$ and $\ell(t)$ are piecewise continuous functions and are assumed to be unknown. The authors verify that the detectability and identifiability of system (5.26) is the same as those for the following system, referred to as the Korn-reduced system,

$$x'(t) = \tilde{A}x(t) + \tilde{B}u(t), \tag{5.27}$$
$$y(t) = \tilde{C}x(t) + \tilde{D}u(t), \tag{5.28}$$

where,

$$\tilde{A} = \begin{bmatrix} 0 & I \\ -M^{-1}(\mathcal{L}_{gg} - \mathcal{L}_{g1}\mathcal{L}_{11}^{-1}\mathcal{L}_{1g}) & -M^{-1}D_g \end{bmatrix},$$

$$\tilde{B} = \begin{bmatrix} F_\delta & 0 \\ M^{-1}F_\omega - M^{-1}\mathcal{L}_{g1}\mathcal{L}_{11}^{-1}F_\theta & 0 \end{bmatrix}, \tilde{C} = \begin{bmatrix} C_\delta - C_\theta \mathcal{L}_{11}^{-1}\mathcal{L}_{1g} & C_\omega \end{bmatrix},$$

$$\tilde{D} = \begin{bmatrix} -C_\theta \mathcal{L}_{11}^{-1}F_\theta & L \end{bmatrix}, x(t) = \begin{bmatrix} \delta(t) \\ \omega(t) \end{bmatrix},$$

To detect and identify cyber attacks, there are two kinds of monitors that are widely used. A static detector, also referred to as Bad Data Detector, is an algorithm that checks for attacks at specific discrete times by comparing measurements at these times with the typical measurements. A dynamic detector, on the other hand, monitors the measurements based on a continuous time signal $y(t)$, $t \in \mathbb{R}^+$, that is checking at every time instant t. The following results offer conditions on identifiability and detectability of attacks by each type of detector.

Theorem 5.2 (Theorem 3.2 in Pasqualetti et al. (2011)). *For system (5.26) and attack set K, the following are equivalent,*
i. the attack set K is undetectable by a static detector,
ii. there exists an attack mode $u_K(t)$ such that for some $\delta(t), \omega(t)$ and every $t \in \mathbb{N} \cup \{0\}$,

$$\tilde{C}\begin{bmatrix} \delta(t) \\ \omega(t) \end{bmatrix} + \tilde{D}u_K(t) = 0. \tag{5.29}$$

Proof: Denoting $\tilde{C}^\dagger y(t) = \begin{bmatrix} \hat{\delta}(t) \\ \hat{\omega}(t) \end{bmatrix}$, it is known (proved in Liu et al. (2011)) that an attack is detected by a static detector if and only if $r(t) = y(t) - \tilde{C}\begin{bmatrix} \hat{\delta}(t) \\ \hat{\omega}(t) \end{bmatrix}$, called the residual, is nonzero for some $t \in \mathbb{N}$, where $\tilde{C}^\dagger$ represents the conjugate transpose of matrix $\tilde{C}$. Here, we have for all $t \in \mathbb{N}$, and some $\delta(t), \omega(t), u_K(t)$,

$$r(t) = y(t) - \tilde{C}\begin{bmatrix} \hat{\delta}(t) \\ \hat{\omega}(t) \end{bmatrix} = y(t) - \tilde{C}\tilde{C}^\dagger y(t)$$

$$= (I - \tilde{C}\tilde{C}^\dagger)y(t) = (I - \tilde{C}\tilde{C}^\dagger)\left(\tilde{C}\begin{bmatrix} \delta(t) \\ \omega(t) \end{bmatrix} + \tilde{D}u_K(t)\right).$$

Thus, for all $t \in \mathbb{N}$, the attack set K is undetectable if and only if equation (5.29) holds.

□

Furthermore, the following result is proved for the identifiability of the attack set by a static detector.

Theorem 5.3 (Theorem 3.3 in Pasqualetti et al. (2011)). *For system (5.26) and attack set K, the following are equivalent,*
i. the attack set K is unidentifiable by a static detector,
ii. there exists an attack set R, with $|R| \leq |K|$ and $R \neq K$ and attack modes $u_K(t), u_R(t)$ such that at every $t \in \mathbb{N} \cup \{0\}$, and some $\delta(t), \omega(t)$,

$$\tilde{C}\begin{bmatrix}\delta(t)\\ \omega(t)\end{bmatrix} + \tilde{D}(u_K(t) - u_R(t)) = 0. \quad (5.30)$$

Proof: Recall from Lemma 5.2 that for an attack set K to be unidentifiable, one needs the condition, $y(x_K, u_K, t) = y(x_R, u_R, t)$ to hold for all $t \in \mathbb{R}^+$ and some attack modes, $u_K(t), u_R(t)$ and initial conditions, $x_K, x_R \in \mathbb{R}^{2n+m}$. Since system (5.26) is linear, this condition is the same as, $y(x_K - x_R, u_K - u_R, t) = 0$, which is exactly equation (5.30) and hence we obtain the equivalence relation.

□

Similarly, as for dynamic detectors, the following theorems are achieved.

Theorem 5.4 (Theorem 3.4 in Pasqualetti et al. (2011)). *For system (5.26) and attack set K, the following are equivalent,*
i. attack set K is undetectable by a dynamic detector,
ii. there exists an attack mode $u_K(t)$, such that for some $\delta(0), \omega(0)$, and for all $t \in \mathbb{R}^+ \cup \{0\}$,

$$\tilde{C}e^{\tilde{A}t}\begin{bmatrix}\delta(0)\\ \omega(0)\end{bmatrix} + \tilde{C}\int_0^t e^{\tilde{A}(t-\tau)}\tilde{B}u_K(\tau)d\tau = -\tilde{D}u_K(t). \quad (5.31)$$

Proof: Note that (5.27) is a linear differential equation and its solution is given by,

$$x(t) = x(0)e^{\tilde{A}t} + \int_0^t e^{\tilde{A}(t-\tau)}\tilde{B}u_K(\tau)d\tau, \quad (5.32)$$

where, $x(0) = \begin{bmatrix}\delta(0)\\ \omega(0)\end{bmatrix}$. According to Lemma 5.1, for the set K to be undetectable, we need $y(\bar{x}(0), u_K, t) = y(\tilde{x}(0), 0, t)$ for some initial states, $\bar{x}(0) := \begin{bmatrix}\bar{\delta}(0)\\ \bar{\omega}(0)\end{bmatrix}, \tilde{x}(0) :=$

$\begin{bmatrix} \tilde{\delta}(0) \\ \tilde{\omega}(0) \end{bmatrix}$, which by linearity is, $y(\bar{x}(0) - \tilde{x}(0), u_K, t) = 0$. Using (5.28) and (5.32), this condition is equivalent to (5.31).

□

With the same reasoning as in the proof of Theorem 5.3, the following result was shown by applying the linear property of $y(t)$.

Theorem 5.5 (Theorem 3.5 in Pasqualetti et al. (2011)). *For system (5.26) and attack set K, the following are equivalent,*
i. attack set K is unidentifiable by a dynamic detector,
ii. there exists an attack set R, with $|R| \leq |K|$ and $R \neq K$, such that for some $u_K(t), u_R(t)$ and $\delta(0), \omega(0)$ and for all $t \in \mathbb{R}^+$,

$$\tilde{C}e^{\tilde{A}t}\begin{bmatrix} \delta(0) \\ \omega(0) \end{bmatrix} + \tilde{C}\int_0^t e^{\tilde{A}(t-\tau)}\tilde{B}(u_K(\tau) - u_R(\tau))d\tau = -\tilde{D}(u_K(t) - u_R(t)). \quad (5.32)$$

□

For more results on attack detection, we refer the reader to Chen, Kar, and Moura (2017), Chen et al. (2018), Hoehn and Zhang (2016), Teixeira, Pérez, Sandberg, and Johansson (2012), and Teixeira, Shames, Sandberg, and Johansson (2012).

5.5 CONCLUSION

As our reliance on the internet to complete transactions increases, the measures required to protect information and guard computer networks from attacks need to be enhanced. In this chapter, by explaining results currently in the literature, we intended to offer some insight to the reader in the area of research regarding cyber security and help him/her gain a better understanding of the various differential equations models that have been studied. Two types of models were described that have been commonly used in the research community to control and combat cyber attacks and determine information on the spread of computer viruses. We concentrated on models that are based on differential equations and here we note that there are similar articles that use models in terms of stochastic differential equations, in which a term representing the noise in the system is added. "Noise" refers to anything that has an affect on the system but has not been accounted for in the equations of the model. For example, Tornatore, Buccellato and Vetro (2005) consider the stochastic Susceptible-Infected-Recovered (SIR) model without time delay as follows,

$$dS(t) = \big(-\beta S(t)I(t) - \mu S(t) + \mu\big)dt - \sigma S(t)I(t)dW(t),$$
$$dI(t) = \big(\beta S(t)I(t) - (\lambda + \mu)I(t)\big)dt + \sigma S(t)I(t)dW(t),$$
$$dR(t) = (\lambda I(t) - \mu R(t))dt,$$

where the term $\sigma S(t)I(t)dW(t)$ represents the noise. Other similar examples include, Liu, Chen, and Jiang (2016), and Chen and Li (2009).

The stochastic version of the systems modeling cyber attacks have also been studied in the literature. See for example, Chen et al. (2018), Miao, Zhu, Pajic, and Pappas (2014) and Mo et al. (2010). Typically, the following system,

$$x_{k+1} = Ax_k + Bu_k + w_k,$$
$$y_k = Cx_k + v_k,$$

is considered, where w_k is the process noise and v_k is the sensor noise. For background information on stochastic differential equations we refer the reader to Klebaner (2005) and Kuo (2006).

APPENDIX 5.1

5.1.1. Equilibrium points and Stability (see Sections 8.3 and 9.7 in Polking et al. (2018) or Section 9.2 in Boyce and DiPrima (2009)).

For the system,

$$x' = f(x, y),$$
$$y' = g(x, y),$$

equilibrium points are where both $f(x,y)$ and $g(x,y)$ are equal to zero. If (x_0, y_0) is an equilibrium point, then $x(t) = x_0, y(t) = y_0$ is an equilibrium solution. We say that an equilibrium point is locally stable if any solution starting near the equilibrium point, will always stay in an arbitrary small neighborhood of the point. More precisely, in the mathematical language, the equilibrium point y_0 is locally stable if for every $\varepsilon > 0$, there is a $\delta > 0$, such that if $y(t)$ is a solution with $|y(0) - y_0| < \delta$, then $|y(t) - y_0| < \varepsilon$ for all $t > 0$. The equilibrium point is locally asymptotically stable if every solution starting near the equilibrium point will eventually approach the equilibrium point as time is set to go to infinity. That is, the equilibrium point is stable and there is a $\beta > 0$, such that each solution, $y(t)$ with $|y(0) - y_0| < \beta$, approaches to y_0 as $t \to \infty$. An equilibrium point is globally stable if every solution converges to y_0 regardless of where it starts in the domain. Furthermore, it is unstable if it is not stable and moves away from the equilibrium point.

5.1.2. Phase Portrait (see Section 9.1 in Boyce and DiPrima (2009) or Section 9.3 in Polking et al. (2018)).

For the system, $\vec{y}' = A\vec{y}$, one may determine the shape of solutions near the origin by determining the eigenvalues and eigenvectors of matrix A. In the case of a 2×2 matrix A, if both eigenvalues are real valued with opposite signs then the solution curves form a shape called a saddle. If both eigenvalues are positive and real, then the origin is called a Nodal Source and if both are negative and real then it is called a Nodal Sink. In the case of complex eigenvalues, if the eigenvalues have no real part, then the phase portrait is called a Center. If there is a real part and it is positive, it is a Spiral Source and if the real part is negative, the origin is a Spiral Sink. A phase portrait example of a Saddle, Nodal Sink, Center and Spiral Sink are given below in Figures 5.1-5.4, respectively. The phase portraits for Nodal Source and Spiral Source have the same structure as the Nodal Sink and Spiral Sink, respectively, with the difference of having arrows pointing away from the origin. As for stability, the Saddle, Nodal Source and Spiral Source are unstable, since they go to infinity as t tends to infinity and the Nodal Sink along with the Spiral Sink are asymptotically stable. Moreover, the Center is stable.

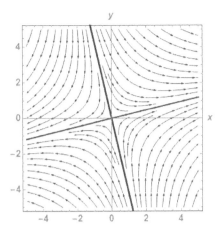

FIGURE 5.1 Saddle.

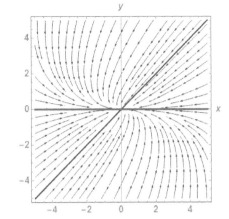

FIGURE 5.2 Nodal Sink.

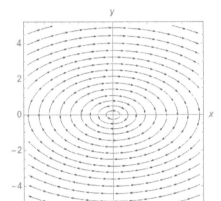

FIGURE 5.3 Center.

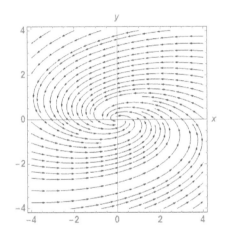

FIGURE 5.4 Spiral Sink.

5.1.3. Trace-Determinant Diagram (see Section 9.4 in Polking et al. (2018)).

The shape of the phase portrait near the origin for the system, $\vec{y}' = A\vec{y}$, may also be determined by the use of the TD diagram. Let $T = \text{trace}(A), D = \det(A)$, then along the same lines as the method explained in 5.1.2 of the Appendix, we have,

$$\det \begin{bmatrix} a_{11} - \lambda & a_{12} \\ a_{21} & a_{22} - \lambda \end{bmatrix} = \lambda^2 - T\lambda + D = 0, \tag{5.33}$$

giving $\lambda = \frac{T \pm \sqrt{T^2 - 4D}}{2}$. Thus, if $T^2 - 4D > 0$ then we have two real eigenvalues, if $T^2 - 4D = 0$ we have a repeated eigenvalue, and $T^2 - 4D < 0$ indicates complex eigenvalues.

To determine the specific sign of the eigenvalues, note that from the characteristic equation,

$$(\lambda - \lambda_1)(\lambda - \lambda_2) = 0,$$

which yields,

$$\lambda^2 - (\lambda_1 + \lambda_2)\lambda + \lambda_1 \lambda_2 = 0.$$

Hence, compared with equation (5.33), we obtain, $T = \lambda_1 + \lambda_2$ and $D = \lambda_1 \lambda_2$ and thus knowing the signs of T and D, are sufficient to determining the signs of the eigenvalues.

5.1.4. Phase Portrait by Jacobian Matrix (see Section 10.1 in Polking et al. (2018)). For the system,

$$x_1' = f_1(x_1, x_2, ..., x_n),$$
$$x_2' = f_2(x_1, x_2, ..., x_n),$$
$$\vdots$$
$$x_n' = f_n(x_1, x_2, ..., x_n),$$

the Jacobian matrix is defined as,

$$\begin{bmatrix} \frac{\partial f_1}{\partial x_1} & \frac{\partial f_1}{\partial x_2} & \cdots & \frac{\partial f_1}{\partial x_n} \\ \frac{\partial f_2}{\partial x_1} & \frac{\partial f_2}{\partial x_2} & \cdots & \frac{\partial f_2}{\partial x_n} \\ \vdots & \vdots & \vdots & \vdots \\ \frac{\partial f_n}{\partial x_1} & \frac{\partial f_n}{\partial x_2} & \cdots & \frac{\partial f_n}{\partial x_n} \end{bmatrix}.$$

Let $J(x_0, y_0)$ be the Jacobian matrix at the equilibrium point, (x_0, y_0), with $x = x_0, y = y_0$. Then to obtain the phase portrait at (x_0, y_0), we may apply the method offered by the TD diagram, described in 5.1.2 of the Appendix using, $J(x_0, y_0)$ instead of the A matrix in system, $\vec{x}' = A\vec{x}$.

5.1.5. La Salle's Invariance Principle (See La Salle (1976)).

A function $V(x)$ is positive definite if $V(x) > 0$ for all $x \neq x_0$ and $V(x) = 0$ for $x = x_0$. $V(x)$ is radially unbounded if $V(x) \to \infty$ as $\|x\| \to \infty$. A set Ω is called an invariant set, if $x(t_0) \in \Omega$ implies $x(t) \in \Omega$, for all $t \geq t_0$. Consider the nonlinear equation, $x' = f(x)$ and let $V(x)$ be a continuously differentiable positive definite function, such that $V'(x) \leq 0$ on $\Gamma = \{x : V(x) \leq c\}$ for some constant $c > 0$ and let M be the largest invariant set in $\{x : V'(x) = 0\} \cap \Gamma$. If $M = \{x_0\}$, then the equilibrium point, x_0 is asymptotically stable. If in addition, $V(x)$ is radially unbounded and the above conditions hold for any constant $c > 0$, then x_0 is globally asymptotically stable. Therefore, to prove that a point, x_0 is globally asymptotically stable by La Salle's Invariance Principle, one needs to find a Lyapunov function, $V(x)$ that is continuously differentiable and positive definite and for all x in the domain, $V'(x) \leq 0$ with $V'(x) = 0$ only when $x(t) = x_0$. For more information on this method, we refer the reader to Sections 2.2 and 2.4 in Kamaleddin and Nikravesh (2013), Section 10.7 in Polking et al. (2018) and Section 9.6 in Boyce, DiPrima (2009). Also for the definition of invariant set see Section 1.8 in Hale (1969).

5.1.6. Hurwitz Condition (see Bellman and Kalaba (1964)). The equation,

$$a_0 x^n + a_1 x^{n-1} + ... + a_n = 0,$$

with a_0 positive and all coefficients being real, has only roots with negative real parts if and only if the value of a_1 and all the determinants,

$$\Delta_2, \Delta_3, ..., \Delta_n,$$

are positive, where,

$$\Delta_\lambda = \begin{vmatrix} a_1 & a_3 & a_5 & \cdots & a_{2\lambda-1} \\ a_0 & a_2 & a_4 & \cdots & a_{2\lambda-2} \\ 0 & a_1 & a_3 & \cdots & a_{2\lambda-3} \\ \cdot & \cdot & \cdot & \cdots & \cdot \\ \cdot & \cdot & \cdot & \cdots & a_\lambda \end{vmatrix}.$$

5.1.7. Input-to-State Stability (see Sontag (2008)). A K_∞ function, $\alpha : \mathbb{R}^+ \to \mathbb{R}^+$, is a function that is continuous, strictly increasing, and unbounded with $\alpha(0) = 0$. A KL function, $\beta(r,t) : \mathbb{R}^+ \times \mathbb{R}^+ \to \mathbb{R}^+$, is a K_∞ function with respect to r and each fixed $t \in \mathbb{R}^+$ and satisfies $\lim_{t \to \infty} \beta(r,t) = 0$. Consider the following general form of system in control theory,

$$x'(t) = f(x(t), u(t)),$$
$$y(t) = h(x(t)),$$

where $x(t) \in \mathbb{R}^n$ is the state of the system, $u(\cdot) : [0, \infty) \to \mathbb{R}^m$ is the input and $y(t) \in \mathbb{R}^p$ is the output. Then the system has input-to-state stability (ISS), if there exists a K_∞ function γ and a KL function β such that,

$$\|x(t)\| \leq \beta(\|x(0)\|, t) + \gamma(\|u(t)\|_\infty), \tag{5.34}$$

for all solutions, where $\|u(t)\|_\infty = \operatorname{ess\,sup}_{s \in [0,t)} \|u(s)\|$. Namely, for each $t \in \mathbb{R}^+$, equation (5.34) needs to hold for all initial conditions $x(0)$ and inputs $u(\cdot)$.

5.1.8. Notation for Examples 1b, 2, 3 in Castillo-Chavez et al. (2002).
Λ: Birth Rate,
μ: Death Rate,
d_i: Death Rate per capita due to disease by Strain i,
$r, \tilde{r}$: Treatment Rates per capita,
$\sigma \leq 1$: Reduction in Infection Rate because of Treatment,
p: Proportion of Individuals who completed the Treatment,
q: Proportion of Individuals who did not complete the Treatment.
β: Transition Rate from Susceptible to Infected,
γ: Transition Rate from Infected to Susceptible,
k: Transition Rate from Exposed to Infected,

5.1.9. Notation for Li (2018).
b: Birth and Death Rates,

β: Transmission Rate,
γ: Fraction of Infected that Recovers per unit time,
γI: Transition Rate from Infected to Recovered.

5.1.10. Notation for Gan et al. (2012).
δ: Rate at which external Susceptible computers get connected to the internet; the same rate is used for internal computers becoming disconnected from the internet,
βSI: Transition Rate from Susceptible to Infected, where $\beta > 0$,
$\alpha_1 I$: Transition Rate from Susceptible to Recovered,
α_2: Transition Rate from Recovered to Susceptible,
γ_1: Transition Rate from Infected to Recovered,
γ_2: Transition Rate from Infected to Susceptible.

5.1.11. Notation for Toutonji et al. (2012).
β: Contact Rate,
μ: Replacement Rate,
θ: Dysfunctional Rate,
α: Transition Rate from Exposed to Infected,
ψ_1: Transition Rate from Vulnerable to Secured,
ψ_2: Transition Rate from Exposed to Secured,
γ: Transition Rate from Infected to Secured,
ϕ: Transition Rate from Secured to Vulnerable.

5.1.12. Notation for Ren et al. (2012).
d: Rate at which External Computers Join the network,
r: Intrinsic Growth Rate of Susceptible computers,
k: Carrying Capacity of Logistic Growth of Susceptible computers,
ε: Recovery Rate,
λ: Transition Rate from Susceptible to Infected once connected to an Infected.

5.1.13. Notation for Xiao et al. (2017).
μ: Both the Rate at which Susceptible computers join the network and the Crash Rate of the computers,
β: Transition Rate from Susceptible to Exposed,
η: Transition Rate from Exposed to Infected,
ε: Transition Rate from Exposed to Recovered,
ξ: Transition Rate from Infected to Quarantined,
φ: Transition Rate from Quarantined to Recovered,
γ: Transition Rate from Infected to Recovered.

REFERENCES

Amin, S., Litrico, X., Sastry, S., & Bayen, A. (2012). Cyber security of water SCADA systems-Part I: Analysis and experimentation of stealthy deception attacks. *IEEE Transactions on Control Systems Technology*, 21(5), 1963-1970.

Amin, S., Litrico, X., Sastry, S., & Bayen, A. (2012). Cyber security of water SCADA systems-Part II: Attack detection using enhanced hydrodynamic models. *IEEE Transactions on Control Systems Technology*, 21(5), 1679-1693.

Anton, H., & Rorres, C. (2005). *Elementary linear algebra: Applications version* (9th ed.). Hoboken, NJ: John Wiley & Sons, Inc.

Bellman, R., & Kalaba, R. (1964). *Selected papers on mathematical trends in control theory*. Mineola, NY: Dover Publications.

Boyce, W. & DiPrima, R. (2009). *Elementary differential equations and boundary value problems* (9th ed.). Hoboken, NJ: John Wiley and Sons, Inc.

Castillo-Chavez, C., Feng, Z., & Huang, W. (2002). On the computation of R_0 and its role on global stability. In C. Castillo-Chavez, S. Blower, P. Driessche, D. Kirschener, & A. Yakubu (Eds). *Mathematical approaches for emerging and re-emerging infectious diseases: Models, methods and theory* (Vol. 125) (pp. 229-250). New York, NY: Springer IMA Volumes in Mathematics and Its Applications.

Chen, G., & Li, T. (2009). Stability of stochastic delayed SIR model. *Stochastics and Dynamics*, 9(2), 231-252.

Chen, Y., Kar, S., & Moura, J. (2017). Dynamic attack detection in cyber-physical systems with side initial state information. *IEEE Transactions on Automatic Control*, 62(9), 4618-4624.

Chen, Y., Kar, S., & Moura, J. (2018). Optimal attack strategies subject to detection constraints against cyber-physical systems. *IEEE Transactions on Control of Network Systems*, 5(3), 1157-1168.

Diekmann, O., Heesterbeek, J., & Metz, J. (1990). On the definition and the computation of the basic reproduction ratio R_0 in models for infectious diseases in heterogeneous populations. *Journal of Mathematical Biology*, 28(4), 365-382.

Ding, D., Han, Q., Xiang, Y., Ge, X., & Zhang, X. (2018). A survey on security control and attack detection for industrial cyber-physical systems. *Neurocomputing*, 275, 1674-1683.

Farraj, A., Hammad, E., & Kundur, D. (2018). A cyber physical control framework for transient stability in smart grids. *IEEE Transactions on Smart Grid*, 9(2), 1205-1215.

Foroush, H., & Martínez, S. (2013). On multi-input controllable linear systems under unknown periodic DoS jamming attacks. In *2013 Proceedings of the Conference on Control and its Applications* (pp. 222-229). Society for Industrial and Applied Mathematics.

Gan, C., Yang, X., Liu, W., Zhu, Q., & Zhang, X. (2012). Propagation of computer virus under human intervention: a dynamical model. *Hindawi Publishing Corporation.* 1-8.

Hale, J. (1969). *Ordindary differential equations.* New York, NY: John Wiley & Sons, Inc.

Han, X., & Tan, Q. (2010). Dynamical behavior of computer virus on internet. *Applied Mathematics and Computation,* 217(6), 2520-2526.

Hoehn, A., & Zhang, P. (2016). Detection of covert attacks and zero dynamics attacks in cyber-physical systems. In *2016 American Control Conference (ACC)* (pp. 302-307). IEEE.

Kamaleddin, S. & Nikravesh, Y. (2013). *Nonlinear systems stability analysis: Lyapunov-based approach.* Boca Raton, FL: CRC Press: Taylor & Francis Group.

Klebaner, F. (2005). *Introduction to stochastic calculus with applications* (2nd ed.). London, England: Imperial College Press (ICP).

Kuo, H. (2006). *Introduction to stochastic integration.* New York, NY: Universitext Springer.

Larson, R. (2013). *Elementary linear algebra* (7th ed.). Boston, MA: Brooks/Cole Cengage Learning.

La Salle, J. (1976). *The stability of dynamical systems* (Vol. 25). Philadelphia, PA: Society for Industrial and Applied Mathematics Regional Conference Series in Applied Mathematics.

Li, M. (2018). *An introduction to mathematical modeling of infectious diseases* (Vol. 2). Cham, Switzerland: Springer Nature.

Liu, Q., Chen, Q., & Jiang, D. (2016). The threshold of a stochastic delayed SIR epidemic model with temporary immunity. *Physica A,* 450, 115-125.

Liu, Y., Ning, P., & Reiter, M. (2011). False data injection attacks against state estimation in electric power grids. *ACM Transactions on Information and System Security,* 14(1), 13.

Lu, A., & Yang, G. (2018). Input-to-state stabilizing control for cyber-physical systems with multiple transmission channels under denial of service. *IEEE Transactions on Automatic Control,* 63(6), 1813-1820.

Manandhar, K., Cao, X., Hu, F., & Liu, Y. (2014). Detection of faults and attacks including false data injection attacks in smart grid using Kalman filter. *IEEE Transactions on Control of Network Systems,* 1(4), 370-379.

Martcheva, M. (2015). *An introduction to mathematical epidemiology* (Vol. 61). New York, NY: Springer Texts in Applied Mathematics.

Miao, F., Zhu, Q., Pajic, M., & Pappas, G. (2014). Coding sensor outputs for injection attacks detection. In *53rd IEEE Conference on Decision and Control,* (pp. 5776-5781). IEEE.

Mishra, B., & Jha, N. (2010). SEIQRS model for the transmission of malicious objects in computer network. *Applied Mathematics and Computation,* 34, 710-715.

Mishra, B., & Pandey, S. (2011). Dynamic model of worms with vertical transmission in computer network. *Applied Mathematics and Computation,* 217(21), 8438-8446.

Mo, Y., Garone, E., Casavola, A., & Sinopoli, B. (2010). False data injection attacks against state estimation in wireless sensor networks. In *49th IEEE Conference on Decision and Control,* (pp. 5967-5972). IEEE.

Pasqualetti, F., Dörfler, F., & Bullo, F. (2011). Cyber-physical attacks in power networks: models, fundamental limitations and monitor design. In *2011 50th IEEE Conference on Decision and Control and European Control Conference* (pp. 2195-2201). IEEE.

Pasqualetti, F., Dörfler, F., & Bullo, F. (2013). Attack detection and identification in cyber-physical systems. *IEEE Transactions on Automatic Control,* 58(11), 2715-2729.

Persis, C., & Tesi, P. (2015). Input-to-state stabilizing control under denial of service. *IEEE Transactions on Automatic Control,* 60(11), 2930-2944.

Polking, J., Boggess, A., & Arnold, D. (2018). *Differential equations with boundary value problems* (2nd ed.). New York, NY: Pearson Modern Classic.

Ren, J., Yang, X., Zhu, Q., Yang, L., & Zhang, C. (2012). A novel computer virus model and its dynamics. *Nonlinear Analysis: Real World Applications,* 13(1), 376-384.

Roberto, J., Piqueira, C., & Araujo, V. (2009). A modified epidemiological model for computer viruses. *Applied Mathematics and Computation,* 213(2), 355-360.

Sá, A., Carmo, L., & Machado, R. (2017). Covert attacks in cyber-physical control systems. *IEEE Transactions on Industrial Informatics,* 13(4), 1641-1651.

Sridhar, S., Hahn, A., & Govindarasu, M. (2011). Cyber physical system security for the electric power grid. *Proceedings of the IEEE,* 100(1), 210-224.

Sontag, E. (2008). Input to state stability: basic concepts and results. In P. Nistri, & G. Stefani (Eds). *Nonlinear and Optimal Control Theory* (pp. 163-220). Berlin, Germany: Springer Lecture Notes in Mathematics.

Teixeira, A., Pérez, D., Sandberg, H., & Johansson, K. (2012). Attack models and scenarios for networked control systems. In *Proceedings of 1st International Conference on High Confidence Networked Systems* (pp. 55-64).

Teixeira, A., Shames, I., Sandberg, H., & Johansson, K. (2012). Revealing stealthy attacks in control systems. In *2012 50th Annual Allerton Conference on Communication, Control and Computing* (pp. 1806-1813). IEEE.

Tornatore, E., Buccellato, S., & Vetro, P. (2005). Stability of a stochastic SIR system. *Physica A*, 354, 111-126.

Toutonji, O., Yoo, S., & Park, M. (2012). Stability analysis of VEISV propagation modeling for network worm attack. *Applied Mathematical Modelling*, 36(6), 2751-2761.

Weiss, H. (2013). The SIR model and the foundations of public health. *Materials Matematics*, 0001-17.

Xiao, X., Fu, P., Dou, C., Li, Q., Hu, G., & Xia, S. (2017). Design and analysis of SEIQR worm propagation model in mobile internet. *Commununications in Nonlinear Science and Numerical Simulation*, 43, 341-350.

CHAPTER 6

Network Science

Elie Alhajjar

CONTENTS

6.1	Introduction	207
6.2	Basics of Graph Theory	209
	6.2.1 Graphs and Subgraphs	210
	6.2.2 Graph Metrics	211
	6.2.3 Centrality Measures	213
6.3	Network Models	214
	6.3.1 The Erdős-Renyi Model	214
	6.3.2 The Watts–Strogatz Model	216
	6.3.3 The Barabási-Albert Model	217
6.4	Cybersecurity Applications	218
	6.4.1 Cyber-Terrorism	219
	6.4.2 Attack Graphs	221
	6.4.3 Insider Threat	222
	6.4.4 Digital Viruses	226
6.5	Implications for Students and Early Career Practitioners	229

6.1 INTRODUCTION

Networks are ubiquitous. The existence of society depends upon a variety of complex networks covering different domains. In the physical realm, they include highways, railroads, the air transportation network, the global shipping network, power grids, water distribution networks, supply networks, global financial networks, telephone systems, and the internet. In the biological realm, they include genetic expression networks, metabolic networks, ant colonies, food webs, river basins, and the global ecological web of Earth itself. In the social realm, they include governments, businesses, universities, social clubs, churches, school systems, and military organizations.

Despite society's intricate dependence on networks, there still exists a huge gap between what we need to know about networks to ensure their smooth functioning and the current state of our scientific knowledge. Many physical networks, such as

DOI: 10.1201/9780429354649-6

the global communication and transportation networks, have very advanced technological implementations, yet their behavior in critical conditions cannot be predicted reliably. The same holds for biological and social networks, in which scientists are continuously trying to understand how they operate.

Networks are built on top of one another. Consider social networks as a common example: they are built on information networks, which in turn are built on communications networks that operate using physical networks for connectivity. More broadly, they are interactive and mutually interdependent systems that together constitute a much larger system. There is a desperate need today for a research field that offers the fundamental knowledge necessary to design large and complex networks in such a way that their behaviors can be predicted prior to building them; network science offers such promise.

As the modeling of networks employs a variety of techniques, network science is considered an interdisciplinary field that combines ideas from areas like mathematics, physics, biology, computer science, statistics, and social sciences to name a few. The field has benefited enormously from the wide range of viewpoints brought to it by practitioners from different disciplines, but it has also suffered from the dispersion of human knowledge about networks across the scientific community.

The last couple of decades have boosted the study of large networked systems due to the ever-increasing availability of large datasets and computer power for their storage and manipulation. In particular, mapping projects of the World Wide Web and the physical Internet offered the first chance to study the topology of large complex networks. Indeed, large complex networks arise in a vast number of natural and artificial systems. The brain consists of many interconnected neurons, ecosystems consist of species whose inter-dependency can be mapped into food webs, social systems may be represented by graphs describing various interactions among individuals, and large infrastructure systems, such as power grids and transportation networks are critical to our modern society, etc. A central result in the study of these large networks is that they are generally characterized by complex topologies and very heterogeneous structures. These features usually find their signature in connectivity patterns statistically characterized by heavy tails and large fluctuations, scale-free properties, and non-trivial correlations, such as high clustering and hierarchical ordering.

However, the structure of the network is only a starting point. The use of the word 'connectedness' in a complex system framework usually refers to two related issues. One is connectedness at the level of structure in the sense of who is linked to whom, and the other is connectedness at the level of behavior, as in the fact that each individual's actions have implicit consequences on the outcomes of the whole system. This means that, in addition to a language for discussing the structure of networks, a framework is needed for reasoning about behavior and interaction in network contexts.

Due to our reliance on IT technology in private and professional life, our systems of control are becoming more and more dependent upon sophisticated and subsequently more complex devices, which are rendering them extremely vulnerable to targeted attacks. In the era of the development of information technologies, a new form of terrorism has become an important threat to modern society. This type of threat was coined the name 'cyber-terrorism' by Denning.

In terms of the connection between network science and cybersecurity, one of the main challenges herein is to identify malicious events in a given network, study their behavior, and deploy the right actions to protect the network. The current work aims to describe this connection and shed some light on potential future directions at the intersection of the two fields. After this short introduction, Section 6.2 defines the basics of graph theory and some tools for the statistical characterization and classification of large networks. We focus on the minimum requirements needed to establish a solid ground for the remainder of the chapter. We record enough reference material for the pursuit of further exploration of this extremely rich mathematical field of study. Section 6.3 is devoted to a brief presentation of the various network models and the general approaches used for their analysis. We discuss three main network models: the Erdős-Renyi model, the Watts-Strogatz model, and the Barabási-Albert model. These are the fundamental models in network science, but they are not the only ones! There have been a lot of developments upon them, and several variations exist within each one of them. In addition to that, we refrain from mentioning every seminal paper that advanced the field due to space constraints. For a deeper and more thorough investigation of the original works, the textbooks by Barabási and Pósfai (2016), Barrat, Barthelemy, and Vespignani (2008), Easley and Kleinberg (2010), Jackson (2008), and Newman (2010) and the references therein serve as a great starting point. Section 6.4 provides a (somewhat subjective, by no means exhaustive) list of applications of network science concepts in cybersecurity instances; it is based on the author's areas of interest and expertise. We conclude the chapter by addressing some final notes and research directions in Section 6.5.

6.2 BASICS OF GRAPH THEORY

In this section, the basic notions and notations needed to describe networks are provided. Needless to say, the expert reader can freely skip this section and use it later as a reference.

The terms *graphs* and *networks* are used indistinctly in the literature. The only nuance is that the term *graph* usually refers to the abstract mathematical concept of nodes and edges, while the term *network* refers to real-world objects in which nodes represent entities of some system and edges represent the relationships between them.

We adopt a natural framework for the rigorous mathematical description of networks, namely graph theory. Graph theory is a vast field of mathematics that can be traced back to the seminal work of Leonhard Euler in solving the Konigsberg bridges problem in 1736. For readers interested in pursuing a deeper study of graph theory, West (2000) and Agnarsson and Greenlaw (2006) are great references.

FIGURE 6.1 A graph and its adjacency matrix.

6.2.1 Graphs and Subgraphs

Let $V = \{v_1, v_2, \ldots, v_n\}$ be a finite set of elements and $V \times V$ the set of all ordered pairs $\{v_i, v_j\}$ of elements of V. A relation on the set V is any subset $E \subseteq V \times V$, where $\times$ denotes the Cartesian product. A *simple undirected graph* is a pair $G = (V, E)$, where V is a finite set of *nodes* (or *vertices*) and E is a relation on V such that $\{v_i, v_j\} \in E$ implies $\{v_j, v_i\} \in E$ and $v_i \neq v_j$; that is, G has no *loops*. The elements of E are called *edges* or *links*; they are denoted as $E = \{e_1, e_2, \ldots, e_m\}$. In general, $n = |V|$ and $m = |E|$, where $|.|$ means the cardinality of a set.

When the orientation of the edges in E matters, $G = (V, E)$ is called a *directed graph*, or a *digraph*. In other words, $\{v_i, v_j\} \in E$ does not necessarily imply that $\{v_j, v_i\} \in E$. Graphically, the orientation of an edge is usually depicted by an arrow that indicates the direction of the edge. Given a graph $G = (V, E)$, a *subgraph* of G is a graph $G' = (V', E')$ such that $V' \subseteq V$ and $E' \subseteq E$. That is, all the vertices in G' are vertices in G and all the edges in G' are edges in G.

In this chapter, we are interested in weighted graphs. A *weighted graph* is a quadruple $G = (V, E, W, \phi)$ where V and E are as above, $W = \{w_1, w_2, \ldots, w_s\}$ is a set of *weights* (i.e. real numbers), and $\phi \colon E \to W$ is a *weight function*, that is, a surjective mapping that assigns a weight to each edge. When the weight of each edge is 1, we call G an *unweighted graph* or simply a graph. If an edge e joins two nodes v_i and v_j, then we say that v_i and v_j are *adjacent* and they are *incident* to e.

The *adjacency matrix* $A = (a_{ij})_{i,j=1}^{n}$ of a weighted graph G is an $n \times n$ array defined as

$$a_{ij} = \begin{cases} \phi(\{v_i, v_j\}) & \text{if } \{v_i, v_j\} \in E \\ 0 & \text{otherwise.} \end{cases}$$

Note that for a simple undirected graph, the adjacency matrix is symmetric, and the entries on the main diagonal are all equal to zero. Figure 6.1 shows an example of a simple graph with five vertices along with the corresponding adjacency matrix. We will use this graph as a working example in the remainder of the section. For directed graphs, the adjacency matrix is not necessarily symmetric due to the direction of the edges.

For a graph $G = (V, E)$, the maximum number of edges is $m = \frac{n(n-1)}{2}$, and a graph satisfying this maximum is called a *complete graph*. A *bipartite graph* is a simple undirected graph $G = (V_1 \sqcup V_2, E)$ such that the vertices in V are divided into two disjoint subsets and there are no edges between vertices in the same subset.

A *k-regular* graph is a graph in which each node is connected to k other nodes. In the graph of Figure 6.1, four edges are missing to make it a complete graph on five vertices with a total of ten edges. Moreover, if the edge $\{v_1, v_2\}$ is removed, then the graph becomes 2-regular as each vertex would then have degree two.

In a graph $G = (V, E)$, a *path* from a node v_i to a node v_j is a collection of ordered vertices $\{v_i, v_{i+1}, \ldots, v_{j-1}, v_j\} \subseteq V$ and a collection of ordered edges $\{(v_i, v_{i+1}), (v_{i+1}, v_{i+2}), \ldots, (v_{j-1}, v_j)\} \subseteq E$. The *length* of a path is the number of edges traversed along the path. A *shortest path*, or a *geodesic path*, from node a v_i to a node v_j, is a path of shortest length. A *cycle* is a closed path, i.e. a path in which $v_i = v_j$. In Figure 6.1, $\{v_0, \{v_0, v_1\}, v_1, \{v_1, v_3\}, v_3, \{v_3, v_4\}, v_4\}$ is a path of length three between the vertices v_0 and v_4, but it is not a shortest path! In order to get from node v_0 to node v_4, the shortest path (length two) is $\{v_0, \{v_0, v_2\}, v_2, \{v_2, v_4\}, v_4\}$.

We say that a graph is *connected* if there is a path between any pair of nodes in the graph. A *component* of a graph is a connected subgraph. A *tree* is a connected graph that has no cycles. One can easily derive that for a tree, $m = n - 1$, there is a unique path between any two given nodes. Equivalently, the deletion of any edge breaks a tree into disconnected components. In the case there is a parent node, or *root*, from which the whole tree arises, then it is called a *rooted tree*. The nodes at the bottom that are connected to only one other node are called *leaves*. Clearly, the graph in Figure 6.1 is connected (one component), and the deletion of the edges $\{v_1, v_2\}$ and $\{v_3, v_4\}$ makes it into a tree rooted at node v_0.

6.2.2 Graph Metrics

The simplest characteristic of a node v_i is its *degree*, which is defined as the number of nodes adjacent to it. It is usually denoted by k_i, and it is not hard to see that in a graph with n vertices

$$m = \frac{1}{2} \sum_{i=1}^{n} k_i. \tag{6.1}$$

In terms of the adjacency matrix, the degree of a node v_i can be expressed as

$$k_i = \sum_{j=1}^{n} a_{ij}. \tag{6.2}$$

In the graph of Figure 6.1, the degrees of the vertices v_0, v_1, v_2, v_3, v_4 are $2, 3, 3, 2, 2$, respectively. From a weighted point of view, one can say that the weighted degrees of the vertices v_0, v_1, v_2, v_3, v_4 are $5, 19, 31, 11, 22$, respectively. The latter calculation corresponds to the sum of the respective rows (or equivalently, columns) of the adjacency matrix.

For directed graphs, the distinction is made between the *in-degree* k_i^{in} and the *out-degree* k_i^{out} of a node v_i. The former is the number of edges pointing in the direction of v_i, while the latter is the number of edges going out of v_i. The *average degree* of an undirected/directed graph is given by

$$\langle k \rangle = \frac{1}{n} \sum_{i=1}^{n} k_i = \frac{2m}{n}, \tag{6.3}$$

$$\langle k^{in} \rangle = \frac{1}{n} \sum_{i=1}^{n} k_i^{in} = \langle k^{out} \rangle = \frac{1}{n} \sum_{i=1}^{n} k_i^{out} = \frac{m}{n}. \quad (6.4)$$

The *diameter* d of a graph $G = (V, E)$ is the maximum shortest path length in G. It simply measures the minimum number of edges needed to connect the two most distant nodes in G. The *average shortest path length* is the average value over all the possible pairs of vertices in the graph. As mentioned in the previous section, the maximum number of edges in a graph is $m = \frac{n(n-1)}{2}$ (complete graph). The *density* D of a graph is the ratio of the number of edges to the maximum such number, i.e. $D = \frac{2m}{n(n-1)}$. We say that a graph is *sparse* if $D \ll 1$ and *dense* if $D \sim 1$. It is worth noting that real-world networks are sparse in general, a feature that implies that their adjacency matrices have entries that are mostly 0. As an illustration, the graph in Figure 6.1 has diameter $d = 2$ and density $D = 0.6$. The shortest path lengths between all pairs of nodes in the graph are $1, 1, 2, 2, 1, 1, 2, 2, 1, 1$; this implies that the average shortest path length is $\frac{14}{10} = 1.4$.

Let $G = (V, E)$ be an undirected graph. The *degree distribution* p_k is the probability that a randomly chosen node has degree k. If we let n_k be the number of nodes of degree k, then $p_k = \frac{n_k}{n}$ and clearly $\sum_{k=0}^{\infty} p_k = 1$. In Figure 6.1, the graph has three nodes of degree 2, and two nodes of degree 3. Hence, the degree distribution is simply $p_2 = \frac{3}{5} = 0.6$, $p_3 = \frac{2}{5} = 0.4$, and $p_k = 0$ for all other values of k. This allows us to define the *t-th moment* of the degree distribution as

$$\langle k^t \rangle = \sum_{k=0}^{\infty} k^t p_k. \quad (6.5)$$

The case $t = 1$ is simply the average degree or the first moment of p_k (see Equation 6.3). When dealing with directed graphs, it is natural to consider two degree distributions: $p_{k^{in}}$, the probability that a randomly selected node has in-degree k^{in}, and $p_{k^{out}}$, the probability of a randomly selected node has out-degree k^{out}.

The *clustering coefficient* of a node measures the likelihood that the adjacent vertices to this node are connected to each other. There are two different definitions for the clustering coefficient in the literature, so the comparison of such coefficients among different graphs must use the same measure. The first definition, often referred to as the *local clustering coefficient*, is the ratio of the number of edges between the neighbors of the node and the maximum number of such possible edges. More precisely, for a node v_i of degree $k_i > 1$:

$$C_i = \frac{2m_i}{k_i(k_i - 1)}, \quad (6.6)$$

where m_i is the number of edges between the neighbors of v_i. If $k_i \leq 1$, we set $C_i = 0$. The *average clustering coefficient* of a graph is simply given by

$$\langle C \rangle = \frac{1}{n} \sum_{i=1}^{n} C_i. \quad (6.7)$$

The second definition, often referred to as the *global clustering coefficient*, is the ratio of the number Δ of triangles in the graph (cycles of length 3) to the number of connected triples (paths of length 2). In mathematical terms,

$$C_\Delta = \frac{6 \times \Delta}{\sum_{i=1}^{n} k_i(k_i - 1)}. \tag{6.8}$$

It is worth mentioning that both measures are normalized and bounded between 0 and 1. As a quick application, the reader is encouraged to verify that the clustering coefficients of the nodes in Figure 6.1 satisfy the following: $C_0 = 1$, $C_1 = C_2 = \frac{1}{3}$, and $C_3 = C_4 = 0$. This implies that the average clustering coefficient of the graph is $\langle C \rangle = \frac{1}{5}(1 + \frac{1}{3} + \frac{1}{3} + 0 + 0) = \frac{1}{3}$. On the other hand, there is only one triangle in the graph, namely $\{v_0, v_1, v_2\}$, so the global clustering coefficient is $C_\Delta = \frac{6(1)}{6+6+2+2+2} = \frac{1}{3}$.

6.2.3 Centrality Measures

By definition, *centrality* aims to capture the notion of 'importance' of a node in a network. There are plenty of centrality measures in the literature and efficient algorithms to compute them for large networks. We will discuss below some of the most commonly used ones.

Perhaps the most natural centrality measure for a node in a graph is simply its degree or *degree centrality*, i.e. the number of nodes adjacent to it. It is a local indicator of a node's importance and does not take into consideration the global characteristics of the graph.

For two nodes v_i and v_j, we denote by d_{ij} as the distance between them. In other words, d_{ij} is the length of the shortest path from v_i to v_j. The *closeness centrality* measures the mean distance from a node to other nodes in the graph. For a vertex v_i, it is defined as

$$c_i = \frac{n}{\sum_{j=1}^{n} d_{ij}}. \tag{6.9}$$

This definition exhibits a technical problem when the graph in hand has more than one component. In this situation, Equation 6.9 is adjusted in the following form

$$c_i = \frac{1}{n-1} \sum_{i \neq j} \frac{1}{d_{ij}}. \tag{6.10}$$

The *betweenness centrality* of a node measures the extent to which this node lies on paths between other nodes. The betweenness centrality of a vertex v_i is given by

$$b_i = \sum_{s \neq v_i \neq t \in V} \frac{\sigma_{st}(v_i)}{\sigma_{st}}, \tag{6.11}$$

where σ_{st} is the total number of shortest paths from node s to node t and $\sigma_{st}(v_i)$ is the number of those paths that pass through v_i. The betweenness centrality is normalized by dividing Equation 6.11 by the number of pairs of vertices not including v_i, which is $\frac{(n-1)(n-2)}{2}$.

As an example, the nodes v_0 and v_1 in Figure 6.1 are considered. It is straightforward to see that $d_{01} = d_{02} = 1$, and $d_{03} = d_{04} = 2$. Equation 6.9 implies that the closeness centrality of node v_0 is $c_0 = \frac{5}{1+1+2+2} = \frac{5}{6}$. To compute the betweenness centrality of node v_1, notice that the only shortest paths that pass through v_1 are the ones from v_0 to v_3, and from v_2 to v_3. Hence, by Equation 6.11 and the normalizing factor, $b_1 = \frac{1}{6}(\frac{1}{1} + \frac{1}{2}) = \frac{1}{4}$.

For a graph G with adjacency matrix A, consider the equation $Ax = \lambda x$ where x is a vector and λ is a scalar. We call λ an *eigenvalue* of A and x the corresponding *eigenvector*. By known results from linear algebra, the matrix A admits n different eigenvectors. The *eigenvector centrality* of a node v_i is the i-th entry of the leading eigenvector of A which, by the Perron-Frobenius theorem (see Newman (2010)), is guaranteed to have all its entries non-negative.

All measures mentioned above are defined in the context of undirected networks. When dealing with directed networks, these definitions need to be reformulated accordingly. Other centrality measures in the literature include Katz centrality, PageRank, hubs and authorities, etc. Due to the technicality of the arguments involved therein, we omit the details and refer the reader to Newman (2010).

6.3 NETWORK MODELS

In this section we examine some of the most widely used models of network structure. Such models have two main functions. On the one hand, they aim to mimic, in a simplified form, the emergence and evolution of real networks in order to shed some light on the mechanism responsible for their formation. On the other hand, generating synthetic networks with preset properties help test the impact of selected network characteristics on the network's behavior.

It is quite impossible to include in one section all existing models together with the multiple variations on each of them. For the sake of brevity, we focus our attention herein on three types of network models: the Erdős-Renyi model, the Watts-Strogatz model, and the Barabási-Albert model.

6.3.1 The Erdős-Renyi Model

A *random graph* is a network model in which the values of certain features are fixed, but the network is otherwise random. It is named an *Erdős-Renyi model* after the two Hungarian mathematicians who founded the field of random graph theory in the 1960s. Lacking any information, the simplest action is to connect pairs of nodes at random with a given connection probability p.

There are two ways to define a random network:

1. Fix n nodes, then choose m distinct pairs of them uniformly at random from all possible pairs and connect them with an edge. We call this model $G(n, m)$.

2. Fix n nodes, then place an edge between each distinct pair with independent probability p. We obtain the $G(n, p)$ model in which the number of edges is not fixed.

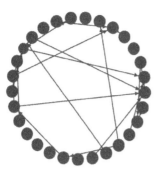 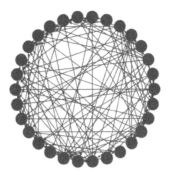

FIGURE 6.2 Two random graphs on 30 vertices.

We focus on the second model and explore some of its key network characteristics. We omit the proofs as well as the algebraic computations for the ease of the reader. All the missing mathematical details can be found in Newman (2010).

For illustration purposes, Figure 6.2 shows two different random graphs with 30 nodes: the one on the left is generated with a probability $p = 0.05$, and the one on the right is generated with a probability $p = 0.25$. It is trivial to see that the higher the connection probability is, the denser the graph becomes.

The probability that a random network has exactly m links is given by

$$p_m = \binom{\frac{n(n-1)}{2}}{m} p^m (1-p)^{\frac{n(n-1)}{2} - m}. \tag{6.12}$$

This is clearly a *Binomial distribution*, from which we can infer the expected number of links and the average degree in a random graph:

$$\langle m \rangle = \frac{pn(n-1)}{2}, \tag{6.13}$$

$$\langle k \rangle = p(n-1). \tag{6.14}$$

Recall that the degree distribution is the probability that a randomly chosen node has degree k. The degree distribution of a random network follows the Binomial distribution; it is dependent on n and p as follows:

$$p_k = \binom{n-1}{k} p^k (1-p)^{n-1-k}. \tag{6.15}$$

In the limit of large n, the Binomial distribution can be approximated by the *Poisson distribution*:

$$p_k = e^{-k} \frac{\langle k \rangle^k}{k!}. \tag{6.16}$$

The Binomial distribution and the Poisson distribution have similar properties. They both have a peak around $\langle k \rangle$; if we increase p the network becomes denser. Moreover, the probability p controls the width of the distribution: the denser the network, the wider the distribution. Hence, the larger are the differences in the degrees.

While the Poisson distribution is only an approximation to the degree distribution of a random network, we adopt its form for p_k due to its analytical simplicity. Most importantly, its properties are independent of the number of nodes and depend only on a single parameter, namely the average degree $\langle k \rangle$.

For any vertex, the probability that any two of its neighbors are also connected to each other is given by the connection probability p. Therefore, the average clustering coefficient is equal to

$$\langle C \rangle = p = \frac{\langle k \rangle}{n-1}. \tag{6.17}$$

Random networks exhibit what we call the 'small world property.' It is rooted in the fact that the number of nodes at distance d from a node v increases exponentially with d. The diameter of a random graph can then be approximated by

$$d \sim \frac{\ln n}{\ln \langle k \rangle}. \tag{6.18}$$

Finally, we briefly discuss the evolution of a random network, in particular the emergence of the *giant component*. Trivially, the largest component of the graph has one node when $p = 0$ and n nodes when $p = 1$. When $1 < \langle k \rangle < \ln n$, it can be shown that the giant component has $\sim n^{\frac{2}{3}}$ nodes and grows to $\sim n$ when $\langle k \rangle \gg \ln n$.

6.3.2 The Watts–Strogatz Model

Erdős-Renyi models are characterized by the vanishing of their average clustering coefficient when n is large. However, many real-world networks are highly clustered while at the same time maintaining a small average distance between vertices. Watts and Strogatz (1998) proposed a model that interpolates between ordered lattices (large clustering coefficient) and purely random networks (small average path length). See Newman (2010) for a detailed exposition.

The model is defined as follows. Consider a ring of nodes, each node being connected to its previous and next neighbors. Now rewire each edge to a randomly chosen node, with probability p. We distinguish two extreme cases:

1. When $p = 0$, no edges are rewired, and we retain the original circle.

2. When $p = 1$, all edges are rewired, and we obtain a random graph.

The degree distribution of the *Watts-Strogatz model* can be computed analytically, and as p gets closer to 1, it reduces to a Poisson distribution. It turns out that short path lengths develop in this model for very small values of the probability p in the sense that only a small fraction of the edges need to be rewired for the path lengths to become short.

In summary, the Watts-Strogatz model allows the tuning of the clustering coefficient within the framework of static random graph theory. In addition to that, it gives a plausible justification of the high clustering coefficients observed in real-world networks.

6.3.3 The Barabási-Albert Model

The insight behind this new model is the fact that in most real-world networks, new edges are not located at random but tend to connect to vertices which already have a high degree. Before explaining the details of the *Barabási-Albert model*, we take a look at scale-free networks first.

By definition, a *scale-free network* is a network whose degree distribution follows a power law $p_k \sim Ck^{-\gamma}$ for some constant C that can be determined via the normalization condition: $\sum p_k = 1$ in the discrete case and $\int p_k dk = 1$ in the continuous case. We call γ the *degree exponent* of the distribution.

The key difference between a random network and a scale-free network is rooted in the different shape of the Poisson and the power-law functions: in a random network most nodes have comparable degrees, and hence hubs are forbidden, while hubs are expected in scale-free networks. Moreover, the more nodes a scale-free network has, the larger are its hubs

$$k_{max} = k_{min} n^{\frac{1}{\gamma-1}}, \qquad (6.19)$$

where k_{max} and k_{min} are the maximum and minimum degree in the network, respectively.

Concerning the small-world property, distances in a scale-free network are smaller than those observed in an equivalent random network. More precisely, the average distance in a scale-free network depends on the number of nodes n and the degree exponent γ as follows

$$\langle d \rangle \sim \begin{cases} const. & \gamma = 2 \\ \ln \ln N & 2 < \gamma < 3 \\ \frac{\ln N}{\ln \ln N} & \gamma = 3 \\ \ln N & \gamma > 3 \end{cases} \qquad (6.20)$$

In order to understand the reason why hubs are the main difference between a random network and a scale-free network, we need to justify the emergence of the scale-free property therein. The Barabási-Albert model offers such an explanation, based on two characteristics:

1. *Growth*: start with n_0 nodes, and at each time step add a new node with e edges that connect the new node to e already existing nodes. We assume $e \leq n_0$.

2. *Preferential Attachment*: the probability that a vertex v_i acquires a new edge is proportional to its degree k_i through

$$\Pi(k_i) = \frac{k_i}{\sum_j k_j}. \qquad (6.21)$$

The dynamical rate equation governing the evolution of $k_i(t)$ can then be formally obtained by considering that the degree growth rate of the vertex v_i increases proportionally to the probability $\Pi(k_i)$ that an edge is attached to it. In the case when edges only come from the newborn vertices, the rate equation becomes

$$\frac{dk_i}{dt} = e\Pi(k_i) = e\frac{k_i}{\sum_j k_j}; \qquad k_i(t_i) = e. \tag{6.22}$$

From the differential equation above, we solve for the degree dynamics of the network. We use the resulting expression to find the degree distribution, as well as the average distance and the average clustering coefficient of the Barabási-Albert model. We omit the mathematical derivations and summarize the findings

$$p_k \sim 2e^2 k^{-3}, \tag{6.23}$$

$$\langle d \rangle \sim \frac{\ln n}{\ln \ln n}, \tag{6.24}$$

$$\langle C \rangle \sim \frac{(\ln n)^2}{n}. \tag{6.25}$$

Finally, it is of fundamental importance to understand that growth and preferential attachment are jointly needed to generate Barabási-Albert models; hence if one of them is absent, either the scale-free property or stationarity is lost.

6.4 CYBERSECURITY APPLICATIONS

There are many different layouts for the general network science research process; we focus here on the one adopted by Borner, Sanyal, and Vespignani (2007).

A network science study typically starts with a hypothesis or research question. Next, an appropriate dataset is collected or sampled, represented, and stored in a format amenable to efficient processing. Subsequently, network measurements are applied to identify features of interest that lead to analyzing and/or modeling the system. Given the complexity of networks and the results obtained, the application of visualization techniques for the communication and interpretation of results is important. Interpretation frequently ensues further refinement of parameter values or algorithms and rerunning of sampling, modeling, measurement, and visualization stages.

There is a major difference between *network analysis* and *network modeling*. The former aims at the generation of descriptive models that explain and describe a certain system, while the latter attempts to design process models that not only reproduce the empirical data but can also be used to make predictions.

It is very hard to collect all the literature that connects cybersecurity to network science and vice versa. In this section, we provide a subjective and by no means exhaustive list of endeavors that lie at the intersection of these two fields. We divide the section into four topics: cyber-terrorism, attack graphs, insider threat, and digital viruses.

6.4.1 Cyber-Terrorism

The National-Security-Archive (2018) defines *cyberspace* as a global domain within the information environment consisting of the interdependent network of information systems infrastructures including the internet, telecommunications networks, computer systems, and embedded processors and controllers. Within this framework, a *cyber attack* is an attack, via cyberspace, targeting an enterprise's use of cyberspace for the purpose of disrupting, disabling, destroying, or maliciously controlling a computing environment/infrastructure, destroying the integrity of the data, or stealing controlled information. In other words, it is a deliberate exploitation of computer systems, digitally-dependent enterprises, and networks to cause harm.

As in the case of universal definitions of terrorism, there is no consensus about defining *cyber-terrorism*. Denning defines cyber-terrorism as the convergence of terrorism and cyberspace. It is generally understood to mean unlawful attacks and threats of attack against computers, networks, and the information stored therein when done to intimidate or coerce a government or its people in furtherance of political or social objectives. Further, to qualify as cyber-terrorism, an attack should result in violence against persons or property, or at least cause enough harm to generate fear. Attacks that lead to death or bodily injury, explosions, plane crashes, water contamination, or severe economic loss would be examples. Serious attacks against critical infrastructures could be acts of cyber-terrorism, depending on their impact.

Terrorists' activities via cyberspace include creating websites/blogs, communication via email, discussion via chat rooms, e-transactions (e-commerce/e-banking), using search engines to collect data and find information, phishing/hacking, viruses, malicious code, etc. As an example of such activities, we consider the website www.anshar.net created by Noordin Mohammed Top in Indonesia. Established for propaganda purposes, this website published successful terrorist attacks, recruited fellow prospective 'soldiers,' and distributed training material for agents.

From an online presence in cyberspace grew a terrorist cell, the *Noordin network*. Over the years, this network claimed responsibility for many terrorist attacks, such as the JW Marriott Hotel bombing in Jakarta (2003), the Australian Embassy bombing in Jakarta (2004), the Bali bombing (2005), and the JW Marriott and Ritz Carlton bombings in Jakarta (2009). Details of these attacks are provided in Everton (2009).

Alhajjar and Russell (2019) and Alhajjar and Morse (2020) propose the concept of *layered network collapse* to study the characteristics of the Noordin network, a prototype of what is called *dark networks*. Informally, given a network formed by a set of layers, the idea is to 'collapse' the layers into only one such layer without losing the overall network information. This is done via a prescribed 'rule' that dictates the way in which the collapsing phenomenon happens.

In our case, the network consists of 139 actors (nodes). The edges in each layer correspond to the contributions of the actors within the following attributes: organizational affiliation, classmate-ship, internal communication, kinship ties, training

TABLE 6.1 Centrality measures for the Noordin network

Agent	Degree Centrality	Eigenvector Centrality	Betweenness Centrality
Azhari	0.1040	0.3502	0.0610
I. Darwish	0.0591	0.1968	0.0496
A. Rofiq	0.0551	0.2043	0.0521
Ubeid	0.0576	0.2049	0.0496
A. Sangkar	0.0515	0.1988	0.0506

events, recruiting ties, business affiliation, operations, friendship ties, religious affiliation, logistical places, mentor ties, and meeting attendance. The data was first collected and published in Everton (2009, 2012) and Roberts and Everton (2011).

Table 6.1 summarizes the results obtained after applying three centrality measures to the collapsed network. It is no surprise that Mohammed Noordin Top was the central player in the terrorist network! However, other main players needed to be unveiled in order to dismantle the operational aspect of the group. We omit Mohammed Noordin Top and record the next five players with the highest scores in Table 6.1.

More broadly, one can ask about the similarities and differences between terrorist networks and criminal networks. In general, terrorists carry out highly coordinated attacks and maintain a high level of secrecy and trust between each other. Therefore, each individual in the network is generally important and well-connected to other agents in the same network; they operate in a tightly-knit team structure.

On the other hand, criminals operate day-to-day smuggling and trading operations. Therefore, they require low-level agents managed by high-level dealers. Low-level workers have little influence and communication within the network and rely on their chain of command; they operate similarly to a corporation in a hierarchical structure.

Alhajjar, Fameli, and Warren (2020) validate the above hypotheses by studying a handful of publicly available datasets that pertain to criminal and terrorist organizations. In their work, 16 different networks are analyzed within the metrics defined in Section 6.2. A sample of the analytical results is provided in Table 6.2, where two networks are compared: a heroin cell in New York City and a terrorist cell in the Philippines.

Terrorist networks are characterized by low average distance, high eigenvector centrality, high average degree, high network density, high global clustering coefficient, and low betweenness centrality. On the contrary, criminal networks have high average distance, low eigenvector centrality, low average degree, low network density, low global clustering coefficient, and high betweenness centrality.

TABLE 6.2 Comparison between a heroin cell and a terrorist cell

Network Metric	Heroin Dealing	Philippines Bombing (2000)
Diameter	4.00	4.00
Average distance	2.36	1.45
Density	0.12	0.58
Average Degree	4.58	8.63
Global Clustering Coefficient	0.22	0.74
Eigenvector Centrality	0.32	0.64

6.4.2 Attack Graphs

Jha, Sheyener, and Wing (2002) define an *attack graph* as a succinct representation of all paths through a system that ends in a state where an intruder has successfully achieved his/her goal. Security analysts use attack graphs for detection, defense, and forensics. When evaluating the security of a network, it is not enough to consider the presence or absence of isolated vulnerabilities. A security analyst must take into account the effects of interactions of local vulnerabilities and find global vulnerabilities introduced by interconnections.

Attack graphs are used to determine if specific goal states can be reached by attackers attempting to penetrate computer networks from initial starting points. They are a collection of scenarios showing how a malicious agent can compromise the integrity of a target system.

Attack graphs are a natural application of a broader concept, namely *scenario graphs*. Formally, an attack graph is a tuple $G = (S, \tau, S_0, S_s, L)$, where:

S is a set of states,
$\tau \subseteq S \times S$ is a transition relation,
$S_0 \subseteq S$ is a set of initial states,
$S_s \subseteq S$ is a set of success states,
L is a labeling of states with a set of propositions true in that state.

There are multiple benefits for the analysis of attack graphs; we mention a few of them here. First, they help determine where to position new Intrusion Detection System (IDS) components for best coverage or where to target security upgrades. Second, they shed light on how to predict the changes in overall network vulnerability that would occur if new exploits of a certain type became available. Third, they identify worst-case scenarios and prioritize defenses accordingly. Finally, they can be used to assess the effectiveness of defense strategies employed while the attack on the network is in progress.

A path in $G = (S, \tau, S_0, S_s, L)$ is a sequence of states $s_1, s_2, \ldots, s_t$ such that $s_i \in S$ and $(s_i, s_{i+1}) \in \tau$ for $i = 1, \ldots, t$. We focus on the paths that start at some initial state $s_0 \in S_0$ and end at some final state $s_f \in S_s$. The number of such paths in an attack graph is a fundamental measure for what we call *network exposure*. Aside from this number, several other metrics play an equally important role in exploring the vulnerabilities in a given system network, e.g. the number of nodes in the graph, the length of the longest/shortest path, the central nodes in the graph, etc.

Attack graphs have the potential of enhancing both heuristic and probabilistic correlation approaches. Since the graph describes all possible attacks, an IDS can match individual alerts to attack edges in the graph. Matching successive alerts to individual paths in the attack graphs dramatically increases the likelihood that the network is under attack.

Attack graphs enable an administrator to perform several kinds of analyses to assess security needs: marking the paths in the attack graph that an IDS might detect, determining where to position new IDS components for best coverage, exploring trade-offs between different security policies and different software/hardware configurations, and identifying the worst-case scenarios in order to prioritize defensive actions accordingly.

As highlighted in the previous paragraphs, attack graphs are an important tool to mathematically and visually represent the sequence of events that might lead to a successful cyber attack. Likewise, attack trees and fault trees are a popular method that can aid cyber-attack perception. Lallie, Debattista, and Bal (2020) describe the fundamental theory of cyber attacks as well as how influential elements of a cyber attack are represented in attack graphs and attack trees. The authors present empirical research aimed at analyzing more than 180 attack graphs and attack trees to identify how they illustrate cyber attacks in terms of their visual syntax. In particular, the survey points to the lack of a standard method of representing attack graphs or trees, and urges the research community to work towards such standardization. In a nutshell, it summarizes the plethora of works that have been done in the field of attack graphs and provides a systematic comparison between them, while contrasting the challenges in this domain.

6.4.3 Insider Threat

Cappelli, Moore, and Trzeciak (2012) define a *malicious insider* as a current or former employee, contractor, or business partner who (i) has or had authorized access to an organization's network, system, or data; (ii) has intentionally exceeded or intentionally used that access in a manner that negatively affected the confidentiality, integrity, availability, or physical well-being of the organization's information, information systems, or workforce.

Insider attacks within an organization include:

1. Low-tech attacks, such as modifying or stealing confidential or sensitive information for personal gain.

2. Theft of trade secrets or customer information to be used for business advantage or to give to a foreign government or organization.

3. Technically sophisticated crimes that sabotage the organization's data, systems, or network.

4. Workplace violence incidents that lead to loss of life or injuries.

For the purpose of this chapter, we focus on insider threats revolving around cyber means. More precisely, we aim to address the following question: in a risk analysis scenario, how much damage can be caused by a potential malicious insider, given the attack resources and the network assets?

Many systems contain layered security or what is commonly referred to as defense-in-depth, where valuable assets are hidden behind different layers or secured in numerous ways. Agnarsson, Greenlaw, and Kantabutra (2016a, 2016b) define a model for cybersecurity systems that uses defense-in-depth/layered-security approaches. We explain the technical details in the next paragraphs.

A *cybersecurity model (CSM)* is a tuple $M = (T, C, P)$, where $T = (V, E)$ is a rooted tree at r with n non-root vertices, $C = \{c_1, c_2, ..., c_n\}$ is a multiset of penetration cost, and $P = \{p_1, p_2, ..., p_n\}$ is a multiset of prizes. A *Security System (SS)*, denoted by (T, c, p), with respect to a cybersecurity model $M = (T, C, P)$ is given by two bijections $c : E(T) \to C$ and $p : V(T)\backslash\{r\} \to P$. A *System Attack (SA)* in a security system (T, c, p) is given by a subtree $\tau \subseteq T$ that contains the root r of T.

Given a security system (T, c, p), the authors define the cost and the prize of a system attack $\tau \subseteq T$ as

$$cost(\tau, c, p) := \sum_{e \in E(\tau)} c(e), \qquad (6.26)$$

$$prize(\tau, c, p) := \sum_{v \in V(\tau)} p(v). \qquad (6.27)$$

Fix a budget $B \in \mathbb{R}$. In a system attack, one of the main goals is to maximize the total prize p constrained by the budget B:

$$prize^*(B, c, p) := max\{prize(\tau, c, p) : \text{for all } \tau \subseteq T \text{ with } cost(\tau, c, p) \leq B\}. \quad (6.28)$$

In simple terms, the model at hand consists of a rooted tree in which the vertices represent a set of assets with their respective values, and the edges represent the road map to go from one asset to another together with the penalty associated to each step. Hence, a system attack is an instance where a subset of assets gets compromised via the routes between them. The prize of such an attack is the sum of the values of the compromised assets, and the cost is the sum of the penalties afforded to acquire them. Since no attacker has an infinite amount of resources, he/she tries to maximize the total prize of the attack while making use of his/her finite budget.

The rationale for this model is simple: in most systems, valuable assets are hidden behind different layers, such as antivirus software, intrusion detection systems, firewalls, encryption, etc. There is an associated cost to break into each level, such as money spent, time invested, penalties for denial, etc. On the other hand, there is an associated prize with every valuable asset that is stolen or compromised. The attacker's ultimate goal is to maximize the total prize given a fixed budget, whereas the defender wants to optimize his or her cost allocation to prevent the compromise of his or her valuable assets.

Alhajjar (2019) adopted this cybersecurity model to analyze the risk of an insider threat instance in cyberspace. We assume the attacker controls the root r of a tree T that represents any type of online network between agents. Our plan follows three steps:

Step 1: Find all rooted subtrees of the original tree; each system attack corresponds to a rooted subtree $\tau \subseteq T$.

Step 2: Find the cost and the prize of each subtree; each system attack is assigned a cost and a prize.

Step 3: Pick the subtree with the highest prize, whose cost is less than B. The system attack satisfying these constraints is the optimal one, i.e. the one causing the most profit to the attacker and the most damage to the defender.

Mathematically, we encode each subtree $\tau \subseteq T$ by a characteristic vector $X(\tau)$, i.e. a vector that has 1 in position i if the subtree contains the vertex v_i and 0 otherwise. Equations 6.26 and 6.27 then become:

$$cost(\tau, c, p) = X(\tau).C(T) \qquad prize(\tau, c, p) = X(\tau).P(T), \qquad (6.29)$$

where $C(T)$ and $P(T)$ denote the cost and the prize vectors of the tree T, respectively, and '.' is the dot product of vectors.

Figure 6.3 shows a tree T rooted at r with cost vector $C(T) = (1, 1, 1, 1, 1, 2)$ and prize vector $P(T) = (10, 2, 10, 3, 10, 40)$. The reader can think of an attacker positioned at node r, who is trying to infiltrate the network with assets worth either 2, 3, 10, or 40 units and with penetration costs worth either 1 or 2 units. It is not hard to see that, based on the attacker's budget (i.e. resources),

$$prize^*(B, c, p) = \begin{cases} 10\lfloor B \rfloor & 0 \leq B < 4 \\ 10\lfloor B \rfloor + 5 & 4 \leq B \leq 7 \\ 75 & B > 7 \end{cases} \qquad (6.30)$$

While there is no general formula to find the optimal prize of an attack, the problem can be solved in a case-by-case study depending on the network in question. Keep in mind that the essential goal of the attacker is to compromise as many assets as possible with the limited resources he/she has access to. Thus, finding the optimal prize is crucial for a defender to evaluate worst-case scenarios and for an attacker to cause maximum damage. Figure 6.4 shows the optimal subtree for the budget $B = 5$, in which the optimal prize is found to be $prize^*(5, c, p) = 55$ by Equation 6.30.

In a similar fashion yet a different context, a complex cyber-physical network refers to a new generation of complex networks whose normal functioning significantly relies on tight interactions between their physical and cyber components.

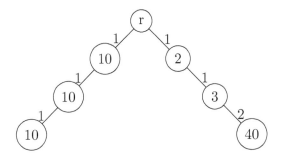

FIGURE 6.3 A rooted tree with 6 non-root vertices.

Wen, Yu, Yu, and Lu (2017) outline the cybersecurity of complex cyber-physical networks and suggest several security enhancing techniques. For a complex cyber-physical network, besides enhancing its cybersecurity, another important issue is related to its functioning, i.e. how to control the networks' collective behaviors for better optimization performance or completing various cooperative tasks.

Synchronization is one of the most fascinating collective behaviors in complex networks, which is a timekeeping behavior in which the states of all nodes in the network converge to the same trajectory. If the network is unable to self-synchronize to a desirable trajectory, then the pinning synchronization problem arises. The control objective of pinning synchronization is to make the states of all nodes in the considered network synchronize to a predesigned trajectory. Wen et al. (2017) address the security control problem of complex networks subject to malicious attacks, with emphasis on secure pinning synchronization control of complex networks in the presence of attacks on nodes. The problem is modeled as a directed tree rooted at the node describing the target system in the original network. A smart control center is embedded in the network for real-time monitoring, control, and operational decision-making; it simply takes the role of detecting the status of each node and activating the repair work.

In the same realm of complex cyber-physical networks, Wan, Cao, Chen, and Huang (2017) study the observer-based security control problem for complex dynamical systems in cyber-physical networks. They investigate the consensus tracking problem of dynamic networks under a cyber-attack scenario, which is assumed to impact the communication channels for the controllers and the observers in an independent way. To this end, an effective algorithm for selecting the feedback gain matrices and coupling strengths for the controllers and observers is designed. To guarantee that the consensus tracking can be achieved, sufficient conditions with parameters solved by two optimization problems are derived.

The problem is again modeled as a rooted tree in which the nodes represent different agents in the network. After the attacks have been detected by the network, a repairing mechanism is activated until the communication topology is recovered. Combining the information of the distributed observers, a control protocol for regulating the followers to track the leader (root of the tree) is then designed. Conditions for guaranteeing the tracking objective under frequent attacks are derived by

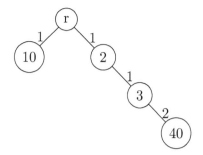

FIGURE 6.4 An optimal subtree for $B = 5$.

utilizing properly designed *Lyapunov functions* on networks. The reader is encouraged to refer to Wan et al. (2017) for the technical definitions and results.

There are plenty of issues and challenges about cybersecurity assessment, enhancing cybersecurity, and security control of complex networks that need to be addressed in the future. Specifically, in the Internet of Things (IOT), each node in various network infrastructures is meant to be sensed and controlled remotely across existing communication networks. For complex cyber-physical networks with a high number of nodes, the task of ensuring the cybersecurity in complex cyber-physical networks within the context of big data or cloud computing remains a challenging yet interesting issue.

6.4.4 Digital Viruses

It is common belief that digital viruses have long existed since the advent of computational devices. Malicious actors usually tend to cause harm to computers, destroy information stored in electronic devices, and/or benefit from stealing such information. We briefly discuss two types of digital viruses to highlight their connection to network science: computer viruses in Newman, Forrest, and Balthrop (2002) and mobile phone viruses in Wang, Gonzalez, Menezes, and Barabasi (2013).

The primary vehicle for transmission of computer viruses is electronic mail (email). Viruses typically arrive on a computer as an attachment to an email message which, when activated by the user, sends further copies of itself to other recipients. The email addresses of these other recipients are usually obtained by examining an email 'address book,' a file in which the user's correspondents' email addresses are stored.

From this address book, we can form a directed graph in which the nodes are computer users, and we place an edge from node v_i to node v_j if user j's email address appears in user i's address book. Newman et al. (2002) analyzed data gathered from a large university computer system serving $\sim 28,000$ users, then reduced that number to $\sim 16,000$ active users.

We now summarize the important conclusions inferred from the study. Technical and experimental details are omitted for the sake of brevity. The curious reader is encouraged to consult Newman et al. (2002) and its citations.

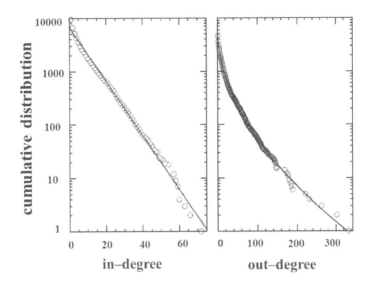

FIGURE 6.5 The in-degree and out-degree distributions of the network (see Newman et al. (2002)).

1. The average degree of a vertex is ∼3.38, which is both the average in-degree and out-degree.

2. The in-degree and out-degree distributions are faster decaying than the power law distribution (see Figure 6.5). More specifically, the in-degree distribution is exponential, and the out-degree distribution is stretched exponential.

3. In directed graphs, *reciprocity* measures the fraction of edges between vertices that point both ways. The network in hand has a reciprocity of 0.23, which is a much higher value than the one expected in random networks.

4. A *strongly connected component* is a subset of vertices in which each vertex can be reached from every other vertex. Typically, the network has one *giant strongly connected component (GSCC)*, which contains a significant fraction of the entire network, as well as a number of smaller strongly connected components. The network studied here has a giant component of size ∼54% of the total network.

5. In the targeted versus random node removal setting, it is shown that randomly removing nodes from the GSCC is an ineffective strategy to combat infection. On the other hand, the identification and the removal of the vertices most responsible for the spread of viruses (e.g. nodes with highest degree) increase the immunity of the remaining vertices.

We expand here on the conclusions enumerated above to put them in the right context. First, an average out-degree greater than one induces spread in any type of virus. Hence, the network of computer users is dense enough for the spreading phenomenon. An effective strategy to mitigate the spread of email viruses would be decreasing the average degree of the network to a threshold below one. Second, the degree distributions follow the preferential attachment introduced in Section 6.3.3. A plausible explanation of this observation is that individuals who already have large address books would be more likely to add to them than individuals who do not. Third, the value of the network reciprocity means that if there is an edge from vertex v_i to vertex v_j, then there is a 23% probability that there will also be an edge from v_j to v_i. This strongly suggests that the observed value is not the result of a random association of vertices; there is a high chance that you have a person in your address book if they have you in theirs, and vice versa. Fourth, it is the giant component that is responsible for large-scale virus epidemics: a virus outbreak that starts with a single vertex will become an epidemic if and only if that vertex falls in the giant component. In other words, the number of vertices infected in such an epidemic is equal at least to the size of the giant component.

This begs the following question: how can one defend against such virus outbreaks? The most common virus prevention strategies focus on random 'vaccination' of computers using antivirus software. However, it is shown in the same study that randomly removing nodes from the GSCC is an ineffective strategy to combat infection. On the contrary, the identification and the removal of the vertices most responsible for the spread of viruses (e.g. nodes with highest degree) has a great potential of increasing the immunity of the remaining vertices.

Likewise, the fast-growing market for smart phones coupled with their almost constant online presence makes these devices the new targets of viruses. Mobile phone viruses are able to self-replicate and spread quickly. Similarly to their biological counterparts, they can spread based on physical proximity when they use Bluetooth communication. Like computer viruses, they can spread by either targeting individuals in the address books of the infected phones (topological behavior), or by randomly selecting contacts/phone numbers to be contacted (scanning behavior).

Wang et al. (2013) investigate the interplay between spreading behaviors employed by Multimedia Message Services (MMS) viruses and the ability that phone providers have to look for anomalies based on messaging volume as a function of time of the day and day of the week. They use a dataset collected by a mobile phone provider for billing and operational purposes during a period of twelve weeks. MMS viruses can spread at different rates, and this rate can be the difference between their success or failure.

The main finding of the paper is that given enough time, sophisticated viruses can infect a large fraction of susceptible phones without being detected by phone providers. This is based on the topology and the dynamics of the so-called *call graph*, defined as follows. Every mobile user is a node, and two users are connected by an undirected link if there had been at least one reciprocated pair of phone calls between them. This procedure eliminates a large number of one-way calls, most of which correspond to single events.

Based on these results, phone providers need to improve their monitoring ability and install necessary patches on infected phones quickly so that MMS viruses can be better restrained. In order to avoid the costly impact of major outbreaks, they must acquire a good understanding of the network formed from connections between users so they can deploy smart anomaly detection schemes and prevent mobile phones from becoming the next platform for virus writers.

In summary, the study of digital viruses infecting computers and smart phones represents an increasingly important application of epidemic phenomena. Computer viruses display just as much diversity as biological viruses; depending on the nature of the virus and its spreading mechanism, the relevant contact network can differ dramatically. MMS viruses exploit the social network behind mobile communications, compared to Bluetooth viruses that take advantage of physical proximity. The networks underlying the spread of both computer and mobile viruses have been shown to be scale-free: they do not behave like random networks (see Section 6.3).

6.5 IMPLICATIONS FOR STUDENTS AND EARLY CAREER PRACTITIONERS

This chapter provides an overview of the field of network science and some of its countless applications to cybersecurity. After a brief introduction about the importance of these two fields, basic notions from graph theory are introduced as graphs are considered the building blocks of all real-world networks. Such notions include adjacency matrices, diameter, paths, connected component, trees, degree, clustering coefficient, centrality measures, etc. Following the prerequisite material, the reader discovers the three main network models in the literature: the Erdős-Renyi model, the Watts-Strogatz model, and the Barabási-Albert model. The focus in Section 6.3 is on pinpointing the differences between the three models, especially when it comes to the corresponding degree distributions.

Section 6.4 of the chapter is devoted to showcasing a handful of the plethora of applications of network science in cybersecurity. The reader should be warned that the choice of network science applications in cybersecurity is selective and subjective at the same time. Due to the timely relevance of both these areas of research, the number of publications and case studies that connect them is expected to grow continuously in the near and remote future. We end this section with some implications on future research that might impact students and early career researchers.

Real-life networks may appear as physical or non-physical networks. Examples of the first kind include servers/routers/computers and the wires that connect them, cities/towns/villages and the roads that connect them, etc. Other examples include websites and the links between them, people and their social connections, scientists and their citation history, etc. In some domains, it is relatively easy to gain access to a complete network dataset, such as social network studies of small groups. However, the acquisition of a complete network dataset is impossible in many applications due to technical and resource constraints.

Whenever a practitioner is collecting data, if the full dataset is hard to reach, sampling methods are used. These methods are usually based on the features of the nodes and/or the links, or even on the general structure of the network. Care must be taken in order to avoid introducing any bias in the data; such an encounter is dependent on the domain in which the data is being collected. Once the full network or the sampled version is mapped, the next step is to conduct basic measurements for the characterization of the network at hand (see Section 6.3). Local structure can be inferred from metrics like the clustering coefficient, statistical properties can be inferred from the degree distribution, and eventually the type of the network will be revealed through the detailed analysis of its structural and functional properties.

Most modeling techniques rely on the static nature of networks, i.e. the stationary properties for which the probability distribution is derived. However, many networks are not static but evolve over time, such as the creation of a social relation or the addition of a new hyperlink. The dynamic evolution of networks can be generally modeled by formally introducing a time variable t that indicates the changes of the network metrics in terms of time. In this situation, the number of nodes and links vary with time, which constitutes a crucial pillar to identify general growth mechanisms in the network at large.

Finally, there remains a lot to be done at the intersection of network science and cybersecurity. Due to the vast amounts of data being collected daily, there is no shortage of networks to be analyzed! Major gaps looming in these fields of research include: (i) developing new network models to better capture the behavior of real-world networks, (ii) benchmarking more network metrics that could help decipher unknown properties, (iii) crowdsourcing for open and publicly available code ready to be employed, and (iv) reaching out to general communities about the importance of the use of network analysis in their daily cybersecurity tasks.

ACKNOWLEDGMENTS

A part of this chapter was written during a visit to the New England Complex Systems Institute (NECSI) and the Massachusetts Institute of Technology (MIT) in Cambridge, MA. The author would like to thank both of these institutes for their hospitality. Moreover, the author acknowledges the support of the Intelligent Cyber-Systems and Analytics Research Lab (ICSARL) and the Network Science Center at the United States Military Academy in West Point, NY.

REFERENCES

Agnarsson, G. and Greenlaw, R. (2006). *Graph theory: Modeling, applications, and algorithms*. Prentice-Hall, Inc.

Agnarsson, G., Greenlaw, R., and Kantabutra, S. (2016a). On cyber attacks and the maximum-weight rooted-subtree problem. *Acta Cybernetica*, 22(3):591–612.

Agnarsson, G., Greenlaw, R., and Kantabutra, S. (2016b). The structure of rooted weighted trees modeling layered cybersecurity systems. *Acta Cybernetica*, 22(4):25–59.

Alhajjar, E. (2019). Insider threat risk analysis based on rooted trees. *Presentation at the Inaugural Workshop of Army Behavioral and Social Scientists, Fort Belvoir, VA*.

Alhajjar, E., Fameli, R., and Warren, S. (2021). Are terrorist networks just glorified criminal cells? *Northeast Journal of Complex Systems (NEJCS)*, 3(1), Art. 1.

Alhajjar, E. and Morse, S. (2020). A layered network collapse method in cyber terrorism. *15th International Conference on Cyber Warfare and Security, Norfolk, VA*, pages 18–22, XII.

Alhajjar, E. and Russell, T. (2019). A case study of the noordin network. *Proceedings of the SBP-BRIMS conference, Washington, DC*.

Barabási, A.-L. and Pósfai, M. (2016). *Network Science*. Cambridge University Press.

Barrat, A., Barthelemy, M., and Vespignani, A. (2008). *Dynamical Processes on Complex Networks*. Cambridge University Press.

Börner, K., Sanyal, S., and Vespignani, A. (2007). Network science. In Cronin, Blaise (Eds.), *Annual Review of Information Science Technology, chapter 12*. American Society for Information Science and Technology. 41:537-607.

Cappelli, D. M., Moore, A. P., and Trzeciak, R. F. (2012). The cert guide to insider threats: How to prevent, detect, and respond to information technology crimes. In *Addison-Wesley Professional, 1st edition*.

Dawood, H. A. (2014). Graph theory and cyber security. In *2014 3rd International Conference on Advanced Computer Science Applications and Technologies*, pages 90–96. IEEE.

Denning, D. E. Cyberterrorism: Testimony before the special oversight panel on terrorism committee on armed services us house of representatives. *Focus on Terrorism*, 9.

Easley, D. and Kleinberg, J. (2010). *Networks, crowds, and markets*. Cambridge University Press.

Everton, S. F. (2009). Network topography, key players and terrorist networks. *Terrorist Networks 1, 12-19*.

Everton, S. F. (2012). *Disrupting dark networks*. Cambridge University Press.

Jackson, M. O. (2008). *Social and Economic Networks*. Princeton University Press, USA.

Jha, S., Sheyner, O., and Wing, J. (2002). Two formal analyses of attack graphs. In *Proceedings 15th IEEE Computer Security Foundations Workshop. CSFW-15*, pages 49–63. IEEE.

Lallie, H. S., Debattista, K., and Bal, J. (2020). A review of attack graph and attack tree visual syntax in cybersecurity. *Computer Science Review*, 35:100219.

National-Research-Council (2005). *Network science*. The National Academies Press, Washington, DC.

National-Security-Archive (2018). https://nsarchive.gwu.edu/news/cybervault/2018-09-19/cyber-glossary. Last accessed November 30, 2020.

Newman, M. E. J. (2010). *Networks: An introduction*. New York: Oxford University Press.

Newman, M. E. J., Forrest, S., and Balthrop, J. (2002). Email networks and the spread of computer viruses. *Physical Review E*, 66(3):035101.

Roberts, N. and Everton, S. F. (2011). Strategies for combating dark networks. *Journal of Social Structure 12, 1-32*.

Wan, Y., Cao, J., Chen, G., and Huang, W. (2017). Distributed observer-based cybersecurity control of complex dynamical networks. *IEEE Transactions on Circuits and Systems I: Regular Papers*, 64(11):2966–2975.

Wang, P., González, M. C., Menezes, R., and Barabási, A.-L. (2013). Understanding the spread of malicious mobile-phone programs and their damage potential. *International Journal of Information Security*, 12(5):383–392.

Wen, G., Yu, W., Yu, X., and Lü, J. (2017). Complex cyber-physical networks: From cybersecurity to security control. *Journal of Systems Science and Complexity*, 30(1):46–67.

West, D. B. (2000). *Introduction to graph theory*. Upper Saddle River: Prentice Hall.

CHAPTER 7

Operations Research

Paul L. Goethals

Natalie M. Scala

Nathaniel D. Bastian

CONTENTS

7.1	Introduction		233
7.2	Decision Analysis		234
	7.2.1	Decision Strategies	235
		7.2.1.1 Attack Trees	236
	7.2.2	Utility Functions	237
	7.2.3	Value Functions	239
	7.2.4	Group Decision Making	240
	7.2.5	Other Approaches	241
	7.2.6	Case Study: Cyber Attack on a Voting Process	241
7.3	Mathematical Optimization		244
	7.3.1	Linear Optimization	245
	7.3.2	Nonlinear Optimization	247
	7.3.3	Integer Programming	248
	7.3.4	Other Approaches	248
	7.3.5	Case Study: Network Interdiction	250
7.4	Stochastic Process Modeling		251
	7.4.1	Markov Processes	252
	7.4.2	Queueing Models	254
	7.4.3	Other Approaches	258
	7.4.4	Case Study: Malware Spread	259
7.5	Conclusion		261

7.1 INTRODUCTION

In the last eighty years, the discipline of operations research has been used extensively to provide analytical evidence for research outcomes in the mathematical sciences. Problems of interest may involve variations of cost-benefit analyses, the measurement of attributes or features, differing levels of uncertainty, and constrained or unconstrained frameworks. The analyses are not always concerned with generating unique optimal solutions; often, an array of near-optimal solutions may be more appropriate depending on the assumptions, factors, and objectives of a particular study. Research work may also be more aligned toward investigating how complex processes behave either in a static or dynamic manner. Sensitivity analyses may accompany outcomes to provide an indication of the robustness of different solutions. In summary, the assortment of methods, tools, and analytics in operations research is vast and growing. Emerging fields such as data science and techniques such as machine learning will fuel continued growth in this mathematical discipline for many years to come.

Interestingly, the domain of cyber theory and applications is also growing, and the products and processes in this realm have qualities and characteristics that warrant using methods in operations research. Cyber attacks can be modeled using hierarchical threat structures and may involve competing decision strategies from both an organization or individual and the adversary. Network traffic flow, intrusion detection and prevention systems, interconnected human-machine interfaces, and automated systems all require higher levels of complexity in modeling and possess inherently random sub-processes. Attributes, such as cyber resiliency, network adaptability, security capability, and information technology flexibility require the measurement of multiple characteristics, many of which may involve both quantitative and qualitative interpretations. For nearly every organization that is invested in some cybersecurity practice, decisions must be made that involve the competing objectives of cost, risk, and performance.

This chapter will highlight only a small portion of the operations research space. The sub-discipline of *decision analysis* is first described, which involves the modeling or measuring of actions, factors, or responses using mathematics to aid in decision making. The sub-discipline of *mathematical optimization* is then discussed, which involves solving for the most desirable outcome or solutions given a constrained or unconstrained framework. The last broad topic, techniques in *stochastic process modeling* are provided to account for the random or uncertain behavior of cyber processes and where insight and greater forecasting power may be achieved. Along the way, the focus will be on the mathematics associated with these sub-disciplines of operations research. Finally, at the conclusion of this chapter, research implications and extensions are offered to the examiner who desires to carry these practices forward in cyber research.

7.2 DECISION ANALYSIS

Decision analysis is an important area within operations research. Related methods and tools can be probabilistic or deterministic and typically address risk, sequential decision making, group decision making, and/or quantifying qualitative data. Decision analysis is useful for scoping decision problems and even includes several accessible methods to enable those without mathematical backgrounds to make data-driven decisions. These methods are especially useful for research and development, capital, or high consequence strategic decisions, especially when the decision process involves qualitative or incomplete information. Decision analysis methods are mostly used in industry, especially by senior management, for these reasons. This section identifies some of the most used decision analysis methods and then details an example for the use of utility with cyber attack trees.

7.2.1 Decision Strategies

Decision analysis methods assist in the evaluation of tradeoffs in decision making and frequently incorporate human elements or qualitative data into the process. The models consider alternatives and events. Alternatives are the courses of action or strategies from which the decision maker ultimately chooses; the decision maker has control. Events (also called states of nature or outcomes) are conditions that may occur in the future over which the decision maker has little or no control. The events under consideration in a decision analysis problem should be mutually exclusive and collectively exhaustive. Payoffs are also typically examined, which are some sort of quantitative reward or cost if a given alternative and event combination comes to fruition in the future.

Managers and leaders may need to make a decision that affects the future, takes on some risk, but has to be made now, while not knowing what the future will bring (i.e., imperfect information). When the probability of future events is unknown or equally likely, decision strategies such as average payoff, aggressive, and conservative are used. The choice of strategy begins with an assessment of the decision maker's risk appetite: risk neutral will apply the average payoff strategy; risk seeking will apply the aggressive strategy; and risk averse will apply the conservative strategy. Alternatives, events, and payoffs are arranged in a table, and the best-case decision is chosen in alignment with the strategy at hand. For example, the average payoff, or risk neutral strategy, averages the payoffs for each alternative. The alternative with the largest average payoff is chosen for a decision with a maximum objective; the alternative with the smallest average payoff is chosen for a minimum objective. For the aggressive strategy, decision rules known as minimin and maximax are used for the minimum and maximum objective, respectively. For the conservative strategy, the minimax rule is used for the minimum objective, and the maximin rule is used for the maximum objective. Further discussion and examples can be found in Goodwin and Wright (2004) and Evans (2016).

When the probability of events is known or can be estimated, expected monetary value (EMV) is used to evaluate decision alternatives, regardless of the decision maker's risk appetite. The formula for EMV of an alternative a_j, is:

$$\text{EMV}(a_j) = \sum_{i=1}^{n} p_i x_{ij},$$

where n is the number of events that could happen in the future, p_i is the probability of event i occurring and x_{ij} is the payoff for event i and alternative j. Alternatives, events, and payoffs are organized into a decision tree, where event nodes are represented by circles, decision nodes are represented by squares, and branches with probability connect the nodes. Payoffs are placed at the terminal ends of the branches, and the EMV is used to identify the best-case decision by working backward from the end of the tree and calculating EMV at every event (circle) node. The maximum or minimum alternative is chosen at each decision (square) node, in alignment with the problem objective. See Goodwin and Wright (2004) and Evans (2016) for examples of decision trees.

Because EMV and decision trees do not consider risk up front, risk profiles and evaluations of the optimal decision are done after the tree calculations; an example is in Evans (2016). Such considerations are necessary, as the decision maker's risk appetite must be considered. EMV operates under the mathematical assumption of repeated trials, or repeated decisions, similar to expected value. In reality, decisions are made only once and typically have large consequences. Therefore, the long run average that is essentially calculated by the EMV is not necessarily reflective of risk, and implications must be considered to align the optimal decision with the decision maker's risk tolerance.

7.2.1.1 Attack Trees

Attack trees are graphical representations of a security problem and identify all of the ways a system can be attacked. Once the structure of the tree is identified along with all methods to attack the system, mitigations can then be designed; the tree assists in identifying and motivating the need for those mitigations. Furthermore, when resources are scarce or the attack tree is complex, evaluation of the tree, through risk analysis and utility theory, can help to identify return-on-investment for mitigations as well as the worst-case or highest probability threats that need attention. Attack trees were first proposed by Schneier (1999) and are now widely used in practice. The methodical design of the tree helps to understand the attacker's ability, motivation, and goals, which, in turn, generate intelligent security countermeasures (Schneier, 1999); the nodes of the tree include all immediate, necessary, and sufficient causes for the occurrence of the overall attack goal of the tree (Du and Zhu, 2013).

The design of an attack tree is formal and takes a logical, deliberate approach. First, a brainstorm of all potential attacks or methods of attacks is done, breaking each type of attack down to the lowest level or single actions that must be taken. This decomposition of more complex thoughts or actions enumerates all threats and aids in both the understanding of the full scope of threat, and the type of countermeasures needed. The visual design of the tree is standard in the literature and uses three shapes connected by straight lines to form a hierarchy of actions; the hierarchy terminates at the lowest level, or single actions to take to attack a system. The three standard shapes are (i) circles to denote terminal or leaf nodes, the lowest level, single actions; (ii) rounded triangles to denote OR relationships, where at least one leaf node in the next lowest hierarchy level must be executed to attack the system; and (iii) flat-bottom ovals to denote AND relationships, where all leaf nodes in the next lowest hierarchy level must be executed together to attack the system. Attack trees can be complex and involve hundreds or thousands of

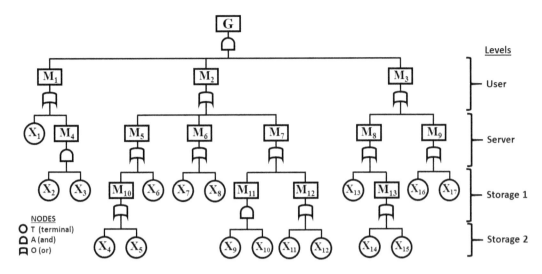

FIGURE 7.1 Example of attack tree depth corresponding to network architecture difficulty.

nodes. They can be drawn in a hierarchy to visualize the attack space or listed in hierarchical form. To list, each branch is given a number, and an extension to that numerical code is made as the nodes on the branches move to lower levels of the hierarchy.

Let G represent an attacker's goal and $M_1, M_2, \ldots, M_q$ represent the attacker sub-goals, with $X_1, X_2, \ldots, X_r$ representing the array of events that can occur in a tree. The jth attack sequence S_j is a combination of k different events, or terminal leaf nodes, $S_j = \{X_{j1}, X_{j2}, \ldots, X_{jk}\}$, one of multiple scenarios that can occur toward reaching an attacker's goal G. The assessed difficulty for an attacker in reaching their goal may be related to the hierarchical structure, whereby increased levels of network architecture correspond with greater attack difficulty. For example, a phishing attack with the goal of acquiring stored data may have a structure similar to that observed in Figure 7.1, where the depth of the tree relates to physical systems. An attack tree may also relate to the costs $C_1, C_2, \ldots, C_l$ associated with various scenarios S_j. For example, it may be more costly for a cyber attacker in terms of attribution to acquire one set of data over another; in this manner, differentiation exists in attack probabilities among the various branches of the tree, as shown in Figure 7.2. Finally, a tree may involve weighting schemes based upon a defender's priorities or the amount of loss incurred due to attacking a particular system.

7.2.2 Utility Functions

The utility of an alternative or course of action is, generally speaking, a numerical score that identifies the attractiveness of the alternatives to the decision maker, when the decision involves risk and uncertainty (Goodwin and Wright, 2004). If the decision space is deterministic (without uncertainty or probability), the score is referred to as value instead of utility. To define utility, consider all potential payoffs

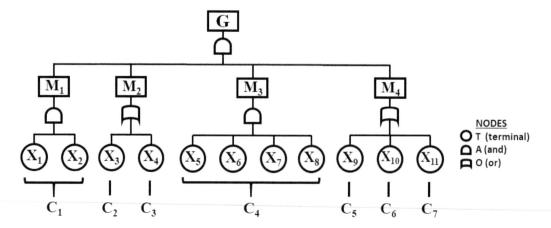

FIGURE 7.2 Attack tree variation in cost according to scenario.

of the decision problem, ordered from largest to smallest. The best payoff p^+ receives $u(p^+) = 1$, and the worst payoff p^- receives $u(p^-) = 0$. Utilities for intermediate payoffs are elicited from the decision maker by considering hypothetical lotteries; the most common approach for these lotteries is probability equivalence (Goodwin and Wright, 2004). To assess utility using this approach for an intermediate payoff x_i, the decision maker is offered a hypothetical choice between receiving x_i for certain or entering a lottery where they may receive p^+ or p^- with specified probabilities; the probabilities are varied until the decision maker is indifferent between the lottery and x_i. The process is repeated until all intermediate payoffs i are assessed, and $0 < u_i(x_i) < 1 \; \forall \; i$ by definition.

Graphing the utilities for all payoffs reveals a decision maker's utility function and risk appetite. The graph is designed such that payoffs are on the x-axis and utilities are on the y-axis. Functions that graph as convex convey a risk seeking attitude, while functions that graph concave convey a risk averse attitude. Risk neutral attitudes graph as straight lines and represent the decision maker's preference as the EMV (Goodwin and Wright, 2004). S-shaped curves denote a combination of risk seeking and risk averse appetites, depending on payoff.

Utility functions may come in many different forms. One commonly used utility function is the n-attribute additive utility function shown below:

$$u(\mathbf{x}) = \sum_{i=1}^{n} w_i u_i(x_i), \text{ where weights } 0 \leq w_i \leq 1 \text{ and } \sum_{i=1}^{n} w_i = 1$$

Another common utility function is the n-attribute multiplicative model, which has the form:

$$u(\mathbf{x}) = \prod_{i=1}^{n} u_i(x_i)^{w_i}, \text{ where weights } 0 \leq w_i \leq 1 \text{ and } \sum_{i=1}^{n} w_i = 1$$

The single-attribute exponential utility function is also frequently found in the literature:

$$u_i(x_i) = 1 - e^{-rx_i}, \text{ where } r \text{ is the risk aversion coefficient.}$$

Many multi-attribute utility functions assume independence of the functions. Without mutual utility independence, the assessment or interpretation of utility functions can be complex; see Keeney and Raiffa (1976) for a discussion. Finally, when considering the nature of a cyber attack, varying attitudes toward risk and differing payoffs and losses from the attacker and defender perspectives may be considered. Extensions of how these functions may be used to model such events are clear.

7.2.3 Value Functions

Utility functions are in contrast to value functions, as value functions do not consider risk and probability. Techniques such as multiple objective decision analysis (MODA) assess value in deterministic problem spaces through value functions. Attributes that are to be considered as part of the decision are arranged in a value hierarchy, and weighted averages are used to calculate preference. Value functions are elicited for the attributes, and alternatives are evaluated against the functions. Further discussion can be found in Keeney and Raiffa (1976) and Parnell, et al. (2013). An example of MODA applied to cybersecurity can be found in Scala and Goethals (2020).

One popular multi-objective approach that utilizes linear or non-linear value functions is the use of the desirability function approach. The method, first introduced by Harrington (1965) and later modified by Derringer and Suich (1980) as well as Derringer (1994), has been used widely in the latter part of the 20th century plus the last twenty years to identify optimal parameters for processes. Suppose we have a total of m responses or characteristics, where the qth response of interest, for $q = 1, 2, ..., m$, is denoted as y_q. The method calls first for estimating each of the responses with fitted response surface functions $\hat{y}_1(\mathbf{x}), \hat{y}_2(\mathbf{x}), ..., \hat{y}_m(\mathbf{x})$, where $\mathbf{x} = x_1, x_2, ..., x_n$ represents a set of factors influencing each response. A transformation of each estimated response is then performed resulting in individual desirability functions $d[\hat{y}_1(\mathbf{x})], d[\hat{y}_2(\mathbf{x})], ..., d[\hat{y}_m(\mathbf{x})]$, where $0 \leq d[\hat{y}_q(\mathbf{x})] \leq 1$, based upon whether the objective is to maximize or minimize the individual response. This transformation depends on the identification of objective values $\tau_1, \tau_2, ..., \tau_m$, as well as lower and upper constraints ℓ_q and u_q, respectively, and shape parameters $\gamma_1, \gamma_2, ..., \gamma_m$ for each response. Finally, given pre-defined weights $w_1, w_2, ..., w_m$, corresponding to the priority of each response in the problem, a composite desirability function D in the form of a geometric mean is maximized. This multi-objective formulation is represented as follows:

$$\text{Max } D = \{d[\hat{y}_1(\mathbf{x})]^{w_1} \cdot d[\hat{y}_2(\mathbf{x})]^{w_2} \cdots d[\hat{y}_m(\mathbf{x})]^{w_m}\}^{1/\sum_{q=1}^{m} w_m}$$

$$\text{s.t. } d[\hat{y}_q(\mathbf{x})] = \begin{cases} 0 & \text{if } \hat{y}_q(\mathbf{x}) > u_q \\ \left(\frac{\hat{y}_q(\mathbf{x}) - u_q}{\tau_q - u_q}\right)^{\gamma_q} & \text{if } \tau_q \leq \hat{y}_q(\mathbf{x}) \leq u_q, \\ 1 & \text{if } \hat{y}_q(\mathbf{x}) < \tau_q \end{cases}$$

if the objective is to minimize the qth response, and

$$d[\hat{y}_q(\mathbf{x})] = \begin{cases} 0 & \text{if } \hat{y}_q(\mathbf{x}) < \ell_q \\ \left(\frac{\hat{y}_q(\mathbf{x}) - \ell_q}{\tau_q - \ell_q}\right)^{\gamma_q} & \text{if } \ell_q \leq \hat{y}_q(\mathbf{x}) \leq \tau_q, \\ 1 & \text{if } \hat{y}_q(\mathbf{x}) > \tau_q \end{cases}$$

if the objective is to maximize the qth response

With this construct, individual and composite desirability functions with a value equal to one are considered ideal, whereas values equal to zero are considered undesirable.

The Analytic Hierarchy Process (AHP) is another approach to deterministic value and uses pairwise comparisons of the attributes on the value hierarchy with respect to the decision goal as well as pairwise comparisons of the alternatives with respect to the attributes. The comparisons are synthesized via linear algebra and eigenvalues; a full discussion of the AHP can be found in Saaty (1990) and Vargas (1990). The decision analysis community has argued over the use of MODA-based value functions versus the AHP; the technical savvy of the decision maker as well as the problem design should guide the decision of which method to use.

7.2.4 Group Decision Making

When decision makers are in groups, the AHP offers methods to aggregate group decisions into a single judgment that can be used to calculate the preference alternative. Dispersion of group judgments should be first calculated (Saaty and Vargas, 2007). Then if dispersion is low, the geometric mean or a weighted geometric mean should be used to aggregate the judgments (Aczél and Saaty, 1983; Aczél and Alsina, 1987). If dispersion is high, the decision makers should consider revising their judgments. If they are unable or unwilling to do so, then an aggregation based on principal components should be used (Scala et al., 2016).

For utility and value judgments, behavioral techniques are preferred for aggregation. Simple mathematical averages can be used when every group member is unbiased and any random error in a group member's judgment is independent of the errors of other members; this is rarely the case in practice (Goodwin and Wright, 2004). Weighted averages also can be used, assigning weight relative to the importance of each group member's assessment; the problem is accurately assessing the weights based on group skill (Goodwin and Wright, 2004). Therefore, behavioral methods are typically better to use; an example is the Delphi method. In this method, decision makers submit their initial assessments to a proctor or facilitator, who then shares the judgments with the group without attribution. Discourse follows, the decision makers resubmit assessments, and the process repeats until consensus is reached. A summary of the Delphi method can be found in Goodwin and Wright (2004), and the full treatment is available in Helmer-Hirschberg

(1967). Biases also need to be considered when obtaining decision-maker assessments; Goodwin and Wright (2004) provide a tutorial on overcoming bias in elicitation.

7.2.5 Other Approaches

Another popular technique in decision analysis is the use of an influence diagram, which is a graphical depiction of a decision problem. All factors and forces are represented to gain an understanding of the decision space. Like decision trees, circles and squares are used to depict events (no control) and decisions (control), respectively. Influence diagrams may be used to inform the construction of decision trees as the diagram summarizes the dependencies that exist among events and decisions (Goodwin and Wright, 2004). Examples of influence diagrams can be found in Scala, Rajgopal, and Needy (2010) and Alves et al. (2020). Affinity diagrams are a closely related method that assists with identification of attributes when creating value hierarchies; Value Focused Thinking (VFT) can also be used to brainstorm and define a decision problem. A discussion on VFT can be found in Keeney (1992; 2008) and Parnell et al. (2013); an example of VFT and affinity diagrams is shown in Scala and Pazour (2016).

7.2.6 Case Study: Cyber Attack on a Voting Process

To illustrate decision analysis techniques, consider an investigation into cybersecurity related to voting systems. Figure 7.3 depicts an attack tree for manipulating votes cast via optical scanners, a type of electronic voting equipment; the figure is an adaptation of an attack tree created by the University of South Alabama, in partnership with the Election Assistance Commission (EAC) in 2009. Voters using optical scanners at polling places darken bubbles on paper ballots next to the names of their choice candidates; the paper ballots are then fed through a large scanning machine, which counts the votes. Votes are then stored on removable media or transmitted via the internet to a central server. Therefore, to attack the voting data, either malware has to be used or the data has to be modified on the storage medium. Then, for using malware as an attack to be employed, either a method needs to be selected and filtered, a ballot definition file is attacked, device tallies are attacked, or tabulation software is attacked. Three methods can be selected and filtered, so each are identified on the next level of the hierarchy

First, three independent attributes are defined: (i) the cost of performing an attack, x_1, (ii) the technical difficulty in performing an attack, x_2, and (iii) the level of difficulty in discovering an attack, x_3. These attributes and the corresponding approach are adapted from Du and Zhu (2013). Then, an ordinal one-to-five scale is defined for each attribute to denote the relative cost or difficulty of each, with five denoting the highest cost or most difficulty. Using this scale, the terminal nodes X_i are individually evaluated for x_1, x_2, and x_3 by a set of subject-matter experts. The Delphi method may then help to reach consensus in assessing value, while minimizing group power dynamics. Since the scores of each attribute do not

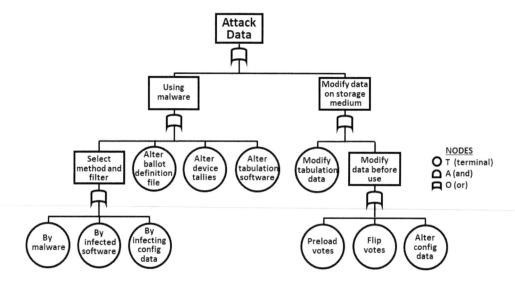

FIGURE 7.3 Attack tree for voting data, adapted from EAC (2009).

depend on the ordinal scale values of the other attributes, the condition of mutual utility independence is met.

To assess the occurrence probability of the events in each tree, a series of calculations is then performed. Since the attributes are inversely proportional to utility, they are converted to a [0, 1] scale using a factor c, with c/x; for this case study, with an ordinal one-to-five scale, $c = 0.2$. Using a general multi-attribute additive utility function:

$$u(\vec{\mathbf{x}}) = w_1 u(x_1) + w_2 u(x_2) + w_3 u(x_3), \text{ where } \sum_{i=1}^{3} w_i = 1,$$

an occurrence is calculated for each node X_i. With the equal weighting of attributes being employed for this case study, the scores in Table 7.1 result. Higher occurrence probabilities require more attention and stronger mitigations from decision makers and information technology professionals who defend the network for which the attack tree is defined. Following Table 7.1, in this example, nodes X_1, X_2, and X_3 all have the highest occurrence likelihoods of 0.083. A complete and detailed example of this approach can be found in Scala et al. (2021).

To determine the total occurrence probability of attack across the entire tree, a fault tree approach is taken. Considering Figure 7.3 and simplifying the discussion, we define 'using malware' as M_1, 'modify data on storage medium' as M_2, 'select method and filter' as M_3, and 'modify data before use' as M_4. Start at the bottom of the tree and calculate the occurrence likelihood of the nodes based upon additive probability formulas. For instance, since M_3 is an OR node, the probability of the union of three events is used to calculate the occurrence likelihood at M_3, whereby

TABLE 7.1 Scoring of attributes and occurrence probability

Node	Vulnerability	x_1	x_2	x_3	Scaled x_1	Scaled x_2	Scaled x_3	Occurrence Probability
X_1	By malware	2	4	2	0.100	0.050	0.100	0.083
X_2	By infected software	2	4	2	0.100	0.050	0.100	0.083
X_3	By infecting config data	2	4	2	0.100	0.050	0.100	0.083
X_4	Alter ballot definition file	3	4	2	0.067	0.050	0.100	0.072
X_5	Alter device tallies	3	4	2	0.067	0.050	0.100	0.072
X_6	Alter tabulation software	3	4	2	0.067	0.050	0.100	0.072
X_7	Modify tabulation data	3	3	2	0.067	0.067	0.100	0.078
X_8	Preload votes	3	3	4	0.067	0.067	0.050	0.061
X_9	Flip votes	3	4	2	0.067	0.050	0.100	0.072
X_{10}	Alter config data	3	3	3	0.067	0.067	0.067	0.067

we assume independence for all events in the tree space:

$$\begin{aligned}P(M_3) &= P(X_1 \cup X_2 \cup X_3)\\ &= P(X_1) + P(X_2) + P(X_3) - P(X_1 \cap X_2) - P(X_1 \cap X_3) - P(X_2 \cap X_3)\\ &\quad + P(X_1 \cap X_2 \cap X_3)\\ &= P(X_1) + P(X_2) + P(X_3) - P(X_1)P(X_2) - P(X_1)P(X_3) - P(X_2)P(X_3)\\ &\quad + P(X_1)P(X_2)P(X_3)\end{aligned}$$

For this case study, the total occurrence likelihood of attacking data is $P(G) = 0.539$, where $P(M_1) = 0.385$, $P(M_2) = 0.250$, $P(M_3) = 0.230$, and $P(M_4) = 0.187$. If the attack tree is known to have shared sub-trees or dependencies among the nodes, alternative techniques for calculating these occurrences will have to occur (Kordy et al., 2014).

This example examined an attack tree with singular leaf nodes and OR relationships as attack scenarios. However, in an AND structure, attack scenarios would have to incorporate multiple attack events occurring at the same time. To illustrate an AND node, Figure 7.4 shows an attack tree on election evidence via a post-election audit. Deliberately modifying data requires two actions in order to be successful: (i) replacing paper tape with fraud and (ii) rewriting data on media. So, in this case, for an attack on election evidence to be executed, any of the following events can occur: {destroy election artifacts}, {mishandle election artifacts}, {add new fraudulent evidence}, {replace paper tape with fraud, rewrite data on media}, {unintentional modify}, {modify deliberately by computer}, {unintentional modify by computer}, {modify via malware attack}, {modify via malware at creation}. The AND node identifies that both actions related to 'deliberately modify' have to occur together in order to execute the attack.

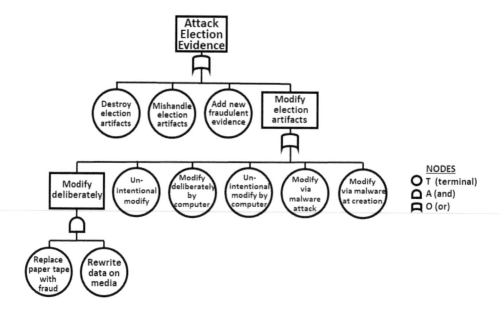

FIGURE 7.4 Attack tree for election evidence, adapted from EAC (2009).

To calculate the probability of such events, the occurrence probability of the leaf nodes would need to be multiplied together (joint probability) to find the probability of the AND attack scenario. For example, in Figure 7.4, the probability for the one AND scenario is the product of the occurrence likelihoods for the two terminal nodes, namely {replace paper tape with fraud} and {rewrite data on media}. Then, when calculating the total probability of the attack tree, the probability of the intersection of events at AND nodes is used in the evaluation.

7.3 MATHEMATICAL OPTIMIZATION

Mathematical optimization is a sub-field of operations research concerned with the theory, solution methods, and modeling techniques for finding the extrema of functions on sets defined by linear and/or nonlinear constraints (equalities and inequalities) in a finite-dimensional vector space. In more practical terms, mathematical optimization is concerned with finding a best decision, according to some criterion, out of a (typically much) larger set of decisions. The respective mathematical model will involve decision variables (one for every elementary decision, often called a solution), constraints (limitations on some combination of decision variables), and an objective function that measures the quality of the solution. Mathematical optimization has been widely accepted because of its ability to model important and complex decision problems across many application domains, including cyber.

In the following sections, we introduce linear optimization, the fundamental modeling variant of mathematical optimization and demonstrate the application of mathematical optimization for cyber research by providing a problem related to network flow interdiction.

7.3.1 Linear Optimization

Linear optimization, which is one of the key branches of mathematical optimization, seeks to minimize or maximize a linear function while satisfying a set of linear constraints (equalities and/or inequalities). The conception of the general class of optimization problems, which is a landmark event in the history of mathematics, and its applications that enabled the ability to solve general systems of linear constraints, is credited to George B. Dantzig for his work around 1947 while developing a deployment, training, and logistics planning tool for the United States Air Force Comptroller (Dantzig, 1948). Since then, there has been a plethora of contributions to the field of linear optimization in terms of both theoretical developments and computational advancements, along with many new and emerging application areas.

We begin with formulating the general Linear Program (LP), where $c_1x_1+c_2x_2+\cdots+c_nx_n$ is the linear *objective function* to be minimized (often denoted by z); $c_1, c_2, \ldots, c_n$ are the *cost coefficients*, and $x_1, x_2, \ldots, x_n$ are the *decision variables*. The following mathematical programming formulation represents the construct of the general LP:

$$\begin{aligned}
\text{Min} \quad & z = c_1x_1 + c_2x_2 + \cdots + c_nx_n \\
\text{s.t.} \quad & a_{11}x_1 + a_{12}x_2 + \cdots + a_{1n}x_n \leq b_1 \\
& a_{21}x_1 + a_{22}x_2 + \cdots + a_{2n}x_n \leq b_2 \\
& \vdots \quad + \quad \vdots \quad + \cdots + \quad \vdots \quad \vdots \\
& a_{m1}x_1 + a_{m2}x_2 + \cdots + a_{mn}x_n \leq b_m \\
& x_1, x_2, \ldots, x_n \geq 0
\end{aligned}$$

In this formulation, the linear inequality $\sum_{j=1}^{n} a_{ij}x_j \leq b_i$ represents the ith constraint, where a_{ij} are called the *technological coefficients*. The right-hand-side vector, which is the column vector whose ith component is b_i, represents the minimal requirements to be satisfied by the constraints. Finally, the non-negativity constraints are represented by $x_j \geq 0 \; \forall j = 1, \ldots, n$. A set of values of the decision variables $x_1, x_2, \ldots, x_n$ that satisfies all of the linear constraints is known as a feasible solution; sets of these points form the feasible region. Thus, solving a general LP is the problem of finding an optimum feasible solution such that a given linear objective function z is minimized subject to a system of linear constraints (Bazaraa et al., 2011).

Consider the following LP problem:

$$\begin{aligned}
\text{Max} \quad & z = 30x_1 + 10x_2 \\
\text{s.t.} \quad & x_1 + 2x_2 \leq 16 \\
& -x_1 + \tfrac{1}{3}x_2 \leq 0 \\
& x_1 + \tfrac{1}{2}x_2 \leq 8 \\
& x_1, x_2 \geq 0
\end{aligned}$$

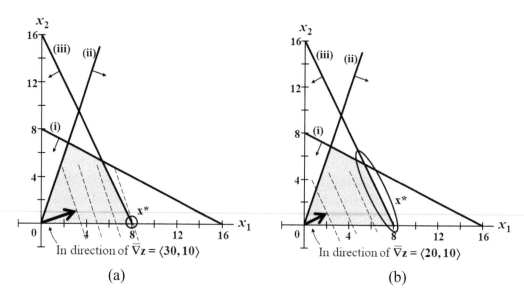

FIGURE 7.5 Feasible region illustration with (a) unique and (b) infinite optimal solutions.

In this simplified example, we have two decision variables x_1 and x_2, and the linear objective function to be maximized is $30x_1 + 10x_2$. The linear constraints define the feasible region, such that the optimization problem is to find a point (feasible solution) in the feasible region having the smallest possible objective function value. Given this two-dimensional construct where the first three constraints are labeled (i), (ii), and (iii), we can observe the solution graphically as shown in Figure 7.5(a). In this illustration, we observe the shaded feasible region, the gradient of the objective function z found by $\vec{\nabla} z = \langle z_{x_1}, z_{x_2} \rangle$, and the resulting level curves of z that identify the maximum being at $\mathbf{x}^* = (8, 0)$, where $z = 240$.

It is noted that with a slight change to the objective function, i.e., changing $c_1 = 30$ to $c_1 = 20$, the gradient vector for the objective function becomes $\vec{\nabla} z = \langle 20, 10 \rangle$, and we find an infinite number of feasible solutions along the boundary of the feasible region, as illustrated by a circle around the $\mathbf{x}^*$ line in Figure 7.5(b). This method of analyzing how changes to problem variables affect the solution space is called *sensitivity analysis*; if small changes result in significant changes to the solution, the problem lacks robustness. The concept of sensitivity analysis is always of great interest to the operations research analyst, as it relates directly to the foundation by which decisions may be made.

In order to represent an optimization problem as an LP, there are several assumptions implicit within the mathematical formulation. First, the objective contribution and resource usage of each activity is proportional to the level of that activity (i.e., proportionality). Second, every function (objective function and constraints) is the sum (i.e., linear combination) of the individual contributions of the activities (i.e., additivity). Third, the decision variables can take any fractional (non-integer) values (i.e., divisibility). Finally, the value assigned to each parameter (cost and/or technological coefficient) is a known deterministic constant (i.e., certainty) (Bazaraa et al., 2011).

If a single optimal solution exists for a particular LP problem, then the optimal solution exists at an extreme point (i.e., corner point) of the feasible region. These extreme points (a geometric notion) correspond to basic feasible solutions (an algebraic characterization); in other words, a point within the feasible region is a basic feasible solution if and only if it is an extreme point. Since a LP problem having a finite optimal objective function value has an optimal solution at an extreme point, an optimal basic feasible solution can always be found for such a problem. In order to solve LP problems, Dantzig (1963) developed the first computationally viable solution technique known as the Simplex Method, which traverses only extreme point solutions of the feasible region in search of continually improving objective function values; when value can no longer improve, the algorithm terminates. The simplex method also discovers when the feasible region is empty and whether the optimal objective function value is unbounded. To learn more about the Simplex Method and more mathematical details of LP, please refer to Bazaraa et al. (2011) and Bertsimas and Tsitsiklis (1997).

7.3.2 Nonlinear Optimization

If the objective function or any one of the constraints in the construct of an optimization problem are nonlinear, we may refer to this construct as a Nonlinear Program (NLP). The desirability function model formulation provided in section 7.2.3 of this chapter is an example of an NLP. Another NLP is provided below to illustrate the graphical nature of the two-dimensional variant. Consider the following problem:

$$\text{Min } z = 30x_1 + 10x_2$$
$$\text{s.t.} \quad x_1^2 + x_2^2 \leq 16$$
$$-x_1 + \tfrac{1}{4}x_2 \leq 0$$
$$(x_1 - 2)^2 + (x_2 - 2)^2 \leq 4$$
$$x_1, x_2 \geq 0$$

In Figure 7.6(a), we can observe the corresponding shaded feasible region, the gradient of the objective function z, and the resulting level curves of z. Since the gradient of a function always points in the direction of maximum increase, we proceed in the direction of the negative gradient vector for a minimization problem. Of note, the last constraint in the NLP dictating that x_1 and x_2 be greater than or equal to zero is not necessary for this problem as the third constraint forces the solution into the first quadrant. In this instance, we call the nonnegativity constraint a *nonbinding constraint*. The optimal solution can be calculated either by substitution between the second and third constraints or via a computer algebra system, the result being $\mathbf{x}^* = (0.255479, 1.02192)$, where $z = 17.8836$. Note that if we were to suggest that either x_1, x_2, or both had to be negative, then no solution would result as indicated in Figure 7.6(b).

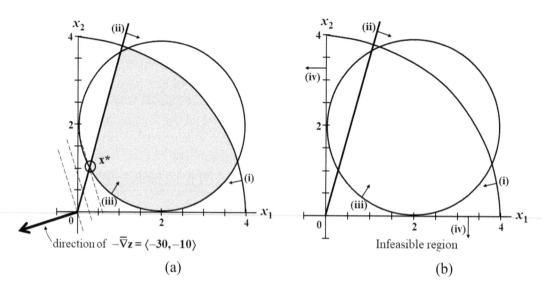

FIGURE 7.6 Feasible region illustration with (a) unique solution and (b) no solutions.

7.3.3 Integer Programming

The previous sections illustrated cases when solutions may be continuous or discrete as well as examples of infinite or no solutions. In cyber research, the variables of interest may be file packets, the number of systems infected, the number of network false alarms, or even a quantity of people. When the variables **x** are restricted to integers only, integer programming is appropriate. Consider the linear and nonlinear two-dimensional examples shown previously but with the variables restricted to integers only. The smallest convex set that contains the solutions in a space is known as the *convex hull*; in Figure 7.7, the convex hull for these two-dimensional examples is shaded and includes all of the feasible points for the solution. It is noted that while the optimal solution **x*** did not change in the linear case (see Figure 7.7(a)), it did change in the nonlinear case (Figure 7.7(b)), the new optimal solution being **x*** = $(1, 1)$. Finally, if some of the variables in an integer programming problem are not discrete, it is referred to as a mixed-integer programming problem.

7.3.4 Other Approaches

In general terms, when we extend our model to n-dimensions, any of the constrained optimization problems discussed previously in this chapter can be written in the form:

$$\begin{aligned} \text{Max } & \mathbf{c}^T \mathbf{x} \\ \text{s.t. } & \mathbf{A}\mathbf{x} \leq \mathbf{b} \\ & \mathbf{x} \geq 0, \end{aligned} \quad (7.1)$$

where **c** and **b** are vectors of known coefficients, **A** is a matrix of coefficients for the left-side of the constraint framework, and **x** represents the variables that are sought for solving the problem. An important concept in linear optimization is the

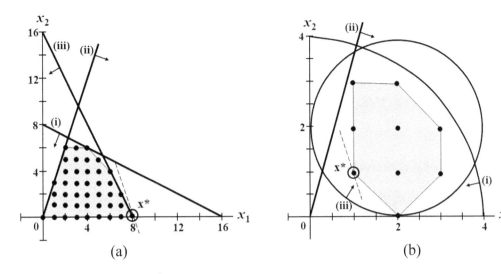

FIGURE 7.7 Convex hull illustrations with identified feasible points.

presence of a dual problem corresponding to a given primal problem. For instance, the *dual* of Equation 7.1 would have the form:

$$\text{Min } \mathbf{b}^T\mathbf{y}$$
$$\text{s.t. } \mathbf{A}^T\mathbf{y} \geq \mathbf{c}$$
$$\mathbf{y} \geq 0.$$

The dual problem can offer insights into the feasibility of the solution space, aid in analyzing the sensitivity of a given problem, and in some cases, may be easier to solve than a given primal problem.

The ability to solve many of these problems can depend also on the presence of the n-dimensional convex hull, the nature of the matrix $\mathbf{A}$, and/or certain conditions on the objective function or constraints. Solution techniques for mathematical optimization are also aligned toward the class of the specific problem; we have shown examples of the linear, nonlinear, and integer programming classes. If $\mathbf{x}$ represents both discrete and continuous variables, a mixed-integer program may be used to formulate the problem. If the objective function or constraints utilize random variables in their construct, a stochastic optimization problem is used. If the aim of the researcher is to acquire a degree of robustness in the solution with levels of uncertainty in the decision space, robust optimization can be applied. Finally, goal programming involves simultaneously optimizing multiple objectives, given a decision variable space and supporting constraints. For solution techniques related to several of these classes, specifically with regard to constrained optimization, see Gould and Leyffer (2003), Rader (2010), or Garcia and Pena (2018).

While we have focused our discussion on constrained optimization problems, the aim of the researcher may be to solve problems in an unconstrained framework. Given the objective of minimizing or maximizing an n-dimensional function $f(\mathbf{x})$ without constraints, solution techniques may involve iterative evaluations of function values or use of derivatives in searching for optimality. Methods for solving unconstrained optimization problems can be found in Fasano (2010), Djordjevic (2019), and Stanimirovic et al. (2020).

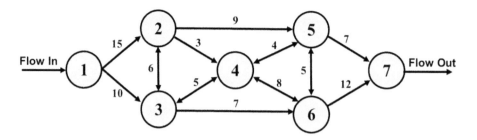

FIGURE 7.8 Network flow diagram.

7.3.5 Case Study: Network Interdiction

Perhaps one of the most common representations for examining or studying the behavior of cyber processes or systems is the network model. It may involve seeking optimal solutions to real-world applications such as communication, energy, or transportation networks. It may also entail investigating how connections facilitate the spread of malware or the flow of information. Modeling network flow enables the researcher to gain a better understanding of a system's behavior; when coupled with optimization tools, improvements to flow or architecture may be identified.

When faced with a network attack or outage of some kind, a defender's objective is frequently to maximize flow through the system. In an energy production problem, this may be transitioning power from a supplier to where it is most in need. In a communications security environment, maximum flow indicators provide locations for optimal sensor placement in the form of intrusion detection systems or botnet defenses. As an example, consider the network flow diagram in Figure 7.8 made up of nodes and directed arcs.

Let X_{ij} represent the flow of a discrete commodity from node i to node j, where the constants along each arc indicate the throughput capacity ϕ_{ij} between nodes. If we let F represent the total flow into the system, i.e. $\sum X_{ij}$, and consider the flow balance at each of the nodes as constraints, the network in Figure 7.8 can be represented by the following linear integer program:

Maximize F
subject to:
$X_{12} + X_{13} - F = 0$ (origin of flow)
$X_{12} + X_{32} - X_{23} - X_{24} - X_{25} = 0$ (node 2 constraint)
$X_{13} + X_{23} + X_{43} - X_{32} - X_{34} - X_{36} = 0$ (node 3 constraint)
$X_{24} + X_{34} + X_{54} + X_{64} - X_{43} - X_{45} - X_{46} = 0$ (node 4 constraint)
$X_{25} + X_{45} + X_{65} - X_{54} - X_{56} - X_{57} = 0$ (node 5 constraint)
$X_{36} + X_{46} + X_{56} - X_{64} - X_{65} - X_{67} = 0$ (node 6 constraint)
$X_{57} + X_{67} - F = 0$ (destination of flow)
$X_{12} \leq 15,\ X_{13} \leq 10,\ X_{23} \leq 6,\ X_{24} \leq 3,\ X_{25} \leq 9,$ (capacity)
$X_{32} \leq 6,\ X_{34} \leq 5,\ X_{36} \leq 7,\ X_{43} \leq 5,\ X_{45} \leq 4,$
$X_{46} \leq 8,\ X_{54} \leq 4,\ X_{56} \leq 5,\ X_{57} \leq 7,\ X_{64} \leq 8,$
$X_{65} \leq 5,\ X_{67} \leq 12$
$X_{ij} \geq 0, X_{ij} \in \mathbb{Z}$ (nonnegativity, integers)

With this construct, the objective is to maximize flow through the network. The resulting maximum flow is 19 units, which is obtained by the flow scheme $\{X_{12} =$

$9, X_{13} = 10, X_{23} = 3, X_{24} = 3, X_{25} = 6, X_{32} = 3, X_{34} = 3, X_{36} = 7, X_{43} = 0, X_{45} = 4, X_{46} = 5, X_{54} = 2, X_{56} = 3, X_{57} = 7, X_{64} = 1, X_{65} = 2, X_{67} = 12\}$. For this network, it is noted that flow is at its capacity for $X_{13}, X_{24}, X_{36}, X_{45}, X_{57}$, and X_{67}. Further analysis on the effect of increasing capacity between nodes is typically of interest to security practitioners.

For network interdiction, contrary to the defender's objective, an attacker's objective may involve identifying the most beneficial connections to target. In this role, the intent is to minimize the total flow in the network by reducing the throughput capacity on various arcs. Let Y_{ij} represent the proportion of the flow between node i and node j that is removed from the attacker. Most attacks have an associated cost in terms of technical difficulty or discovering difficulty. Let c_{ij} be the cost to attack the arc between node i and node j, whereby the capacity of the attacker is less than some overall constant ω. We can think of ω as providing a limit to the number of arcs that an attacker can influence in a network. Looking at the same network in Figure 7.8, this problem can be formulated by considering the dual problem to the defender's problem, where γ_{ij} represents the dual variables for each of the capacity constraints and ψ_n represents the dual variables for each of the node constraints. We seek to minimize $z = \sum \phi_{ij} b_{ij}$, where $b_{ij} = (1 - Y_{ij})\gamma_{ij}$. Suppose that c_{ij} for the attacker is analogous to the throughput capacity for each arc; then, we have the following linear program:

$$\begin{aligned}
\text{Minimize } z = & 15b_{12} + 10b_{13} + 6b_{23} + 3b_{24} + 9b_{25} + 6b_{32} \\
& + 5b_{34} + 7b_{36} + 5b_{43} + 4b_{45} + 8b_{46} + 4b_{54} \\
& + 5b_{56} + 7b_{57} + 8b_{64} + 5b_{65} + 12b_{67}
\end{aligned}$$

subject to:
$15Y_{12} + 10Y_{13} + 6Y_{23} + 3Y_{24} + 9Y_{25} + 6Y_{32}$ (attacker cost)
$+ 5Y_{34} + 7Y_{36} + 5Y_{43} + 4Y_{45} + 8Y_{46} + 4Y_{54}$
$+ 5Y_{56} + 7Y_{57} + 8Y_{64} + 5Y_{65} + 12Y_{67} \leq \omega$
$b_{ij} + Y_{ij} + \psi_j - \psi_i \geq 0$ (arc constraints)
$0 \leq Y_{ij} \leq 1$ (defined proportion)
$\psi_1 = 1, \psi_7 = 0, \psi_k \in \{0, 1\}$ for $k = 2, \ldots, 6$ (dual variable constraints)
$b_{ij} \geq 0$ (nonnegativity)

By the nature of the proportion, this is a mixed-integer programming problem. An interesting study is to examine the attacker's objective with varying ω. For $\omega = 5$, we have $b_{57} = 0.286$, $y_{57} = 0.714$, with $z = 14$. This suggests focusing solely on reducing the capacity of the arc between node 5 and 7 to roughly 29% of its original capacity. For $\omega = 10$, we have $b_{67} = 0.75$, $y_{57} = 1$, $y_{67} = 0.25$, with $z = 9$. This suggests that it would be best for the attacker to completely eliminate the arc between node 5 and node 7 and reduce the arc between node 6 and node 7 to 75% of its capacity. The attractiveness of the connection between node 6 and node 7 is clear – any added attacker capacity will likely target this bridge. For additional reading on network interdiction problems, see Smith and Song (2019).

7.4 STOCHASTIC PROCESS MODELING

As mentioned previously, in the cyber realm, most sub-processes may be viewed as inherently random in nature. For instance, given a set of conditions, there exists some array of probabilities that governs the spread of malware from one system to another, the frequency disruption of a network, the behavior of a botnet, or the attack likelihood for an organization. We may also be more interested in how these events or sub-processes evolve over time to denote trends and forecast equilibria. When randomness or uncertainty is present, researchers will frequently use stochastic process modeling to study or investigate outcome tendencies. These mathematical models enable one to observe the changes in either discrete or continuous time using probabilistic rules.

Let X represent a random variable and t represent an index as a subset of the interval $[0, \infty)$. The set of random variables $\{X_t\}$, where $t \in [0, \infty)$ is referred to as a *stochastic process*. In most stochastic modeling scenarios, t is represented as time; using this construct, we may consider time as a countable set of numbers as in a discrete-time process or as an infinite set corresponding to a continuous-time process. Chapter 10 provides the reader with an overview of fundamentals in probability that may govern the behavior of random variables. Several probability distributions are also presented, which may be used to model either discrete-time or continuous-time processes. Perhaps one of the most popular modeling techniques in stochastic process modeling utilizes the Markov model, named after the mathematician that founded stochastic process theory in the nineteenth and twentieth centuries and discussed in greater detail in the following paragraphs.

7.4.1 Markov Processes

A Markov chain is a stochastic process whereby the outcomes follow the memoryless property; the probability of being in a future state only depends on the current state and not on any previous outcomes. Given a set of states $j_0, j_1, \ldots, j_n, j_{n+1}$, in mathematical terms and using conditional probability notation, the memoryless property states:

$$P(X_{n+1} = j_{n+1} | X_n = j_n, \ldots, X_1 = j_1, X_0 = j_0) = P(X_{n+1} = j_{n+1} | X_n = j_n)$$

If p_{jk} is the probability of going from state j to state k in one iteration, we can represent all transitions between states for an n-state process using a transition matrix $\mathbf{P}$:

$$\mathbf{P} = \begin{bmatrix} p_{11} & p_{12} & \cdots & p_{1n} \\ p_{21} & p_{22} & \cdots & p_{2n} \\ \vdots & \vdots & \ddots & \vdots \\ p_{n1} & p_{n2} & \cdots & p_{nn} \end{bmatrix}, \text{ where } p_{jk} \geq 0 \text{ and } \sum_{k=1}^{n} p_{jk} = 1$$

For a Markov chain, researchers will often present an n-state process in the form of a state transition diagram for ease of reference and illustration. Consider a two-state process with transition matrix $\mathbf{P}$ in the form:

$$\mathbf{P} = \begin{bmatrix} 0.3 & 0.7 \\ 0.4 & 0.6 \end{bmatrix}$$

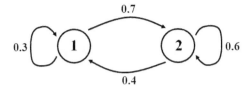

FIGURE 7.9 Example of a state transition diagram.

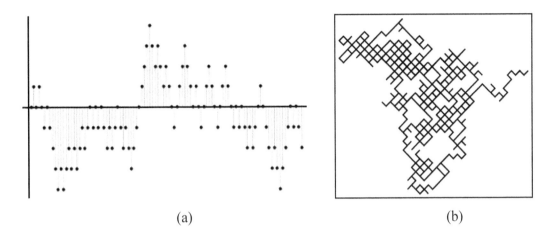

FIGURE 7.10 Illustrations of the symmetric random walk.

The corresponding state transition diagram for this matrix is shown in Figure 7.9.

One of the simplest Markov chains to illustrate is the random walk, often used to describe the motion of molecules. It has also been used by researchers to describe the motion of self-propagating malware between devices in close-proximity networks (Sharma and Gupta, 2016; Thompson and Morris-King, 2018), as well as a sequence of random observations created when using network monitoring software to record a characteristic of interest. If the movement from one transition to another is of equal probability, the random walk is referred to as *symmetric*; otherwise, if the distribution or probabilities are not equal, it is *asymmetric*. Simulated illustrations of the (a) one-dimensional and (b) two-dimensional symmetric random walks are shown in Figure 7.10.

Several terms are defined to differentiate among various Markov processes. If states j and k can transition between one another as in Figure 7.9, they are said to communicate with one another. If two states j and k do not communicate, we should observe $p_{jk} = 0$, $p_{kj} = 0$, or both. A Markov chain is *irreducible* if all states in the state transition diagram are shown to communicate with one another. If there is a probability of returning to a state j after a process starts at state j following a finite length of time, the state is known as *recurrent*; otherwise, the state is *transient*. Moreover, states may be classified as periodic or aperiodic. The period of the jth state among n states is the largest integer d whereby $p_{jj} = 0$ whenever n is not divisible by d. If d for the ith state is greater than one, the state is *periodic*; otherwise, if $d = 1$, the state is *aperiodic*. A Markov chain is said to be aperiodic if all of its states are aperiodic. Finally, a Markov chain is *regular* if its corresponding transition matrix taken to some power contains all positive entries.

Based upon these definitions, there are several special types of Markov chains that offer unique contributions. Regular Markov chains, as well as those that are irreducible, aperiodic, and with a finite number of states, have what is known as a *limiting distribution*. This distribution gives the researcher insights into the long-term behavior of a process. For an n-state process, the limiting distribution can be written as:

$$\boldsymbol{\pi} = \{\pi_1, \pi_2, \ldots, \pi_n\}, \text{ where } \pi_j > 0, \sum_{j=1}^{n} \pi_j = 1,$$

and where π_j represents the probability of being in state j over a long period of time. To find the limiting distribution, solve the system of equations $\boldsymbol{\pi}\mathbf{P} = \boldsymbol{\pi}$ for the probability vector $\boldsymbol{\pi}$.

Another special type of Markov chain is one that contains an *absorbing* state, that is where the probability of reaching a certain state is one. Processes that typically end in one finalized state, such as death or complete loss, can be modeled as absorbing processes; as such, they will have at least one state j where $p_{jj} = 1$ in the diagonal of $\mathbf{P}$. Since the long-term behavior of an absorbing process is known, of greater interest to the researcher is the expected number of visits to each state as well as the expected time across all states before absorption occurs. These results can be identified by examining the elements and rows of the fundamental matrix $\mathbf{F}$. This matrix is found by first rearranging the states in $\mathbf{P}$ so that the absorbing state is first, partitioning $\mathbf{P}$ into an identity matrix $\mathbf{I}$, a matrix of zeroes $\mathbf{0}$, and two remainder matrices $\mathbf{Q}$ and $\mathbf{R}$ as shown below:

$$\mathbf{P} = \left[\begin{array}{c|c} \mathbf{I} & \mathbf{0} \\ \hline \mathbf{R} & \mathbf{Q} \end{array}\right].$$

Then use $\mathbf{Q}$ to find $\mathbf{F} = (\mathbf{I}_n - \mathbf{Q})^{-1}$. One may further find the matrix product $\mathbf{FR}$, which lists the probabilities that an object in an initial non-absorbing state will lead to an absorbing state.

7.4.2 Queueing Models

In operations research, queueing models are traditionally used to study problems where traffic flow is of high interest, such as with vehicular or air transportation, switchboard telephone calls, evacuation drills, vaccine distribution, or urban planning. They may be integrated within large-scale logistic problems to examine optimal flow or the effect of resourcing on different sub-processes. In the relatively recent past, queueing models have been extended to many cyber applications in novel ways. For instance, researchers have modeled the arrival of security incident notifications from various network sensors for information technology management (Bhattacharya et al., 2019), used queueing models to measure cyber resilience against malicious attacks (Fink et al., 2014), and proposed innovative routing algorithms for data transmission to improve blockchain technology (Yang et al., 2018). The use of queueing models to analyze and ultimately improve cyber systems will only likely continue to grow in the future.

Queueing system structures can be represented in various ways and are usually characterized by a finite or infinite-capacity queue, some type of service policy, and some arrangement of one or more servers. When describing various queueing systems frequently observed in research, Kendall notation is commonly used, which was first suggested by Kendall (1953). The long variation of this notation consists of a partition '(i) / (ii) / (iii) / (iv) / (v) / (vi)' whereby each element represents (i) the distribution of arrival times, (ii) the distribution of service times, (iii) the number of servers, (iv) the capacity of the queue, (v) the population size, and (vi) the queueing policy used. Most often, the assumptions of an infinite queue capacity, infinite population size, and a first-in first-out (FIFO) queueing policy reduces the six-element long notation to an abbreviated three-element variation.

Moreover, depending on the problem of interest, the interarrival times between customers (or individual elements) arriving at the queue, as well as the time in service, may be according to a fixed interval or governed by some probability distribution. The distribution most often observed in modeling interarrival or service times is the exponential distribution. Since the distribution of times remaining until an event occurs or is completed is the same regardless of the time that has already passed, the exponential distribution is appropriate, as it is characterized by the memoryless property. Hence, similar to discussion in Section 7.4.1, the distribution is typically referred to as 'Markovian,' with the designation M in queue notation. Alternatively, a general distribution is represented with a G, and D is used to signify the use of a degenerate distribution, which may have fixed or constant times. For instance, the M/M/1 queueing system consists of one server with each individual's or element's arrival and service times according to an exponential distribution, whereby the assumptions described previously for (iv), (v), and (vi) in the notation are considered.

Let λ represent the average number of individuals or elements arriving and μ as the average number being served per time period. The expected fraction of time that a server is busy is known as the server utilization, ρ; for q servers, it is calculated as $\rho = \lambda/(q\mu)$, whereby a queueing system is considered stable for $\rho < 1$. For non-exponentially distributed interarrival and service times, let m_a and m_s be the average arrival and service rate, respectively, with σ_a and σ_s being the corresponding standard deviations for these distributions. Figure 7.11 depicts frequently observed queueing system constructs; the capacity of the queue is considered infinite, except when a k-capacity limit is considered, as in the M/M/1/k case.

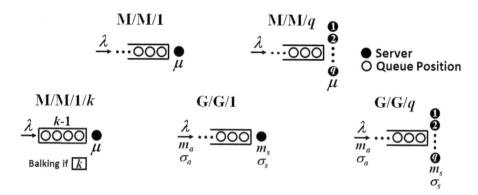

FIGURE 7.11 Traditional queueing model structures with their accompanying notation.

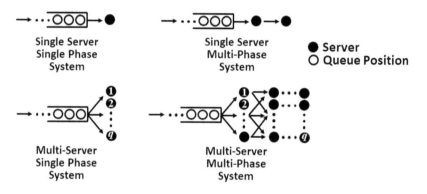

FIGURE 7.12 Various queueing service structures.

In studying queueing systems, there are numerous measures that may be of interest to the researcher. Often, the objective is to examine ways to minimize the average time an individual or element spends in the queue, in service, or in the system. Queue lengths, the average number of individuals or elements in the system, and service reliability may also be a focus when examining their effect on bottlenecks or overall throughput. Finally, the long-term behavior of the system, as with Markov chains, can be an objective. The system outputs from these models are affected by not only the distribution times, and the number of resources or servers available, but also how the queueing system is designed. Individuals or elements may enter one queue with one server, one queue with multiple servers, multiple queues with one server, or multiple queues with multiple servers. They may also require receiving service from multiple stations before exiting the queueing system. Figure 7.12 shows various service structures that one may investigate.

Many calculations of interest for a queueing system with one server are functions of λ, μ, and ρ; we consider the most often used model: M/M/1. The percentage of the time that all servers are busy for this model is given by the server utilization, $\rho = \lambda/\mu$. Let π_n be the probability there are n individuals or elements in the system.

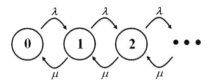

FIGURE 7.13 M/M/1 Markov chain state transition diagram.

Then, the probability no individuals or elements are in the system (i.e., the servers are idle) is

$$\pi_0 = 1 - \frac{\lambda}{\mu} = 1 - \rho.$$

To compute the probabilities of j elements being in the system, we can think of the M/M/1 as a Markov chain transitioning between states where each state represents the number of elements in the system (see Figure 7.13).

When a process is in a steady state at a node j, there is a flow balance; the rate entering j is equivalent to the rate leaving node j. For state 0, the rate in for an M/M/1 is $\mu\pi_1$, and the rate out is $\lambda\pi_0$. Setting these equal to one another and solving for π_1, we have:

$$\pi_1 = \frac{\lambda}{\mu}\pi_0 = \rho(1 - \rho),$$

which is the probability of one element being in the system. Setting the inward flow equal to the outward flow for state 1, and then solving for π_2 using a substitution for π_1, we have:

$$\lambda\pi_0 + \mu\pi_2 = \lambda\pi_1 + \mu\pi_1 \to \lambda\pi_0 + \mu\pi_2 = (\lambda + \mu)\pi_1 \to \lambda\pi_0 + \mu\pi_2 = (\lambda + \mu)\frac{\lambda}{\mu}\pi_0 \to$$

$$\mu\pi_2 = \left(\frac{\lambda^2}{\mu} + \lambda - \lambda\right)\pi_0 \to \mu\pi_2 = \frac{\lambda^2}{\mu}\pi_0 \to \pi_2 = \frac{\lambda^2}{\mu^2}\pi_0 = \rho^2(1 - \rho).$$

This pattern continues to show that $\pi_j = \rho^j(1-\rho)$. It is also noted that $\sum_{j=1}^{n} \pi_j = 1$.

Let L be the average number of elements or individuals in the system. Using the Law of Total Expectation, this is calculated as:

$$L = 0 \cdot P(\text{zero in system}) + 1 \cdot P(\text{one in system}) + 2 \cdot P(\text{two in system}) + \cdots$$

$$= \sum_{j=0}^{\infty} j\pi_j = \sum_{j=0}^{\infty} j\rho^j(1-\rho) = (1-\rho)\sum_{j=0}^{\infty} j p^j = \rho(1-\rho)\sum_{j=0}^{\infty} jp^{j-1}.$$

Appealing to differentiation, since $\frac{d}{d\rho}(\rho^j) = j\rho^{j-1}$, and using the sum of the geometric series, $\sum_{j=0}^{\infty} \rho^j = \frac{1}{1-\rho}$; through substitution, we have:

$$L = \rho(1-\rho)\sum_{j=0}^{\infty}\frac{d}{d\rho}(\rho^j) = \rho(1-\rho)\frac{d}{d\rho}\sum_{j=0}^{\infty}\rho^j = \rho(1-\rho)\frac{d}{d\rho}\left(\frac{1}{1-\rho}\right)$$

$$= \rho(1-\rho)\frac{1}{(1-\rho)^2} = \frac{\rho}{1-\rho}.$$

Let W be the average amount of time spent in the system. Little's law, $L = \lambda W$, can be applied to find a measure strictly in λ and μ:

$$W = \frac{L}{\lambda} = \frac{\rho}{\lambda(1-\rho)} = \frac{\lambda/\mu}{\lambda\left(1 - \lambda/\mu\right)} = \frac{\lambda/\mu}{\lambda - \lambda^2/\mu} = \frac{\lambda}{\lambda\mu - \lambda^2} = \frac{\lambda}{\lambda(\mu - \lambda)} = \frac{1}{\mu - \lambda}.$$

The calculations for L and W can further be broken down to determine the average number of elements in the queue or in service as well as the average time spent in the queue or in service.

In cyber research, queueing models have been developed extensively to model the flow of packets across networks. Yates et al. (2020) provide a comprehensive literature survey for research performed in the design and optimization of systems where this traffic flow is observed. Consider a network router that has data packets arriving at an average rate of 140 packets per second (pps) whereby it takes the router one millisecond to forward them. Assuming that the router can manage an infinite queue of data packets, we can calculate our items of interest. The arrival rate is $\lambda = 140$ pps, and the service rate is one packet per 0.001 seconds or $\mu = 1000$ pps. The percentage of time that the router is busy is $\rho = \lambda/\mu = 0.14$, and the probabilities of zero, one, two, or j data packets in the router are $\pi_0 = 1 - \rho = 0.86$, $\pi_1 = \rho(1-\rho) = 0.12$, $\pi_2 = \rho^2(1-\rho) = 0.02$, and $\pi_j = (0.14)^j(0.86)$, respectively. The average number of packets in this system is $L = \rho/(1-\rho) = 0.16$, i.e., the likelihood of more than one packet in the system is very low. Finally, the average amount of time a data packet spends in this system, both in the queue waiting and being processed by the router, is $W = 1/(\mu - \lambda) = 0.001$ seconds.

7.4.3 Other Approaches

Perhaps one of most used approaches to modeling stochastic processes, aside from those described in this section, are Monte Carlo methods. These techniques involve numerical algorithms that generate statistical estimates of quantities based upon random sampling of a distribution. Depending on the complexity of the problem, the method can require a large computational effort to approximate parameters of interest, utilizing repeated sampling via random number generation. As such, both the analysis of error and unbiased sampling are topics of interest to the researcher using Monte Carlo methods. A sufficient generator is also a requirement; see James and Moneta (2020) for discussion regarding this matter.

Monte Carlo methods utilize a transformation technique to generate random variables in a desired probability distribution; most often, the transformation performed is via the Inverse Transform method. Given a desired probability distribution function $f(y)$ with corresponding cumulative distribution function (cdf) $F(y)$, the inverse cdf, $F^{-1}(y)$, can first be computed. With $0 \leq F(y) \leq 1$, we can then produce a random variable X with the same approximate cdf by generating a uniform distribution on this same interval, i.e., $z \in U(0,1)$, and evaluating the inverse cdf to get $X = F^{-1}(z)$. Complex processes may involve a large number of sub-processes with various parameter distributions, all of which may be estimated using such a technique. When combined with the computational power of today's computers, it is clear to see why Monte Carlo methods may be popular for operations research analysts.

7.4.4 Case Study: Malware Spread

If a wireless network is not properly secured, computer systems connected within the network are more vulnerable to a cyber attack. Any infected computer or device connected to such a wireless network can be used to launch an attack against another system in the network. Various epidemiological models by researchers have shown how malware spreads among users through access to wireless routers (Akritidis et al., 2007; Hu et al., 2009).

Suppose that we have such a chain of computer systems or devices in a vulnerable wireless network. Under a certain array of conditions for proximity and connectivity of the users in the network as well as malware efficiency and complexity, some probabilities of malware infection and survivability can be inferred. Consider a chain of computer systems, each of which interacts with the previous system, and may be infected from that system, which then may infect the next system, and so on, with each transition representing an hour of time. Given this scenario, three states are identified: (i) State 1 - a computer system acquires the malware and loses all functionality, (ii) State 2 - a computer system acquires the malware and is able to maintain functionality, and (iii) State 3 - a computer system in the network does not acquire the malware. The research intent is to model this scenario as a stochastic process so that insights can be gained on the behavior or impact of the malware over a long period of time. To compare outcomes, we provide two cases on the extremes of aggressive and non-aggressive malware, with the transition matrices being given by, respectively:

$$\mathbf{P}_1 = \text{First Computer} \begin{array}{c} \\ 1 \\ 2 \\ 3 \end{array} \begin{array}{c} \text{Second Computer} \\ \begin{array}{ccc} 1 & 2 & 3 \end{array} \\ \begin{bmatrix} 0.05 & 0.00 & 0.95 \\ 0.75 & 0.15 & 0.10 \\ 0.25 & 0.75 & 0.00 \end{bmatrix} \end{array} \quad \mathbf{P}_2 = \text{First Computer} \begin{array}{c} \\ 1 \\ 2 \\ 3 \end{array} \begin{array}{c} \text{Second Computer} \\ \begin{array}{ccc} 1 & 2 & 3 \end{array} \\ \begin{bmatrix} 0.05 & 0.20 & 0.75 \\ 0.05 & 0.20 & 0.75 \\ 0.00 & 0.00 & 1 \end{bmatrix} \end{array}$$

At each hour, an initial number of computer systems is affected or not affected in some way. Of note, a computer in State 1, which loses all functionality, is shown to transition to State 3 with high probability; this makes intuitive sense, as it will no longer be able to acquire the malware. The accompanying state transition diagrams for $\mathbf{P}_1$ and $\mathbf{P}_2$ are provided in Figure 7.14.

Since $\mathbf{P}_1^2$ has all positive elements in its transition matrix, it is regular. To compute its limiting distribution, we solve the system of linear equations for the vector $\boldsymbol{\pi}$ in $\boldsymbol{\pi} \mathbf{P}_1 = \boldsymbol{\pi}$:

$$\begin{bmatrix} \pi_1 & \pi_2 & \pi_3 \end{bmatrix} \begin{bmatrix} 0.05 & 0.00 & 0.95 \\ 0.75 & 0.15 & 0.10 \\ 0.25 & 0.75 & 0.00 \end{bmatrix} = \begin{bmatrix} \pi_1 & \pi_2 & \pi_3 \end{bmatrix} \rightarrow \begin{array}{c} 0.05\pi_1 + 0.75\pi_2 + 0.25\pi_3 = \pi_1 \\ 0.15\pi_2 + 0.75\pi_3 = \pi_2 \\ 0.95\pi_1 + 0.10\pi_2 = \pi_3 \end{array}$$

Once the equations are set to zero, the last equation in the system can be deleted, as it is dependent on the other two. In particular, adding the negative of the first two

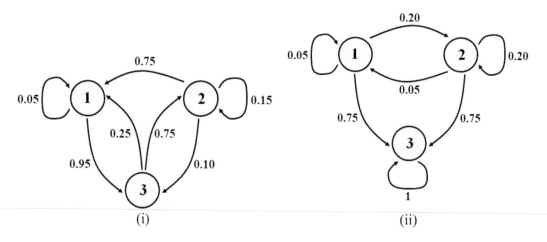

FIGURE 7.14 State transition diagrams for (a) aggressive and (b) non-aggressive spread.

equations results in the third equation. Since π is a probability vector, its elements must sum to one. We are left solving the system of equations; the last step utilizes Gauss-Jordan elimination:

$$\begin{array}{l} -0.95\pi_1 + 0.75\pi_2 + 0.25\pi_3 = 0 \\ -0.85\pi_2 + 0.75\pi_3 = 0 \\ 0.95\pi_1 + 0.10\pi_2 - \pi_3 = 0 \end{array} \rightarrow \begin{array}{l} -0.95\pi_1 + 0.75\pi_2 + 0.25\pi_3 = 0 \\ -0.85\pi_2 + 0.75\pi_3 = 0 \\ \pi_1 + \pi_2 + \pi_3 = 1 \end{array} \rightarrow \pi = \begin{bmatrix} 0.33769 \\ 0.31046 \\ 0.35185 \end{bmatrix}^T.$$

This result suggests that as time progresses, we anticipate close to one-third of the computer systems acquiring the malware and losing complete functionality, with close to 35% of the computer systems failing to acquire the malware.

Since $\mathbf{P}_2$ taken to any power continues to result in zero entries within the transition matrix, it is not regular. The limiting distribution need not be calculated; with $p_{33} = 1$, the third state is identified as an absorbing state, so $\pi = (0, 0, 1)$. In this case, it is often of interest to determine the expected time in a state j or the probabilities associated with going from one state to another at each transition. The matrix $\mathbf{P}_2$ is first rearranged so that the absorbing state is first; then the matrix is partitioned:

$$\mathbf{P}_2 = \begin{bmatrix} 0.05 & 0.20 & 0.75 \\ 0.05 & 0.20 & 0.75 \\ 0 & 0 & 1 \end{bmatrix} \rightarrow \mathbf{P}_2 = \begin{bmatrix} 1 & 0 & 0 \\ 0.75 & 0.05 & 0.20 \\ 0.75 & 0.05 & 0.20 \end{bmatrix} \rightarrow \mathbf{P}_2 = \begin{bmatrix} \mathbf{I} & \mathbf{0} \\ \hline \mathbf{R} & \mathbf{Q} \end{bmatrix}$$

$$= \begin{bmatrix} 1 & 0 & 0 \\ \hline 0.75 & 0.05 & 0.20 \\ 0.75 & 0.05 & 0.20 \end{bmatrix}.$$

With the matrix $\mathbf{Q}$ identified, we can find the fundamental matrix $\mathbf{F}$:

$$\mathbf{F} = (\mathbf{I}_n - \mathbf{Q})^{-1} = \left(\begin{bmatrix} 1 & 0 \\ 0 & 1 \end{bmatrix} - \begin{bmatrix} 0.05 & 0.20 \\ 0.05 & 0.20 \end{bmatrix} \right)^{-1} = \begin{bmatrix} 0.95 & -0.20 \\ -0.05 & 0.80 \end{bmatrix}^{-1}$$

$$= \begin{bmatrix} 1.0667 & 0.2667 \\ 0.0667 & 1.2667 \end{bmatrix}.$$

The first row of **F** tells us that if a system is currently infected without functionality, it can expect to be a little over one hour in this state on average before transitioning to the absorbing state. Similarly, if a system is currently infected and maintaining functionality, it can expect to be a little over 15 minutes in this state on average before transitioning to the absorbing state. Also, on average, since the rows of **F** add up to the same numbers, the expected amount of time in states 1 and 2 are the same, at 1.3333 hours, before the absorbing state is reached, and the malware has lost its effectiveness completely. Finally, we have that

$$\mathbf{FR} = \begin{bmatrix} 1.0667 & 0.2667 \\ 0.0667 & 1.2667 \end{bmatrix} \begin{bmatrix} 0.75 \\ 0.75 \end{bmatrix} = \begin{bmatrix} 1 \\ 1 \end{bmatrix},$$

which is what we would expect for an absorbing chain. The probability of ending up in the absorbing state if a system was originally in either State 1 or State 2 is one.

7.5 CONCLUSION

In this chapter, we have briefly outlined several sub-fields of operations research where contributions to cyber research are made. This is certainly not all inclusive; the discipline of operations research involves much more than what is covered here. Researchers are developing new and innovative descriptive and prescriptive analytics to identify anomalous and potentially malicious network behavior, establish thresholds for managing cyber risk, and provide decision makers with tools to evaluate attributes such as network resilience, reliability, and efficiency. Research is also providing valuable insights regarding the vulnerabilities and gaps within critical infrastructure capabilities.

Methods, tools, and practices in optimization are also providing great benefit in practical areas; for example, providing more efficient strategies for managing or protecting cyber-physical systems in the communication and power industries. Operations research is heavily used in the development of risk approaches for minimizing monetary and non-monetary cost, both of which are having an immense effect on organizations facing cyber attacks today. Researchers utilizing multi-response optimization have sought to minimize objectives such as damage, the number of intrusions, and inner attack surface connectivity, while maximizing objectives like attack difficulty, discoverability, and system responsiveness. Game theoretical optimization approaches have incorporated defender choices based upon different attacker behaviors to result in the most beneficial outcome.

In addition, researchers are using stochastic process modeling to develop descriptive and predictive analytics to help identify what is occurring or forecast what might happen in the future. This relies on a component of randomness, which relates well to many cyber processes characterized by uncertainty, such as malware spread, network failures, data processing, and attacker-defender behavior. Monte Carlo simulation methods are being used to model and compare risk scenarios for computer viruses and phishing attacks, analyze the event frequency for external threats and organizational losses, as well as validate and test cyber-physical system performance. These types of models enable inferences on the long-term behavior or outcomes of cyber processes that would otherwise seem too complex to understand.

Despite the immense benefit and contribution that operations research provides to cybersecurity efforts, there are still many areas where the tools, techniques, and methods of the discipline may be applied. Many experts describe the internet as a domain of transportation, where data files move from one location to another. This analogy opens up great potential contributions from the operations research sub-fields of supply chain management and logistics as well as the broad science of transportation networks. In addition, the future growth in the population of cyber-physical systems will certainly influence methodologies. Data-driven analysis should become more prevalent as a result, providing a stronger foundation for the methods of the discipline. A greater focus on the interconnectedness between the dimensions of cyber space, such as between the individual, the information, as well as the physical and geographic components, is also likely.

While only discussed in some parts of this chapter, computer simulation is used extensively as a mechanism for examining, analyzing, and interpreting complex processes and systems in operations research. All methods and techniques described in this chapter frequently incorporate simulation into studying aspects of a problem. In decision analysis, simulation could examine the effect of various weighting schemes or cost-based parameters on recommended outcomes. In mathematical optimization, it may be the manipulation of the constraint space or consideration of new variables on the optimal solution. In stochastic process modeling, it may be the adjustment of parameters or resources, and their effect on long-term behavior or the results associated with utilizing different distributions for modeling elements. The power of simulation enables the researcher to take a solution under one set of conditions, and then apply an array of other conditions in an experimental framework to observe the effects. In essence, it facilitates performing a complete and thorough sensitivity analysis. So, as the power of computers continue to grow in the future, corresponding advances to operations research efforts in simulation will likely result.

Lastly, one element of the research space that will undoubtedly have a tremendous future influence on the mathematics of operations research in the cyber realm will be the drive toward machine learning. The predominant feature with the models, methods, and tools presented in this chapter is that they are explicitly designed to realistically represent or solve real-world problems. Closing the disparity between the model and what is modeled is an objective of machine learning and also of every mathematician. The science of machine learning offers an opportunity to take predictive models applied in practice and improve their predictive power over time. Combined with the methods of operations research, machine learning can provide a clear advantage in concepts such as cyber awareness, adaptive defense, cyber deception, and intrusion detection and prevention.

REFERENCES

Aczél, J., and Alsina, C. (1987). Synthesizing judgements: A functional equations approach. *Mathematical Modelling*, 9(3-5), 311–320.

Aczél, J., and Saaty, T. L. (1983). Procedures for synthesizing ratio judgements. *Journal of Mathematical Psychology*, 27, 93–102.

Akritidis, P., Chin, W.Y., Lam, V.T., Sidiroglou, S., and Anagnostakis, K.G. (2007). Proximity breeds danger: emerging threads in metro-area wireless networks. *Proceedings of the 16th USENIX Security Symposium*, 323–338, Berkeley, California.

Alves, T. C. L., Liu, M., Scala, N. M., and Javanmardi, A. (2020). Schedulers and schedules: a study in the U.S. construction industry. *Engineering Management Journal*, 32(3), 166–185.

Bazaraa, M. S., Jarvis, J. J., and Sherali, H. D. (2011). *Linear programming and network flows*. Hoboken, New Jersey: John Wiley & Sons.

Bertsimas, D., and Tsitsiklis, J. N. (1997). *Introduction to linear optimization*. Belmont, Massachusetts: Athena Scientific.

Bhattacharya, A., Bopardikar, S., Chatterjee, S., and Vrabie, D. (2019) Cyber threat screening using a queuing-based game-theoretic approach. *Journal of Information Warfare*, 18(4), 37–52.

Dantzig, G. B. (1948). *Programming in a linear structure*. Washington, D.C.: Comptroller of the United States Air Force.

Dantzig, G. B. (1963). *Linear programming and extensions*, Princeton University Press, Princeton, NJ.

Derringer, G. (1994) A balancing act: optimizing a product's properties. *Quality Progress*, 51–58.

Derringer, G., and Suich, R. (1980) Simultaneous optimization of several response variables. *Journal of Quality Technology*, 12(4), 214–219.

Djordjevic, S. S. (2019) Some unconstrained optimization methods. *Applied Mathematics*, edited by Bruno Carpentieri. London, England: IntechOpen Limited.

Du, S., and Zhu, H. (2013). *Security assessment in vehicular networks*. New York City, New York: Springer-Verlag.

Evans, J. R. (2016). *Business analytics: methods, models, and decisions*, 2nd ed. New York City, New York: Pearson Education.

Fasano, G. (2010) Methods for large-scale unconstrained optimization. *Wiley Encyclopedia of Operations Research and Management Science*, edited by James Cochran. Hoboken, New Jersey: John Wiley & Sons, Inc.

Fink, G.A., Griswold, R.L., and Beech, Z.W. (2014) Quantifying cyber-resilience against resource-exhaustion attacks. *Proceedings of the 1st International Symposium on Resilient Cyber Systems*, Denver, Colorado.

Garcia, J., and Pena, A. (2018) Robust optimization: Concepts and applications. *Nature-inspired methods for stochastic, robust and dynamic optimization*, edited by Del Ser, J. and Osaba, E. London, England: IntechOpen Limited.

Goodwin, P., and Wright, G. (2004). *Decision analysis for management judgment, 3rd ed.* Hoboken, New Jersey: John Wiley & Sons, Inc.

Gould, N., and Leyffer, S. (2003) An introduction to algorithms for nonlinear optimization. *Frontiers in Numerical Analysis*, edited by Blowey, J.F., Craig, A.W., and Shardlow, T. Berlin, Germany: Springer-Verlag, pp. 109–197.

Harrington, E.C. (1965) The desirability function. *Industrial Quality Control*, 21(10), 494–498.

Helmer-Hirschberg, O. (1967). Analysis of the future: the Delphi method. Santa Monica, California: RAND Corporation.

Hu, H., Myers, S., Colizza, V., and Vespignani, A. (2009) Wifi networks and malware epidemiology. *Proceedings of the National Academy of Sciences of the United States of America*, 106(5), 1318–1323.

James, F., and Moneta, L. (2020) Review of high-quality random number generators. *Computing and Software for Big Science*, 4(2), 1–12.

Keeney, R. L. (1992). *Value-focused thinking: a path to creative decision making.* Cambridge, Massachusetts: Harvard University Press.

Keeney, R. L. (2008). Applying value-focused thinking. *Military Operations Research*, 13(2), 7–17.

Keeney, R. L., and Raiffa, H. (1976). *Decisions with multiple objectives: preferences and value tradeoffs.* Hoboken, New Jersey: John Wiley & Sons, Inc.

Kendall, D.G. (1953) Stochastic processes occurring in the theory of queues and their analysis by the method of imbedded Markov chain. *The Annals of Mathematical Statistics*, 24(3), 338–354.

Parnell, G. S., Bresnick, T. A., Tani, S. N., and Johnson, E. R. (2013). *Handbook of decision analysis.* Hoboken, New Jersey: John Wiley & Sons, Inc.

Rader, D. J. (2010) *Deterministic operations research: models and methods in linear optimization.* Hoboken, New Jersey: John Wiley & Sons, Inc.

Saaty, T. L. (1990). How to make a decision: the analytic hierarchy process. *European Journal of Operational Research*, 48, 9–26.

Saaty, T. L., and Vargas, L. G. (2007). Dispersion of group judgments. *Mathematical and Computer Modelling*, 46, 918–925.

Scala, N. M., Goethals, P. L., Dehlinger, J., Mezgebe, Y., Jilcha, B., and Bloomquist, I. (2021). Evaluating mail-based security for electoral processes using attack trees. *Risk Analysis*, in press.

Scala, N. M., Rajgopal, J., and Needy, K. (2010). Influence diagram modeling of nuclear spare parts process. *Proceedings of the 2010 Industrial Engineering Research Conference*, Cancun, Mexico.

Scala, N. M., Rajgopal, J., Vargas, L. G., and Keedy, K. L. (2016). Group decision making in the Analytic Hierarchy Process. *Group Decision and Negotiation*, 25, 355–372.

Scala, N. M., and Goethals, P. (2020). A model for and inventory of cybersecurity values: metrics and best practices. In N. M. Scala and J. P. Howard (Eds.), *Handbook of Military and Defense Operations Research*. London, England: CRC Press, 55–81.

Scala, N. M., and Pazour, J. A. (2016). A value model for asset tracking technology to support naval sea-based resupply. *Engineering Management Journal*, 28(2), 120–130.

Schneier, B. (1999). Attack trees. *Dr. Dobb's Journal*. Retrieved from https://www.schneier.com/academic/archives/1999/12/attack_trees.html.

Sharma, K., and Gupta, B.B. (2016) Multi-layer defense against malware attacks on smartphone wi-fi access channel. *Procedia Computer Science*, 78, 19–25.

Smith, J.C., and Song, Y. (2019) A survey of network interdiction models and algorithms. *European Journal of Operational Research*, 283(3), 797–811.

Stanimirovic, P. S., Ivanov, B., Ma, H., and Mosic, D. (2020) A survey of gradient methods for solving nonlinear optimization. *Electronic Research Archive*, 28(4), 1573–1624.

Thompson, B., and Morris-King, J. (2018) An agent-based modeling framework for cybersecurity in mobile tactical networks. *Journal of Defense Modeling and Simulation: Applications, Methodology, Technology*, 15(2), 205–218.

United States Election Assistance Commission (EAC) Advisory Board and United States Election Assistance Commission Standards Board (2009). *Election operations assessment: threat trees and matrices and threat instance risk analyzer (TIRA)*. Retrieved from https://www.eac.gov/assets/1/28/Election_Operations_Assessment_Threat_Trees_and_Matrices_and_Threat_Instance_Risk_Analyzer_(TIRA).pdf

Vargas, L. G. (1990). An overview of the analytic hierarchy process and its applications. *European Journal of Operational Research*, 48, 2–8.

Yang, T., Zhai, F., Liu, J., Wang, M., and Pen, H. (2018) Self-organized cyber physical power system blockchain architecture and protocol. *International Journal of Distributed Sensor Networks*, 14(10), 1–9.

Yates, R.D., Sun, Y., Brown, R., Kaul, S.K., Modiano, E., and Ulukus, S. (2020) Age of information: an introduction and survey. *Cornell University*, retrieved from: https://arxiv.org/abs/2007.08564.

CHAPTER 8

Data Analysis

Raymond R. Hill

Darryl K. Ahner

CONTENTS

8.1	Introduction	268
8.2	Modeling Methods to Support Cyber Research	269
	8.2.1 Visualization Methods	269
	8.2.2 Simulation Modeling and Analysis	271
	8.2.3 Risk Management Frameworks	273
	8.2.4 Cyber Red Teaming	276
	8.2.5 Game Theory for Cyber Security	277
	8.2.6 Markov Chains for Cyber Security	280
8.3	Data Analytic Methods	281
	8.3.1 Classification of Methods	282
	8.3.2 Supervised-Predictive Tools	284
	8.3.2.1 Regression Analysis	284
	8.3.2.2 Logistic Regression	285
	8.3.2.3 Tree Methods	287
	8.3.3 Supervised-Clustering Methods	288
	8.3.3.1 Discriminant Analysis	288
	8.3.3.2 Support Vector Machines	290
	8.3.4 Unsupervised-Clustering Methods	292
	8.3.4.1 Multidimensional Scaling	292
	8.3.4.2 k-Nearest Neighbors	293
	8.3.4.3 k-Means Clustering	294
	8.3.5 Unsupervised-Dimension Reduction Methods	294
	8.3.5.1 Principal Components Analysis	294
	8.3.5.2 Factor Analysis	295
	8.3.6 What Has Not Been Discussed	296
8.4	Concluding Remarks	297

DOI: 10.1201/9780429354649-8

8.1 INTRODUCTION

The modern world is an electronically connected world. However, electronic connections carry risk. That risk involves the loss of the connections as well as the exploitation of the connection by those with often times nefarious intent. Defeating these attacks against our connected devices via simply disconnecting those devices is not a viable option. Thus, we have the need to exploit cyber research opportunities to advance our methods and understanding of how to protect our electronic systems in our connected world. This chapter supports this cyber research objective by focusing on the role of modeling and analytics, and in particular data analytics, as it supports and advances research and operations in cyber warfare. More specifically, we focus on some of the modeling and data analytical, or machine learning, methods used in the quantitative analytical aspects of cyber research. We then present some of the various quantitative methods available to the cyber researcher. First, we want to set the stage for our cyber discussion by defining some common terms, some of which may already be familiar to some readers.

We live in a cyber age. We interact with and depend upon computers and information technology (IT) constantly. This reliance on cyber assets creates shortfalls when the technology fails and creates particularly severe consternation when those failures are the result of malicious actions.

Cyber attack is the term used to categorize the malicious actions against an IT system. Cyber attacks can affect just about everything, financial systems, our national infrastructure, our personal systems like phones, computers, cars, even the systems in our homes.

Cyber warfare is the collective term used to classify the use of technology to attack IT systems. The instruments within the cyber warfare domain are varied. A computer virus is a mobile computer program that can infect a computer causing all sorts of problems, from degrading computer performance all the way to completely incapacitating the system. Other computer programs can attack the IT system to allow unauthorized access to the system after which a host of hostile actions can occur.

Cyber warfare involves offensive as well as defensive components. An offensive component is meant to inflict some desired harm on a targeted IT system. A defensive component is meant to prevent the opposing offensive components from being successful in their offensive efforts.

> *Cyber defense includes the non-real time analysis of historical data and includes actuarial-like predictions of future events.* (Herring & Willett, 2014)

The cyber domain, with all its useful as well as its malicious components is replete with data. Data governs everything, and given the highly technological aspect of the cyber domain, one can obtain data on just about any action that occurs in the domain. However, data is not quite the same as information; we need to exploit the data via modeling and analysis to obtain the information needed to take any informed actions necessary. Thus, it is useful for those embarking on research in cyber and cyber warfare to have an appreciation of data and how to analyze that data to gain insight for subsequent decision making actions.

Our focus in this chapter is on modeling and data analytics for cyber. Both modeling and data analytics naturally require data. Of course there are myriad issues associated with data, enough to warrant a complete discussion on its own. However, we are going to sidestep the issue with data, obtaining the data, storing the data, verifying and validating the data as well as updating the data, to keep our focus on the analytical models and methods available to exploit that data and in the process produce the information needed for actions.

8.2 MODELING METHODS TO SUPPORT CYBER RESEARCH

The cyber domain is hard to understand; it is large and complex. One really needs to understand the domain to gain particular insights into defensive and offensive effects. Models are used to gain insight into large complex systems. This section introduces a sample of modeling approaches used to understand the cyber domain.

8.2.1 Visualization Methods

Visual methods are a valuable tool to the cyber researcher and operator. The ability to display data from disparate sources and in a variety of forms provides the cyber analyst a wealth of data from which to deduce necessary information and insight. The interpretation of the data is not necessarily automated, but rather interpretation is left to the analyst. This presents challenges and offers advantages.

A key challenge in using visual methods for the display of a wealth of information is avoiding information overload. A question driving research in the human factors/human computer interaction domain is how to present disparate yet abundant forms of information in a manner conducive to effective decision making. The potential solutions are as vast as the number of researchers in the field. Despite presentation challenges, these visually-focused decision support systems remain valuable because of the advantage they afford.

An advantage offered by these data rich, visually dense displays is that human decision makers are wonderfully adept at making sense out of all the data. With experience, the human decision maker learns to focus on the salient features of the data and fairly quickly make very accurate assessments. The human decision maker is also a wonderful pattern recognizer, as we can associate patterns in the display that are of interest and note changes in the patterns expected.

There are a variety of useful visual methods. Dynamic plots allow the user to see trends and spot spikes. Statistical process control principles may be applied to provide limits on observed behavior to get indications of dynamic processes deviating from their expected behavior. Fairly simple rules can be added to a visual display to provide specific alerts or warnings. In addition, displays can quickly move between data sources, combine or stack data from various sources, and much more.

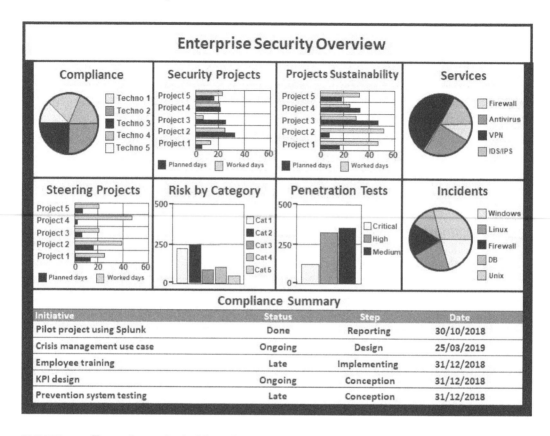

FIGURE 8.1 Example static dashboard visualization.

Goodall and Sowul (2009) is a nice example of research in visual display frameworks useful for the cyber domain. In addition, their work discusses other frameworks that are available. Knaflic (2015) is a easy to follow and comprehensive guide to developing effective graphics and visuals. The text not only features many examples but provides meaningful theory and methodology, so a user can actually implement effective graphics.

However, it is often the case that gaining a deeper understanding of IT processes subject to cyber attack may require using models of the systems or processes to provide insight beyond that attainable via purely visual methods. Insight sought from these models vary. Common objectives include trying to predict cyber attack occurrence or success, testing various cyber offensive or defensive strategies, or even trying to predict the range of impacts that may result from cyber attacks.

The example in Figure 8.1 (B2B, 2015) is a great example of a dashboard involving lots of visual components. Note the use of varied information with a variety of presentation methods, including pie charts, bar charts and histograms. In addition, the bottom of the display contains examples of using simple rules to provide specific information for the user. The challenge, of course, is how to present this amount of information while retaining comprehension of some overall status.

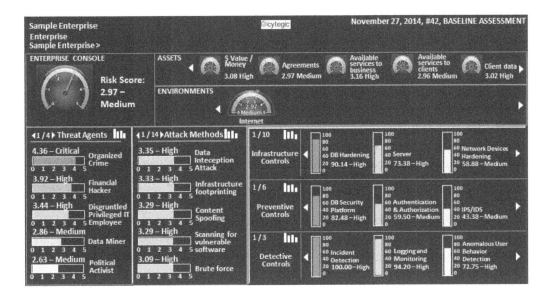

FIGURE 8.2 Example dynamic dashboard visualization.

The dashboard in Figure 8.2 (Indik, 2020) depicts more of a real-time display. The dashboard features a lot of information and is quite busy. However, despite the grey scale figures used here, for the implemented version the designers have used color coding and odometer dials as a means to draw user attention to those items deemed important. This dashboard would likely be quite appropriate for a real-time monitoring and control system.

8.2.2 Simulation Modeling and Analysis

Simulation modeling has long been a favored analytical approach, particularly in military settings (Hill & Miller, 2017). Simulation is a computer-based, or digital, representation of the physical system or process under examination. The simulation model abstracts the important features of the system or process of interest, represents those important features in a digital model designed to mimic the system or process, and then produces data representative of the output from the system or process of interest under the specified simulation conditions. Simulation is quite adept with complex systems or processes, when considering scenarios that might not be reasonably tested in an actual setting, or when trying to gain understanding into conceptual systems or processes.

There are multiple paradigms, or modeling approaches, in the simulation domain. These include:

- Discrete-event simulation

- Agent-based simulation, to include complex-adaptive agent simulation

- Spreadsheet or Monte-Carlo sampling methods

In a *discrete-event simulation* (DES), simulation time moves in discrete steps. Events are defined to capture the abstraction of the system or process, and the occurrence of these events happen at discrete points of time. Part of the simulation logic is to schedule when that event is to occur. An ordered list of events is maintained, and the simulation time advances to the next event time once a current event completes processing. A common graphical depiction of a DES involves items, or entities, moving around a network structure. Kuhl, Sudit, Kistner, and Costantini (2007) use discrete-event simulation to simulate, and evaluate, cyber attacks against computer networks. Since DES is adept at modeling network structures and the flow of items through that network, the DES paradigm is quite popular for modeling the network structures found in the cyber domain.

The *Petri net* is another network modeling method focused on the connections between elements of the network and the transactions between those elements. Petri nets are often instantiated as a DES. Chen, Sancez-Aarnoutse, and Buford (2011) use the Petri net paradigm to examine cyber attacks made against a smart grid.

An *agent-based simulation* model is a particular type of DES in which the entities, or agents, within the model abstraction are provided an increased level of autonomy or intelligence. Entities in DES typically follow defined sequences of processes and do not interact in the simulated environment. This means the entities do not have mechanisms to observe conditions or behaviors in the simulation and react to what is observed. The agent-based simulation entities are provided capabilities to sense aspects of their simulation environment and adjust their behaviors accordingly. Weimer, Miller, and Hill (2016) provide a concise primer on the methods of agent-based simulation focused on general defense applications, while Kotenko and Ulanov (2005) provide an example where the agent modeling paradigm is used to examine and gain insight into denial of service attacks.

A *Monte-Carlo*, or *sampling simulation* is useful when time passage is not relevant for the simulation behaviors. For instance, consider a financial model built based on a set of specified inputs. The model is calculated and returns some answer, let's say return on investment. The answer depends upon the inputs, but what if the inputs are actually uncertain and any particular combination are just one instance? A Monte Carlo method would sample the distributions of the input parameters generating many input instances, record the corresponding answers, and return not only average answers but also the risk surrounding that answer based on the empirical distributions obtained during the sampling. These approaches are quite useful in financial planning, risk management, inventory analysis, and functional evaluation. Metropolis (1987) provides a history of the Monte Carlo method starting with its early roots as a classified defense analytical method.

There are valid reasons why simulation should be a tool in every cyber researcher's toolbox. The simulation is designed to digitally mimic the essential characteristics of the system under study. Simulation fidelity refers to the level of detail modeled. For instance, an IT simulation model may represent detailed packet traffic in a network or a network switch. Another IT simulation model may focus on the communication links foregoing the physical-level details of the network. Simulation scope refers to the breadth of the system model; maybe one model is a single

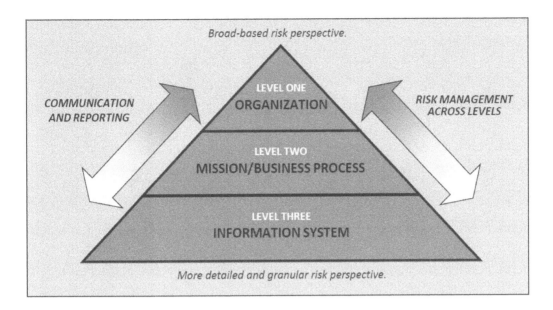

FIGURE 8.3 Organizational wide risk management framework.

computer while another model provides a representation of a complete network. The incredible breadth of simulation possibilities makes it a flexible, valuable analytical tool. Its ability to incorporate uncertainty in the system abstraction, and its ability to generate a wealth of simulated data as a result of running a very large number of scenarios means simulation models are quite useful in gaining insight into cyber warfare and are a valuable tool for the cyber researcher. As Hill and Miller (2017) point out, simulation has been a tool of choice for the military for a very long time and has had and will continue to have similar success in the cyber domain.

8.2.3 Risk Management Frameworks

Risk management frameworks (RMF) are methods that help identify, assess, and manage cyber risks of critical system infrastructures for cyber defense. The two main publications that cover the details of RMF are National Institute of Standards and Technology (NIST) Special Publication 800-37, 'Guide for Applying the Risk Management Framework to Federal Information Systems,' and NIST Special Publication 800-53, 'Security and Privacy Controls for Federal Information Systems and Organization.' An RMF is a structured process that integrates information security and risk management activities and is purposefully designed to be technology neutral so that the methodology can be applied to any type of information system without modification. As shown in Figure 8.3, the RMF is equally applicable to a broad range of systems at three levels of an organization, namely at the organization level, mission or business process level, and information systems level.

Regardless of the level at which the RMF is applied, the NIST RMF includes steps to Categorize, Select, Implement, Assess, Authorize, and Monitor, which

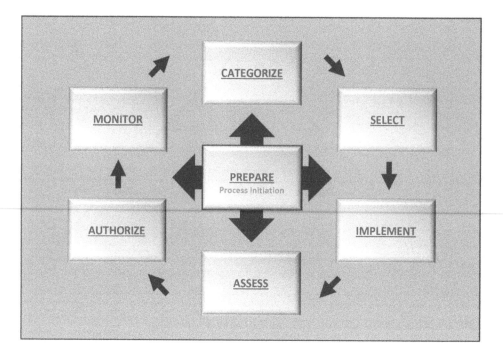

FIGURE 8.4 NIST risk management framework.

are implemented in a cycle, as illustrated within the RMF cycle in Figure 8.4. The main philosophy of RMFs are to balance system stakeholder and user requirements with controls. Controls are restrictions or safeguards appropriate for achieving the particular cybersecurity or privacy objective. To accomplish this, we begin with the Prepare step in Figure 8.4. The Prepare step ensures adequate risk management preparation at the organizational and system levels. The security and privacy activities associated with the Prepare step help avoid an implementation of RMF that becomes too costly and demands too many skilled security and privacy professionals. These associated activities lead to an organization having comprehensive, enterprise-wide governance as well as the appropriate resources in place to avoid an overly burdensome implementation of RMF. Some of the activities in the Prepare step are explained for the organization and system level.

At the organization level:

- Provide an organization-wide forum to consider all sources of risk
- Designate a Senior Accountable Official for Risk Management
- Ensure an effective security program is established
- Assess ongoing organization-wide security and privacy risk
- Align information security management processes with strategic, operational, and budgetary planning processes

At the system level:

- Implement an enterprise architecture strategy that facilitates effective security and privacy solutions

- Determine placement of system within the enterprise architecture

- Identify the types of information to be processed, stored, and transmitted by the system

An architecture is a formal description and representation of a system, organized in a way that supports reasoning about the structures and behaviors of the system. The system level understanding is critical to understand and prepare for RMF implementation. Many of these activities have been assigned to, or coordinated by, Chief Data Officers whose position has recently been formed within organizations.

We begin the more detailed discussion of the RMF process with the Categorize step in Figure 8.4 that informs the risk management processes and tasks by determining the adverse impact to operations and assets, individuals, and other organizations to the loss of confidentiality, integrity, and availability of the system and the information processed, stored, or transmitted by those systems. In this step, the worst-case scenario is explored to help define the criticality and sensitivity of the system and the scenario's adverse impact on the system. The next step, the Select step, identifies baseline security controls and provides tailoring guidance and supplements controls as needed based on risk assessment. The Implement step applies security controls within the system architecture using sound systems engineering practices and determines security configuration settings. The Assess step determines the control effectiveness by determining whether the controls were implemented correctly and whether they are performing as intended. The Authorize step determines the risk and, if acceptable, authorizes the operation of the system. If the system is authorized to operate, the RMF process has determined that the risk of operating the system is acceptable and appropriate controls have been put in place to manage risk to the system. Finally, the Monitor step continuously tracks changes to the system that may affect security controls and reassesses control effectiveness.

Some typical types of controls include:

- Access control, who is allowed to use the system

- Awareness and training, ensuring proper use of the system

- Security Assessment and Authorization, formalizing an instance of the Assess and Authorize steps

- Configuration Management, managing changes to the system

- Identification and Authentication, ensuring only authorized users

- System and Information Integrity

Within the larger RMF construct and within the application of various controls, there are opportunities to leverage modeling and data analytical methods to improve cyber defense effectiveness as it relates to the ability to detect cyber threats, the speed of responding to identified threats, and the management of scarce resources. Many areas of research have arisen out of placing the larger cyber defense problem within this construct.

8.2.4 Cyber Red Teaming

Cyber Red Teaming is an advanced form of assessment that can be used to identify cyber weaknesses in a variety of systems. It is especially beneficial when the target system is still in development and designers can readily affect improvements. The Red Team acts as an adversary attempting to penetrate the system security measures and manipulate the system in some nefarious way. This action is a cyberattack. Such a cyberattack may be any type of offensive maneuver that targets computer information systems, infrastructures, computer networks, or personal computer devices. Some common types of cyberattacks are Denial-of-Service (DoS), phishing, password attacks, Structured Query Language (SQL) injection attack, and malware attacks. By correcting system vulnerabilities or otherwise accounting for these successful cyberattacks, red teaming leads to system cyber security improvements. Red Teaming is a type of exploratory wargaming where systems are tested and refined without the need for actual hostilities. Several components need to be present to have an effective war game. First, a clear objective of the wargame needs to be established. For instance, the objective of a war game can be to explore and develop decision alternatives, to improve decision making abilities, to explore shortcomings of a systems process, or to better understand the events that lead to a particular outcome, among others. Other components included in wargaming that can add to the effectiveness of Red Teaming are:

- Red Team: In cyber Red Teaming, this team is often a highly skilled team, often certified by an appropriate agency, that possesses the tools and capabilities to attack a system.

- Blue Team: In cyber Red Teaming, this team should consist of the actual individuals who operate the system and employ the RMF controls.

- White Team (Adjudicator): This neutral person or team make a formal judgement of an effect of an attack by the Red Team and subsequent application of a control by the Blue Team. Cyber Red Teaming often involves an attack on an actual system and the White team may need to assess the outcome of such an attack by acting as a referee. Additionally, the White Team may provide overall support to the wargaming activity.

- Dynamics: Cyber Red Teaming often involves free play by each side, both the Red Team and Blue Team. Alternatively, a more structured deliberate turn-by-turn approach may be used that has both teams making decisions simultaneously. Dynamics also involves the observability of the system. Since the White Team observes the actions of the Blue Team and Red Team, the White Team typically has a clearer picture of the state of the system at any moment. Conversely, the Red Team and Blue Team may have imperfect information of the state of the system at any moment of the game. It is important that data regarding the knowledge of the Blue Team, Red Team, and White Team at any time is maintained during the Cyber Red Teaming exercise.

- Measures of Merit: This measure defines to what degree an objective has been achieved. Measures of Merit may include measures of performance of a system and measures of effectiveness, which are measures designed to correspond to accomplishment of system mission objectives and achievement of the desired system result.

Cyber Red Teaming can benefit greatly from improved modeling using simulations, metrics, and data analytics.

8.2.5 Game Theory for Cyber Security

Game theory is a study of strategic decision under competition. More formally, it is the study of mathematical models of conflict and cooperation between intelligent, rational decision-makers (Tambe, 2011; Wu, Shiva, Roy, Ellis, & Datla, 2010). The term *rational* implies that players are making choices to satisfy their best self interests. Each situation and player has a portfolio of actions (called the action profile) that they can perform and the associated payoffs (or utility) for those actions. In the Theory of Rational Choice, a person would evaluate their action profiles and then pick that profile to maximize their utility. Formally we define the function $u()$ as the utility function for a player and a and b as two different action profiles. If a player acts rationally, then they would pick action profile a if

$$u(a) > u(b) \tag{8.1}$$

Notice that we need not know the actual function $u()$ in order to determine which action profile is used given that the player acts rationally. It suffices for a player to know which action profile he or she prefers over another. This does make intuitive sense. For example, I might prefer product A over product B because it has a certain feature, but I have not necessarily evaluated with mathematical precision exactly by how much I prefer product A over product B.

The *Nash equilibrium* is the steady state of the game. If a game has a Nash equilibrium, then each player has an incentive to retain their decision over changing it to a different decision. Unlike a decision that is simply based on personal (think greedy decision making) incentives, a Nash equilibrium takes into account each player's potential decisions. Thus, for a player to determine the optimal and steady state decision strategy, a player should ask him or herself: Given my opponent's decision X, what should my decision be? If we let $A_{r,s,i} = D_i(P_r | P_s = d_{s,i})$ represent the optimal action profile player P_r can make, given P_s action profile is $d_{s,i}$, where

	P2 Confess (C)	P2 Quiet (Q)
P1 Confess (C)	2, 2	0, 3
P1 Quiet (Q)	3, 0	1, 1

FIGURE 8.5 Payoff matrix in prisoner's dilemma.

r represents Player 1 and s represents Player 2, then the set of Nash equilibria ($S_N E$) can be represented as:

$$S_{NE} = (\cup_{i=1}^{n}\{A_{r,s,i}, d_{2,i}\}) \cap (\cup_{i=1}^{n}\{A_{r,s,i}, d_{1,i}\}) \qquad (8.2)$$

If $S_{NE} \neq 0$; then there exists an action profile for each player such that if both players choose their respective action profile, there is no incentive to deviate from this action profile. Let $|S_{NE}|$ be the cardinality (the number of elements) of the set S_{NE}. If $|S_{NE}| > 1$ then multiple Nash equilibria exist; however, Nash equilibria need not be equally preferable. Depending on the player utility function or payoff values, an action profile that is also a Nash equilibrium could be better than another action profile that is also a Nash equilibrium. Further analysis is needed after S_{NE} has been obtained to decide on which action profile to focus.

We will illustrate how to obtain the set of Nash equilibria with a game called the Prisoner's Dilemma. The Prisoners Dilemma deals with two suspects, *P1* and *P2*. After their arrest they are kept separate, so they cannot corroborate their stories. Both have the option to confess or remain quiet, and both would like to minimize the time they would have to serve. If both stay quiet, the police can only get them for a lesser charge of 1 year. However, if only one of them confess then he or she who confesses will receive no sentence, while the other will get the full sentence of 3 years. Finally, if both confess, then both still receive a higher sentence of 2 years. This is summarized in the following payoff matrix in Figure 8.5.

If each player simply acted in their own self-interest, (that is, trying to maximize their utility by minimizing their time) each would confess trying to obtain no sentence at all. However in acting in their own self-interest, they receive a 2-year sentence each. Now if each player tried to maximize their utility by taking the action profile of their opponent into account, we obtain the following decisions:

- Given that *P1* chooses Quiet, *P2* chooses Quiet
- Given that *P1* chooses Confess, *P2* chooses Quiet

- Given that *P2* chooses Quiet, *P1* chooses Quiet

- Given that *P2* chooses Confess, *P1* chooses Quiet

Utilizing Equation 8.2, our viable set of Nash equilibria becomes:

$$(\{Q,Q\} \cap \{C,Q\}) \cup (\{Q,Q\}\{Q,C\}) = \{(Q,Q)\} \tag{8.3}$$

Therefore the Nash equilibrium is for both to choose 'Quiet' with cardinality 1. Therefore the optimal solution is the only Nash equilibrium and will result in only a 1-year sentence for both suspects.

Game theory provides a mathematical model of the interactions between players and has recently gained increased prevalence for cyber applications (Chung, Kamhoua, Kwiat, Kalbarczyk, & Iyer, 2016; Do et al., 2017; Shiva, Roy, & Dasgupta, 2010; Sokri, 2020; K. Wang et al., 2016; Y. Wang, Wang, Liu, Huang, & Xie, 2016). Similar to the prisoner's dilemma problem, these game theoretical models consist of a collection of participants, called players, who are each defined by a set of actions and strategies. These actions and subsequent strategies make the critical assumption that players act rationally, that is they assume that a perfectly rational player could justifiably play [a strategy] against multiple, similarly perfectly rational opponents (Tambe, 2011; Wu et al., 2010). In practice, this means the game players will behave in an expected manner so these models do not consider the 'unexpected' behavior sometimes found in reality.

Games are expressed in normal form or extensive form. Normal form games restrict each player to act simultaneously and are usually expressed in the form of a matrix. The solution to these games is the best payoff for each player such that neither player can do any better by deviating from their strategy. This means that given one player chooses to play a specific strategy, the other player chooses the best response to that strategy such that their utility, or benefit, is at least as good as all other responses. Thus, the best response need not be unique. The strategy played is known as the Nash equilibrium. The Nash equilibrium may not be unique and in some cases may not exist. However, in a mixed strategy game where each players strategies are played probabilistically, at least one Nash equilibrium will always exist.

Extensive form games add a temporal aspect to the game instead of requiring simultaneous player actions over a single turn or period of time. Actions are taken by players sequentially over either an infinite amount of time or until a terminal point is reached. These games are sometimes referred to as *Stackelberg games*.

Yuan, Sun, and Liu (2016) present a two-level Stackelberg game in order to evaluate the actions taken by a resilient control system as part of a cyber-physical system in the event of a DoS or distributed DoS attack. Each level represents the layer of the cyber-physical system in which each Stackelberg game is being played. The first, or internal, level takes place at the cyber layer where the players are the intrusion detection system (IDS) and the cyber attacker. The second, or external, level occurs at the physical layer between the external disturbance on the system and the controller.

To find an optimal defensive response strategy, Jiang, Tian, Zhang, and Song (2008) propose a two-player, zero-sum stochastic game where the players are defined as the attacker and defender. The state space is defined by the level of system privileges the attacker has obtained at any given point in the game, with transitions occurring based on the action taken by the attacker. Since the attacker is continuously gaining privileges from the defenders system, the utilities for each player are additive inverses based on the amount of cost to the defender, considering operational costs, cost of damage, any residual costs, and the cost effect on the system based on the defenders action.

Musman and Turner (2018) present a minimax game where each player seeks to minimize the maximum utility of the opponent. This game, called the Cyber Security Game, provides a preparatory evaluation of a systems ability to withstand an attack by considering qualities such as the purpose, the entire range of hazards, and all possible attacks. A suite of tools is used to analyze the quality of the entire system and the degree of risk incurred given strategic investments in the development of the system. Chapter 11 provides an in-depth discussion of various game theoretic methods.

8.2.6 Markov Chains for Cyber Security

Markov Chains are stochastic processes defined by $\mathbf{X} = \{X_n, n = 0, 1, 2, \ldots\}$ and finite n, when $X_n = i$ the stochastic process is defined to be in state i at time n. Let P_{ij} be the probability of moving, or transitioning, from state i to state j and

$$P\{X_{n+1} = j | X_n = i, X_{n-1} = i_{n-1}, \ldots, X_1 = i_1, X_0 = i_0\} = P_{ij} \quad (8.4)$$

for states $i_0, i_1, \ldots, i_{n-1}, i$, and j and $n \geq 0$ Ross (1996). Thus, any future state X_{n+1}, given we know the past states $X_0, X_1, \ldots, X_{n-1}$ as well as the present state X_n is independent of the past states and depends only on the present state (Ross, 1996).

If the transition probability

$$P\{X_{n+1} = k | X_n = j\} = P\{X_1 = k | X_0 = j\} \text{ for } n = 1, 2, \ldots,$$

then it is stationary and can be denoted by p_{jk}. Note that being probabilities, $0 \leq p_{jk} \leq 1$ for all j, k and as the process must make a transition into some state, $\sum_k p_{jk} = 1$ for all j. The matrix $\mathbf{P} = (p_{jk})$ denotes the one-step transition probabilities P_{jk} for all j, k. $\mathbf{A}$ is called a stochastic matrix if $\sum_k a_{jk} = 1$ and $0 \leq a_{jk} \leq 1$.

For Markov Chains, a class is a set of states that are all mutually reachable. In general, a Markov Chain can have many classes, but often will have one class where every state is reachable from every other state. If this is the case, we call the Markov Chain irreducible. Additionally, if every state can reach every other state in a finite number of transitions, then the Markov Chain is positive recurrent. Finally, let $\{n_j\}$ be the set of numbers such that $p_{jj}^{(n_j)} > 0$. Let d_j be the greatest common divisor of $\{n_j\}$, d_j is called the period of j. d_j is the same for every state in the same class as j. That is, period is a class property. If $d_j = 1$, state j is aperiodic.

If a Markov chain is aperiodic irreducible positive recurrent (called an ergodic Markov chain), then $\lim_{N \to \infty} P_{jk}^{(n)} = \pi_k$, where π_k is the long run proportion of time spent in state k. This provides a means to solve ergodic Markov chains by solving the system of equations given in vector form by $\pi = \pi P$.

Gore, Padilla, and Diallo (2017) provide a model for cyber security using Markov chains in order to determine the likelihood of identified attack paths through a defenders network while considering eight predetermined factors. These attack paths indicate potential weak points in a defenders network. As with the methodology presented by Musman and Turner (2018) using game theory, their approach is preemptive in nature and is limited in usability during a cyber attack event.

Larkin (2019) combines game theory and Markov chains to develop an operational network security problem consisting of 20 tactical normal form games that provides an assessment of the resiliency of a cyber defenders network. Each tactical game informs transitional probabilities of a discrete-time Markov chain over an attacker-defender state space. The Markov chain provides an assessment of the conditional path through the operational problem with an expected cost of damage to the defender network. The solutions of the tactical sub-games and, in turn, the operational problem, allow assessment of the risk of various network functions and project the effects of network improvement resource allocation decisions via an integer program. Network resiliency is enhanced against potential malicious external actors through improved resource allocation of these scarce resources that is informed by the Markov models.

This methodology is particularly useful from the perspective of the cyber defender. The methodology provides a means to effectively decompose an attack and corresponding defense into tactical sub-games, provides an assessment of the networks resilience to include assessing resiliency from the current state, and provides a means of increasing resiliency if additional resources or technologies become available.

8.3 DATA ANALYTIC METHODS

This section delves into specific techniques used in data analytics.

Data analytics involves the extraction of meaning from data. In our view, the basic language of data analytics is statistics. This is not a new nor universally accepted concept. This view grew in part out of our reading the musings in the seminal work by Tukey (1962) when referring to analysis, and its growing importance to the field of applied statistics. Thus, our discussion here on the methods of data analytics is easily viewed as a discussion on the statistical methods useful to the cyber research.

Jensen (2020) posited, 'Statistics = Analytics' while comparing the roles and views of statisticians and data analysts. With the data in hand, statistical techniques, or processing algorithms, can be applied to generate the insights needed for management and leadership. Our focus on data analytics for cyber research takes on this statistical flavor; we will overview some of the statistical, or machine learning, methods useful for analytics.

However, a slight divergence of discussion is warranted to establish that we adhere to the concept of 'Statistics $\neq$ Data Science.' Statistics and analytics focus on the methods. A statistical point of view examines the workings of the underlying algorithms in addition to the interpretation of the information generated by the algorithm. As Breiman (2001) discusses, this is the culture focused on using the data to make inferences regarding the underlying process, which can be described by some mathematical model. The data analytical point of view is more focused on the use of the algorithm as a black box to generate the prediction information that is then interpreted for managerial and leadership insight. This corresponds to the second culture mentioned in Breiman (2001), where the algorithmic model does not require, or provide, insight into an unknown process generating the data. We see data science as an expansion of the data analytics point of view to include the data engineering, data management, data scraping, and data collection challenges. The data science focus must consider how to collect the data for analysis, and more importantly, consider how to continue to collect, store, manipulate, and make available that data needed for the analytics. We do not address these data science attributes here other than to acknowledge this difference between statistics, data analytics, and data science, as well as the need for all skill sets to be involved.

8.3.1 Classification of Methods

There are a variety of ways to characterize data analytical methods. A common way to classify the methods are one of

- Descriptive

- Prescriptive

- Predictive

Jensen (2020) lists the above methods in a model focused on the types of questions to answer in an analysis. Kulachi, Frumosu, Khan, Ronsch, and Spooner (2020) use similar categories in their recent work recounting their data science experiences.

A descriptive method helps understand the data. For instance, visual or graphing methods are common. Examples include scatter plots, histograms, and box-whisker plots, each of which provide insight into the distributional aspects of the underlying data. Just as common are summary methods, such as computing means, variances, or constructing distributional comparisons. Other descriptive methods describe how to map the dataset of interest into fewer dimensions or how to classify the data based on attributes. Such methods also provide an element of prediction when new data are added. Some of these methods are further discussed below.

Prescriptive methods produce strategies to attain some defined goal, or set, of goals. For instance, a simulation study may provide strategies to best counter some anticipated cyber attack. Similar strategies may emanate from other modeling approaches as well, such as optimization, probabilistic modeling, or decision analytical methods. The analysis of data from past events, and subsequent actions, may prescribe strategies yielding the best outcome based on trends or profiles found in the past data.

Predictive methods provide information pertaining to future or as yet unobserved events. Various interested parties, which in the defense realm are often leadership, may want to know the impact of a massive cyber attack against infrastructure assets. Leadership may wish to know where the next attacks might occur. In each case, the analyst provides insight into future or as yet unrealized events. The data analytic approaches may involve forecasting based on past events, predicting results based on some set of characteristics, or classifying outcomes based on similarities to past, known events. Further discussion on descriptive, predictive, and prescriptive methods is given in Chapter 9.

Modern data analytics tend to use the following taxonomy for the various tasks:

- Prediction

- Classification

- Dimension reduction

From a data analytics perspective, prediction is of prime importance. Modern datasets can be quite large; there is, in fact, a vast literature on 'big data' (see Kulachi et al. (2020) for a recounting of recent experiences). With such large datasets, the data analysts tend to ignore the underlying data generation mechanisms to focus on prediction accuracy. The statistical view focuses on the data as a sample from some underlying population distribution and draws inferences from the analysis to understand that underlying data generation mechanism.

Classification methods focus on using the characteristics of the current dataset to create homogeneous partitions of the data. This partitioning forms the basis for subsequent predictions. New data are classified based on the parameters learned and used to create the partitions. New data are then associated with a partition and the particular characteristics of the members of that partition. In the cyber domain, data on past attacks can be exploited to classify the severity or possible outcomes of some new or anticipated attack.

Dimension reduction methods take data with many dimensions, or characteristics, and effectively describe that same data with fewer characteristics. The resulting reduced dimension dataset is used to build simpler, more tractable and sometimes more effective models. A cyber attack dataset may have a large number of characteristics. Comprehending all these dimensions may be daunting to the cyber analyst. A dimension reduction method can be used to identify a subset of fewer yet important characteristics to which the analyst can focus their analytical and monitoring efforts and still gain insight on the full dataset.

Modern data analytics also classify methods by whether they use what is called *supervised* or *unsupervised* learning. An algorithm estimates, or learns, undefined parameters through mechanisms such as direct calculation, iterative improvement, or some heuristic search process. The two learning method classifications differ

TABLE 8.1 Classification of methods discussed

	Predictive	Classification	Dimension Reduction
Supervised	Regression Logistic Regression Trees	Discriminant Analysis Support Vector Machine	
Unsupervised		K-Means k-Nearest Neighbor Multidimensional Scaling	Factor Analysis Principal Components

based on whether or not response data are used in the model parameter estimation process.

A *supervised learning algorithm* uses response data to guide its parameter search process. *Regression analysis* uses the response data to find the parameter values that minimize some loss function, such as the mean square error between model predictions and the actual response data. *Time series models* use the response data to find the model parameters used to predict future data points. A *genetic algorithm* returns parameter values that the algorithm deems best attain some defined measure guiding the search.

An *unsupervised learning algorithm* does not use response data in the model parameter estimation process. For example, a clustering algorithm might use the characteristics of the independent data to find which points can be grouped together and how many different groups effectively describe the input data. Dimension reduction methods exploit the relationships among the data characteristics to find some smaller dimension or set of important characteristics. Since the response data are not used, these are classified as unsupervised.

Naturally, there can also be cases wherein the method uses both supervised and unsupervised approaches. Frumosu and Kulahci (2018) examine such hybrid approaches.

Table 8.1 displays methods discussed in the remainder of this chapter in terms of method type and manner of supervision.

8.3.2 Supervised-Predictive Tools

Supervised-predictive tools employ the past data, response and characteristics, to form a model useful for predicting outcomes based on particular characteristic values.

8.3.2.1 Regression Analysis

Regression analysis techniques model the linear relationship between the values of a response variable and some collection of predictor variables. Regression analysis assumes the data follow some underlying true mathematical model form and we can use the data collected to estimate that underlying model. The general form of

the linear model is:
$$Y = \beta_0 + \beta X + \epsilon \quad (8.5)$$
where β is a vector of unknown (but learned) regression coefficients, X is a $(n \times k)$ matrix of n observations and k predictor variables, ϵ is the independent error component usually assumed to follow a normal distribution with a zero mean and some unknown, but constant, variance, and Y is the response variable. The k predictors can be separate characteristics or functions of other characteristics, such as interaction or nonlinear terms.

The collected data, Y_i, are used to estimate Equation 8.5 yielding
$$\hat{Y} = b_0 + BX + e \quad (8.6)$$
where $\hat{Y}$ is the estimated response, b_0 is the estimate of β_0 and the matrix B is the estimate of matrix β in Equation 8.5 and e is the estimate of the error component in Equation 8.5, commonly referred to as the residual component, or the residuals. A common estimation approach is based on a least-squares criteria where the estimated parameters of the model are set such that the sum of the squared estimated errors, defined as $\sum_{i=1}^{n}(Y_i - \hat{Y}_i)^2$, is minimized.

Analysts are generally concerned with two aspects of a regression model. First how well the model predicts the values used to estimate the model. As Equation 8.5 depicts, data are assumed comprised of the output from an underlying model (with unknown parameter values) and error. The variance of the response data are explained by the estimated model and the estimated error. Naturally, we want a model to accurately describe the data, and that accuracy is measured in terms of the response variability explained. A model that explains most of the variability is a good fitting model. Two measures of model fit are commonly used: the R^2 value and the R^2_{adj} value, both of which fall between 0 (really poor fit) and 1 (perfect fit). Each measure is a ratio involving the variance explained to the total variance in the response.

The second aspect, more important for inference purposes, is the adequacy of the model. Model adequacy focuses on the assumptions of the error term in Equation 8.5 as realized in the estimated form of Equation 8.6. The residuals, $e_i = Y_i - \hat{Y}_i$, are examined for relative normality, constancy of variance, and a general lack of patterns explained by characteristics currently excluded from the model. An adequate model enables inference regarding the statistical properties of the model and the variety of statistical tests found in regression modeling.

With an adequate well-fitting model, analysts can make inferences regarding the importance of the predictor variables with respect to the response variable as well as make predictions of the response given new values of the predictor variables.

8.3.2.2 Logistic Regression

When the response data are binary, such as 0-1, false-true, miss-hit, failure-success, then the regression model takes on a special form called the *logistic regression*. In a logistic regression we want a model whose output provides a probability associated with one of the binary responses as a function of some set of predictor variables. This requires some manipulation of the binary response.

With binary outcomes, the probability of 'success' is π_i. Convention is that a success is coded as a 1. Naturally, the range of $\pi_i \in [0, 1]$. We can transform this range as

$$\frac{\pi_i}{1 - \pi_i} \in [0, \infty] \quad (8.7)$$

and finally

$$\eta_i = \ln \left[\frac{\pi_i}{1 - \pi_i} \right] \in [-\infty, \infty]. \quad (8.8)$$

The final transformation, Equation 8.8, gives the range we generally prefer in regression. Unlike linear regression in which the model is estimated using closed form, least squares results, the logistic regression is estimated numerically. The usual linear regression assumes normality of the data. A more general setting treats normality as just one case of a number of data distributions. This general setting is the class of general linear models, for which linear regression is a case. Logistic regression is in the class of generalized linear models but based on a binomial distribution of data. The function fit is then

$$\eta_i = X^t \hat{\beta}. \quad (8.9)$$

The resulting function maps the particular characteristics of a data point to the likelihood of that point being a success (or a failure if using the complement).

Figure 8.6 shows notional missile data from Montgomery, Peck, and Vining (2012). On the left are the data for some missile speed (the predictor variable) along the y-axis and the outcome of 0 (hit) and 1 (miss) is along the x-axis. Clearly,

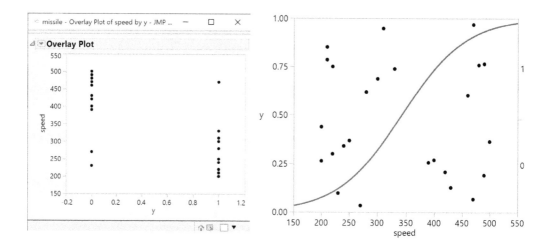

FIGURE 8.6 Notional logistic regression, data on the left, function on the right. On the left, y represents the binary response outcome. On the right, y represents the probability of success given the predictor variable value of speed.

data are hard to use intuitively as there is not a clear cut pattern even though there does appear to be a favoring of hits at higher speeds. The data as plotted shows the binary nature of the results. Examining the left side plot rotated a quarter turn shows the overlap of 0 and 1 results at the intermediate speeds. This common situation complicates the prediction process.

The resulting logistic equation is displayed in the plot on the right side of Figure 8.6. The x-axis is the predictor variable speed, while the response output y in this case is the probability of success (i.e., a value of 1) for that predictor variable value. This plot shows the ability of the logistic regression model to differentiate the misses and hits and provide a functional probability of hit for a given value of speed.

The estimated function also provides the log-odds ratio. While regression coefficients provide a increase in the response per unit increase in the dependent variable (with others held constant), the log-odds in logistic regression provide the increase in the probability of success given the unit increase in the dependent variable (again with all others held constant).

8.3.2.3 Tree Methods

A *decision tree* recursively partitions data based on characteristics and responses of the data such that each partition is increasingly homogeneous in the response data. Each sub-partition refines the level of classification and adds complexity. A very simple tree is shown in Figure 8.7. The tree building process is often called a *recursive binary splitting approach* since each partition is further divided into two sub-partitions. A tree could be developed that partitions data on every characteristic; such a tree on non-trivial data is generally considered to suffer from over fitting, merely repeating the input data and not providing much useful insight. In practice, trees are pruned which involves reducing the depth of the tree by removing less important splits. These less complex trees are then used to classify the existing data as well as new data that requires processing. A *regression tree* partitions based on models of the response at each node in the tree. The regression tree prediction is based on a quantitative model of the data in the partition associated with each tree leaf. A *classification tree* uses data characteristics to create the recursive splits using the responses in each final partition to return the classification prediction.

Trees suffer from the greedy approach used to create them; the algorithm partitions the tree based on some 'best at this time' metric. This causes the tree to be built using very similar steps each time, even when the training set used to build the tree differs. Two ensemble-based extensions to a single tree model are available to boost prediction accuracy: (i) bagging or bootstrapping, and (ii) the random forest.

A *bootstrap forest* generates many trees from the data set by defining a variety of training data sets based on re-sampling the original data. Each of these trees are used to develop a prediction all of which are then combined to return the ensemble prediction. This is referred to as a bootstrap forest or a 'bagging' approach. However, since each tree is built using the best partition approach at each step,

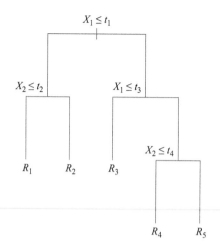

FIGURE 8.7 Example of a simple regression tree.

each tree is likely to have a similar structure due to similar sequences in the recursive splitting process. A *random forest* extends the bootstrapping method by randomly selecting the partitioning variables in each step. This results in multiple trees that differ in structure due to the random component and differ in content due to the bootstrapped training data set. Thus, the random forest provides the predictive power of the tree with the robust prediction quality of a diverse set of such trees.

Ramirez (2018) used such forests to classify acoustic signals. For cyber, a trained tree can be used to quickly, and quite accurately, classify incoming attacks.

8.3.3 Supervised-Clustering Methods

Clustering methods are used to find groupings among the data. Generally, discriminant functions are used as the primary mechanism to assign data to groups or clusters. As shown in Figure 8.6, the logistic regression has clustering capabilities for the two-group problem.

8.3.3.1 Discriminant Analysis

In a discriminant analysis, linear functions separate the predictor data into groups based on their responses, with a high degree of classification accuracy. Consider Figure 8.8 from Rencher and Christensen (2012). In this two-dimensional case, we seek a linear function that separates the data accurately. This is an example of a two-group discriminate problem. The reader will note some similarity with a logistic regression model as a two-group discriminator.

This two group problem reduces to funding the linear function

$$Z = \hat{b}X \tag{8.10}$$

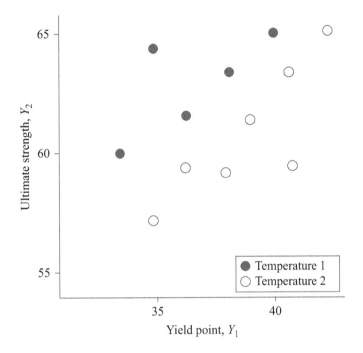

FIGURE 8.8 Example of a two-group discriminant problem. The axes are plotted using the predictor variables Y_1 and Y_2.

with $\hat{b}$ found such that it maximizes a distance function

$$\Delta = \frac{(b^t \mu_1 - b^t \mu_2)^2}{b^t \Sigma b} \qquad (8.11)$$

where X is the matrix of characteristics for the n observations, b is the estimate of the linear function coefficients, μ_i are the respective means of each group, and Σ is the pooled sample covariance matrix.

For multi-level, non-binary responses, the method extends to finding multiple discriminant functions using multiple means, also assuming a common variance. For example, consider a multivariate set of nine predictor variables used to classify four groups of wolves (Morrison, 1990). Figure 8.9 depicts the results of a discriminant analysis.

The nine dependent variables are weighted and combined to yield four discriminant functions or canonical dimensions. Figure 8.9 depicts the data as defined in the first two canonical dimensions. The dots represent each of the data points plotted in terms of their first two canonical scores. The circles represent the groupings of the four responses based on all four canonical scores. The original dimensions of the data are represented with the nine axes in the approximate middle of the figure. The lower case letters within the circles are the four classifications of the data in the dataset.

We note that the independent variables, $X_1, \ldots, X_9$ generate the four discriminant functions defining the canonical axes used to best describe the separating planes between each of the groups.

290 ■ Mathematics in Cyber Research

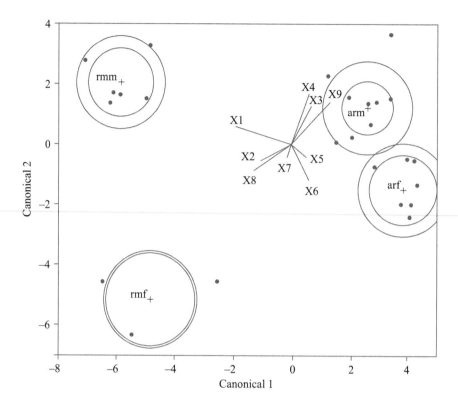

FIGURE 8.9 Example of a multi-group discriminant problem.

In cyber research, historical data on types of attacks can be used to develop the discrimination functions to categorize some new attack. Correct categorization can then lead to timely and correct actions in response to the attacks.

One might notice some overlap among the groupings in Figure 8.9. When no separating hyperplane between groups exists, we can use a more flexible analysis method called *support vector machines*.

8.3.3.2 Support Vector Machines

Support vector machine (SVM) algorithms are among the newer and more powerful classification methods. There are many cases where linear classifiers fail to adequately discriminate. For instance, consider the two groupings in Figure 8.10. This nonlinear boundary is better suited than a linear function to classify between the grouping of the data delineated by the different gray scales. In higher dimensions, we are looking for hyperplanes, or rather separating or supporting hyperplanes to accomplish the discrimination with the ability to have those hyperplanes come in as nonlinear functions when necessary.

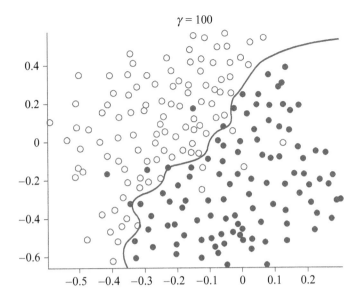

FIGURE 8.10 Two-group discrimination with nonlinear function via a support vector machine.

The support vector machine algorithm uses nonlinear discrimination functions such as seen in the Figure 8.10. Solving an optimization problem to find this nonlinear function is fairly complex, but using some interesting modifications to the data, becomes quite tractable.

Just as in discriminant analysis, a hyperplane of the form:

$$f(x) = WX + b \tag{8.12}$$

where the matrix W provides the weights applied to the data elements of X, separates the groups. The hyperplane is situated to be some distance from the points in each group. Making this distance a hard constraint yields

$$\frac{y_i[(WX) + b]}{||W||} \geq \Delta \tag{8.13}$$

known as the hard margin classifier. In practice, this needs to be relaxed and this involves adding slack. Normalizing Equation 8.13 and adding the slack leads to the quadratic optimization problem used to obtain the classifier.

$$\text{Min} \quad \frac{||W||^2}{2} + C \sum_{i=1}^{n} \phi_i \tag{8.14}$$

subject to:

$$y_i[(WX) + b] \geq 1 - \xi_i \tag{8.15}$$

$$\xi_i \geq 0 \tag{8.16}$$

Quite often the classifier must take on a nonlinear form. Finding the correct nonlinear form via an optimization as shown is not a tractable problem. However, if we can project the data into a higher dimension such that a linear classifier exists in the higher dimension, then the previous solution approach works. Kernel functions project the data into a higher dimension. In this higher dimension, linear separating hyperplanes are calculated. The regularization parameter dictates the classification accuracy of the hyperplane while the gamma parameter determines the nearness of the points to the hyperplane that are defining that hyperplane. Once calculated, the data are reduced back to the original dimension. The linear hyperplanes in the higher dimension are quite often nonlinear surfaces when projected back into the original, lower dimension space. The line in Figure 8.10 is the nonlinear discrimination function found via this support vector machine approach.

8.3.4 Unsupervised-Clustering Methods

A clustering algorithm examines data characteristics to create a set of groups containing points with closely related characteristics. Since no response data associated with the characteristics are used in the clustering phase, the algorithms are considered unsupervised.

8.3.4.1 Multidimensional Scaling

The *multidimensional scaling approach* examines how 'near' points are in multidimensional space and maps those points onto a two-dimensional (or in some cases three dimensional) chart retaining the relative distance between the points. Because the original data might involve many variables, and the multidimensional scaling algorithm maps that high dimension distance between points into a two or three-dimensional map while maintaining the relative distances among the data, multidimensional scaling can also be viewed as a dimension reduction method. However, its use is primarily to uncover similarities among data thus we keep it here as a clustering method.

For example, suppose the data have ten characteristics. It is easy to plot the data in three dimensions at a time and observe how close the points are to one another. However, there are 120 of these plots to examine and that is a lot to try and keep straight. The multidimensional scaling plot gives 1 plot from which to infer which observations are most similar, least similar, etc.

Manly and Alberto (2017) present an often used example of counts of voting agreement among 15 New Jersey Congressmen regarding environmental issues. The dimensions are the X and Y axes in two-dimensional space and the groups show agreement. Party affiliation is depicted by the letter in the parentheses. Clearly, there are party lines for the voting, and a single Republican that aligned with the Democrats in terms of agreement. Levi and Williams (2013) use MDS to clarify survey data while examining the issue of cybercrime reduction. Lee, Yu, Dargahi, Conti, and Bianchi (2018) use MDS to detect cloning in an Internet of Things scenario.

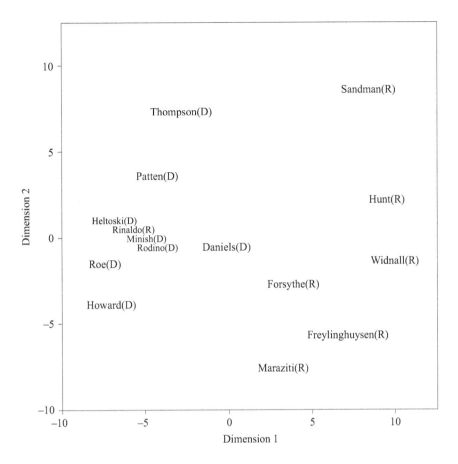

FIGURE 8.11 Voting agreement among 15 New Jersey Congressmen with respect to environmental issues.

8.3.4.2 k-Nearest Neighbors

The *nearest neighbor algorithm* is a conceptually simple classification method. The challenge is this: given a database of past observations, responses and characteristics, classify and predict some new data. The solution is to first query the database finding those k points whose characteristics put it nearest to the new point of interest. Nearness can be defined in a variety of ways. An effective and commonly used measure is the Euclidean distance measure. Then, with this cluster of k points, determine the anticipated, or predicted, response of the new data by some majority vote of the corresponding responses of the points in the cluster or average response value for quantitative data.

The approach is simple to implement but does come with some drawbacks. First, the entire database of historical data is maintained and used for each new data point query. There is also the need to define the distance measure used to find the nearest neighbors of the new data point and what is meant by close. There is the rule, or majority vote, by which the new data response is predicted based on the responses from among the points in the k-sized cluster. Finally, there is the question of how to produce the predicted response in the case of quantitative response data.

A point of debate is whether this method is actually a supervised method. Because the response is not used to define the clusters but is used to generate the prediction, the method is considered here as an unsupervised method.

8.3.4.3 k-Means Clustering

Another popular and easy to understand clustering algorithm is the k-means clustering approach. Given a collection of quantitative observations, the goal is to compute clusters of 'similar' points.

The algorithm is iterative and works by first randomly defining k arbitrary points, or centroids, within the dataset. The k clusters are then defined by assigning each data point to the centroid to which it is closest. Once all data are assigned to clusters, the data in each cluster are used to create an updated set of centroids. Based on the new centroids, the data are re-evaluated for cluster membership and the resulting centroids again updated. The iterative approach continues until the centroid definition converges to some final estimates.

A new data point is assigned to that cluster whose centroid is closest to the data point. The new data point is then associated with that cluster. Using the algorithm does require defining the number of clusters k and the distance function to use. Drawbacks of the approach includes its sensitivity to outlier points and possible problems encountered when the initial random centroids are poorly selected.

8.3.5 Unsupervised-Dimension Reduction Methods

Many datasets are large not only in terms of number of observations but also in the number of characteristics associated with each data point. Many of the characteristics are highly related. In such situations, these related characteristics can be replaced by a representative characteristic or some new aggregate characteristic. This reduces the dimension of the problem, often quite drastically, leading to better descriptive models and better understanding of the data.

8.3.5.1 Principal Components Analysis

The underlying assumption in *principal component analysis* (PCA) is that the p characteristics of the data can be weighted to align with some p unknown principal components, as in:

$$PC_1 = w_{1,1}X_1 + \ldots + w_{1,p}X_p$$
$$\vdots$$
$$PC_p = w_{p,1}X_1 + \ldots + w_{p,p}X_p$$

but not all of these p principal components are useful. Is there a reduced set of components that explain the data to a sufficient degree? Geometrically, find the $m < p$ principal axes of the p-dimensional cloud of data to sufficiently explain the data.

Mathematically, find the principal component that explains the most variance. Once found, find the second principal component that explains most of the remaining variance. Continue the process until m such components are found. When the data are summarized using the covariance or correlation matrix, the iterative process reduces to the eigenvalue problem.

It is well known that the eigenvalues of the covariance matrix sum to the variance of the data. Thus, the largest eigenvalues related to those eigenvectors reflect the principal axes of the data and explain the most variance. The principal components are derived from the eigenstructure of either the covariance or correlation matrix associated with the data set. Typically the value of m is set to be those number of eigenvalues that sum to a percentage of the total variance, for example, 80% of the total variance.

		England	N Ireland	Scotland	Wales
Original example from Mark Richardson's class notes Principal Component Analysis	Alcoholic drinks	375	135	458	475
	Beverages	57	47	53	73
What if our data have way more than 3-dimensions? Like, 17 dimensions?! In the table is the average consumption of 17 types of food in grams per person per week for every country in the UK.	Carcase meat	245	267	242	227
	Cereals	1472	1494	1462	1582
	Cheese	105	66	103	103
	Confectionery	54	41	62	64
	Fats and oils	193	209	184	235
	Fish	147	93	122	160
The table shows some interesting variations across different food types, but overall differences aren't so notable. Let's see if PCA can eliminate dimensions to emphasize how countries differ.	Fresh fruit	1102	674	957	1137
	Fresh potatoes	720	1033	566	874
	Fresh Veg	253	143	171	265
	Other meat	685	586	750	803
	Other Veg	488	355	418	570
	Processed potatoes	198	187	220	203
	Processed Veg	360	334	337	365
	Soft drinks	1374	1506	1572	1256
	Sugars	156	139	147	175

FIGURE 8.12 Example 17-dimensional dataset on UK eating patterns.

For example consider the 17-dimensional data regarding eating habits in the UK as presented by Powell and Lehe (2020). Figure 8.12 displays the data employed. Figure 8.13 is the data reduced to two dimensions via principal components indicating the similarities between the eating habits of the countries.

8.3.5.2 Factor Analysis

The *factor analysis model* (FA) is similar to but fundamentally distinct from the principal components model. Instead of generating principal components as linear combinations of predictor variables, factor analysis represents the predictor variables themselves as linear combinations of new variables called factors. Thus, we consider

$$X_1 = v_{1,1}CF_1 + \ldots + v_{1,p}CF_p$$
$$\vdots$$
$$X_p = v_{p,1}CF_1 + \ldots + v_{p,p}CF_p$$

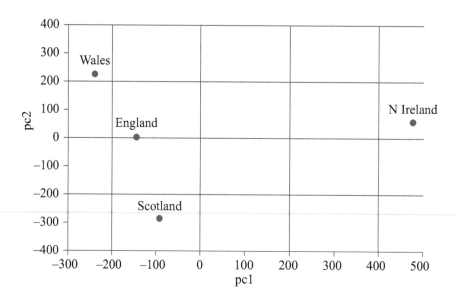

FIGURE 8.13 UK data reduced to two-dimensional distance representation via principal components.

The unknown loadings on the common factors are found by factorizing the $(p \times p)$ correlation matrix of the data into an $(m \times p)$ loading matrix, Λ'. Then,

$$\Sigma = \Lambda\Lambda' + \Psi \qquad (8.17)$$

where the Ψ diagonal matrix provides the unique component of Σ not explained by the $\Lambda\Lambda'$ matrix. The loadings give the insight into how each characteristic is explained by the common factors. The loadings provide insight into how to combine the data to obtain the aggregate factors. As with PCA, the loadings in FA are derived using the eigenvalues and eigenvectors of the covariance or correlation matrix of the data.

Unfortunately, the loadings are not unique so there are actually various transformations, or rotations, that are used to clarify the interpretations of the factor loadings. Default transformations implemented by software packages are usually sufficient to promote factor analysis interpretation.

The key takeaway is that the factor analysis gives useful information on how to create the m aggregate characteristics to reduce the problem dimension, while the principal components model provides insight into which characteristics to drop due to their strong relationship with other characteristics.

8.3.6 What Has Not Been Discussed

Any book chapter is necessarily limited. This chapter is no exception. This chapter examined a representative sample of data analytical methods aligned with the classification provided in Table 8.1. The discussion was not complete. For instance, there are other classification methods, much more than the three discussed here. For example, the *Naïve Bayes* method or the use of neural networks for classification. There are also a wide variety of pattern recognition algorithms, none of which we discuss here. Those pattern recognition algorithms include neural networks, wavelet models, and a variety of deep-learning network models. For instance, Butt (2018) used neural networks for cyber attack detection. Bishop (2016) is a well cited source text on this topic. In general, there are a plethora of excellent books, papers, and available tutorials on each of the subjects. For instance the encyclopedia by Sammut and Webb (2011) on the variety of machine learning topics, the introductory book by Manley and Alberto (2017) and the more advanced text by James, Witten, Hastie, and Tibshirani (2013).

8.4 CONCLUDING REMARKS

The Department of Defense (DoD) has been using analytics for a long time, probably since its inception. Analytics have been employed to support conflict probably as long as there has been human conflict (see Hill and Miller (2017) for a similar discussion on military simulation). The cyber domain is quickly emerging as a key domain of interest to the military and one that warrants full consideration to protect it from exploitation or debilitation.

There are many facets to the cyber attack or cyber protection, so there are ample opportunities to conduct cyber research and have that research be meaningful and crucial. For instance, how can we improve pattern recognition or image processing algorithms to run quicker while providing improved accuracy. How do we go about examining the high volume, high velocity data associated with our cyber monitoring systems to accurately highlight intrusion attempts. Finally, how can we exploit our variety of quantitative modeling methods to actually identify the intruders attempting to access our systems. Underscoring cyber research, whether that research is focused on software or hardware, detailed transmission protocols or application protection, will be modeling, data, and the analysis of data to gain insight useful to the research.

This chapter provides two insights. First, there are an array of analytical methods that are applicable to understanding the cyber domain. Second, there are many algorithms out there for the analysis of data. This chapter discussed common ways to classify the algorithms and provided an overview of a representative sample of those.

Hopefully, this chapter provides impetus to the new cyber researcher to expand beyond their particular areas of interest to gain knowledge and insight into the tools of data analytics as it has and will continue to be of importance in the cyber domain.

ACKNOWLEDGMENTS

This work was supported by the Office of the Secretary of Defense, under the STAT COE.

Disclaimer: The views expressed in this chapter are those of the author and do not reflect the official policy or position of the United States Air Force, Department of Defense or the U.S. Government.

REFERENCES

B2B. (2015). *Cytegic monitors cyber security threats in real time.* Retrieved 22 Sep 2020, from https://www.b2bnn.com/2015/02/cytegic-monitors-cyber- security-threats-real-time/cytegicdashboard/

Bishop, C. M. (2016). *Pattern recognition and machine learning.* New York: Springer.

Breiman, L. (2001). Statistical modeling: The two cultures. *Statistical Science*, 16(3), 199–231.

Butt, S. A. (2018). *Cyber data anomoly detection using auto-encoder neural networks* (Unpublished master's thesis). Air Force Institute of Technology, Wright-Patterson AFB, United States.

Chen, T. M., Sancez-Aarnoutse, J. C., & Buford, J. (2011, December). Petri net modeling of cyber-physical attacks on smart grid. *IEEE Transactions on Smart Grid*, 2(4), 741–749.

Chung, K., Kamhoua, C. A., Kwiat, K. A., Kalbarczyk, Z. T., & Iyer, R. K. (2016, January). Game theory with learning for cyber security monitoring. In *17th International Symposium on High Assurance Systems Engineering (HASE)* (p. 1–8). Piscataway, New Jersey.

Do, C. T., Tran, N. H., Hong, C., Kamhoua, C. A., Kwiat, K. A., Blasch, E., & Iyengar, S. S. (2017). Game theory for cyber security and privacy. *ACM Computing Surveys (CSUR)*, 50, 1–37.

Frumosu, F. D., & Kulahci, M. (2018). Big data analytics using semi-supervised learning methods. *Quality and Reliability Engineering International*, 34, 1413–1423.

Goodall, J. R., & Sowul, M. (2009). Viassist: Visual analytics for cyber defense. In *Proceedings of the 2009 Conference On Technologies for Homeland Security* (pp. 143–150). Piscataway, New Jersey.

Gore, R., Padilla, J., & Diallo, S. (2017). Markov Chain Modeling of Cyber Threats. *Journal of Defense Modeling and Simulation: Applications, Methodology, Technology*, 14(3), 233–243. doi: 10.1177/1548512916683451

Herring, M. J., & Willett, K. D. (2014). Active cyber defense: A vision for real-time cyber defense. *Journal of Information Warfare*, 13(2), 46–55.

Hill, R. R., & Miller, J. O. (2017). A history of united states military simulation. In *Proceedings of the 2017 Winter Simulation Conference* (pp. 346–364). Piscataway, New Jersey.

Indik. (2020). *Cyber security example.* Retrieved 22 Sep 2020, from https://www.indik-dashboard.com/en/examples/cybersecurity_example3

James, G., Witten, D., Hastie, T., & Tibshirani, R. (2013). *An introduction to statistical learning with applications in R.* New York, NY: Springer.

Jensen, W. A. (2020). Statistics = analytics? *Quality Engineering*, 32(2), 133–144.

Jiang, W., Tian, Z., Zhang, H., & Song, X. (2008, April). A Stochastic Game Theoretic Approach to Attack Prediction and Optimal Active Defense Strategy Decision. In *2008 IEEE International Conference on Networking, Sensing and Control* (p. 648–653). doi: 10.1109/ICNSC.2008.4525297

Knaflic, C. N. (2015). *Storytelling with data: A data visualization guide for business professionals.* John Wiley & Sons.

Kotenko, I., & Ulanov, A. (2005). Agent-based simulation of ddos attacks and defense mechanisms. *International Scientific Journal of Computing*, 4(2), 113–123.

Kuhl, M. E., Sudit, M., Kistner, J., & Costantini, K. (2007). Cyber attack modeling and simulation for network security analysis. In *Proceedings of the 2007 Winter Simulation Conference* (pp. 1180–1188). Piscataway, New Jersey.

Kulachi, M., Frumosu, F. D., Khan, A. R., Ronsch, G. O., & Spooner, M. P. (2020). Experiences with big data: Accounts from a data scientist's perspective. *Quality Engineering*, 32(4), 529–542.

Larkin, M. T. (2019). *A stochastic game theoretical model for cyber security* (Unpublished master's thesis). Air Force Institute of Technology, Wright-Patterson AFB, United States.

Lee, P., Yu, C.-M., Dargahi, T., Conti, M., & Bianchi, G. (2018). Mdsclone: Multidimensional scaling aided clone detection in internet of things. *IEEE Transactions on Information Forensics and Security*, 13, 2031–2046.

Levi, M., & Williams, M. L. (2013). Multi-agency partnerships in cybercrime reduction: Mapping the uk information assurance network cooperation space. *Information Management & Computer Security*, 21(5), 420–443.

Manley, B. F. J., & Alberto, J. A. N. (2017). *Multivariate statistical methods: A primer* (4th ed.). Boca Raton: CRC Press.

Manly, B. F. J., & Alberto, J. A. N. (2017). *Multivariate statistical methods: A primer, fourth edition.* CRC Press.

Metropolis, N. (1987). The beginning of the Monte Carlo method. *Los Alamos Science*, 125–130.

Montgomery, D. C., Peck, E. A., & Vining, G. G. (2012). *Introduction to linear regression analysis*, (Fifth ed.). Hoboken, NJ: John Wiley & Sons.

Morrison, D. F. (1990). *Multivariate statistical methods* (Third ed.). New York, NY: McGraw Hill.

Musman, S., & Turner, A. (2018). A Game Theoretic Approach to Cyber Security Risk Management. *Journal of Defense Modeling and Simulation: Applications, Methodology, Technology*, 15(2), 127–146. doi: 10.1177/1548512917699724

Powell, V., & Lehe, L. (2020). *Principal component analysis*. Retrieved 16 Jan 2020, from http://setosa.io/ev/principal-component-analysis/

Ramirez, R. C. (2018). *Characterization of ambient noise* (Unpublished master's thesis). Air Force Institute of Technology, Wright-Patterson AFB, United States.

Rencher, A. C., & Christensen, W. F. (2012). *Methods of multivariate analysis* (Third ed.). John Wiley & Sons.

Ross, S. M. (1996). *Stochastic Processes* (2nd ed.). University of California, Berkeley: John Wiley & Sons, Inc.

Sammut, C., & Webb, G. I. (Eds.). (2011). *Encyclopedia of machine learning*. new York: Springer.

Shiva, S., Roy, S., & Dasgupta, D. (2010, April). Game theory for cyber security. In *Proceedings of the Sixth Annual Workshop on Cyber Security and Information Intelligence Research* (p. 1-4). New York, NY: Association for Computing Machinery.

Sokri, A. (2020). Game theory and cyber defense. In *Games in management science* (p. 335–352). Springer, Cham.

Tambe, M. (2011). *Security and Game Theory: Algorithms, Deployed Systems, Lessons Learned.* New York: Cambridge University Press. doi: 10.1017/CBO9780511973031

Tukey, J. W. (1962). The future of data analytics. *The Annals of Mathematical Statistics*, 33(1), 1–67. doi: 10.1080/10618600.2017.1384734

Wang, K., Du, M., Yang, D., Zhu, C., Shen, J., & Zhang, Y. (2016). Game-theory-based active defense for intrusion detection in cyber-physical embedded systems. *ACM Transactions on Embedded Computing Systems (TECS)*, 16, 1–21.

Wang, Y., Wang, Y., Liu, J., Huang, Z., & Xie, P. (2016, June). A survey of game theoretic methods for cyber security. In *IEEE First International Conference on Data Science in Cyberspace (DSC)* (p. 631–636). Piscataway, New Jersey.

Weimer, C. W., Miller, J. O., & Hill, R. R. (2016). Agent-based modeling: An introduction and primer. In *Proceedings of the 2016 Winter Simulation Conference* (pp. 65–79). Piscataway, New Jersey.

Wu, Q., Shiva, S., Roy, S., Ellis, C., & Datla, V. (2010). On Modeling and Simulation of Game Theory-Based Defense Mechanisms Against DoS and DDoS Attacks. *Proceedings of the 2010 Spring Simulation Multiconference on - SpringSim '10*, 10. Retrieved from http://dl.acm.org/citation.cfm?id=1878537.1878703 doi: 10.1145/1878537.1878703

Yuan, Y., Sun, F., & Liu, H. (2016). Resilient Control of Cyber-Physical Systems Against Intelligent Attacker: A Hierarchal Stackelberg Game Approach. *International Journal of Systems Science*, 47(9), 2067–2077. Retrieved from http://dx.doi.org/10.1080/00207721.2014.973467 doi: 10.1080/00207721.2014.973467

CHAPTER 9

Statistics

Nita Yodo

Melvin Rafi

CONTENTS

9.1	Introduction	303
9.2	Fundamental Elements of Statistics	305
	9.2.1 Organization, Properties, and Characteristics	305
	9.2.2 Applications in Cyber Research	307
9.3	Descriptive Statistics	309
	9.3.1 Summary Tables	309
	9.3.2 Graphical Displays	309
	9.3.3 Numerical Descriptive Measures	313
	9.3.4 Statistical Software Packages	315
	9.3.5 Applications in Cyber Research	316
9.4	Inferential Statistics	317
	9.4.1 Estimation Methods	317
	9.4.2 Hypothesis Testing	318
	9.4.3 Applications in Cyber Research	319
9.5	Correlation and Regression Analysis	320
	9.5.1 Correlation Analysis	321
	9.5.2 Regression Analysis	322
	9.5.3 Applications in Cyber Research	324
9.6	Research Implications and Extensions: Cyber Resilience Research	325
	9.6.1 Research Implication on Cyber Resilience	326
	9.6.2 Potential Extensions	327
9.7	Conclusions	328

9.1 INTRODUCTION

Cyber threats are difficult to predict, measure, and manage. Obtaining every single piece of cyber-related data in a very complex cyberspace seems to be an impossible challenge. By their very nature, cyber threats are fast-changing, borderless, random, and the collected observation data on cyber activities are often vague or incomplete. A 2007 study conducted by the University of Maryland concluded that hackers

DOI: 10.1201/9780429354649-9

attack, on average, every 39 seconds, or 2,244 times a day (Cukier, 2007). With advancements in technology, there is no doubt that the frequency of cyber attacks has never been greater than at the present time.

Software companies that provide cybersecurity software and services publish their own annual cyberthreat reports. Kaspersky Lab released a report on financial cyber threats that 889,452 users of Kaspersky Lab solutions were attacked by banking Trojans in 2018, which is equal to an increase of 16% compared to 2017 (Kaspersky, 2019). Symantec (now NortonLifeLock Inc.), in their February 2019 internet security threat report, revealed big increases in cyber threats compared to the previous year, such as a 56% increase in web attacks, a 33% increase in mobile ransomware, and a 78% increase in supply chain attacks (among other statistics) (Symantec, 2019). Cyber attacks are not only increasing in frequency but also in magnitude of impact. In 2018, there were 351,937 complaints received by the Federal Bureau of Investigation (FBI)'s Internet Crime Complain Center (IC3). Cyber attacks were responsible for a total of $2.7 billion in financial losses. On average, there were 954 internet crime complaints received every day (FBI, 2019). For unsuccessful or non-harmful cyber attacks, these incidents typically went unreported.

Statistics is a branch of the mathematical science that involves the collection, analysis, and interpretation of data. Although accurate information on cyber incidents is often challenging to obtain, statistics have been widely used to make conclusions in cybersecurity threats, cybercrimes, cyber attacks, and other related cyber applications. Statistics is a field of study that transforms observation data into common knowledge. Since a subject matter may be better understood through the availability of some observation data, statistical learning is essential for determining proper methods to collect the data, employing the correct analyses, and effectively presenting the results. Two main reasons why statistical methods are crucial in cyber research include: (1) statistics can be used as a guide to learn from data and navigate around common problems that could lead to making improper conclusions; (2) Decisions and opinions are often made based on observational data, and it is crucial to reevaluate the quality of analyses that others have presented. In addition to data interpretation, statistics are critical to discoveries, decisions, and predictions made from data.

This chapter on statistical methods in cyber research is organized into seven sections. Section 9.1 gives a background on properties of cyber threats that are random and unknown and how statistics fits into cyber research. The fundamental element of statistics and the difference between qualitative and quantitative data are introduced in Section 9.2. In Section 9.3, descriptive statistics, which include describing data graphically and numerically, are presented. Section 9.4 reviews two standard methods of doing inferential statistics. Section 9.5 details regression and correlation analysis to statistically study the relationships of two variables.

In today's world, cybersecurity alone is no longer sufficient to overcome the evolving magnitude and frequency of future cyber attacks. New cyber threats are emerging every day, and the impacts of these threats are often unknown. In order to cope with these uncertainties, the concept of resilience is adapted to cyber research. Cyber resilience, in general, involves the ability of a system to withstand, respond, and overcome the impact of a cyber threat. The implications of statistical methods on the current cyber resilience research efforts will be detailed at the end of this chapter in Section 9.6, which discusses research implications and extensions.

9.2 FUNDAMENTAL ELEMENTS OF STATISTICS

9.2.1 Organization, Properties, and Characteristics

In a nutshell, the science of statistics can be commonly categorized into two major branches. *Descriptive statistics* are statistics used to organize, summarize, and describe raw observations in the form of common information. Common information refers to information that the majority of people can understand, relate to, or share. *Inferential statistics* are statistics used to make an inference or draw conclusions about a more extensive data set (population data) from a smaller data set (sample data).

In addition to descriptive and inferential statistics, probability is another branch of statistics. While statistics are often based on analyzing data from past events, probability deals with estimating the likelihood of future events using a numerical scale ranging between 0 (impossibility) and 1 (certainty). Predictions for future events are often based on the analysis of the past or present observed events. Consequently, while the scope of this chapter focuses only on statistical methods, it is worthwhile to note that probability and statistics typically go hand in hand.

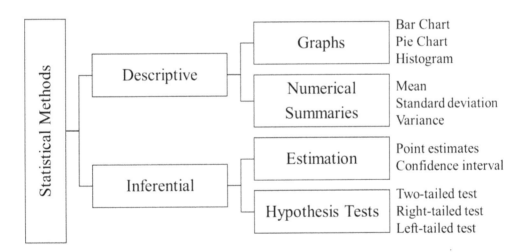

FIGURE 9.1 Statistical methods in a nutshell.

The field of inferential statistics enables one to make educated guesses about the numerical characteristics of large groups. The logic of sampling gives a way to test conclusions about the large group using only a small portion of its members. There are some basic terminologies in statistical methods:

(1) A *population* is the target, problem, or subject of interest. A population is generally represented as a more extensive set of data. It is the entire pool of data from which a statistical *sample* is drawn.

(2) A *sample* refers to the subset of the population that is being analyzed, typically represented in a smaller data set. A sample is drawn from a population. There are many sampling methods to select a sample set from a population; random and cluster sampling are two common sampling methods (Acharya, Prakash, Saxena, & Nigam, 2013). In random sampling, each sample has equal probability to be selected, whereas cluster sampling selects a predefined group in a population as the samples (Figure 9.2).

(3) A *variable* is a characteristic or property of interest from the population or sample.

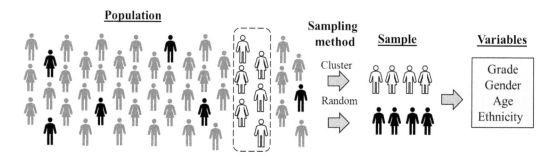

FIGURE 9.2 Graphical representation for a population, a sample, and variables.

Prior to analyzing and making inferences from a dataset, understanding the different types of data is crucial in statistical analysis. Data is the raw information from which statistical interpretation will be based. There are certain statistical measurements that are only appropriate for specific data sets. Thus, correct conclusions can be achieved by employing correct measurements in statistical analysis. In general, there are two types of data, qualitative data and quantitative data (Figure 9.3).

Qualitative data deals with descriptions. It can be represented by words, symbols, and pictures, but not by numbers. Qualitative data can be observed, but cannot be directly measured. It is also referred to as categorical data. Information consisting of qualitative data can be sorted by category, but not by number. There are two general types of qualitative data: nominal data and ordinal data.

(1) *Nominal data* is used for labeling or naming purposes, without any type of quantitative value. There is no intrinsic ordering from highest to lowest in nominal data. All variables have the same significance. For example, gender (male, female), color (red, blue, etc.), or region (north, south, east, west).

(2) *Ordinal data* is used for ordering purposes based on their relative position on a scale. Ordinal data can be sorted from highest to lowest, or vice versa. However, calculations cannot be done with ordinal data as they only show sequence instead of value. For example, rank (1st, 2nd, 3rd), a letter grade (A, B, C), or quality (good, average, bad).

The main difference between nominal and ordinal data is that nominal data is qualitative data that cannot be placed in any order, while ordinal data is qualitative data that can be ordered.

Quantitative data is represented by numbers and is also referred to as numerical data. It is often used to answer questions such as 'how many,' 'how much,' or 'how often.' Quantitative data is measurable and can be quantified arithmetically. There are two general types of quantitative data: discrete data and continuous data.

(1) *Discrete data* involves only whole numbers or integers as the measure. It cannot be further divided into any finer details or fractions. For example, the number of people affected, the number of passwords exposed, or the number of phishing emails received.

(2) *Continuous data* involves almost any numeric value between two numbers. It can be divided into finer details and can include fractions and decimal numbers. Continuous data can be converted into different units of measurements. For example, length (inches, feet, etc.), time (hour, minute, seconds, etc.), or temperature (Celsius, Fahrenheit, Kelvin, etc.).

Depending on the physical representation of the data, most discrete and continuous data can take both positive and negative values. Positive and negative values are often used to identify a pattern, such as whether a trend is increasing or decreasing. In monetary terms, a positive value typically means gain, while a negative value means loss.

9.2.2 Applications in Cyber Research

In cyber research, collecting, understanding, and interpreting data is an essential step prior to conducting any statistical analysis. The fundamental elements of statistics have been used widely in cyber research applications. Some examples are described below.

- *Cyberbullying.* Hinduja and Patchin (2010) published a study on cyberbullying. The goal of the study was to determine if cyberbullying is related to suicidal ideation among youths. The *population* in their study was the entire youth population in the United States. This study was conducted with a random *sample* of 1,963 middle-schoolers. The data was collected from a survey distributed to approximately 30 middle schools. There were 1,963 responses (data) collected. The *variables* from the population of interest were grade (6th-8th), gender (male/female students), age (10-16 years old), and ethnicity/race (Caucasian, African, Hispanic, etc.). A graphical representation of a population, samples, and variables is shown in Figure 9.2.

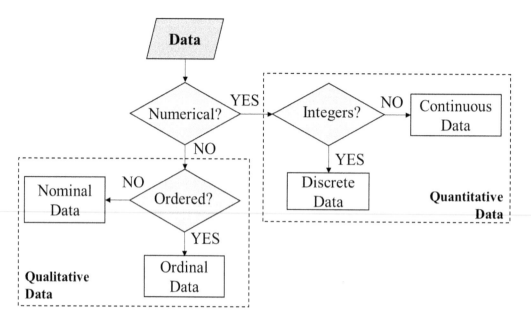

FIGURE 9.3 A flow chart to identify different types of data.

- *Cyber risk measurement.* Facchinetti, Giudici, and Osmettti (2020) published a book chapter on cyber risk measurement with ordinal data when quantitative data of cyber loss data is not available. In their study, they accounted for business lines, cyber attack techniques (0-day, distributed denial of service (DDoS), malware, phishing/social engineering, phone hacking, SQL (or code) injection, and others), and a three-level of severity. Both business lines and cyber attack techniques are considered to be qualitative nominal data, whereas the levels of severity in low, medium, and high are considered to be qualitative ordinal data. In their case study, they employed a sample data of 1,127 cyber attacks observed in 2017. The frequency of the cyber attacks is quantitative data. Based on their case study, they concluded that 0-day attacks and phone hacking are associated with low severity, DDoS and SQL injection are associated with medium severity, and malware and phishing/social engineering are related to high severity.

- *Cybersecurity threats analysis.* Kerzhener, Tan, and Fosse (2015) proposed a model-based systems engineering method to analyze cybersecurity threats on networked cyber-physical systems (CPS). They introduced seven qualitative attributes that can be determined for a vulnerability. These attributes are access vector (local, adjacent network, network), access complexity (high, medium, low), authentication (multiple, single, none), confidentiality (none, partial, complete), integrity (none, partial, complete), availability (none, partial, complete), and gain access (none, user, admin). The first six attributes are employed from the common vulnerability scoring system (CVSS), which is a standard for assessing vulnerability in computer system security. Based on the seven attributes, Kerzhener et al. (2015) determined a quantitative 'network criticality of vulnerability' measure with a quantitative score from 0 to 10, where 10 represents that the network is extremely vulnerable to intrusion.

9.3 DESCRIPTIVE STATISTICS

Descriptive statistics involves describing, summarizing, and presenting the characteristics of a data set. The objective of descriptive statistics is to transform the data into information that can be easily understood. The information from data can be presented in tabular, graphical, or numerical format.

9.3.1 Summary Tables

Both qualitative and quantitative data can be summarized in a tabular format, where data is presented in rows and columns. A frequency distribution table is often used to obtain quantitative measures from qualitative data. A frequency distribution or table of frequencies is a table of qualitative data with the corresponding frequencies in each category or class. Some quantitative measures that can be obtained from a frequency table are:

(1) *Frequency (or absolute frequency)* refers to the number of times that a value appears. Frequency is represented as f_i, where $i=1, 2,...,n$ represents each category or class from the qualitative data. The sum of the absolute frequencies is equal to the total number of data, N.

(2) *Relative frequency* (rf_i) is the result of dividing the frequency for each class with the total number of data. The sum of the relative frequencies is equal to 1 (or 100% if the percentage is used for relative frequency). Relative frequency is often used to observe the proportion of each class of data with respect to the total number of data.

(3) *Cumulative frequency* (cf_i) is the sum of absolute frequency in class i and the absolute frequency from its previous classes. When $i=1$, the cumulative frequency is equal to its associated frequency, $cf_1 = f_1$.

(4) *Cumulative relative frequency* (crf_i) is the result of dividing the cumulative frequency by the total number of data, N. When $i=1$, the cumulative relative frequency is equal to its relative frequency, $crf_1 = rf_1$. Similar to rf_i, crf_i takes the value between 0 and 1 (or 0% and 100%).

9.3.2 Graphical Displays

When dealing with a large amount of data, the use of graphical representations can make complex information easier to view and interpret. Graphical images are exhibited through various tables, charts, or graphs. The types of data and the information to be displayed often determine which graph is appropriate for use. Graphical displays for qualitative (or categorical) data are often combined with quantitative measures to enable information clarity—for example, when studying trends or proportions. Some graphs (Figure 9.4) that are commonly used for qualitative data are:

(1) *Bar graph:* A bar graph can be represented vertically or horizontally depending on the presentation of the data. The categories (classes or groups) of the qualitative data are represented by bars. The height or length of each bar can either be the category frequency, relative frequency, or percentage, which typically refers to a quantitative measure. Bar graphs can be displayed as single, stacked, or grouped. It should be noted that some platforms, such as Microsoft Excel make a distinction between bar charts and column charts (or vertical bar charts). Data in a bar chart is often presented horizontally with the data frequencies on the x-axis and the categories on the y-axis, whereas data in a column chart is presented vertically with categories on the x-axis and data frequencies on the y-axis.

(2) *Line graph:* A line graph is similar to a bar graph where the categories of qualitative data are represented by lines instead of bars. A line graph is appropriate only when both the x- and y-axes display ordinal data, rather than nominal data. Although bar graphs can also be used in this situation, line graphs are generally better for studying a trend or comparing changes over time.

(3) *Pie Chart:* A pie chart is represented as a circular graph (a pie) that has been cut into several slices. Each slice of the pie represents a different category of qualitative data. The size of each slice is proportional to the category of the relative frequency. A pie chart is best used to showcase a scenario in which proportion is the primary information to be displayed.

(4) *Pareto chart:* A Pareto chart consists of a bar graph with the categories (the bars) arranged by height in descending order from left (higher frequency) to right (lower frequency) and a line graph in the secondary axis representing the information of cumulative relative frequency.

For quantitative (numerical) data, graphical displays are often used to realize frequency distributions among the data. Minimal data processing is typically necessary to construct a quantitative graphical display. Some graphical displays that are commonly used for quantitative data are:

(1) *Histogram:* A histogram is fundamentally a bar graph that displays the frequency distribution of the numerical data. Each bar represents data ranges. The height of the bar indicates how much data or the count of the data that falls in that range. A general method for constructing a histogram is as follows.
Step 1: All the data is partitioned into equal classes (or bins). There are some recommended rules for the number of classes and the width (Birgé & Rozenholc, 2006), such as Sturges' rule (Scott, 2009).

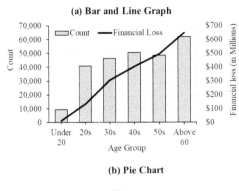
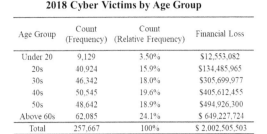
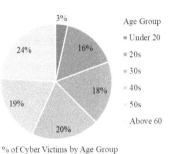
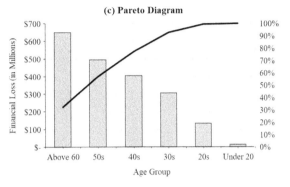

FIGURE 9.4 Various graphical displays for qualitative data adapted from the IC3 2018 Internet Crime Report, page 16 (FBI, 2019).

Step 2: Count the number of observations (or frequencies) that fall in each class.

Step 3: Plot the resulting frequency distribution as a bar chart.

There are different shapes of histograms, as shown in Figure 9.5. Some of the common histogram patterns are associated with either one or a combination of the following properties.

- *Symmetric / Asymmetric:* A histogram is said to be symmetric if it possesses two identical shapes and sizes when divided in the center of the histogram. An asymmetric histogram is indicated by a distribution peak that is off from the center of the histogram.

- *Unimodal / Bimodal / Multimodal:* A unimodal histogram has one single peak, while a bimodal histogram has two peaks. A histogram that has more than two peaks is often called a multimodal histogram.

- *Uniform:* The frequency distribution in a uniform histogram is almost equal for all the classes or bins.

- *Bell-shaped:* A bell-shaped histogram is an asymmetric unimodal histogram that has a peak point in the middle. In statistics, this bell-shaped histogram is also known as the 'normal distribution.'

- *Skewed:* For asymmetric histograms, skewness is a measure of the asymmetry of the data distribution. A negatively skewed histogram indicates that the tail (the end of the distribution) is on the left side, whereas a positively skewed histogram suggests that the tail is on the right side.

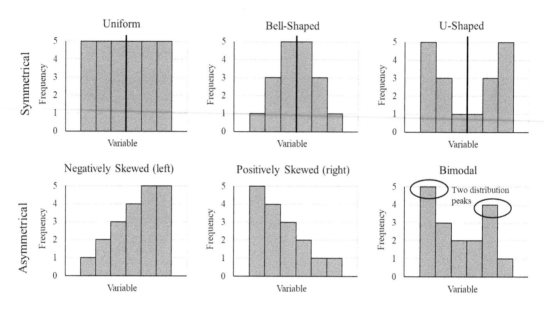

FIGURE 9.5 Shapes of histogram associated with its frequency distributions.

(2) *Stem-and-Leaf plot:* A stem-and-leaf plot displays individual quantitative data in a graphical format. If the data is not too large and complex, a stem-and-leaf plot can preserve the detail of individual observations. The general idea is to split the data into a 'stem' portion and a 'leaf' portion. There are several ways to split the data up. For example, continuous data represented by the quantity 30.5 can be split at the decimals as 30 for the 'stem' portion and 5 for the 'leaf' portion.

(3) *Dot plot:* Similar to the histogram and stem-and-leaf plot, a dot plot can be used to present a frequency distribution by using 'dots.' A dot plot is only suitable for small to moderately sized data sets.

(4) *Frequency polygon graph:* Frequency polygons serve the same purpose as histograms. In addition to absolute frequency, relative frequency or cumulative frequency can be displayed in a frequency polygon. A cumulative frequency polygon is also known as an ogive. The difference between a frequency polygon and a Pareto chart is that the variables in frequency polygons are not arranged based on the frequency.

(5) *Box-and-whisker plot:* Unlike other graphical displays, a box-and-whisker plot displays other numerical properties such as minimum value, lower quartile,

mid-quartile (median), upper quartile, and maximum value. A box plot is useful for identifying outliers and for comparing distributions. It should be noted that a box plot and a box-and-whisker plot can be used interchangeably.

The general graphical displays for quantitative data can be found in Figure 9.6. More information on how to properly construct graphical displays for quantitative data can be found in (Mendenhall & Sincich, 2016).

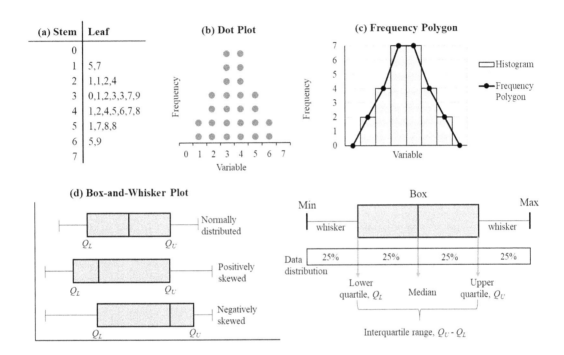

FIGURE 9.6 Various graphical displays for quantitative data.

9.3.3 Numerical Descriptive Measures

Numerical descriptive measures are a quantitative value or a number computed from a dataset to describe: (1) the center of the distribution; (2) the spread of data; and (3) the relative position of a data (an observation) within its set (Mendenhall & Sincich, 2016). A measure of central tendency is a value that attempts to describe the center of the distribution. Three standard measures for central tendency are mean, median, and mode.

(1) *Mean (average)* is the average value of the measurements. The population mean is denoted as μ, while the sample mean is denoted as $\bar{x}$. The equation for calculating mean value is shown in Table 9.1.

(2) *Median* is the middle number when the measurements are arranged in ascending or descending order. If the number of measurements n is odd, there will only be one middle value. If the number of measurement n is even, there

TABLE 9.1 Different symbols for population and sample

Numerical Measures	Population	Sample
Number of data points or measurements	N (population size)	n (sample size)
Mean (arithmetic)	μ	$\bar{x} = \dfrac{\sum_{i=1}^{n} x_i}{n}$
Variance	$\sigma^2 = \dfrac{\sum_{i=1}^{n}(x_i - \mu)^2}{n}$	$s^2 = \dfrac{\sum_{i=1}^{n}(x_i - \bar{x})^2}{n}$
Standard deviation	$\sigma = \sqrt{\sigma^2}$	$s = \sqrt{s^2}$

will be two middle values. Thus, the median is typically taken as the mean of the two middle measurements when n is even.

(3) *Mode* is the value that occurs most frequently in the measurements; this is the value with the most frequency.

A measure of data variation measures the spread of data points (observations) in a data set (or measurement). Standard measures for data variation are range, variance, and standard deviation.

(1) *Range / Interquartile range:* A range is the difference between the largest and the smallest value in a measurement. An interquartile range is the difference between the upper and lower quartiles (Figure 9.6).

(2) *Variance:* The variance is the average squared deviation of data from the mean. The variance for a population (σ) and a sample (s) are denoted by different symbols. The equations for variance are shown in Table 9.1.

(3) *Standard deviation:* The standard deviation is the average deviation of data from the mean and is denoted by σ for a population and by s for a sample. The standard deviations σ and s are respectively equal to the square root of population variance and sample variance (Table 9.1).

The relative position of data within the set can be measured by quartiles. A data set must be ordered from the smallest to the largest value to calculate quartiles. Quartiles divide ordered data into four parts. This is shown in Figure 9.6, which shows three quartiles values: lower quartile, middle quartile, and upper quartile.

(1) *Lower quartile (Q_L)* is also known as the first quartile (Q_1) or the 25th empirical quartile. Q_L is a value that marks where 25% of the data is below this point or value.

(2) *Middle quartile* is also known as the second quartile (Q_2) or mid-quartile. A middle quartile is fundamentally the median of the data. Mid-quartile marks where 50% of the data lies below this point.

(3) *Upper quartile (Q_U)* is also known as the third quartile (Q_3) or the 75th empirical quartile. Q_U marks where 75% of the data lies below this point.

Quartiles are often useful when detecting outliers. Outliers are observations or data points that are outside the range of most data, indicating unusually small or large values in a dataset. Typically, outliers are located at the furthest ends of a distribution.

9.3.4 Statistical Software Packages

Various statistical software packages are available. Each offers the user different feature sets, such as ease-of-use, interaction through a graphical user interface (GUI), or the ability to analyze statistics programmatically through coding. Listed below are five popular software packages for statistical analysis. Please note that this list of software is not exhaustive.

Microsoft Excel: Microsoft Excel is an excellent option for beginners to get started with statistics. It is widely accessible to many individuals and companies. Although Microsoft Excel is a spreadsheet program, it can be utilized for simple statistical analyses, such as generating statistical charts and calculating simple statistical properties.

IBM's SPSS (Statistical Package for the Social Sciences): Generally known just by SPSS, this is popular with global companies and industries. It offers a more superior ability to perform statistical analysis compared to Microsoft Excel and can handle large data sets. SPSS has built-in data manipulation and an option to automate analysis. It allows batch processing to carry out advanced statistical analysis for extensive and complex data.

MATLAB, by MathWorks: MATLAB, by itself, is an analytical platform and programming language that is widely used by researchers and scientists. MATLAB thus requires some foundational programming skills compared to Microsoft Excel or SPSS. MATLAB offers a specific statistical analysis toolbox for advanced statistical analysis. In addition, MATLAB also provides a lot of other advanced toolboxes from different research areas. Although MATLAB may not offer ease-of-use for beginners in statistics, it has a lot of flexibility in terms of interfacing statistical analysis with real-world applications and diverse research areas.

R: R is a free software package for statistical computing and graphics. It can handle relatively complex data analysis and generate various graphical displays. R requires a certain degree of programming skill, and it allows users to add additional functionality through programming of new functions. Compared to MATLAB, R does not offer as many interface options with other research areas.

Minitab: Minitab software offers a range of both basic and relatively advanced statistical analysis. Some commands can be executed through a GUI or through scripts. Although more sophisticated statistical analysis can be performed with scripted commands, the programming skill requirements are minimal in Minitab as compared to MATLAB or R.

9.3.5 Applications in Cyber Research

The cyber domain utilizes descriptive statistics in many applications. Presenting information in descriptive statistics can make data more easily understandable by the general public. As an example, Table 9.2 shows the number of cyber-related complaints received by the IC3 over the last five years, where the information is presented as a frequency table. Comparing the information in the 'Year' and 'Frequency of Complaints' columns, a trend of increasing cyber complaints over the year can be seen from the table. In addition, other statistical measures can be calculated from the frequency column, such as relative frequency, cumulative frequency, and cumulative relative frequency.

TABLE 9.2 The Internet Crime Complaint Center (IC3) complaint statistics from 2014-2018

Year, i	Frequency of Complaints* f_i	Relative Frequency $rf_i = f_i/N$	Cumulative Frequency $cf_i = f_i + f_{i-1}$	Cumulative Relative Frequency $crf_i = cf_i/N$
2014	269,422	0.178 (17.8%)	269,422	0.178
2015	288,012	0.191 (19.1%)	557,434	0.369
2016	298,728	0.198 (19.8%)	856,162	0.567
2017	301,580	0.200 (20.0%)	1,157,742	0.767
2018	351,937	0.233 (23.3%)	1,509,679	1.000
Total	N = 1,509,679	1.000 (100%)	–	–

*Source: The Internet Crime Complaint Center (IC3) 2018 Internet Crime Report, page 5.

In addition to the table formats, graphical displays such as dashboards, control charts, performance indicators, and metrics in the cyber domain are frequently developed with descriptive statistics as a basis for monitoring. Dashboards are widely used within information technology across many organizations as a graphical key performance indicator of a process and heavily utilize descriptive statistics measures in the background. Bar graphs, line graphs, and pie charts are often included as part of a dashboard display. Statistical process control (SPC) charts, often associated with quality control applications, also employ descriptive statistics measures in detecting anomalies or outliers (deviations from the mean).

Three commonly used statistical charts in cyber research are the Shewhart chart, cumulative sum (CUSUM) chart, and the exponentially weighted moving average (EWMA) chart. Sklavounos, Leondakianakos, and Edoh (2019) employed an EWMA chart to evaluate different types of cyber intrusion (DDoS, User to Root (U2R), Root to Local (R2L), and Probe) in single or multiple attacking manners. Bouyeddou, Harrou, Sun, and Kadri (2017) employed all three charts as a comparative study to detect SYN flood cyber attacks, a type of DoS attack. Their results showed that the Shewhart chart is suitable for detecting high-intensity cyber attacks, whereas CUSUM and EWMA are more suitable for low-intensity cyber attacks.

The impact of cybersecurity dangers can be measured and understood more by the general public when presented numerically. Many federal and non-federal organizations publish monthly or yearly cybersecurity statistics in terms of the average number of cyber attacks per day, the average time needed to identify data lost in a data breach, or the average cost of a data breach. These numbers are descriptive statistics. Similarly, a study in 2007 (mentioned also in the Introduction) that concluded an average frequency of 2,244 hacker attacks a day (Cukier, 2007) and the numbers reported on the 2018 FBI crime reports are some examples of numerical measures of descriptive statistics in the cyber domain.

In cyber research, statistical software packages can be employed to efficiently generate statistical figures and calculate statistical properties of a dataset. Choosing the right software to use in cyber research will depend on the nature of the research question that one has to answer. SPSS is often used in the social studies aspect of cyber research (for example, in studying the impact of cyberbullying (Turan, Polat, Karapirli, Uysal, & Turan, 2011)) or in cybersecurity awareness among parents (Ahmad et al., 2018). It should be kept in mind that regardless of the statistical software package used in statistical analysis, the results of statistical analysis depend heavily on how the data is collected, processed, and analyzed. Although the most advanced statistical software package may be utilized, the results may not be valid if the data used to perform statistical analysis is not collected in a correct or ethical manner.

Statistical software packages are often integrated with simulation and programming software packages. Although statistical software packages have good theoretical properties, on their own, they may not be sufficient to conduct in-depth cyber research. This is often because there are other aspects to consider in addition to statistical analysis, such as simulating a network intrusion or modeling the responses to cyber attacks. The seamless integration of statistical analysis and system modeling within MATLAB/Simulink has been widely leveraged for modeling cyber-physical systems (Al Faruque & Ahourai, 2014; Levy, Raviv, & Baker, 2019).

9.4 INFERENTIAL STATISTICS

Inference about a population parameter can be statistically made in one of two ways: (1) estimating the value of the population parameter (typically a point or an interval); or (2) hypothesis testing to validate a hypothesis about the value of a setting or scenario.

9.4.1 Estimation Methods

There are generally two types of estimation methods: point estimation and interval estimation. The 'hat' symbol typically denotes estimation in statistics. A point estimator is a numerical estimate based on the measurement contained in a sample. The sample mean $\bar{y}$ of a data set is the point estimator for the population mean $\hat{\mu}$. Similarly, sample variance s^2 is the point estimator for the population variance $\hat{\sigma}^2$. Another way to estimate the value of the population parameter is to use the interval estimation, which is the range of numbers in which a population parameter lies, considering the margin of error. The interval in the statistic is commonly known as a confidence interval. More information on estimation methods can be found in Mendenhall and Sincich (2016).

9.4.2 Hypothesis Testing

Hypothesis testing is a method of making a decision about one or more population parameters. Two datasets are compared in hypothesis testing, one from sampling and the other from an idealized model. There are several general elements in a statistical test, as follows (Mendenhall & Sincich, 2016):

(1) *The null hypothesis*, H_o, is the statement being tested. Typically, the null statement is the statement 'there is no significant difference between the two data sets that are being tested."

(2) *Alternative (or research) hypothesis*, H_a or H_1, is the statement being tested against the null hypothesis, which states 'there is a significant difference between the two datasets that are being tested.' Table 9.3 shows three forms of the alternative hypothesis, the two-tailed test $\neq$, lower (left)-tailed test $<$, and upper (right)-tailed test $>$.

(3) A *test statistic* is employed to decide whether the null hypothesis, H_o, should be accepted or rejected. Test statistics correspond to a type of distribution. Common test statistics are the z-test, t-test, chi-squared (χ^2) test, and F-test, which respectively correspond to the normal distribution, Student's t-distribution, chi-square distribution, and F-distribution. The selection of an appropriate test statistic to be employed depends on the purpose of the test statistic, whether the test statistic is used to test for a one-sample test, a two-sample test, or a paired test. More details on the various test statistics for hypothesis testing can be found in Dixon and Massey (1951) and Mendenhall and Sincich (2016).

(4) A *significance level*, α, is a probability threshold below which the null hypothesis, H_o, will be rejected. Common significance level values are 10%, 5%, and 1%; these are respectively equal to the 90%, 95%, and 99% confidence levels (1-α).

(5) *P-value*, the probability value as a result of the test statistic assuming the null hypothesis, H_o, is true. P-values in Table 9.3 are $-z_\alpha$, z_α, $-z_{\alpha/2}$, or $z_{\alpha/2}$.

(6) The *rejection region* is the value of the test statistic that will imply the rejection of the null hypothesis, H_o. Rejection regions are indicated as the shaded areas in Table 9.3.

(7) The *conclusion* is the decision made on whether to accept or reject the null hypothesis, H_o.

TABLE 9.3 Hypothesis testing variations based on the alternative hypothesis

Hypothesis Testing	The Null Hypothesis, H_o The Alternative Hypothesis, H_a	Graphical Representation
Two-tailed test	$H_o: \mu = \mu_o$ $H_a: \mu \neq \mu_o$	
Left-tailed test	$H_o: \mu \geq \mu_o$ $H_a: \mu < \mu_o$	
Right-tailed test	$H_o: \mu \leq \mu_o$ $H_a: \mu > \mu_o$	

Since a statistical test can result in one of only two outcomes (rejecting or not rejecting the null hypothesis), the test conclusion is also subjected to only two types of error, that is *Type I and Type II error* (Table 9.4). Type I error is rejecting the null hypothesis even though it is true. An example of a Type I error can be illustrated by a false alarm in cyber intrusion or threat detection, where the alarm goes off despite no threat actually existing. The probability of making a Type I error is denoted by the symbol $\alpha = p(\text{Reject } H_o | H_o \text{ true})$. Type II error is *not* rejecting the null hypothesis even though it is false. Type II error can be illustrated by a missed detection scenario, where the alarm did not go off when there are actual threats. The likelihood of making a Type II error is indicated by the symbol $\beta = p(\text{Accept } H_o | H_o \text{ false})$.

TABLE 9.4 General conclusion and consequences of a hypothesis testing

Decision (Conclusion from hypothesis test)	The "True" State of Nature	
	$H_o = true$ (No cyber threats)	$H_a = true$ (Cyber threats exist)
Reject H_o **Alarm**	Type I error (α) **False Alarm**	Correct decision **No error**
Not rejecting H_o **No Alarm**	Correct decision **No error**	Type II error (β) **Missed detection**

9.4.3 Applications in Cyber Research

Statistical hypothesis testing has been widely applied in cyber research to test various hypotheses, such as:

- *Trust hypothesis test.* Pavlovic (2015) employed the hypothesis testing approach to test the trustworthy behavior of a system. A system S was considered to have a set of four observable behaviors $B = \{a, b, c, d\}$. A trustworthy system should demonstrate an acceptable behavior (a) at least 98% of the time, no more than 0.5% of the time the system may be blocked (b) and crash (c), and no more than 1% of the time the system may have delayed (d) functions. The null hypothesis states that the system S behaves according to the trustworthy probability distribution, Pr_o: $\{a = 0.98, b = 0.005, c = 0.005, d = 0.01\}$. If the test statistic result indicates that the null hypothesis should not be rejected, this means one should continue trusting the system S. On the other hand, the system S is deemed to be untrustworthy if the null hypothesis should be rejected.

- *Bad data detection.* Hypothesis testing was employed in cybersecurity analysis for detecting bad data in the electric power system (Teixeira, Amin, Sandberg, Johansson, & Sastry, 2010). The null hypothesis H_o, was meant for testing if no bad data exists, and the alternative hypothesis H_a was for testing if bad data did exist.

- *Deception attack monitoring.* The hypothesis testing approach was also implemented to diagnose cyber-attacks in a cyber-physical system. In the study conducted by Kwon, Liu, and Hwang (2013), hypothesis testing was used to test if a residual from a fault detection algorithm exceeded a certain threshold. If there are no faults detected, the null hypothesis should be accepted. Otherwise, the alternative hypothesis should be accepted if the algorithm declares a fault in the system that may be related to a cyber attack.

9.5 CORRELATION AND REGRESSION ANALYSIS

An important application of statistics is to be able to estimate the relationship between two or more quantitative variables. This can fundamentally be achieved by correlation and regression analysis. The *strength* of the relationship between two quantitative variables can be estimated through a correlation analysis. A high correlation indicates that the two variables have a strong relationship, whereas a low correlation is associated with a weak relationship between the two variables. A correlation is often assumed to follow a linear line. Thus, correlation analysis is often related to linear regression analysis (Franzese & Iuliano, 2019). Regression analysis is a statistical approach that involves predicting the value of a dependent variable (response) based on the known value of one or more independent variables. Both correlation and regression analysis are part of the fundamental methods behind modern machine learning algorithms.

9.5.1 Correlation Analysis

The strength of the linear relationship between two variables is measured with the correlation coefficient, r, which varies between -1 and 1. ± 1 indicates perfect correlation, $+1$ indicates perfect positive correlation, and -1 indicates perfect negative correlation. 0 indicates no correlation between the two variables; i.e., the two variables do not seem to be related to each other at all. Figure 9.7 illustrates various correlation relationships between two variables and their associated correlation coefficient.

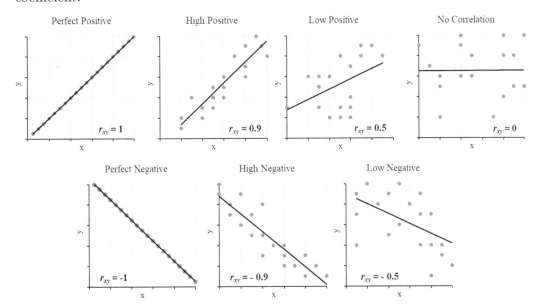

FIGURE 9.7 Various correlation and regression relationships.

The coefficient correlation can be obtained with Pearson's product-moment correlation coefficient method, which is calculated by dividing the covariance of variables x and y by the product of their standard deviations.

$$r_{xy} = \frac{Cov(x,y)}{\sqrt{s_X^2 \; s_Y^2}} = \frac{\int_{i=1}^{n}(x_i - \overline{x})(y_i - \overline{y})}{\sqrt{\int_{i=1}^{n}(x_i - \overline{x})^2}\sqrt{\int_{i=1}^{n}(y_i - \overline{y})^2}} \quad (9.1)$$

If the two variables of interest are ordinal or associated with a ranking, the rank correlation coefficient can be calculated. A rank correlation measures the relationship between two variables' rankings. Spearman's correlation (Zar, 1972) and Kendall's tau (Kendall, 1948) are examples of some rank correlation measures.

9.5.2 Regression Analysis

Both correlation and regression analysis quantify the relationship between two variables. One distinct difference between correlation and regression is that a correlation analysis produces a quantitative statistical value (correlation coefficient), whereas a regression analysis produces a quantitative equation, known as a regression model. Generally, a regression model can be represented as:

$$y_i = f(x_i, \beta) + \varepsilon_i \tag{9.2}$$

where y_i is the dependent variable, x_i is the observed data or independent variables used to predict model y_i, β is an unknown parameter, and ε_i is the random error. The goal of regression is to estimate the function $f(x_i, \beta)$ that closely fits the observed data. The function $f(x_i, \beta)$ can take many forms. Broadly, the function $f(x_i, \beta)$ is either a linear or a non-linear function, which is further associated with a linear regression or a non-linear regression model, as shown in Figure 9.8.

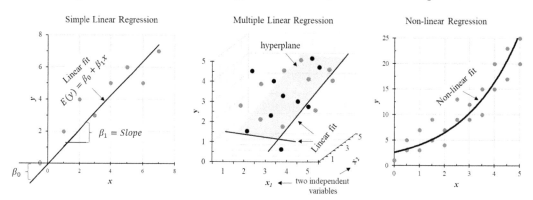

FIGURE 9.8 Various regression models.

A *linear regression* model uses a linear (or a line) approach to describe the relationships between dependent and independent values. This line is illustrated in each subfigure in Figure 9.7 and Figure 9.8. In the scenario involving only one dependent variable y and one independent variable x, such as in Figure 9.7, the regression model is called a simple linear regression model. The general form of the simple linear regression model is:

$$y = \beta_0 + \beta_1 x + \varepsilon \tag{9.3}$$

where β_0 and β_1 are the unknown parameters of the model. β_0 is the y-intercept and β_1 represents the slope or the coefficient of x (Figure 9.8). If the ε term is expected or assumed to be zero, $E(\varepsilon)=0$, then the expected value or the mean value of y, $E(y)$ is:

$$E(y) = \beta_0 + \beta_1 x \qquad (9.4)$$

When there is more than one independent variable involved in x_i, where $i = 1, 2, \ldots, n$, the regression model is known as a multiple linear regression model. The lines in multiple linear regression are represented as a hyperplane. Figure 9.8 shows a hyperplane between two independent variables in multiple linear models. The general form of a multiple linear regression model is:

$$y = \beta_0 + \beta_1 x_1 + \beta_2 x_2 + \cdots + \beta_n x_n + \varepsilon \qquad (9.5)$$

The least-squares method is often employed to fit linear regression models. The estimation of the unknown parameters $\widehat{\beta}_0$ and $\widehat{\beta}_1$ for simple linear regression models can be obtained by:

$$\text{Slope: } \widehat{\beta}_1 = \frac{\int_{i=1}^{n} (x_i - \overline{x})(y_i - \overline{y})}{\int_{i=1}^{n} (x_i - \overline{x})^2} = \frac{Cov(x,y)}{s_x^2} \qquad (9.6)$$

$$\widehat{\beta}_1 = r_{xy} \frac{s_y}{s_x}$$

$$y - \text{intercept: } \widehat{\beta}_0 = \overline{y} - \widehat{\beta}_1 \overline{x} \qquad (9.7)$$

where r_{xy} is the correlation coefficient for variable x and y, s_x^2 is the sample variance of x, s_x and s_y are the standard deviations of x and y, and $\overline{x}$ and $\overline{y}$ are the average of x_i and y_i, respectively. The slope in a simple linear regression model $\widehat{\beta}_1$ is related to its correlation coefficient. When $\widehat{\beta}_1$ is positive; this also denotes that there is a positive correlation between x and y (r_{xy} is positive). Similarly, when there is a negative correlation between x and y (r_{xy} is negative), the slope $\widehat{\beta}_1$ is also negative.

If the relationship between x and y is not linear, the regression model is called a non-linear regression model. The estimation of a non-linear regression model is often approached using the approximation method. The least-squares method can be employed to fit the non-linear regression model. However, the non-linear least squares method must be applied iteratively until the convergence is achieved. In some scenarios, it is possible to linearize the non-linear functions, such as exponential and logarithmic functions. Thus, the ordinary least squares method can be employed to estimate the unknown parameter in non-linear regression models without an iterative process.

Goodness-of-fit is a measure of how good a statistical model fits the observation data. Generally, a regression model is deemed to have a good fit if the difference between the observed values and the predicted values are small. The coefficient of determination, R-squared (R^2), is one measure to quantify the goodness-of-fit in a regression model. The measure of R^2 normally ranges between 0 and 1 (or 0% and 100%). The closer the R^2 value is to 1, the better the regression model fits the data. If R^2 equals 1, the regression model fits the data perfectly. For simple linear regression, R^2 equals the square root of the sample correlation coefficient r^2 when an intercept is included. The most general equation for R^2 is:

$$R^2 = 1 - \frac{SS_{error}}{SS_{total}} \tag{9.8}$$

$$SS_{error} = \int_{i=1}^{n} (y_i - \hat{y})^2 \tag{9.9}$$

$$SS_{total} = \int_{i=1}^{n} (y_i - \overline{y})^2 \tag{9.10}$$

Whether linear or non-linear, a regression model is often treated as a predictive model. Once a regression model is developed, multiple scenarios of predicting the response y can be carried out by varying x. Note that the accuracy of the predicted values depends on the accuracy of the developed fitted regression model. More information on linear and non-linear regression analysis and model fitting can be found in Bates and Watts (1988), Graybill (1976), Haitovsky (1973), Ratkowsky and Giles (1990), and Younger (1979).

9.5.3 Applications in Cyber Research

Correlation analysis and linear regression analysis have many practical uses in cyber research. Most applications fall into either: (1) analyzing relationships; or (2) predicting/forecasting based on the estimation (fitted) regression model.

In cyberbullying applications, Erdur-Baker (2010) employed multiple regression analyses to study the correlation between cyberbullying and traditional bullying. It was found that a bully in a traditionally physical environment and cyber environment were correlated, but this did not apply to the victims. Ayas and Deniz (2014) predicted a moderate positive correlation between cyberbullying and psychological systems in elementary students. Şahin (2012) concluded also a significant correlation of becoming a cyber victim due to loneliness among adolescents. Barlett and Chamberlin (2017) examined cyberbullying patterns with age. It was found that the relationship between age and cyberbullying followed a quadratic function, where the exposure to cyberbullying increased from youth to emerging adulthood and then decreased toward the later stages of life.

In cyber prediction applications, a cyber attacker model profile (CAMP) was proposed to determine the root causes of cybercrimes and predict cyber attacks that were based on correlation and regression analysis (Watters, McCombie, Layton, & Pieprzyk, 2012). Lavrova and Pechenkin (2015) employed both correlation and regression analysis to detect potential known and unknown security events in Internet-of-Things (IoT) applications. In their study, correlation analysis was used to find the interconnections in data collected from IoT devices. Regression analysis was utilized to determine the forms (fitted model) of the analytic interconnections between IoT devices. Further, they also proposed an event correlation method to investigate the relationship between two security incidents.

The relationships of the data presented in Table 9.2 and Figure 9.4 can be further interpreted by their coefficient of correlation r, coefficient of determination R^2, and

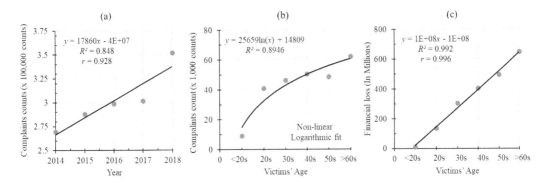

FIGURE 9.9 Correlation and regression models for data presented in Table 9.2 and Figure 9.4.

regression model, which can be obtained with correlation and regression analysis. The results are shown in Figure 9.9. Figure 9.9(a) shows the relationship between the frequency of complaints and the year based on the data from Table 9.2. The data were fitted with a simple linear regression model. It can be further interpreted that a positive linear correlation exists between the year and the number of cyber complaints received. Based on the fitted model, it is to be expected that the number of cyber-related crime complaints will continuously grow in the future. Figure 9.9(b) shows a non-linear relationship between complaint counts and the victims' age. Figure 9.9(c) shows a very strong positive relationship between financial losses and the victims' age. All the data used in Figure 9.9 are obtained from the 2018 FBI-IC3 report (FBI, 2019).

9.6 RESEARCH IMPLICATIONS AND EXTENSIONS: CYBER RESILIENCE RESEARCH

Statistics have been widely used in cyber research and in the application of cybersecurity in general. Various collected data can be transformed into general knowledge with statistical methods that have been employed to track the trend of cybercrimes (FBI, 2019; Kaspersky, 2019; Symantec, 2019), intrusion, fault, and attack detections (Kwon et al., 2013; Pavlovic, 2015; Teixeira et al., 2010), quantify relationships (Ayas & Deniz, 2014; Barlett & Chamberlin, 2017; Şahin, 2012), or to make predictions (Lavrova & Pechenkin, 2015; Watters et al., 2012). Accurate statistics on cyber incidents are often difficult to obtain as many cyber-related incidents go unreported. One of the reasons for this is no global standard for reporting cybercrimes. As the world becomes more cyber-oriented due to the advancement of the internet, the occurrence of cyber attacks will undoubtedly increase in regularity. Although cybersecurity practices are continuously improving with the emergence of new cyber threats, cybersecurity, by itself, still falls short in overcoming all forms of cyber attacks. Keeping pace with the increasing magnitude of cyber attacks is a major challenge for everyone. In addition to advancing cybersecurity, the concept of cyber resilience should be adopted to ensure the continuity of functions and operations in the presence of cyber attacks.

9.6.1 Research Implication on Cyber Resilience

Cyber resilience is a growing concept in the area of cyber safety and security. Cyber resilience incorporates a resilience perspective into cyber-related activities. The objective of cyber resilience is to promote the stability and the continuation of an entity (or entities) operating under the negative influences of cyber-related events (Linkov & Kott, 2019). Smart and connected engineered systems (autonomous systems, IoT, or cyber-physical systems), critical infrastructure (power, transportation, or telecommunications), businesses or enterprises, and governmental agencies are some examples of the entities that may benefit from cyber resilience. In order to realize cyber resilience, the concept of resilience should be: (1) accurately defined and measured; (2) assessed under various scenarios; and (3) enhanced in the implementation.

- *Cyber resilience definition and measures.* Björck et al. (2015) defined cyber resilience as the ability to deliver a continuous intended outcome despite adverse cyber events. Statistical methods are often employed in quantifying resilience. A number of resilience metrics for cyber systems have been identified and assessed using quantitative and qualitative measures by Linkov et al. (2013). If the cyber events are associated with a specific application, a resilience metric from the application itself, such as in energy systems (Roege, Collier, Mancillas, McDonagh, & Linkov, 2014; Yodo & Arfin, 2020), infrastructure systems (Afrin & Yodo, 2019; Francis & Bekera, 2014), transportation systems (Cox, Prager, & Rose, 2011; Ganin et al., 2019), or other general engineered systems (Yodo & Wang, 2016) can be considered or modified to measure cyber resilience in a specific application.

- *Cyber resilience assessment.* Statistic- and probabilistic-based methods, such as the Bayesian network approach (Wu, Yin, & Guo, 2012; Xie, Li, Ou, Liu, & Levy, 2010; Yodo, Wang, & Zhou, 2017) and machine learning approaches (Buczak & Guven, 2015; Lamba & Dutta, 2019; Xin et al., 2018), can be employed to assessed cyber resilience. Model-based or pure simulation approaches can be adopted to evaluate cyber resilience, such as using generic simulation software such as Simulink (Al Faruque & Ahourai, 2014; Kerzhner et al., 2015; Yodo & Wang, 2018). Other qualitative approaches to assess cyber resilience can also be implemented. For example, Khan, Al-Saher, and Rauf (2015) proposed a cyber resilience engineering framework (CREF) to measure cyber resilience from different aspects (attacks and failures) and at different levels (pro-active, resistive, and reactive). Walker et al. (2002) proposed a four-step resilience analysis system that focuses on long-term sustainability and considers stakeholders as an integral part of the system.

- *Cyber resilience enhancement.* In general, resilience enhancement can be divided into four categories based on the disruptive event scenario: (1) prior to an event; (2) during an event; (3) after an event (short-term); and (4) after an event (long-term). Depending on the application, some strategies that have been explored to improve resilience in general that can also be integrated to improve cyber resilience in practice include: fault detection and system monitoring employed as a prior-to-an-event strategy (Ding, Han, Xiang, Ge, & Zhang, 2018; Gupta & Kulariya, 2016), resilient control systems developed to minimize loss during an event and promote efficient restoration immediately thereafter (Bou-Harb et al., 2017; Rafi & Steck, 2013; Rafi, Steck, & Rokhsaz, 2012; Yodo, Wang, & Rafi, 2017), and cyber resilience management frameworks applied to ensure sustainable resilience after an event (CISCO, 2017; Refsdal, Solhaug, & Stølen, 2015; Siegel, Sagalow, & Serritella, 2002). Ratasich et al. (2019) summarized the state of the art of existing work on anomaly detection, fault-tolerance, self-healing, and other methods applicable to achieve resilience in an IoT application.

9.6.2 Potential Extensions

Statistical methods are often used as fundamental methods in various scientific research and social studies. Some of these research areas have grown into their own specialized disciplines. In addition to the concept of cyber resilience, cyber research may also benefit from statistical methods within the following disciplines:

- *Machine learning, deep learning*, and *data mining;* This is a subfield of computer science or data analytics that formulates algorithms in order to make predictions or learn from data. Regression analysis and hypothesis testing are some fundamental statistical methods behind various advanced algorithms in this area. Machine learning is often applied for cybersecurity intrusion detection and situational awareness (Buczak & Guven, 2015; Xin et al., 2018).

- *Reliability engineering.* Reliability is a measure of the ability of a system or component to perform its required functions under stated conditions for a specified period of time. Statistical methods are often used to determine the degradation prediction, remaining useful life (RUL), warranty period, or mean time to failure (MTTF). Advances in reliability engineering can be leveraged for cyber-physical systems to ensure the continuation of functions in the presence of cyber attacks (Marashi, Sarvestani, & Hurson, 2017).

- *Risk analysis, assessment, and management.* In addition to reliability, risk analysis is an assessment approach that is often used to find the root cause of failure (Cherdantseva et al., 2016; Ralston, Graham, & Hieb, 2007). After a root cause of cyber risk has been found, a proper management approach can be implemented to reduce cyber risks and increase cyber resilience with an effective cyber risk management approach (Öğüt, Raghunathan, & Menon, 2011). This area would potentially be of interest to large organizations. There are numerous cybersecurity management policies and standards that have been developed by various experts in academia, industry, and government. For example, the National Institute of Standards and Technology (NIST) cybersecurity risk management involves five functions: identify, protect, detect, respond, and recover (Barrett, 2018; US-CERT, n.d). Cisco Systems, Inc.'s cybersecurity risk management consists of three macro pillars: executive management (strategy), operations, and tactical (technology) (Cisco, 2017).

- *Actuarial science.* Mathematical and statistical methods are employed in actuarial science to assess risk in the insurance and finance industries. In addition, actuarial science also deals with developing protection plans (insurance coverage) to cover misfortunes posed by an unforeseen event. The evaluation of risks in actuarial science that deals with a lot of uncertainty can be leveraged to mitigate cyber risks. Further, strategic insurance coverage plans to protect against cyber risk can be developed (Young, Lopez Jr, Rice, Ramsey, & McTasney, 2016).

- *Social and demography statistics.* In social studies, statistical methods are utilized to study human behavior in physical, social, or cultural environments. Social statistics have been widely employed in cyberbullying applications (Ayas & Deniz, 2014; Erdur-Baker, 2010; Şahin, 2012). It can be further expanded to study social interaction in the cyber environment. Together with demographic studies, social statistics can be applied to evaluate the migration of cyber-related events among the dynamic population that changes over time and space. This area is useful in practice to track cyber threats that originate from social networking sites (Gharibi & Shaabi, 2012; Xu, Yang, Cheng, & Lim, 2014).

The reader should note that the disciplines identified above are not exhaustive. There are a wide range of disciplines in socio-economic, psychology, computer science, and general scientific research that benefit cyber research. The application of statistical methods from diverse channels of multi-disciplinary research hopes to aid in overcoming new cyber threats.

9.7 CONCLUSIONS

This chapter has introduced the concept of applying fundamental statistical methods to cyber research. Descriptive statistics generally deals with organizing, summarizing, and presenting data. Inferential statistics involve interpreting a larger dataset from a smaller dataset, testing a hypothesis, and assessing relationships among two or more variables. Although accurate statistics on cyber-related incidents can be difficult to obtain, this chapter has presented several fundamental statistical methods that may be utilized to make educated observations and extrapolations in cyber research. This is especially important moving into the future, as the world becomes increasingly connected and the Internet-of-Things becomes ubiquitous. In order to keep up with the growing risk of cyber threats, there is a need for a cultural shift to incorporate a cyber resilience perspective into cyber research, so as to strengthen cybersecurity.

REFERENCES

Acharya, A. S., Prakash, A., Saxena, P., and Nigam, A. (2013). Sampling: Why and how of it. *Indian Journal of Medical Specialties*, 4(2), 330–333.

Afrin, T., and Yodo, N. (2019). Resilience-based recovery assessments of networked infrastructure systems under localized attacks. *Infrastructures*, 4(1), 11.

Ahmad, N., Asma'Mokhtar, U., Fauzi, W. F. P., Othman, Z. A., Yeop, Y. H., and Abdullah, S. N. H. S. (2018). Cyber Security Situational Awareness among Parents. Paper presented at the *2018 Cyber Resilience Conference (CRC)*.

Al Faruque, M. A., and Ahourai, F. (2014). A model-based design of cyber-physical energy systems. Paper presented at the *2014 19th Asia and South Pacific Design Automation Conference (ASP-DAC)*.

Ayas, T., and Deniz, M. (2014). Predicting the exposure levels of cyber bullying of elementary students with regard to psychological symptoms. *Procedia-Social and Behavioral Sciences*, 116, 4910–4913.

Barlett, C. P., and Chamberlin, K. (2017). Examining cyberbullying across the lifespan. *Computers in Human Behavior*, 71, 444–449.

Barrett, M. P. (2018). Framework for improving critical infrastructure cybersecurity, Version 1.1. Retrieved from https://www.nist.gov/cyberframework.

Bates, D. M., and Watts, D. G. (1988). *Nonlinear regression analysis and its applications (Vol. 2)*: Wiley New York.

Birgé, L., and Rozenholc, Y. (2006). How many bins should be put in a regular histogram. *ESAIM: Probability and Statistics*, 10, 24–45.

Björck, F., Henkel, M., Stirna, J., and Zdravkovic, J. (2015). Cyber resilience-fundamentals for a definition. Paper presented at the *WorldCIST (1)*.

Bou-Harb, E., Lucia, W., Forti, N., Weerakkody, S., Ghani, N., and Sinopoli, B. (2017). Cyber meets control: A novel federated approach for resilient cps leveraging real cyber threat intelligence. *IEEE Communications Magazine*, 55(5), 198–204.

Bouyeddou, B., Harrou, F., Sun, Y., and Kadri, B. (2017). Detecting SYN flood attacks via statistical monitoring charts: A comparative study. Paper presented at the *2017 5th International Conference on Electrical Engineering-Boumerdes (ICEE-B)*.

Buczak, A. L., and Guven, E. (2015). A survey of data mining and machine learning methods for cyber security intrusion detection. *IEEE Communications Surveys and Tutorials*, 18(2), 1153–1176.

Cherdantseva, Y., Burnap, P., Blyth, A., Eden, P., Jones, K., Soulsby, H., and Stoddart, K. (2016). A review of cyber security risk assessment methods for SCADA systems. *Computers and Security*, 56, 1–27.

CISCO. (2017). Cybersecurity management program. Retrieved from https://www.cisco.com/c/dam/en/us/products/collateral/security/cybersecurity-management-programs.pdf.

Cox, A., Prager, F., and Rose, A. (2011). Transportation security and the role of resilience: A foundation for operational metrics. *Transport Policy*, 18(2), 307–317. doi:DOI 10.1016/j.tranpol.2010.09.004.

Cukier, M. (2007). Study: Hackers attack every 39 seconds. Retrieved from https://eng.umd.edu/news/story/study-hackers-attack-every-39-seconds.

Ding, D., Han, Q.-L., Xiang, Y., Ge, X., and Zhang, X.-M. (2018). A survey on security control and attack detection for industrial cyber-physical systems. *Neurocomputing*, 275, 1674–1683.

Dixon, W. J., and Massey Jr, F. J. (1951). Introduction to statistical analysis. McGraw Hill.

Erdur-Baker, Ö. (2010). Cyberbullying and its correlation to traditional bullying, gender and frequent and risky usage of internet-mediated communication tools. *New Media and Society*, 12(1), 109–125.

Facchinetti, S., Giudici, P., and Osmetti, S. A. (2020). Cyber risk measurement with ordinal data. *Statistical Methods and Applications*, 29(1), 173–185.

FBI. (2019). Internet crime center 2018 internet crime report. Retrieved from https://pdf.ic3.gov/2018_IC3Report.pdf.

Francis, R., and Bekera, B. (2014). A metric and frameworks for resilience analysis of engineered and infrastructure systems. *Reliability Engineering and System Safety*, 121, 90–103.

Franzese, M., and Iuliano, A. (2019). Correlation Analysis. Elsevier.

Ganin, A. A., Mersky, A. C., Jin, A. S., Kitsak, M., Keisler, J. M., and Linkov, I. (2019). Resilience in Intelligent Transportation Systems (ITS). *Transportation Research Part C: Emerging Technologies*, 100, 318–329. doi:https://doi.org/10.1016/j.trc.2019.01.014.

Gharibi, W., and Shaabi, M. (2012). Cyber threats in social networking websites. arXiv preprint arXiv:1202.2420.

Graybill, F. A. (1976). *Theory and application of the linear model*. Duxbury Press North Scituate, MA.

Gupta, G. P., and Kulariya, M. (2016). A framework for fast and efficient cyber security network intrusion detection using apache spark. *Procedia Computer Science*, 93, 824–831.

Haitovsky, Y. (1973). *Regression estimation from grouped observations*. Hafner Press.

Hinduja, S., and Patchin, J. W. (2010). Bullying, cyberbullying, and suicide. *Archives of Suicide Research*, 14(3), 206–221.

Kaspersky. (2019). Financial cyberthreats in 2018. Retrieved from https://securelist.com/financial-cyberthreats-in-2018/89788/.

Kendall, M. G. (1948). Rank correlation methods. Griffin.

Kerzhner, A. A., Tan, K., and Fosse, E. (2015). Analyzing cyber security threats on cyber-physical systems using Model-Based Systems Engineering. Paper presented at the *AIAA SPACE 2015 Conference and Exposition*.

Khan, Y. I., Al-Shaer, E., and Rauf, U. (2015). Cyber resilience-by-construction: modeling, measuring and verifying. Paper presented at the *Proceedings of the 2015 Workshop on Automated Decision Making for Active Cyber Defense*.

Kwon, C., Liu, W., and Hwang, I. (2013). Security analysis for cyber-physical systems against stealthy deception attacks. Paper presented at the *2013 American Control Conference*.

Lamba, A., and Dutta, N. (2019). SR-MLC: Scalable Resilience Machine Learning Classifiers Approach in Cyber Security. *International Journal of Current Research*, 11, 3283–3290.

Lavrova, D., and Pechenkin, A. (2015). Applying correlation and regression analysis to detect security incidents in the internet of things. *International Journal of Communication Networks and Information Security*, 7(3), 131.

Levy, M., Raviv, D., and Baker, J. (2019). Data center simulations deployed in MATLAB and Simulink using a cyber-physical systems lens. Paper presented at the *2019 IEEE 9th Annual Computing and Communication Workshop and Conference (CCWC)*.

Linkov, I., Eisenberg, D. A., Plourde, K., Seager, T. P., Allen, J., and Kott, A. (2013). Resilience metrics for cyber systems. *Environment Systems and Decisions*, 33(4), 471–476.

Linkov, I., and Kott, A. (2019). Fundamental concepts of cyber resilience: Introduction and overview. In *Cyber resilience of systems and networks* (pp. 1–25): Springer.

Marashi, K., Sarvestani, S. S., and Hurson, A. R. (2017). Consideration of cyber-physical interdependencies in reliability modeling of smart grids. *IEEE Transactions on Sustainable Computing*, 3(2), 73–83.

Mendenhall, W. M., and Sincich, T. L. (2016). *Statistics for engineering and the sciences.* Chapman and Hall/CRC.

Öğüt, H., Raghunathan, S., and Menon, N. (2011). Cyber security risk management: Public policy implications of correlated risk, imperfect ability to prove loss, and observability of self-protection. *Risk Analysis: An International Journal*, 31(3), 497–512.

Pavlovic, D. (2015). Towards a science of trust. Paper presented at the *Proceedings of the 2015 Symposium and Bootcamp on the Science of Security.*

Rafi, M., and Steck, J. (2013). Response and recovery of an MRAC advanced flight control system to wake vortex encounters. Paper presented at the *AIAA Infotech@ Aerospace Conference*, Boston (August 2013).

Rafi, M., Steck, J., and Rokhsaz, K. (2012). A microburst response and recovery scheme using advanced flight envelope protection. Paper presented at the *AIAA Guidance, Navigation, and Control Conference and Exhibit*, Minneapolis, MN.

Ralston, P. A., Graham, J. H., and Hieb, J. L. (2007). Cyber security risk assessment for SCADA and DCS networks. *ISA Transactions*, 46(4), 583–594.

Ratasich, D., Khalid, F., Geissler, F., Grosu, R., Shafique, M., and Bartocci, E. (2019). A roadmap toward the resilient internet of things for cyber-physical systems. *IEEE Access*, 7, 13260–13283.

Ratkowsky, D. A., and Giles, D. E. (1990). Handbook of nonlinear regression models.

Refsdal, A., Solhaug, B., and Stølen, K. (2015). Cyber-risk management. In *Cyber-Risk Management* (pp. 33–47). Springer.

Roege, P. E., Collier, Z. A., Mancillas, J., McDonagh, J. A., and Linkov, I. (2014). Metrics for energy resilience. *Energy Policy*, 72, 249–256.

Şahin, M. (2012). The relationship between the cyberbullying/cybervictmization and loneliness among adolescents. *Children and Youth Services Review*, 34(4), 834–837.

Scott, D. W. (2009). Sturges' rule. *Wiley Interdisciplinary Reviews: Computational Statistics*, 1(3), 303–306.

Siegel, C. A., Sagalow, T. R., and Serritella, P. (2002). Cyber-risk management: technical and insurance controls for enterprise-level security. *Information Systems Security*, 11(4), 33–49.

Sklavounos, D., Leondakianakos, A., and Edoh, A. (2019). Statistical process control method for cyber intrusion detection (DDoS, U2R, R2L, Probe). *International Journal of Cyber-Security and Digital Forensics*, 8(1), 82–89.

Symantec. (2019). Internet security threat report (ISTR). Retrieved from https://www.symantec.com/content/dam/symantec/docs/reports/istr-24-2019-en.pdf.

Teixeira, A., Amin, S., Sandberg, H., Johansson, K. H., and Sastry, S. S. (2010). Cyber security analysis of state estimators in electric power systems. Paper presented at the *49th IEEE Conference on Decision and Control (CDC)*.

Turan, N., Polat, O., Karapirli, M., Uysal, C., and Turan, S. G. (2011). The new violence type of the era: Cyber bullying among university students: Violence among university students. *Neurology, psychiatry and brain research*, 17(1), 21–26.

US-CERT. (n.d). Cybersecurity framework. Retrieved from https://www.us-cert.gov/ccubedvp/cybersecurity-framework.

Walker, B., Carpenter, S., Anderies, J., Abel, N., Cumming, G., Janssen, M., Lebel, L., Norberg, J., Peterson, G., and Pritchard, R. (2002). Resilience management in social-ecological systems: a working hypothesis for a participatory approach. *Conservation Ecology*, 6(1).

Watters, P. A., McCombie, S., Layton, R., and Pieprzyk, J. (2012). Characterising and predicting cyber attacks using the Cyber Attacker Model Profile (CAMP). *Journal of Money Laundering Control*, 15(4), 430–441.

Wu, J., Yin, L., and Guo, Y. (2012). Cyber attacks prediction model based on Bayesian network. Paper presented at the *2012 IEEE 18th International Conference on Parallel and Distributed Systems*.

Xie, P., Li, J. H., Ou, X., Liu, P., and Levy, R. (2010). Using Bayesian networks for cyber security analysis. Paper presented at the *2010 IEEE/IFIP International Conference on Dependable Systems and Networks (DSN)*.

Xin, Y., Kong, L., Liu, Z., Chen, Y., Li, Y., Zhu, H., Gao, M., Hou, H., and Wang, C. (2018). Machine learning and deep learning methods for cybersecurity. *IEEE Access*, 6, 35365-35381.

Xu, Y. C., Yang, Y., Cheng, Z., and Lim, J. (2014). Retaining and attracting users in social networking services: An empirical investigation of cyber migration. *The Journal of Strategic Information Systems*, 23(3), 239–253.

Yodo, N., and Arfin, T. (2020). A resilience assessment of an interdependent multi-energy system with microgrids. *Sustainable and Resilient Infrastructure*, 1–14. doi:10.1080/23789689.2019.1710074.

Yodo, N., and Wang, P. (2016). Engineering resilience quantification and system design implications: A literature survey. *Journal of Mechanical Design*, 138(111408), 1–13.

Yodo, N., and Wang, P. (2018). A control-guided failure restoration framework for the design of resilient engineering systems. *Reliability Engineering and System Safety*, 178, 179–190. doi:https://doi.org/10.1016/j.ress.2018.05.018.

Yodo, N., Wang, P., and Rafi, M. (2017). Enabling resilience of complex engineered systems using control theory. *IEEE Transactions on Reliability*, 1–13. doi:10.1109/TR.2017.2746754.

Yodo, N., Wang, P., and Zhou, Z. (2017). Predictive resilience analysis of complex systems using dynamic Bayesian networks. *IEEE Transactions on Reliability*, 66(3), 761–770.

Young, D., Lopez Jr, J., Rice, M., Ramsey, B., and McTasney, R. (2016). A framework for incorporating insurance in critical infrastructure cyber risk strategies. *International Journal of Critical Infrastructure Protection*, 14, 43–57.

Younger, M. S. (1979). Handbook for linear regression (Vol. 1). Duxbury Press North Scituate, MA.

Zar, J. H. (1972). Significance testing of the Spearman rank correlation coefficient. *Journal of the American Statistical Association*, 67(339), 578–580.

CHAPTER 10

Probability Theory

David M. Ruth

CONTENTS

10.1	Introduction	335
10.2	Probability Fundamentals	336
	10.2.1 Sample Spaces and Events	336
	10.2.2 Axioms of Probability	336
	10.2.3 Conditional Probability	339
	10.2.4 Random Variables	340
	10.2.5 Discrete Distributions	341
	10.2.5.1 Bernoulli(p) and Binomial(n, p)	343
	10.2.5.2 Hypergeometric(m, n, N)	344
	10.2.5.3 Geometric(p)	345
	10.2.5.4 Negative Binomial(r, p)	345
	10.2.5.5 Poisson(λ)	346
	10.2.6 Continuous Distributions	347
	10.2.6.1 Uniform(a, b)	349
	10.2.6.2 Normal(μ, σ)	349
	10.2.6.3 Exponential(λ)	350
	10.2.6.4 Gamma(α, λ)	351
	10.2.6.5 Chi-squared(ν)	351
	10.2.7 Expectation and Variance	352
	10.2.7.1 Expectation	352
	10.2.7.2 Variance	353
10.3	Probability Models in Cyber Research	354
	10.3.1 Bayes' Rule	354
	10.3.2 Markov Chains	355
	10.3.3 Information Entropy	357
10.4	Conclusions	358

10.1 INTRODUCTION

"It is remarkable that a science, which commenced with the consideration of games of chance, should be elevated to the rank of the most important subjects of human knowledge." (Laplace, 1812, p.195)

DOI: 10.1201/9780429354649-10

While Pierre-Simon Laplace's claim that probability ranks among the most important subjects of human knowledge may be debated, there is no doubt that probability is of vital importance to a wide array of fields: statistics, physics, engineering, computer science, biology, medicine, meteorology and oceanography, finance, and political science to name just a few. Specifically, probability theory plays a key role in understanding problems in the cyber domain, especially with regard to cybersecurity risk. This chapter highlights probability fundamentals in a cyber context and provides tools for those conducting cyber research to solve problems involving randomness and uncertainty.

10.2 PROBABILITY FUNDAMENTALS

"...[T]he theory of probabilities is at bottom only common sense reduced to calculus; it makes us appreciate the exactitude that which exact minds feel by a sort of instinct without being able ofttimes to give a reason for it." (Laplace, 1812, p.196)

10.2.1 Sample Spaces and Events

Probability is built upon a notion of *sets* that consist of all the possible outcomes of some experiment. Before the experiment is performed the outcome is unknown, but after the experiment is performed some particular result is the realized outcome.

Definition 10.1. The *sample space* S of an experiment is the set consisting of all possible outcomes of the experiment. Any subset $E \subseteq S$ is called an *event*. We say the E event *occurred* if the outcome realized from the experiment is an element of E. A sample space may be finite, countably infinite, or uncountable.

Example 10.1. Flip a coin 2 times and record the face that shows on each flip, in the order each face is observed. The finite sample space is $S = \{HH, HT, TH, TT\}$, and the event described by 'the first flip was heads' is $E = \{HH, HT\}$.

Example 10.2. Flip a coin until the first head is observed and record the face that shows on each flip. The countably infinite sample space is $S = \{H, TH, TTH, TTTH, ...\}$, and the event described by 'the first head occurred in less than 3 flips' is $E = \{H, TH\}$.

Example 10.3. Measure and record the temperature of a glass of liquid water in degrees Celsius. Using the interval notation $(a, b) = \{x : a < x < b\}$, the uncountable sample space is $S = (0, 100)$, and the event described by 'the water was cooler than 50°C' is $E = (0, 50)$. (In this example, 0 degree water is assumed solid and 100 degree water is assumed gas.)

10.2.2 Axioms of Probability

One would like to formally quantify notions such as, 'When a coin is flipped twice, what is the chance that at least one head is observed?' To do this, we introduce a set function P that takes an event as its input and assigns a real number between 0 and 1 as its output. The function P satisfies the following *axioms of probability*:

For any sample space S and any event $E \subseteq S$:

Axiom 10.1. $P(E) \geq 0$.

Axiom 10.2. $P(S) = 1$.

For any sequence of disjoint events $E_1, E_2, \ldots$:

Axiom 10.3. $P\left(\bigcup_{i=1}^{\infty} E_i\right) = \sum_{i=1}^{\infty} P(E_i)$.

Axiom 10.1 requires that all probabilities are at least zero, and Axiom 10.2 assigns the value one to the probability that an observed outcome is, in fact, an outcome in the sample space. Axiom 10.3 maintains that for any sequence of mutually exclusive events, the probability of at least one of these events occurring is the sum of the individual event probabilities. This notion of assigning event probabilities between 0 and 1 such that probabilities add in a natural way is quite intuitive, as are several useful results that follow directly from these axioms:

(i) $P(E^c) = 1 - P(E)$. In particular, $P(\emptyset) = 0$ (since $P(\emptyset) = P(S^c)$).

(ii) If $E \subseteq F$, then $P(E) \leq P(F)$.

(iii) $P(E \cup F) = P(E) + P(F) - P(E \cap F)$.

This last result is a special case of a more general result called the *inclusion-exclusion principle*, which is

$$P\left(\bigcup_{i=1}^{n} E_i\right) = \sum_{i=1}^{n} P(E_i) - \sum_{1 \leq i < j \leq n} P(E_i \cap E_j) \\ + \sum_{1 \leq i < j < k \leq n} P(E_i \cap E_j \cap E_k) \\ + \cdots + (-1)^{n-1} P(E_1 \cap E_2 \cdots \cap E_n). \quad (10.1)$$

Example 10.4. Let S be as in Example 10.1, and define probability function P so that each outcome in S has the same probability; that is, $P(\{HH\}) = P(\{HT\}) = P(\{TH\}) = P(\{TT\}) = \frac{1}{4}$. (These are the probabilities that would be assigned if the coin is equally likely to land on heads or tails; this is often referred to as a 'fair coin.') What is the probability of flipping at least one head?

Solution: By Axiom 10.3,

$$P(\text{'at least one head flipped'}) = P(\{HH, HT, TH\})$$
$$= P(\{HH\}) + P(\{HT\}) + P(\{TH\}) = \frac{3}{4},$$

since $\{HH\}$, $\{HT\}$, and $\{TH\}$ are disjoint events.

Example 10.5. Using the probabilities in Example 10.4, what is the probability the first or second flip is a tail?

Solution: By the inclusion-exclusion principle,

$$P(\text{'the first or second flip is a tail'}) = P(\{TH, TT\} \cup \{HT, TT\})$$
$$= P(\{TH, TT\}) + P(\{HT, TT\}) - P(\{TT\})$$
$$= \frac{1}{2} + \frac{1}{2} - \frac{1}{4} = \frac{3}{4}.$$

This same result could, of course, be computed directly as

$$P(\text{'the first or second flip is a tail'}) = P(\{TH, HT, TT\}) = 3/4.$$

Example 10.6. Let H_k denote the event 'exactly k hacking attempts are made on some network system in a specified time period,' and assign probabilities

$$P(H_k) = e^{-2} 2^k / k!$$

for all $k \in \{0, 1, 2, \ldots\}$. (We will see later that this is the probability function associated with the *Poisson distribution*, commonly used to model the occurrence of discrete events in time. In this case these events occur at an average rate of 2 attacks per time period.) To see that $P(S) = 1$, note that $S = \bigcup_{i=0}^{\infty} H_i$ since the events $H_0, H_1, \ldots$ partition S, and so by Axiom 10.3:

$$P\left(\bigcup_{i=0}^{\infty} H_i\right) = \sum_{i=0}^{\infty} P(H_i) = e^{-2} \sum_{i=0}^{\infty} 2^i / i! = e^{-2} e^2 = 1.$$

What is the probability that more than 1 hacking attempt is made in the specified time period?

Solution:

$$P(\text{'more than 1 hacking attempt is made'}) = P\left(\bigcup_{i=2}^{\infty} H_i\right)$$
$$= 1 - P(H_0 \cup H_1)$$
$$= 1 - P(H_0) - P(H_1)$$
$$= 1 - e^{-2} - 2e^{-2} \approx 0.594.$$

Venn diagrams are often used to illustrate probabilities associated with events, with the sizes of regions in the Venn diagram corresponding to relative probability values. Figure 10.1 shows a Venn Diagram with events E_1 and E_2 in sample space S, where event E_1 is more likely to occur than event E_2.

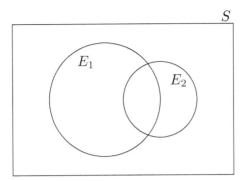

FIGURE 10.1 Venn diagram with $P(S) = 1 > P(E_1 \cup E_2) > P(E_1) > P(E_2) > P(E_1 \cap E_2) > 0$.

10.2.3 Conditional Probability

An essential concept in probability theory is the idea of conditional probability. For example, suppose a particular company might use Firewall 1 or Firewall 2 to enhance the security of a particular network. Let W_k denote the event 'Firewall k is used' for $k \in \{1, 2\}$, and let B denote the event 'a cybersecurity breach occurs.' Figure 10.2 shows a notional Venn Diagram with a sample space S consisting of all possibilities where either firewall is used and whether a breach occurs. The diagram illustrates that a breach is much more likely if Firewall 1 is used, since the relative size of $B \cap W_1$ to W_1 is much larger than relative size of $B \cap W_2$ to W_2. This motivates the following definition:

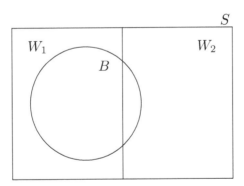

FIGURE 10.2 The conditional probability of a breach (event B) is higher when Firewall 1 is used (event W_1) than when Firewall 2 is used (event W_2), since $P(B \cap W_1)/P(W_1) > P(B \cap W_2)/P(W_2)$ (based on event sizes in the Venn diagram).

Definition 10.2. If E_1 and E_2 are events with $P(E_2) > 0$, then the *conditional probability of E_1 given E_2* is defined as

$$P(E_1|E_2) = \frac{P(E_1 \cap E_2)}{P(E_2)}.$$

Sometimes it is useful to rewrite this expression as

$$P(E_1|E_2)P(E_2) = P(E_1 \cap E_2) = P(E_2|E_1)P(E_1).$$

The left equality can be interpreted as, 'the probability that events E_1 and E_2 occur together is equal to the probability that E_2 occurs times the probability E_1 occurs given E_2 occurs;' similarly for the right equality.

Example 10.7. Using the probabilities in Example 10.4, what is the probability the first or second flip is a tail, given that at least one head is flipped?

Solution:

P('the first or second flip is a tail' | 'at least one head is flipped')
$$= \frac{P(\{HT, TH, TT\} \cap \{HH, HT, TH\})}{P(\{HH, HT, TH\})}$$
$$= \frac{P(\{HT, TH\})}{P(\{HH, HT, TH\})}$$
$$= \frac{1/2}{3/4} = \frac{2}{3}.$$

Definition 10.3. Events E_1 and E_2 are *independent* if $P(E_1|E_2) = P(E_1)$.

An equivalent definition of independence states that E_1 and E_2 are independent if $P(E_1 \cap E_2) = P(E_1)P(E_2)$. This relation follows directly from Definitions 10.2 and 10.3, since for independent events

$$P(E_1 \cap E_2) = P(E_1|E_2)P(E_2) = P(E_1)P(E_2). \tag{10.2}$$

10.2.4 Random Variables

Event notation can be useful but is often cumbersome. For instance, in Example 10.6 we have

$$P(\text{'more than 1 hacking attempt is made'}) = P\left(\bigcup_{i=2}^{\infty} H_i\right),$$

where H_k denotes the event 'exactly k hacking attempts are made on some network system in a specified time period' from before. It might be more natural, however, to let X be the number of hacking attempts made, so then one could write

$$P(\text{'more than 1 hacking attempt is made'}) = P(X > 1).$$

This notation uses the notion of a *random variable* to analyze probabilities associated with quantities of interest. In this particular case, '$X = 0$' corresponds to the event H_0, '$X = 1$' corresponds to the event H_1, and so forth.

Definition 10.4. A *random variable* X is a function that takes an outcome from a sample space S as its input and assigns a real number as its output.

Example 10.8. Let S be as in Example 10.1, and define random variable X by

$$X(s) = \text{number of heads in outcome } s, s \in S.$$

Then $X(HH) = 2$, $X(HT) = X(TH) = 1$, and $X(TT) = 0$.

Example 10.9. Let S be as in Example 10.2, and define random variable X by

$$X(s) = \text{the number of tails observed in outcome } s, s \in S.$$

Then $X(H) = 0$, $X(TH) = 1$, $X(TTH) = 2, \ldots$

Random variables are used routinely to facilitate the expression of probability ideas. For Example 10.1, the idea, 'What is the probability that both flips of a coin will show heads?' can be succinctly expressed as, 'What is $P(X = 2)$?' The expression '$P(X = 2)$' is shorthand for '$P(\{s : X(s) = 2, s \in S\})$,' that is, the probability an event occurs that includes an outcome for which the random variable X takes the value 2. Such shorthand is standard and will be used throughout this chapter.

Random variables can be categorized as *discrete* or *continuous*. The following two sections will introduce these types of random variables as well as some specific named instances that are commonly used. Before doing so, we note that the word 'distribution' is used somewhat generally to describe probabilities associated with a random variable X, and its exact meaning depends on context. For example, the phrase 'the distribution of X' may refer to

- the cumulative distribution function, F_X, or

- the probability mass function, p_X, or the probability density function, f_X, or

- the named family to which X belongs (e.g., 'X has a Normal distribution').

These ideas are all discussed in the following sections. Summary tables for regularly-encountered distributions are in Appendix 10.1.

10.2.5 Discrete Distributions

In many cases, random events of interest involve a countable sample space; for example, the number of errors in a digital signal consisting of 1's and 0's, or the number of hacking attempts made against a network on a given day. Such events can be modeled using *discrete random variables*.

Definition 10.5. X is a *discrete random variable* if it takes on a finite or countably infinite number of possible values.

Definition 10.6. If X is a discrete random variable, then its *probability mass function* (or *p.m.f.*), p_X, is defined by

$$p_X(k) = P(X = k).$$

For a discrete random variable, the set of values k such that $p_X(k) > 0$ is called the *support* of X. If we let A denote the support of X, then all p.m.f.'s have the property that $0 < p_X(k) \leq 1$ for all $k \in A$ and $\sum_{k \in A} p_X(k) = 1$.

Example 10.10. Recalling again the case of flipping a coin twice in Example 10.1, the associated p.m.f. is

$$p_X(0) = \frac{1}{4}, \quad p_X(1) = \frac{1}{2}, \quad p_X(2) = \frac{1}{4},$$

and $p_X(k) = 0$ for $k \notin \{0, 1, 2\}$.

Definition 10.7. If X is any random variable, then its *cumulative distribution function* (or c.d.f.), F_X, is defined by

$$F_X(x) = P(X \leq x).$$

Note that this definition pertains to any type of random variable, and can take positive values for arguments not in the support of X. For instance, in the case of Example 10.10,

$$p_X(1.5) = 0$$

but

$$F_X(1.5) = P(X \leq 1.5)$$
$$= p_X(0) + p_X(1) = \frac{3}{4}.$$

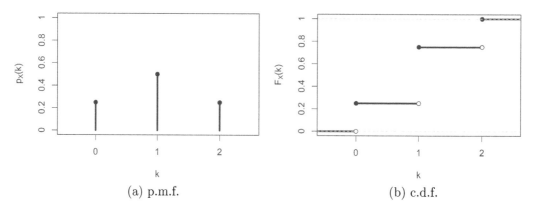

(a) p.m.f.   (b) c.d.f.

FIGURE 10.3 Probability mass and cumulative distribution functions for the number of heads in two coin flips.

Figure 10.3 shows the p.m.f. and c.d.f. for the two coin flip case side by side. Notice that high probability mass is indicated by large p.m.f. values and also by large jumps between c.d.f. values.

10.2.5.1 Bernoulli(p) and Binomial(n, p)

The most basic probability model involves an experiment that can result in a binary outcome such as a success or a failure, where success occurs with probability p. Such an experiment is called a *Bernoulli trial*. A random variable X is said to have a *Bernoulli(p) distribution* if

$$p_X(1) = p, \qquad p_X(0) = 1 - p. \tag{10.3}$$

We use the convention in this chapter of referring to specific distributions by both family name (in this case, Bernoulli) and associated parameter(s) (in this case, p); such a designation completely specifies the distribution function. It is also common in literature to encounter distributions by name only (for example, 'X is a Bernoulli random variable'), leaving the reader to know what parameters are associated with the distribution.

The Bernoulli distribution is a special case of a very important distribution which models the probability of achieving k successes on n trials. A random variable X is said to have a *Binomial(n, p) distribution* if

$$p_X(k) = \binom{n}{k} p^k (1-p)^{n-k} \tag{10.4}$$

for $k \in \{0, 1, 2, \ldots, n\}$; $p_X(k) = 0$ otherwise. This p.m.f. has two parts: $p^k(1-p)^{n-k}$ is the probability of achieving some specific ordered instance of k successes and $n-k$ failures on n independent trials, and the number of distinct ways to order these successes and failures is $\binom{n}{k} = \frac{n!}{k!(n-k)!}$.

Example 10.11. Observe that in the case of flipping a coin twice (with p.m.f. given in Example 10.10),

$$p_X(0) = \frac{1}{4} = \binom{2}{0}(0.5)^0(1-0.5)^2,$$

$$p_X(1) = \frac{1}{2} = \binom{2}{1}(0.5)^1(1-0.5)^1,$$

$$p_X(2) = \frac{1}{4} = \binom{2}{2}(0.5)^2(1-0.5)^0,$$

so $X \sim \text{Binomial}(n=2, p=1/2)$. In other words, X counts the number of heads obtained in 2 independent trials where the head probability is $p = 1/2$.

Example 10.12. *Transmission error.* A string of one hundred 1's and 0's is used to encode some communication for transmission. Suppose the transmission process is such that each character in the string has a 1% chance of being represented incorrectly, independent of the others. What is the probability that more than one character in the string is incorrect?

Solution: Let X = the number of incorrect characters. Then $X \sim$ Binomial($n = 100, p = .01$), and the quantity of interest is

$$P(X > 1) = 1 - p_X(0) - p_X(1)$$
$$= 1 - .99^{100} - 100\,(.01)^1\,(.99)^{99}$$
$$\approx 0.264.$$

10.2.5.2 Hypergeometric(m, n, N)

Another regularly encountered scenario involves the idea of selecting objects from a finite collection without replacement, where each object is classified *a priori* as belonging to one of two groups. For example, the classic 'simple urn' model consists of a container containing $m + n$ marbles, m of which are green and n of which are red; the experiment draws N marbles at random without replacement and records the number of green marbles drawn, k. A random variable X is said to have a *Hypergeometric(m, n, N) distribution* if

$$p_X(k) = \frac{\binom{m}{k}\binom{n}{N-k}}{\binom{m+n}{N}} \qquad (10.5)$$

for $k \in \{\max(0, N-n), \ldots, \min(m, N)\}$; $p_X(k) = 0$ otherwise. Conceptually, this p.m.f. can be expressed as

$$P(k \text{ green balls are drawn}) = \frac{\begin{pmatrix}\text{the number of}\\\text{ways to draw}\\k\text{ green balls}\\\text{from } m \text{ of}\\\text{them}\end{pmatrix}\begin{pmatrix}\text{the number of}\\\text{ways to draw}\\N-k \text{ red balls}\\\text{from } n \text{ of them}\end{pmatrix}}{\begin{pmatrix}\text{the number of ways}\\\text{to draw } N \text{ total balls}\\\text{from } m+n \text{ of them}\end{pmatrix}}. \qquad (10.6)$$

Example 10.13. *Network vulnerability.* Suppose a network consists of 25 computers, 10 of which are vulnerable to cyber attack. An attacker attacks 5 different computers, selected at random. What is the probability at least three vulnerable computers are attacked?

Solution: Let X = the number of vulnerable computers attacked. Then $X \sim$ Hypergeometric($m = 10, n = 15, N = 5$), and the quantity of interest is

$$P(X \geq 3) = p_X(3) + p_X(4) + p_X(5)$$
$$= \frac{\binom{10}{3}\binom{15}{2} + \binom{10}{4}\binom{15}{1} + \binom{10}{5}\binom{15}{0}}{\binom{25}{5}}$$
$$\approx 0.301.$$

10.2.5.3 Geometric(p)

The distribution families introduced so far all have finite support; for example, the number of successes on 10 attempts is an integer between 0 and 10 (i.e., a Binomial($n = 10, p$) random variable). Consider the case where Bernoulli trials with success probability p are repeated until a first success is achieved, and the trial of the first success is recorded. A random variable X is said to have a *Geometric(p)* distribution if

$$p_X(k) = p(1-p)^{k-1} \tag{10.7}$$

for $k \in \{1, 2, \ldots\}$; $p_X(k) = 0$ otherwise. This p.m.f. provides the probability of experiencing $k-1$ consecutive failures, $(1-p)^{k-1}$, multiplied by the probability of a single success on the last trial, p.

Example 10.14. *PIN vulnerability.* There are 10,000 ($= 10^4$) distinct 4-digit Personal Identification Numbers (PINs): 0000, 0001, ..., 9999. Suppose a hacker has breached an online system where users enter a 4-digit PIN to access an account, and the hacker need only correctly guess the user's PIN in order to gain account access. If each new user is equally likely to have any allowable PIN, and the hacker gets one opportunity to guess each user's PIN, what is the probability the hacker first correctly guesses a single PIN within 1000 tries?

Solution: Let $X =$ the trial on which the hacker first correctly guesses a PIN. Then $X \sim$ Geometric($p = 1/1000$), and the quantity of interest is

$$P(X \le 1000) = p_X(1) + p_X(2) + \cdots + p_X(1000)$$
$$= \frac{1}{1000}\left(\frac{999}{1000}\right)^0 + \frac{1}{1000}\left(\frac{999}{1000}\right)^1 + \cdots + \frac{1}{1000}\left(\frac{999}{1000}\right)^{999}$$
$$\approx 0.632.$$

Note that this same probability could be computed alternatively as follows:

$$P(X \le 1000) = 1 - P(X > 1000)$$
$$= 1 - P(\text{the first 1000 guesses are incorrect})$$
$$= 1 - (999/1000)^{1000}$$
$$\approx 0.632.$$

10.2.5.4 Negative Binomial(r, p)

Using the same framework as the Geometric distribution, where independent trials are attempted in succession with success probability p, one might be interested not in how many tries it takes until the first success is achieved, but rather in the number of tries until the rth success is achieved. A random variable X is said to have a *Negative Binomial(r, p) distribution* if

$$p_X(k) = \binom{k-1}{r-1} p^r (1-p)^{k-r} \tag{10.8}$$

for $k \in \{r, r+1, \ldots\}$ and $r \in \{1, 2, \ldots\}$; $p_X(k) = 0$ otherwise. This p.m.f. provides the probability of experiencing exactly $r-1$ successes on $k-1$ tries, which is $\binom{k-1}{r-1} p^{r-1}(1-p)^{k-r}$, multiplied by the probability of a single success on the kth (and final) trial, p. Note that the Geometric(p) distribution is a special case of the Negative Binomial(r, p) distribution, with $r = 1$.

Example 10.15. *PIN vulnerability, revisited.* Under the same conditions as Example 10.14, what is the probability the hacker first correctly guesses three PINs within 1000 tries?

Solution: Let $X =$ the trial on which the hacker first correctly guesses three PINs. Then $X \sim$ Negative Binomial($r = 3, p = 1/1000$), and the quantity of interest is

$$P(X \leq 1000) = p_X(3) + p_X(4) + \cdots + p_X(1000)$$
$$= \binom{2}{2}\left(\frac{1}{1000}\right)^3\left(\frac{999}{1000}\right)^0 + \binom{3}{2}\left(\frac{1}{1000}\right)^3\left(\frac{999}{1000}\right)^1$$
$$+ \cdots + \binom{999}{2}\left(\frac{1}{1000}\right)^3\left(\frac{999}{1000}\right)^{997}$$
$$\approx 0.080.$$

10.2.5.5 Poisson(λ)

In many situations of interest, one observes a phenomenon that occurs at random with some average rate; for example, customers may arrive at an automatic teller machine (ATM) at an average rate of 5 per hour, or severe earthquakes may occur in a certain region at an average rate of 1 every 2 years. These occurrences are generally referred to as 'arrivals' with average rate of occurrence (or 'arrival rate') λ. Under certain conditions, such randomness can be well-modeled using a *Poisson* random variable. A random variable X is said to have a *Poisson(λ) distribution* if

$$p_X(k) = \frac{e^{-\lambda} \lambda^k}{k!} \tag{10.9}$$

for $k \in \{0, 1, \ldots\}$; $p_X(k) = 0$ otherwise.

Example 10.16. *Network attacks.* Suppose a particular network experiences attacks of some form at an average rate of 15 attacks every 24-hour period. Use the Poisson distribution to find the probability that no more than 10 attacks occur in the next 24 hours.

Solution: Let $X =$ the number of attacks in the next 24 hours. Then $X \sim$ Poisson($\lambda = 15$), and the quantity of interest is

$$P(X \leq 10) = p_X(0) + p_X(1) + \cdots + p_X(10)$$
$$= \frac{e^{-15} 15^0}{0!} + \frac{e^{-15} 15^1}{1!} + \cdots + \frac{e^{-15} 15^{10}}{10!}$$
$$\approx 0.118.$$

The Poisson distribution is closely related to the Binomial distribution in the following way: Divide the interval over which Poisson arrivals occur into n subintervals (where n is very large), and suppose a single 'arrival' occurs on each subinterval with probability $p = \lambda/n$, with each arrival occurring independent of others. Let X = the number of arrivals on the entire interval. With respect to the entire interval, $X \sim \text{Poisson}(\lambda)$ with its associated p.m.f.; however, with respect to the n subintervals, $X \sim \text{Binomial}(n, p = \lambda/n)$ with p.m.f.

$$p_X(k) = \binom{n}{k} \left(\frac{\lambda}{n}\right)^k \left(1 - \frac{\lambda}{n}\right)^{n-k}. \tag{10.10}$$

If one were to model Example 10.16 using $n = 1000$ subintervals and success probability $p = 15/1000$, the resulting solution would be

$$P(X \le 10) = P(X = 0) + P(X = 1) + \cdots + P(X = 10)$$
$$= \binom{1000}{0} \left(\frac{15}{1000}\right)^0 \left(1 - \frac{15}{1000}\right)^{1000}$$
$$+ \binom{1000}{1} \left(\frac{15}{1000}\right)^1 \left(1 - \frac{15}{1000}\right)^{999}$$
$$+ \cdots + \binom{1000}{10} \left(\frac{15}{1000}\right)^{10} \left(1 - \frac{15}{1000}\right)^0$$
$$\approx 0.117,$$

which is quite close to the Poisson solution. In fact, it can be shown that the Poisson(λ) p.m.f. computes the pointwise limit of the Binomial($n, p = \lambda/n$) p.m.f. as $n \to \infty$ and $p \to 0$, with $np = \lambda$ held constant. A proof is beyond the scope of this chapter, but we include the result here to highlight that the Poisson approximation to the Binomial can be useful computationally in circumstances where n is large and p is small.

10.2.6 Continuous Distributions

While discrete distributions are used to assign probabilities when an outcome space is countable, a different approach is needed for uncountable sample spaces. For example, one might be interested in how long the wait will be until some next random event occurs, where time is measured on a continuous scale. In such cases, *continuous random variables* are used.

Definition 10.8. X is a *continuous random variable* if the quantity $P(a \le X \le b)$ can be expressed as

$$P(a \le X \le b) = \int_a^b f_X(x)\, dx$$

for some real-valued, non-negative function f_X and for all real numbers a and b with $a < b$. In such case, f_X is called the *probability density function* (or *p.d.f.*) for X.

For a continuous random variable, the set of values x such that $f_X(x) > 0$ is called the *support* of X. All p.d.f.'s have the properties that $f_X(x) \geq 0$ for all x and $\int_{-\infty}^{\infty} f_X(x)\,dx = 1$. We also note here the technical detail that for a continuous random variable, $P(X = a) = 0$ for all real numbers a since

$$P(X = a) = P(a \leq X \leq a) = \int_a^a f_X(x)\,dx = 0. \tag{10.11}$$

Resultingly, one can interchange the use of $<$ and $\leq$ signs when computing probabilities for continuous random variables. (This may <u>not</u> be done in general for discrete random variables.)

One can think of the p.d.f. for X as a function that expresses how probability is 'concentrated' around each value of x (hence the word 'density'). So, if $f_X(x)$ is relatively large for all x on the interval (a, b), then the probability that X will take a value on the interval (a, b) is relatively high compared to other intervals.

The definition for cumulative distribution function (c.d.f.) is unchanged from Definition 10.7; that is, $F_X(x) = P(X \leq x)$. However, for the continuous random variable case c.d.f. values are computed by

$$F_X(a) = \int_{-\infty}^a f_X(x)\,dx \tag{10.12}$$

for all real numbers a. Note then that by the properties of integrals

$$P(a \leq X \leq b) = \int_a^b f_X(x)\,dx = F_X(b) - F_X(a) \tag{10.13}$$

and by the Fundamental Theorem of Calculus, $f_X(x) = F'_X(x)$.

Figure 10.4 shows a generic p.d.f. alongside its associated c.d.f. Notice that the c.d.f. is steeper where the p.d.f. is higher, which is similar to the relationship between the p.m.f. and c.d.f. for a discrete random variable.

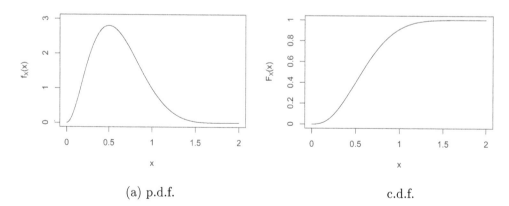

(a) p.d.f.

c.d.f.

FIGURE 10.4 A generic probability density function and its associated cumulative distribution function.

10.2.6.1 Uniform(a, b)

The most basic continuous distribution is the *Uniform distribution*, in which probability density is evenly distributed across some interval (a, b). A random variable X is said to have a *Uniform(a, b) distribution* if

$$f_X(x) = \frac{1}{b-a} \tag{10.14}$$

for $a < x < b$; $f_X(x) = 0$ otherwise. Drawing from the Uniform distribution models the notion of drawing a number 'completely at random' from some interval.

Example 10.17. Let $X \sim \text{Uniform}(a = 0, b = 1)$. Find $P(1/4 < X < 3/4)$.

Solution:

$$P(1/4 < X < 3/4) = \int_{1/4}^{3/4} \frac{1}{1-0} \, dx = 3/4 - 1/4 = 1/2.$$

This result illustrates the general notion that Uniform probabilities are proportionate to sub-interval length.

10.2.6.2 Normal(μ, σ)

Perhaps the most widely-encountered continuous probability distribution is the *Normal distribution*. A random variable X is said to have a *Normal(μ, σ) distribution* if

$$f_X(x) = \frac{1}{\sigma\sqrt{2\pi}} e^{-\frac{1}{2}\left(\frac{x-\mu}{\sigma}\right)^2} \tag{10.15}$$

for $-\infty < x < \infty$. The shape of this density function is the classic bell-shaped curve, symmetric around $x = \mu$. Many natural random phenomena follow this distribution (approximately), and it is also the limiting distribution for several other well-known distributions. Furthermore, a result called the *Central Limit Theorem* says that the sum of a large number of independent and identically distributed random variables has a distribution that is approximately Normal. *This is one of the most remarkable and important results in all of probability theory.*

The parameters μ and σ can be interpreted as the *mean* and *standard deviation* of the Normal distribution; these concepts are explained in Section 10.2.7. Unfortunately, the Normal p.d.f. has no closed-form antiderivative, and so there is no closed form for its c.d.f. by which to compute exact probabilities. Instead, tables are readily available with approximate probabilities for the Normal distribution, and computer-based applications that compute Normal probabilities are commonplace. It is helpful (and necessary if using probability tables) to standardize a Normal random variable as follows:

$$X \sim \text{Normal}(\mu, \sigma) \Rightarrow Z = \frac{X - \mu}{\sigma} \sim \text{Normal}(0, 1). \tag{10.16}$$

The Normal(0, 1) distribution is called the *Standard Normal distribution*, and the c.d.f. for Z is commonly denoted as $\Phi(z)$. Values of $\Phi(z)$ for typical nonnegative values of z are given in Appendix 10.2. For negative values of z, the symmetry of the Standard Normal distribution about $z = 0$ gives

$$\Phi(z) = 1 - \Phi(-z) \quad \text{for all } -\infty < z < \infty. \tag{10.17}$$

Example 10.18. Let $X \sim \text{Normal}(\mu = 1, \sigma = 4)$. Find $P(0 < X < 4)$.

Solution: Let $Z = \frac{X-\mu}{\sigma} = \frac{X-1}{4}$. Then $Z \sim \text{Normal}(0, 1)$ and the quantity of interest is

$$P(0 < X < 4) = P\left(\frac{0-1}{4} < \frac{X-1}{4} < \frac{4-1}{4}\right)$$
$$= P\left(-\frac{1}{4} < Z < \frac{3}{4}\right) = P\left(Z < \frac{3}{4}\right) - P\left(Z < -\frac{1}{4}\right)$$
$$= \Phi(0.75) - \Phi(-0.25).$$

Appendix 10.2 gives

$$\Phi(0.75) \approx 0.7734 \text{ (in row '0.7' and column '.05')}$$

and

$$\Phi(0.25) \approx 0.5987 \text{ (in row '0.2' and column '.05')}$$

so $\Phi(-0.25) \approx 1 - 0.5987 = 0.4013$, and the final result is $P(0 < X < 4) \approx 0.7734 - 0.4013 = 0.3721$.

10.2.6.3 Exponential(λ)

A distribution often used to model the time between the occurrence of two events is the *Exponential distribution*. A random variable X is said to have an *Exponential*(λ) *distribution* if

$$f_X(x) = \lambda e^{-\lambda x} \tag{10.18}$$

for $x > 0$; $f_X(x) = 0$ otherwise. In the context of waiting times, the parameter λ represents the average rate at which phenomena of interest occur. As λ increases, an increasing amount of the area under f (i.e., probability) is associated with smaller values of x.

Example 10.19. *Network attacks, continued.* Consider the network attack scenario in Example 10.16. Assume the time until the first attack of the day occurs has an Exponential distribution. What is the probability on a given day that the first attack occurs after two hours have elapsed?

Solution: Let T = the number of attacks in 24 hours. Then $T \sim \text{Exponential}(\lambda = 15/24)$, and the quantity of interest is

$$P(T > 2) = 1 - P(T \leq 2)$$
$$= 1 - \int_0^2 \frac{15}{24} e^{-15x/24} \, dx$$
$$= 1 - \left(1 - e^{-15(2)/24}\right) \approx 0.287.$$

An interesting relationship exists between the Poisson discrete distribution and the Exponential continuous distribution. Events occurring according to a Poisson process will have the number of event occurrences modeled as a Poisson distribution with parameter λ with the inter-arrival times of those events modeled by the Exponential distribution with parameter λ.

10.2.6.4 Gamma(α, λ)

The Exponential distribution is a special case of the *Gamma distribution*. A random variable X is said to have a *Gamma(α, λ) distribution* if

$$f_X(x) = \frac{\lambda^\alpha}{\Gamma(\alpha)} x^{\alpha-1} e^{-\lambda x} \tag{10.19}$$

for $x > 0$; $f_X(x) = 0$ otherwise. Notice that if $X \sim$ Gamma($\alpha = 1, \lambda$) then $X \sim$ Exponential(λ). One of its many applications is that the Gamma distribution can be used to model the time until the αth event occurs when events occur at an average rate λ and α is a positive integer (in which case $\Gamma(\alpha) = (\alpha - 1)!$).

Example 10.20. *Network attacks, continued again.* Using the same assumptions of Example 10.19, what is the probability on a given day that the third attack occurs after two hours?

Solution: Let $T =$ the number of attacks in 24 hours. Then $T \sim$ Gamma($\alpha = 3, \lambda = 15/24$), and the quantity of interest is

$$P(T > 2) = 1 - P(T \leq 2)$$

$$= 1 - \int_0^2 \frac{(15/24)^3}{2!} x^2 e^{-15x/24} \, dx$$

$$= 1 - \left(1 - \frac{388}{128} e^{-15(2)/24}\right) \approx 0.868.$$

While the integration by parts required to exactly compute the definite integral above would be cumbersome for larger α, probabilities for all the standard distributions discussed in this chapter (and indeed for many others) are easily computed using any standard statistical software application.

10.2.6.5 Chi-squared(ν)

The last distribution in this section is another special case of the Gamma distribution, the *Chi-squared distribution*. A random variable X is said to have a *Chi-squared(ν) distribution* if

$$f_X(x) = \frac{1}{2^{\nu/2} \Gamma(\nu/2)} x^{\frac{\nu}{2}-1} e^{-\frac{1}{2}x} \tag{10.20}$$

for $x > 0$; $f_X(x) = 0$ otherwise. Notice that if $X \sim$ Gamma($\alpha = \nu/2, \lambda = 1/2$) then $X \sim$ Chi-squared(ν). The Chi-squared distribution is widely used in statistical inference and also is used to model probabilities associated with sums of independent squared Normal random variables (such as the distribution of sample variance).

10.2.7 Expectation and Variance

Our discussion up to this point has explored a variety of ways in which probability values can be assigned to events in a sample space. Two important attributes of the probability distribution of a random variable X are its *location*, which provides a measure of the distribution's *center*, and its *dispersion*, which provides a measure of the distribution's *spread*.

10.2.7.1 Expectation

If X is a random variable with support A, the most common measure for the location of the distribution of X is the *expected value of X*, defined as

$$\mu_X = \mathrm{E}[X] = \sum_{k \in A} k \, p_X(k) \tag{10.21}$$

if X is discrete with p.m.f. p_X and this sum exists, and

$$\mu_X = \mathrm{E}[X] = \int_{x \in A} x \, f_X(x) \, dx \tag{10.22}$$

if X is continuous with p.d.f. f_X and this integral exists. The expected value of X is sometimes called the *mean of X*; hence the use of the Greek letter μ. (Note: In this chapter we will subscript μ with the random variable to which it refers; in many cases when a random variable is the only one of interest, μ_X is written simply as μ.)

Expected value in probability is mathematically equivalent to the standard definition for center of mass in a physical setting; in the probability case the 'mass' of interest is probability mass. The phrase 'expected' connotes that $\mathrm{E}[X]$ is representative of a typical value X might take (although it is not necessary that $\mathrm{E}[X]$ is in the support of X).

Example 10.21. Let X be a discrete random variable with p.m.f. $p_X(k) = k/6$, $k \in \{1, 2, 3\}$. Find μ_X.

Solution:
$$\mu_X = 1 \, p_X(1) + 2 \, p_X(2) + 3 \, p_X(3) = 7/3.$$

In this case, X is twice as likely to take on the value 2 than 1, and three times as likely to take on the value 3 than 1, so $\mu = 7/3$ is representative of a typical value (i.e., close to but larger than 2). $\mathrm{E}[X]$ can also be thought of as the long-run average value of observed X's over many random draws of X.

Example 10.22. Let X be a continuous random variable with p.d.f.

$$f_X(x) = \begin{cases} e^{-x}, & x > 0; \\ 0, & \text{otherwise.} \end{cases}$$

Find μ_X.

Solution:
$$\mu_X = \int_0^\infty x\, e^{-x}\, dx = 1.$$

Note that if X is a random variable and g is a real-valued function then we can consider the random variable $Y = g(X)$. A useful property is that

$$\mu_Y = \mathrm{E}[g(X)] = \begin{cases} \sum_{k \in A} g(k)\, p_X(k) & \text{for discrete } X; \\ \int_{x \in A} g(x)\, f_X(x)\, dx & \text{for continuous } X. \end{cases} \quad (10.23)$$

The linearity property for expectation follows immediately; that is, for any constants a and b,

$$\mathrm{E}[aX + b] = a\mathrm{E}[X] + b. \quad (10.24)$$

10.2.7.2 Variance

To measure the dispersion of the distribution of X, one is interested in the expected value of some deviation of X from μ_X. The most common measure is the *variance of X*, defined as

$$\sigma_X^2 = \mathrm{Var}[X] = \mathrm{E}\bigl[(X - \mu_X)^2\bigr]. \quad (10.25)$$

Note that σ_X^2 is non-negative since the random variable $(X - \mu_X)^2$ never takes negative values. Also, since the variance of X measures the square of the expected deviation of X from μ_X, one often expresses dispersion in terms of $\sigma_X = \sqrt{\sigma_X^2}$. This is called the *standard deviation of X*; hence the use of the Greek letter σ. Figure 10.5 illustrates expected value and variance for some generic p.d.f.'s. In Figure 10.5(a), the mean of the solid-lined distribution is less than the mean of the dashed-lined distribution (while the standard deviations are equal); in Figure 10.5(b), the standard deviation of the solid-lined distribution is less than the standard deviation of the dashed-lined distribution (while the means are equal).

A helpful relationship to compute variance follows directly from the linearity of expectation:

$$\begin{aligned}
\sigma_X^2 &= \mathrm{E}\bigl[(X - \mu_X)^2\bigr] = \mathrm{E}\bigl[X^2 - 2\mu_X X + \mu_X^2\bigr] \\
&= \mathrm{E}\bigl[X^2\bigr] - \mathrm{E}[2\mu_X X] + \mathrm{E}\bigl[\mu_X^2\bigr] \\
&= \mathrm{E}\bigl[X^2\bigr] - 2\mu_X \mathrm{E}[X] + \mu_X^2 \\
&= \mathrm{E}\bigl[X^2\bigr] - 2\mu_X^2 + \mu_X^2 \\
&= \mathrm{E}\bigl[X^2\bigr] - \mu_X^2.
\end{aligned}$$

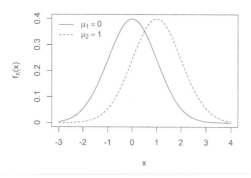
(a) Different mean, same variance

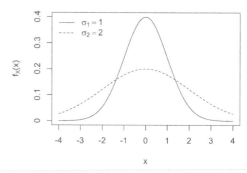
(b) Same mean, different variance

FIGURE 10.5   Effects of differing mean and variance on a p.d.f.

Example 10.23. For Example 10.21, find σ_X.

Solution: $E[X^2] = 1^2 \, p_X(1) + 2^2 \, p_X(2) + 3^2 \, p_X(3) = 6$ and $\mu_X^2 = (7/3)^2 = 49/9$, so

$$\sigma_X^2 = E\left[X^2\right] - \mu_X^2 = 6 - 49/9 = 5/9 \Rightarrow \sigma_X = \sqrt{5}/3.$$

10.3   PROBABILITY MODELS IN CYBER RESEARCH

This section highlights just a few of the many probability applications in cyber research, including Bayes' rule, Markov chains, and information entropy.

10.3.1   Bayes' Rule

An extremely useful probability result for a wide range of problems is *Bayes' Rule*, used to compute conditional probabilities. Bayes' Rule states that for any events $E_1, E_2, \ldots, E_n$, and F,

$$\begin{aligned} P(E_i|F) &= \frac{P(F|E_i)P(E_i)}{P(F)} \\ &= \frac{P(F|E_i)P(E_i)}{P(F|E_1)P(E_1) + P(F|E_2)P(E_2) + \cdots + P(F|E_n)P(E_n)} \end{aligned} \quad (10.26)$$

for all $i \in \{1, 2, \ldots, n\}$. The expansion of the denominator

$$P(F) = P(F|E_1)P(E_1) + P(F|E_2)P(E_2) + \cdots + P(F|E_n)P(E_n) \quad (10.27)$$

is often referred to as the *Law of Total Probability*. The unconditional probability $P(E_i)$ is called the *prior probability*, and the conditional probability $P(E_i|F)$ is called the *posterior probability*. This result is encountered in a variety of cybersecurity applications, either in direct use (as in the following example) or in Bayesian statistical models.

Example 10.24. *Attack pathways.* Suppose a network intrusion can occur via four distinct 'attack pathways:' (i) physical access, (ii) remote and access control, (iii) wireless access, and (iv) web application, with the probability of attacks along these pathways being 0.2, 0.4, 0.3, and 0.1, respectively. An attempted attack along each pathway is successful with probabilities 0.1, 0.05, 0.25, and 0.05. If a successful network intrusion just occurred, what is the probability that the pathway was wireless access?

Solution: Let E_1, E_2, E_3, and E_4, be the events associated with attacks along each respective pathway, and let F be the event 'an attempted attack was successful.' Then the known probabilities are

$$P(E_1) = 0.2, \quad P(F|E_1) = 0.1;$$
$$P(E_2) = 0.4, \quad P(F|E_2) = 0.05;$$
$$P(E_3) = 0.3, \quad P(F|E_3) = 0.25;$$
$$P(E_4) = 0.1, \quad P(F|E_4) = 0.05.$$

So, Bayes' Rule gives

$$P(E_3|F) = \frac{P(F|E_3)P(E_3)}{P(F|E_1)P(E_1) + P(F|E_2)P(E_2) + P(F|E_3)P(E_3) + P(F|E_4)P(E_4)}$$
$$= \frac{0.25(0.3)}{0.1(0.2) + 0.05(0.4) + 0.25(0.3) + 0.05(0.1)}$$
$$= 0.625.$$

Notice that the posterior probability $P(E_3|F) = 0.625$ is higher than the prior probability $P(E_3) = 0.3$. This makes sense since the probability of a successful attack along the wireless access path, $P(F|E_3) = 0.25$, is much higher than the probabilities of successful attacks along the other paths.

10.3.2 Markov Chains

A *Markov model* is a probability model that describes the behavior of a system that is characterized by discrete states and transitions between states. Events consist of transitioning from one state to another, and the probability of moving to some new state depends only on the current state. Formally, this model includes:

- a state space $\mathcal{M} = \{1, \ldots, m\}$;
- random variables $X_n, n \in= \{0, 1, \ldots\}$, each taking values in $\mathcal{M}$;
- transition probabilities $p_{ij} = P(X_{n+1} = j | X_n = i), i, j \in \mathcal{M}$, which are assumed independent of n.

The sequence of random variables $X_1, X_2, \ldots$ is called a *discrete-time Markov chain* (or simply just a *Markov chain*). The property that future state probabilities depend only on present state is called the *Markov property* or *memoryless property*. This property allows the computation of virtually all probabilities of interest associated with state random variables only knowing transition state probabilities and initial state probability. For example, the p.m.f. for the joint event $\{X_0 = i_0, X_1 = i_1, \ldots, X_k = i_k\}$ can be written, using conditional probability and the memoryless property, as

$$P(X_k = i_k, X_{k-1} = i_{k-1}, \ldots, X_0 = i_0)$$
$$= P(X_k = i_k | X_{k-1} = i_{k-1}, \ldots, X_0 = i_0) P(X_{k-1} = i_{k-1}, \ldots, X_0 = i_0)$$
$$= P(X_k = i_k | X_{k-1} = i_{k-1}) P(X_{k-1} = i_{k-1}, \ldots, X_0 = i_0)$$
$$= p_{i_{k-1} i_k} P(X_{k-1} = i_{k-1}, \ldots, X_0 = i_0)$$
$$= \ldots = p_{i_{k-1} i_k} p_{i_{k-2} i_{k-1}} \cdots p_{i_0 i_1} P(X_0 = i_0). \tag{10.28}$$

One can organize the transition probabilities for a Markov chain into a *one-step transition matrix*

$$\mathbf{P} = \begin{pmatrix} p_{11} & p_{12} & \cdots & p_{1m} \\ p_{21} & p_{22} & \cdots & p_{2m} \\ \vdots & \vdots & \ddots & \vdots \\ p_{m1} & p_{m2} & \cdots & p_{mm} \end{pmatrix}, \tag{10.29}$$

and it can be shown by straightforward linear algebra that $P(X_k = j | X_0 = i)$, annotated $\mathbf{P}^k_{i,j}$, is simply the (i, j)-entry of $\mathbf{P}^k$ for $k \in \{1, 2, \ldots\}$, where $\mathbf{P}^k$ is the matrix product of $\mathbf{P}$ taken k times. So then, the matrix $\mathbf{P}^k$ can be thought of as the k-step transition matrix for the Markov chain.

Example 10.25. *Network infection.* Consider a network system (that includes an e-mail system) subject to hacking that consists of state space $\mathcal{M} = \{1, 2, 3, 4\}$ where the states are: (i) 'clean,' (ii) 'virus in the system with risk of infection,' (iii) 'system infected by virus but no critical harm yet done,' and (iv) 'system infected by virus and critical harm done.' Suppose transitions from state to state occur with the Markov property over a succession of discrete time periods, where exactly one transition (possibly to the same state) happens during each period with probabilities as follows:

$p_{11} = 0.95$ (no malicious activity);

$p_{12} = 0.05$ (e-mail containing linked virus is received but not yet opened);

$p_{21} = 0.80$ (virus-linked e-mail deleted without opening link);

$p_{22} = 0.15$ (no action taken on virus-linked e-mail);

$p_{23} = 0.05$ (virus link opened, causing a system infection);

$p_{31} = 0.30$ (infected system restored through virus clean-up operations);

$p_{32} = 0.40$ (infected system restored to risk-of-infection state through appropriate quarantine action);

$p_{33} = 0.10$ (no action taken on infected system);

$p_{34} = 0.20$ (infected system critically harmed);

$p_{44} = 1.00$ ('absorption' in a critically harmed state);

$p_{ij} = 0$ (otherwise).

What is the probability that a clean system reaches a critically harmed state within 10 discrete time periods?

Solution: The one-step transition matrix is

$$\mathbf{P} = \begin{pmatrix} 0.95 & 0.05 & 0 & 0 \\ 0.80 & 0.15 & 0.05 & 0 \\ 0.30 & 0.40 & 0.10 & 0.20 \\ 0 & 0 & 0 & 1 \end{pmatrix},$$

so

$$\mathbf{P}^{10} = \begin{pmatrix} 0.95 & 0.05 & 0 & 0 \\ 0.80 & 0.15 & 0.05 & 0 \\ 0.30 & 0.40 & 0.10 & 0.20 \\ 0 & 0 & 0 & 1 \end{pmatrix}^{10} \approx \begin{pmatrix} 0.9354 & 0.0565 & 0.0031 & 0.0049 \\ 0.9236 & 0.0558 & 0.0031 & 0.0174 \\ 0.7228 & 0.0437 & 0.0024 & 0.2311 \\ 0 & 0 & 0 & 1 \end{pmatrix}.$$

Therefore, $P(X_{10} = 4 | X_1 = 1) = \mathbf{P}^{10}_{1,4} \approx 0.0049$.

The concept of a Markov chain can be extended to continuous time as a *continuous-time Markov chain*; both discrete- and continuous-time Markov chains are employed in a wide variety of cybersecurity applications.

10.3.3 Information Entropy

Yet another important use of probability in cyber applications involves the idea of *information entropy*, or simply *entropy*. Entropy is especially useful in cryptographic applications where the goal is to identify randomness (or lack thereof) in some data. A primary building block for the concept of entropy is *surprise*; that is, if an event E occurs, how surprised are we by its occurrence? A common measure of surprise for event E with $p = P(E)$ is

$$S(p) = \log_2(1/p) = -\log_2 p. \tag{10.30}$$

For example, if $p = 1$ then the surprise when E occurs is $S(1) = 0$. As p decreases from 1 toward 0, $S(p)$ increases without bound. If X is a discrete random variable that takes on values $x_1, x_2, \ldots, x_n$ with probabilities $p_1, p_2, \ldots, p_n > 0$ then the entropy of X, $H(X)$, is defined to be the average surprise experienced upon learning the value of X; that is,

$$H(X) = -\sum_{i=1}^{n} p_i \log_2 p_i. \tag{10.31}$$

Entropy has extensive applications in coding theory. A related idea is *relative entropy*, which is a measure of how different one probability distribution is from another. The *relative entropy* of probability vector $\mathbf{p} = (p_1, p_2, \ldots, p_n)$ to probability vector $\mathbf{q} = (q_1, q_2, \ldots, q_n)$, where all probabilities are nonzero, is defined to be

$$D(\mathbf{p}, \mathbf{q}) = -\sum_{i=1}^{n} p_i \log_2(q_i/p_i). \tag{10.32}$$

Note that if $\mathbf{p} = \mathbf{q}$, then $D(\mathbf{p}, \mathbf{q}) = 0$.

Example 10.26. *Detecting malicious domain names.* Suppose all internet domain names consist of only the characters 'a,' 'b,' 'c,' or 'd,' and suppose also that these letters occur with the frequencies listed in the table below for 'legitimate' and 'malicious' domain names. You observe a sample of domain names from a source that is either legitimate or malicious; the letter frequencies for the sample are also in Table 10.1 below.

TABLE 10.1 Letter frequencies for domain names

Domain names	Characters			
	a	b	c	d
Legitimate	0.10	0.20	0.30	0.40
Malicious	0.25	0.25	0.25	0.25
Observed	0.20	0.30	0.20	0.30

Find whether the relative entropy is higher for the observed sample relative to the expected frequencies for legitimate or malicious domain names.

Solution: Let **p** and **q** be the character probabilities for legitimate and malicious domain names, respectively, and let **r** be the estimated character probabilities from the observed sample. Then

$$D(\mathbf{r},\mathbf{p}) = -0.2\log_2(0.1/0.2) - 0.3\log_2(0.2/0.3) - 0.2\log_2(0.3/0.2)$$
$$- 0.3\log_2(0.4/0.3)$$
$$\approx 0.134$$

and

$$D(\mathbf{r},\mathbf{q}) = -0.2\log_2(0.25/0.2) - 0.3\log_2(0.25/0.3) - 0.2\log_2(0.25/0.2)$$
$$- 0.3\log_2(0.25/0.3)$$
$$\approx 0.029.$$

Since $D(\mathbf{r},\mathbf{q}) < D(\mathbf{r},\mathbf{p})$ (i.e., the observed frequencies are 'closer' to the malicious distribution than to the legitimate distribution), one might suspect the observed batch of domain names are from a malicious source.

10.4 CONCLUSIONS

Probability theory is a vital field of study in mathematics which has abundant applicability to problems in cyber research. Basic notions of sample spaces, events, and probability measure undergird every problem that involves uncertainty. Ideas such as discrete and continuous random variables are very convenient for modeling the uncertainty that arises in frequently encountered settings. Extensions of these concepts, such as Bayes' Rule, Markov chains, and information entropy find many uses in the realm of cyber research and in other areas of study as well. While this chapter introduces probability fundamentals, the field of probability is rich and well-documented; the interested reader is encouraged to explore these ideas in references that provide a more full treatment of the subject. For example, Ross (2019) and Blitzstein and Hwang (2019) are excellent introductions to probability theory for application to any field.

For cyber-specific applications, Edgar and Manz (2017) provide a useful overview of mathematical research methods in cyber research. Liu et al. (2019) and Gore et al. (2017) describe approaches to cybersecurity using Markov chains, and Dudorov et al. (2013) employ probability extensively to model the risk of attack from cyber-crime related malware. Information theory, a heavily probabilistic area of study, will continue to play a prominent role as an application area for cyber. A host of challenges in the realm of secure communications and data system management with privacy assurance are approached using an information-theoretic framework. Specific examples include: securing mobile links in a wireless system, ensuring privacy for biometric data, managing online data repositories, and safeguarding modern power grids that monitor usage data in real time. Schaefer et al. (2017) is a helpful reference to learn more about this emerging field. Other references pertaining to probability applications in cyber research abound.

REFERENCES

Blitzstein, J.K. and Hwang, J. (2019). *Introduction to Probability (Second Edition)*. Boca Raton, FL: CRC Press.

Dudorov, D., Stuples, D., and Newby, M. (2013). Probability analysis of cyber attack paths against business and commercial enterprise systems. *Proceedings of the IEEE European Intelligence and Security Informatics Conference (EISIC)*, 38–44.

Edgar, T.W. and Manz, D.O. (2017). *Research Methods for Cyber Security*. Cambridge, MA: Elsevier Inc.

Gore, R., Padilla, J., and Diallo, S. (2017). Markov Chain modeling of cyber threats. *The Journal of Defense Modeling and Simulation: Applications, Methodology, Technology*, 14(3), 233–244.

Laplace, P.S. (1812). *A Philosophical Essay on Probabilities*, translated from the Sixth French Edition by Truscott, F.W. and Emory, F.L. London, UK: Chapman and Hall, Limited.

Liu, Q., Xing, L., and Zhou, C. (2019). Probabilistic modeling and analysis of sequential cyber-attacks. *Engineering Reports*, 1:e12065. Retrieved from https://doi.org/10.1002/eng2.12065.

Ross, S. (2019). *A First Course in Probability (Tenth Edition)*. New York, NY: Pearson.

Schaefer, R.F., Boche H., Khisti, A., and Poor, H.V. Editors. (2017) *Information Theoretic Security and Privacy of Information Systems*. Cambridge, UK: Cambridge University Press.

APPENDIX 10.1 TABLES OF COMMON DISTRIBUTIONS

10.1.1 Discrete Distributions

Distribution	Parameters	$p_X(k)$	Mean μ	Variance σ^2	Notes
Bernoulli	p	$p^k(1-p)^{1-k}$	p	$\mu(1-p)$	(1)
Binomial	n, p	$\binom{n}{k}p^k(1-p)^{n-k}$	np	$\mu(1-p)$	(1),(2)
Hypergeometric	m, n, N	$\binom{m}{k}\binom{n}{N-k}/\binom{m+n}{N}$	$Np, p=\frac{m}{m+n}$	$\mu(1-p)(\frac{m+n-N}{m+n-1})$	(3)
Geometric	p	$p(1-p)^{k-1}$	$1/p$	$\mu(1-p)/p$	(4)
Negative	r, p	$\binom{k-1}{r-1}p^r(1-p)^{k-r}$	r/p	$\mu(1-p)/p$	(4),(5)
Poisson	λ	$e^{-\lambda}\lambda^k/k!$	λ	μ	(6)

Notes:

(1) The Bernoulli distribution can model the probability of achieving a success ($X=1$) on a single trial with probability p. This is a special case of the Binomial distribution with $n=1$. Support: $k \in \{0,1\}$.

(2) The Binomial distribution can model the probability of achieving k successes on n independent trials. Support: $k \in \{0, 1, \ldots, n\}$.

(3) The Hypergeometric distribution can model the probability of drawing k green marbles from an urn containing m green marbles and n red marbles on a total draw of N marbles without replacement. Support: $k \in \{\max(0, N-n), \ldots, \min(m, N)\}$.

(4) The Geometric distribution can model the probability of first achieving success on trial k after experiencing $k-1$ consecutive failures, where all trials are independent. This is a special case of the Negative Binomial distribution with $r=1$. Support: $k \in \{1, 2, \ldots\}$.

(5) The Negative Binomial distribution can model the probability of first achieving a total of r successes on trial k, where r is a positive integer. Support: $k \in \{r, r+1, \ldots\}$.

(6) The Poisson distribution can model the probability of achieving k successes over some continuous interval of success opportunity when the expected number of successes on that interval is λ. Support: $k \in \{0, 1, \ldots\}$.

10.1.2 Continuous Distributions

Distribution	Parameters	$f_X(x)$	Mean μ	Variance σ^2	Notes
Uniform	$a < b$	$1/(b-a)$	$(a+b)/2$	$(b-a)^2/12$	(1)
Normal	μ, σ	$\frac{1}{\sigma\sqrt{2\pi}} e^{-\frac{1}{2}\left(\frac{x-\mu}{\sigma}\right)^2}$	μ	σ^2	(2)
Exponential	λ	$\lambda e^{-\lambda x}$	$1/\lambda$	μ/λ	(3)
Gamma	α, λ	$\frac{\lambda^\alpha}{\Gamma(\alpha)} x^{\alpha-1} e^{-\lambda x}$	α/λ	μ/λ	(3),(4),(5)
Chi-squared	ν	$\frac{1}{2^{\nu/2}\Gamma(\nu/2)} x^{\frac{\nu}{2}-1} e^{-\frac{1}{2}x}$	ν	2μ	(5)

Notes:

(1) The Uniform distribution can model probabilities associated with drawing a number completely at random from the continuous interval (a, b), where each number is equally likely to be on any subinterval of fixed length. Support: $x \in (a, b)$.

(2) The Normal distribution can model probabilities associated with a variety of natural phenomena and is often referred to as "the bell curve.' It is also the limiting distribution for *sample means* by the Central Limit Theorem. Support: $x \in (-\infty, \infty)$.

(3) The Exponential distribution can model probabilities associated with the time waited (in continuous time) until a first success arrives, where successes arrive at a rate of λ successes per unit time. This is a special case of the Gamma distribution with $\alpha = 1$, and is the continuous analog to the Geometric distribution. Support: $x \in (0, \infty)$.

(4) The Gamma distribution can model probabilities associated with the time waited (in continuous time) until a total of α successes arrive, where successes arrive at a rate of λ successes per unit time and α is a positive integer. (However, f_X is a valid p.d.f. even when $\alpha > 0$ is not an integer.) This is the continuous analog to the Negative Binomial distribution. Support: $x \in (0, \infty)$.

(5) The Chi-squared distribution can model probabilities associated with the sum of the squares of ν independent Standard Normal random variables and can be used to find probabilities associated with *sample variances*. This is a special case of the Gamma distribution with $\alpha = \nu/2$ and $\lambda = 1/2$. Support: $x \in (0, \infty)$.

APPENDIX 10.2 NORMAL PROBABILITIES

Area $\Phi(z)$ under the Standard Normal p.d.f. to the left of z:

z	.00	.01	.02	.03	.04	.05	.06	.07	.08	.09
0.0	0.5000	0.5040	0.5080	0.5120	0.5160	0.5199	0.5239	0.5279	0.5319	0.5359
0.1	0.5398	0.5438	0.5478	0.5517	0.5557	0.5596	0.5636	0.5675	0.5714	0.5753
0.2	0.5793	0.5832	0.5871	0.5910	0.5948	0.5987	0.6026	0.6064	0.6103	0.6141
0.3	0.6179	0.6217	0.6255	0.6293	0.6331	0.6368	0.6406	0.6443	0.6480	0.6517
0.4	0.6554	0.6591	0.6628	0.6664	0.6700	0.6736	0.6772	0.6808	0.6844	0.6879
0.5	0.6915	0.6950	0.6985	0.7019	0.7054	0.7088	0.7123	0.7157	0.7190	0.7224
0.6	0.7257	0.7291	0.7324	0.7357	0.7389	0.7422	0.7454	0.7486	0.7517	0.7549
0.7	0.7580	0.7611	0.7642	0.7673	0.7704	0.7734	0.7764	0.7794	0.7823	0.7852
0.8	0.7881	0.7910	0.7939	0.7967	0.7995	0.8023	0.8051	0.8078	0.8106	0.8133
0.9	0.8159	0.8186	0.8212	0.8238	0.8264	0.8289	0.8315	0.8340	0.8365	0.8389
1.0	0.8413	0.8438	0.8461	0.8485	0.8508	0.8531	0.8554	0.8577	0.8599	0.8621
1.1	0.8643	0.8665	0.8686	0.8708	0.8729	0.8749	0.8770	0.8790	0.8810	0.8830
1.2	0.8849	0.8869	0.8888	0.8907	0.8925	0.8944	0.8962	0.8980	0.8997	0.9015
1.3	0.9032	0.9049	0.9066	0.9082	0.9099	0.9115	0.9131	0.9147	0.9162	0.9177
1.4	0.9192	0.9207	0.9222	0.9236	0.9251	0.9265	0.9279	0.9292	0.9306	0.9319
1.5	0.9332	0.9345	0.9357	0.9370	0.9382	0.9394	0.9406	0.9418	0.9429	0.9441
1.6	0.9452	0.9463	0.9474	0.9484	0.9495	0.9505	0.9515	0.9525	0.9535	0.9545
1.7	0.9554	0.9564	0.9573	0.9582	0.9591	0.9599	0.9608	0.9616	0.9625	0.9633
1.8	0.9641	0.9649	0.9656	0.9664	0.9671	0.9678	0.9686	0.9693	0.9699	0.9706
1.9	0.9713	0.9719	0.9726	0.9732	0.9738	0.9744	0.9750	0.9756	0.9761	0.9767
2.0	0.9772	0.9778	0.9783	0.9788	0.9793	0.9798	0.9803	0.9808	0.9812	0.9817
2.1	0.9821	0.9826	0.9830	0.9834	0.9838	0.9842	0.9846	0.9850	0.9854	0.9857
2.2	0.9861	0.9864	0.9868	0.9871	0.9875	0.9878	0.9881	0.9884	0.9887	0.9890
2.3	0.9893	0.9896	0.9898	0.9901	0.9904	0.9906	0.9909	0.9911	0.9913	0.9916
2.4	0.9918	0.9920	0.9922	0.9925	0.9927	0.9929	0.9931	0.9932	0.9934	0.9936
2.5	0.9938	0.9940	0.9941	0.9943	0.9945	0.9946	0.9948	0.9949	0.9951	0.9952
2.6	0.9953	0.9955	0.9956	0.9957	0.9959	0.9960	0.9961	0.9962	0.9963	0.9964
2.7	0.9965	0.9966	0.9967	0.9968	0.9969	0.9970	0.9971	0.9972	0.9973	0.9974
2.8	0.9974	0.9975	0.9976	0.9977	0.9977	0.9978	0.9979	0.9979	0.9980	0.9981
2.9	0.9981	0.9982	0.9982	0.9983	0.9984	0.9984	0.9985	0.9985	0.9986	0.9986
3.0	0.9987	0.9987	0.9987	0.9988	0.9988	0.9989	0.9989	0.9989	0.9990	0.9990
3.1	0.9990	0.9991	0.9991	0.9991	0.9992	0.9992	0.9992	0.9992	0.9993	0.9993
3.2	0.9993	0.9993	0.9994	0.9994	0.9994	0.9994	0.9994	0.9995	0.9995	0.9995
3.3	0.9995	0.9995	0.9995	0.9996	0.9996	0.9996	0.9996	0.9996	0.9996	0.9997
3.4	0.9997	0.9997	0.9997	0.9997	0.9997	0.9997	0.9997	0.9997	0.9997	0.9998

Numerical values in Appendix 10.2 are rounded to four decimal places and are easily calculated to greater precision using statistical software. The values in this table are replicated in many probability and statistics textbooks; for example, see Ross (2019, p.204).

CHAPTER 11

Game Theory

Andrew Fielder

CONTENTS

11.1	Introduction	364
11.2	Game Theory Basics	364
	11.2.1 Representation	364
	11.2.2 Players	365
	11.2.3 Utilities	366
	11.2.4 Strategies	367
	11.2.4.1 Pure Strategies	367
	11.2.4.2 Mixed Strategies	368
	11.2.5 Dominant Strategies	369
	11.2.6 Equilibria	370
	11.2.7 Strategies in Security Games	370
11.3	Nash Games	371
11.4	Extensive Form Games	373
	11.4.1 Sub-Games	374
	11.4.2 Information Sets	376
	11.4.3 Multi-Stage Games	377
11.5	Stackelberg Games	377
	11.5.1 Strong Stackelberg Equilibria	379
	11.5.2 Weak Stackelberg Equilibria	379
11.6	Practical Operation	379
	11.6.1 Data	380
	11.6.2 Assumptions	380
	11.6.3 Uncertainty	381
	11.6.4 Timing	382
	11.6.5 Solution Design	382
	11.6.6 Scalability	383
11.7	Research Extensions	384
	11.7.1 Operational Security	384
	11.7.2 Privacy	385
	11.7.3 Machine Learning	386
11.8	Conclusions	386

DOI: 10.1201/9780429354649-11

11.1 INTRODUCTION

Decision making in the field of cybersecurity is based on the fundamental understanding of the risk that exists to users, organizations, networks, and systems (Cox, 2009). In order to make appropriate decisions on how to defend from potential attacks, there needs to be an understanding of the threats that are being faced, the impact of successful attacks, and the potential for mitigations.

However, the complexity of many cybersecurity problems makes full understanding of this risk almost impossible, as there are often properties of emergent behaviors that are difficult to capture (Armstrong et al., 2009). In this way, tools and methods for the support of decision making can help improve reasoning for delivering decisions. One such method for supporting decision making is game theory.

Game theory is a formal structured means of making decisions, where it has seen a wide range of applications since it's formal inception by von Neumann and Morgenstern in 1947. The principle of game theory is that there are a number of players competing to get the best result they can from a structured situation. In traditional economics terms, the best result might be profit maximization.

In a game, the players attempt to devise strategies that when played against other players will result in receiving this maximum reward.

Game theory can thus present a natural method for understanding cybersecurity problems. A common means of representing the dynamics of cybersecurity is defining the problem abstractly as attackers and defenders competing to control a system.

While there have been many different approaches for the utilization of game theory for providing decision support for cybersecurity, many use the attacker-defender model in some way (Roy et al., 2010).

This chapter looks to provide an overview of the use of game theory for topics related to the field of cybersecurity. First the basic concepts of game theory are discussed, before discussions of the different types of games that are used to solve cybersecurity problems. The chapter closes with a discussion of the utilization of game theory, firstly considering practical implementation and latterly known areas of application and usage.

11.2 GAME THEORY BASICS

In order to discuss the use of game theory for the purposes of cybersecurity, it is first important to introduce the basic terms used in game theory, and their relationship to cybersecurity.

11.2.1 Representation

There are two general views that are used for representing game theoretic problems: the matrix form, and the tree form. The matrix form presents the information to the decision maker as a look-up matrix, where the players strategies represent the rows and columns of the matrix, and the payoffs are given by the cell in which the strategies of the players intersect. The tree form presents the information as a

TABLE 11.1 Table of common notation

Notation	Definition
N	The set of all players
n_i	Player i
S	The set of all strategies
s_i	Pure strategy i
U	The set of all utility functions
$u_n(s_j)$	The utility function for player n to play strategy i
p_i	The probability of strategy i
σ_i	A mixed strategy for player i
t_i	Target i in a security game
a_i	Attacker action to attack target i
d_i	Defender action to protect target i

decision tree, where actions are selected by players from the root node to the any leaf node, which contains the payoff for the entire path through the decision tree.

In the examples given for two player games with an attacker and defender, the common notation used for the players are a and d respectively.

11.2.2 Players

When defining a game, the first thing that needs to be established are the set of agents that are decision makers in the game.

Game theory defines the players as a set of rational agents who take actions based on a set of available choices. Cybersecurity is often described as a two-player game, with two players, an attacker and a defender, competing for conflicting objectives.

The concept of a rational agent is that the agent or player in game theory terms will always act to their own advantage. Generally, this means acting in a way that will generate the best possible reward for the decisions made in a game. Without this rational agent, game theory does not work.

If an agent is acting to maximize their reward for taking actions, then it is feasible for other agents to predict this. It is this capability to predict actions that is fundamental to game theory. In game theory, the players are always trying to define the best reaction to what they think their opponent is going to do. If a player acts irrationally, then any strategies made by other players would not be a true.

When defining a game, there is a need to define the rules of that game. The rules are defined as the set of the moves that each player can make. Each game can have a completely unique set of moves, and each player need not have moves that reflect the moves of other players.

Classically, game theory will define a number of players, n as a subset of N. In security games with a known set of two players, the attacker is represented by a, and the defender by d.

To illustrate this, take the example of two-players in a simplistic representation.

TABLE 11.2 A basic formulation of a two-player game with an attacker and defender

	a_1	a_2
d_1	1, −1	−5, 5
d_2	−5, 5	1, −1

Table 11.2 gives a basic two-player standard form game in which a defender chooses a target to defend, and an attacker a target to attack. The utilities are defined in the table as the reward first obtained for the defender and the second for the attacker.

In this simple game the defender, d, is charged with protecting 2 targets from the attacker, a. In this example d can either choose to defend target 1 d_1 or target 2 d_2. Likewise, the attacker can choose to attack target 1 or 2 using a_1 or a_2 respectively. In this example, there is no incentive for either player to prefer one target over another.

Defining optimal behavior relies upon a system of belief that each player has about the environment, and that the best outcomes for each player are representative of those with the highest reward. Deviating from this belief means that solutions would not reflect the behavior of players in reality.

In the previous example, there is no benefit to either player for defending or attacking one target over another. Rational behavior dictates then that neither target would be preferred by either player when choosing a move to make. Giving preference to either target would in some games result in a worse performance.

Outside of rational players, an additional player called Nature is also present in some games. The Nature player takes actions with some probability distribution and makes no formal strategic decisions. A Nature player will govern aspects of the game that are outside of the control of any of the players. For cybersecurity, nature might cover attributes such as the success rate of a defensive control or the probability that an exploit is created for a vulnerability.

11.2.3 Utilities

The origin of utility is as a measure of satisfaction, identifying the extent to which an individual is happy with an outcome of an event or situation from a philosophical perspective. The concept of utility has been used to define preference.

Within game theory, the utility of a player, also called payoff, is the reward that a player will receive for making a particular choice within the game. Depending on the game, the utility can take a variety of different forms. In cybersecurity, we often see this in terms of an expected financial loss or as a damage metric.

The purpose of defining utility is to drive the rational decision making of the players. By defining what a player gets for choosing a particular option, an implied ranking or priority is created amongst possible options.

By defining a set of rewards for actions, a player can identify the expected outcome based on their preferences. When set against an indifferent player or nature, a player can calculate the expected utility. This is the probability of each occurrence

TABLE 11.3 A two-player game with differently valued targets

	a_1	a_2
d_1	$1,-1$	$-11,11$
d_2	$-5,5$	$1,-1$

TABLE 11.4 A two-player non-zero-sum game

	a_1	a_2
d_1	$1,-1$	$-11,5$
d_2	$-5,11$	$1,-1$

of nature multiplied by the reward they receive for that option. In decision theory this is called the utility function.

Utility, $u_{i_j}^m$, is the given reward for player i for taking action j, where $u_{i_j}^m \in U$.

Considering a second example in Table 11.3, there is now a difference between the rewards for each of the players based on what is attacked and defended.

In this case, a clear preference can be seen for d to play d_2. As in this case the worst outcome possible would be to get a score of -5, whereas with the other target, d could end up with a worst case score of -11.

The utility function in a two-player game is given as the reward a player gains from taking an action given the actions of all other players. This is given the fact that until each player has taken their action the exact reward is unknowable.

The examples so far have all reflected a form of game called a zero-sum game. These games reflect that the gain by one player must be equal to the loss of the other player. In a cybersecurity scenario, these games are used to represent scenarios where the attacker is maximizing damage rather than having their own unique value for a given activity.

Conversely non-zero-sum games are those where the players differ on the value of the reward for taking an action. This might come from a difference in valuation of the success or failure, or as a factor of cost for the action. Table 11.4 shows that a values the success of attacking target 1 higher than d's perceived loss for having it hit. However, a gains less from attacking target 2, but that scenario would be worse overall for d.

11.2.4 Strategies

The moves that a player makes can be considered on two different levels, the actions that a player actually takes in a game, and the process by which they choose between different sets of actions.

11.2.4.1 Pure Strategies

In game theory, the complete action taken by a player is a pure strategy. This represents the way a player decides to act throughout the game. A strategy might consist of one single choice or action in a simple game, or a set of decisions made. The definition of a strategy has to be with respect to the moves of all other players. This is because sometimes the choices of another player might mean that some choices might no longer be available to a player as a result. In a game such as chess, an opponents move to take your queen means that any future move in the strategy cannot involve moving that piece.

In game theory, it is important to note, that a pure strategy is uninterruptable. Once all players have decided on a strategy to play, then for the purposes of that game, all actions will be carried out.

A pure strategy can be as complex as desired to represent the real choices that must be made in a given scenario. For cybersecurity, this might be a set of steps to protect a target, or a budget allocation for defending a network.

Regardless of the complexity, a pure strategy for a player is given as s_i, and the set of possible strategies is given by:

$$S_i = \{s_i^1, \cdots, s_i^m\} \tag{11.1}$$

Where s_i^m is the mth strategy for player i.

In the previous examples, each player has had two strategies given by d_1, d_2 for d, and a_1, a_2 for a. In the above notation this would give:

$$S_d = \{s_d^1, s_d^2\} \tag{11.2}$$

$$S_a = \{s_a^1, s_a^2\} \tag{11.3}$$

More conventionally S_d and S_a can be written as D and A respectively.

Revisiting the utility of a player, utility, u_i, is defined as the utility of a player given the actions of all the players, and is generally of the form $u_i(s_i, s_{-i})$, where s_i is the strategy played by player i and s_{-i} is the strategy of all other players.

For security scenarios, this is typically then of the form $u_a(s_a, s_d)$ and $u_d(s_d, s_a)$ for the attacker and defender respectively.

In the above example if a player chooses to take action d_2, then the reward will be given based on the action of the attacker. If the attacker plays a_1, then the defender will receive $u_d(d_2, a_1) = -5$.

In this case, d has played the pure strategy d_2, whereas a has played pure strategy a_1.

11.2.4.2 Mixed Strategies

When trying to decide upon which set of actions to take, the best solution might be to always take the same decisions. In cases where a game is repeated a number of times, it is feasible that a single strategy might not yield the best results based on the decisions made by all the players of the game. When there are competing options for possible outcomes based on the possible actions of another player, it can be most optimal to have a set of feasible moves that are best against each of those different options. This set of possible pure strategies defines the mixed strategy.

TABLE 11.5 A two-player game with dominant strategies d_1 and a_2

	a_1	a_2
d_1	2, −2	−5, 5
d_2	1, −1	−11, 11

More formally, a mixed strategy is a probability distribution across all the different pure strategies. When making a decision, a pure strategy is played with the probability in accordance with the distribution of the mixed strategy. This means that a single pure strategy is eventually played, but it is not clear to any competitors what that exact set of actions is, as in classical game theory they will also lock in their strategy at this time.

A mixed strategy, given by σ, for a player i is given as:

$$\sigma_i = (p_i^1, \cdots, p_i^m) \tag{11.4}$$

Where p_i^j represents the probability of player i playing pure strategy j. Each mixed strategy is subject to the constraint:

$$\sum_j p_i^j = 1 \tag{11.5}$$

The utility for a mixed strategy is, therefore, given as $u_i(\sigma_i, \sigma_{-i})$.

Mixed strategies are designed to create flexibility in the way that a player acts. Masking the exact manner in which a player acts is generally needed to cover for potential bad outcomes of otherwise good strategies.

In security games, mixed strategies are vitally important to minimizing expected damage to targets. By being able to change the methods or targets of defense, means that an attacker cant guarantee that any given action will be successful.

11.2.5 Dominant Strategies

A dominant strategy is given as any strategy that for any outcome the strategy performs equal or better than any other strategy. Where dominated strategies exist, they can be removed from the formulation to simplify the game. This is typically done to reduce the time taken to solve a game.

This can be illustrated by the following example:

In Table 11.5, the strategy d_1 dominates strategy d_2. This is given that there is no choice that a can make that would make d_2 a better prospect for d. As such d will always choose to play d_2 such that $p_d^1 = 1$ and thus $p_d^2 = 0$. Similarly, a will always choose to play a_2 as their minimum payoff in this example is 5, which is greater than any reward achieved for playing a_1.

It should be noted that while the example given here is a pure strategy dominating another pure strategy, the same can be applied to a mixed strategy.

11.2.6 Equilibria

With a mixed strategy, there is no guarantee that the result of any given game is going to give a player the desired result. The aim of a mixed strategy is to reduce the risk that each player gets undesirable results.

Masking behavior by selecting from a distribution over strategies only provides benefit, providing there is a rational justification for the distribution. Poorly crafted or unrepresentative mixed strategies can lead to poor performance. It is in the best interest of each player to optimize their behavior to maximize their payoff, however, this needs to be done with respect to their opponents optimal decision making.

The optimal strategy for all of the players is referred to as the equilibrium. As a more formal definition, an equilibrium is the point at which no player is incentivized to deviate from their strategy given by σ^*. This is given as:

$$u_i(\sigma_i^*, \sigma_{-i}) \geq u_i(\sigma_i', \sigma_{-i}) \tag{11.6}$$

Where σ_i' is any alternative strategy to σ_i, and σ_{-i} is the best response strategy amongst all other players.

A game reaching an equilibrium means that for all players, the payoff that each player can expect on average is no worse than any other strategy given that their opponents do not deviate from their optimal strategy.

In some cases there might be multiple equilibrium solutions to the game. These equilibria provide alternative optimal strategies to the players, and might not contain similar features or pure strategies.

11.2.7 Strategies in Security Games

The formulation of the two-player game aligns with the traditional cybersecurity scenario, in which there is a malicious entity attempting to subvert a system (Kiekintveld et al., 2009). As such the traditional model of a two-player game can be applied to this concept, where the two players are defined as an attacker and a defender. This attacker-defender model is core to most game theoretic cybersecurity models.

In its simplest form cybersecurity can be seen as a form of the classic 2x2 problem. In this simple view, an attacker and defender are each presented two targets t_1 and t_2, the defender d must choose which of the two targets to defend at any one time, where the attacker a chooses which of the two targets to attack. The defender seeks to protect the target that the attacker wants to exploit, where the defender gets a positive payoff if they pick the same target as the attacker and a negative payoff if they select the wrong one.

One common formulation of games for security is in the Colonel Blotto game (Roberson, 2006). The Colonel Blotto game is defined as a competitive game, where each player has a number of resources to assign to a number of battlefields. Each player secretly assigns a number of resources to each battlefield, and for each battlefield, the winner is the player that assigned the most resources to that battlefield.

The examples presented so far are the equivalent to the most basic form of the Colonel Blotto game where each player only has one resource available each.

In security, the battlefields of the Colonel Blotto games are used to represent systems, devices, or other resources that might be desirable to an attacker (Guan et al., 2019; Min et al., 2017). A defender has to decide how to allocate their resources such as monitoring or system administrator time to cover these targets from an attacker who has some resource to compromise any of those targets (Gupta et al., 2014).

11.3 NASH GAMES

The traditional simultaneous move game is often called a Nash game in cybersecurity scenarios. It derives its name from the equilibrium that forms the optimal solution. This model of game stems from the decision making of the attacker and the defender, where neither player has an information advantage over the other player.

The traditional Nash game is designed to be a matrix form game, where each complete strategy is represented as a row or column of the matrix. The utilities for each of the players are represented in each of the cells. For any iteration of the game each player picks one of the pure strategies, and receives their reward in the cell at the intersection.

A mixed strategy for a Nash game, is the set of probabilities that a player selects any given row or column as their action for the game. The utilization of mixed strategies is important in games involving security.

The optimal solution to a game is defined as the strategy from which no player can achieve a higher utility by deviating. Simply, this means that the expected payoff of all players is maximized with respect to each of the other players. In this case, the expected payoff is defined as the sum of each utility given the probability that the utility is achieved.

$$u_a = \sum_j p_a^j u_a(\sigma_a, \sigma_d) \tag{11.7}$$

$$u_d = \sum_j p_d^j u_d(\sigma_d, \sigma_a) \tag{11.8}$$

The general form of the Nash equilibrium for the two-player security game can be defined as:

$$max_{\sigma_a} \left(\sum_j p_a^j u_a(\sigma_a, \sigma_d) \right) \tag{11.9}$$

$$max_{\sigma_d} \left(\sum_j p_d^j u_d(\sigma_d, \sigma_a) \right) \tag{11.10}$$

TABLE 11.6 σ^* and expected utilities for numerical examples

Example	Player	σ^*	Expected Utility
Symmetric Payoffs: Table 11.2	d	0.5, 0.5	-2
	a	0.5, 0.5	2
Non-Symmetric Payoffs: Table 11.3	d	0.333, 0.667	-3
	a	0.667, 0.333	3
Non-ZeroSum: Table 11.4	d	0.667, 0.333	-3
	a	0.667, 0.333	3
Domination: Table 11.5:	d	1, 0	-5
	a	0, 1	5

Taking the examples given the optimal strategies can be calculated as shown in Table 11.6.

Taking the first example given in Table 11.2, the Nash Equilibrium can be calculated as $p_d^1 = 0.5$, $p_d^2 = 0.5$, $p_a^1 = 0.5$, and $p_a^2 = 0.5$. The rationale for this is that by both players choosing to play this strategy, they will achieve an average payoff of $u_d^* = -2$ and $u_a^* = 2$.

It is important to note the Nash equilibrium uses the average expected payoff. The average here is the reward expected if the game is played an infinite number of times as opposed to a single shot. This is important, as the players need to be indifferent with regards to the way in which they play.

This can be further explained by taking the example from Table 11.3. Table 11.6 shows that d plays d_1 with $p_1 = 0.333$ and $p_2 = 0.666$, which gives an expected payoff of -3 as:

$$\begin{aligned} u_d(d_1, a_1) p_d(d_1) p_a(a_1) &= 1 * 0.333 * 0.666 = 0.220 \\ u_d(d_1, a_2) p_d(d_1) p_a(a_2) &= -11 * 0.333 * 0.333 = -1.220 \\ u_d(d_2, a_1) p_d(d_2) p_a(a_1) &= -5 * 0.666 * 0.666 = -2.220 \\ u_d(d_2, a_2) p_d(d_2) p_a(a_2) &= 1 * 0.666 * 0.333 = 0.220 \end{aligned} \quad (11.11)$$

To demonstrate the loss of utility a player receives by even a slight deviation, consider the domination example in Table 11.5. In this case the expected payoff for d is -5 by only playing d_1. If there is even a small change in σ for d then their expected utility drops.

$$\begin{aligned} u_d(d_1, a_2) p_d(d_1) p_a(a_2) &= -5 * 0.999 * 1 = -4.995 \\ u_d(d_2, a_2) p_d(d_2) p_a(a_2) &= -11 * 0.001 * 1 = -0.011 \\ u_d(d_1, a_2) p_d(d_1) p_a(a_2) + u_d(d_2, a_2) p_d(d_2) p_a(a_2) &= -4.995 - 0.011 = -5.006 \end{aligned}$$
$$(11.12)$$

In this case d expects to lose an additional 0.006 on average per iteration of the game by deviating from the optimal strategy.

Nash games are most commonly used in security scenarios where there are no observable properties of the game by the attacker. In Nash security games the defender is aware that there is an attacker, and that the attacker has a variety of methods of attack and targets that they can compromise, but no information on their strategy.

In the Nash formulation of a security game, the attacker likewise knows what the defender can do to mediate against their attacks, but does not know the strategy the defender will employ. Games where the attacker has information on the defender's strategy are discussed in later sections.

Nash games are, therefore, defined as games of imperfect information. Perfect information states that all knowledge available at the start of the game is still known at the end of the game. In this context perfect information relates specifically to knowing the strategy of the other player. Since a player in a Nash Game cannot be certain of the strategy another player is using, only predict their likely course of action, a Nash Game is, therefore, a game of imperfect information.

The other form of information considered in games is completeness. This relates to the level of understanding players have of the actions and rewards in the game. A game is said to have complete information if all players have accurate information about the actions and payoffs of the game.

More advanced forms of game theoretic modelling such as Hypergames make use of players with incomplete information (Bennett, 1980). The aim of these games is to not only devise optimal strategies, but to build knowledge of the true nature of the games. To date there has been some limited exploration of Hypergames for cybersecurity (Bakker et al., 2019).

11.4 EXTENSIVE FORM GAMES

To this point, the strategy that a player chooses is selected as a single decision made prior to starting the game. This means that every action that is taken is pre-selected. In reality, not every decision is taken far ahead of time, and so a more responsive form of the game is needed.

The examples presented so far have all been of standard form or represented as strategic games. The matrix representation allows players to easily see the outcomes based on the pure strategies selected.

Regardless of the complexity of the problem, the pure strategy is an encoding of all the actions that are to be taken in the game by a player ignoring timing and sequencing of events. These games normally called standard form games do not always best represent the back and forth nature that comes from a realistic model of decision making.

To address this, there is an alternative method of representing games with multiple decision points. Extensive form games represent in a tree-like structure the decision points that players have available to them.

Taking the example from Table 11.6, the game can be represented in extensive form as demonstrated in Figure 11.1.

The example expresses that while there is an ordered decision, where d plays before a, a has no knowledge of the action taken by the attacker and must pick

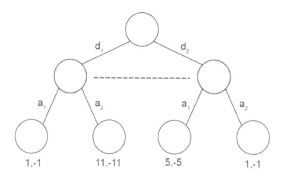

FIGURE 11.1 Basic tree structure for example in Table 11.6.

without knowledge of what their opponent has selected. This restriction makes the extensive form notation operate in the same way as a standard form game.

In an extensive form game, utilities for the game are associated with a payoff at the leaf nodes of the tree. These utilities represent the payoffs for each player for a complete traversal of the tree. Since payoffs are only defined at the leaf of a tree, no payoffs are defined for interim states.

One of the main uses of extensive form games is in the representation of games with a Nature player. In these cases, the optimal decision may change based upon an external influence. Finally, it is important to note that, for complex games, the size of the tree can become an ineffective representation.

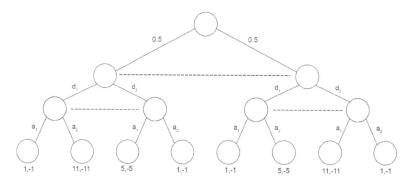

FIGURE 11.2 Basic tree structure with a nature player.

Figure 11.2 shows an example with a nature player. In this case nature plays before either d or a. In this case nature defines which of t_1 and t_2 is more valuable. This changes the utility of the players.

11.4.1 Sub-Games

Each branching path of the game tree in itself can be seen as a game. Each point represents a decision that needs to be made, and requires the representative players to make a choice. With the notion of a player having complete rationality, this

TABLE 11.7 The equivalent strategic form of the game in Figure 11.2.

	a_1	a_2
d_1	1, −1	−8, 8
d_2	−8, 8	1, −1

choice will always be the mathematically optimal choice. A sub-game is defined as any set of decisions that causes a branching of the game tree.

To solve extensive form games, the optimal strategy for each player is the optimal strategy across the whole of the tree. This means solving each of the sub-games. A sub-game equilibria is, therefore, the equilibrium of any individual sub-game within the game tree, where the equilibrium is calculated in the same way as for a strategic game.

The most basic sub-game is the last decision in any set of branches. In the example in Figure 11.1, this relates to each of the decisions made by a. If there is complete knowledge of the state the game is in, then a will always choose the target with a positive utility.

A sub-game can define the utilities as the outcome of additional sub-games, where the structure of the tree has multiple decision points for players.

It is, therefore, possible to solve extended form games by using a process called backwards induction, which iterates back from the last decision to the first decision in the game defining the best possible outcome at each stage (Ben-Porath, 1997). This process, therefore, starts at the leaf nodes of the tree, calculating first the optimal solution for each of the tree edge sub-games. The expected utilities for each of the players taking optimal decisions for that sub-game are then used as the payoffs for the preceding sub-game. This is repeated until an optimal strategy has been found for the node at the root of the game tree.

In the case of the game with a nature player, each branch after the outcome of a nature player will result in a new sub-game. Players will derive their optimal strategy for each of those sub-games. Each of those sub-game strategies can be used as further strategies for the extended game, where the utilities given by sub-game equilibria are weighted in accordance with the probabilities of the Nature player.

Revisiting Figure 11.2, the payoff for d for each of the sub-game is −3, and 3 for a. However, the strategies are not identical for the two sub-games. In the left sub-game d plays d_1 with $p = 0.667$, whereas in the right sub-game they play d_1 with $p = 0.333$. Since d wishes to be indifferent to the expected utility, they will play σ_d^{*left} with $p = 0.5$ and σ_d^{*right} with $p = 0.5$. Overall this results in the equivalent strategy of d_1 with $p = 0.5$ and d_2 with $p = 0.5$. This is because the effective strategic form game can be represented as shown in Table 11.7.

Given that nature plays with $p = 0.5$, this means that the successful states of each branch will give either −5 or −11 for d depending on the play by nature. Since they have equal probability of occurring, the expected utility would be the sum of the utilities multiplied by the probability of the branch being reached. This, therefore, results in an expected value of −8 on each branch. Solving the equilibrium of this game give $\sigma_d^* = (0.5, 0.5)$ and $\sigma_a^* = (0.5, 0.5)$.

11.4.2 Information Sets

In most security games, there will be instances where the outcomes of previous stages are not known to players. In these cases, the players have to make a decision without knowing what their opponent may have already done. The player making a decision, therefore, creates an information set about the possible options that could exist and the payoffs that exist for each of the remaining branches in the game tree (Alpcan and Basar, 2003).

Where there is an information set across all different possibilities at all stages, then the game can be represented as the traditional static game. As no player has gathered specific information on the decisions of other players all plays are being made based on a perceived notion of optimality, which can be derived from the Nash Equilibrium.

Information sets are represented in figures here by the dotted lines that connect the nodes at the same level of the tree.

The number and location of information sets will create different equilibria for the same game tree. Taking the example from the initial tree example in Figure 11.1, it is possible to create a form of the game where there are no information sets, which creates a decision tree.

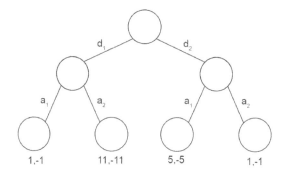

FIGURE 11.3 Tree structure for example in Table 11.6 with an information set for a after d has made their choice.

In the example in Figure 11.3, a has knowledge of the play made by d and so will always choose to then attack the target that has not been protected by d. This means that the utilities for the sub-game are $u_d^1 = -11$ and $u_d^2 = -5$. This is because the attacker will always know to attack the opposite target since they have complete information of the state of the game when making the decision. As a best response to this d, therefore, chooses to play pure strategy d_2 as there is no better reward they can achieve than −5 by changing strategy.

11.4.3 Multi-Stage Games

Multi stage games represent games that are played over a number of stages. Decisions and strategies of each stage may be independent from previous stages.

An example of a cybersecurity multi-stage game was devised by Lye and Wing (1947), who outlined an attacker attempting to achieve one of a handful of goals. In their formulation, an attacker attempts to breach a system by completing a number of steps, at each point a game is played by the players to reach a new state of system compromise. The aim of the players is to maximize a utility associated with the state that the system reaches.

A specific case of a multi-stage game is a repeated game, where the players play the same game multiple times. Each stage of the game allows for the players to make new decisions.

Conventionally, there is an information set associated with the start of each new iteration. This information set defines that the players are aware of the last set of actions taken by the other players.

The strategies that this creates can employ more responsive measures, responding specifically to the result of the previous iteration. A common strategy that is used in iterative prisoners dilemma games is for players to play tit-for-tat, where rather than using a probabilistic strategy, they employ a reactive response to their opponent's moves (Axelrod et al., 1987). More strategic solutions have been developed over tit-for-tat strategies (Nowak and Sigmund, 1993). This kind of strategy cannot be represented using traditional probabilities.

11.5 STACKELBERG GAMES

A specific form of the extended form game is the Stackelberg Game. This game creates a model of a leader and a follower, in which one player, the leader, commits to a strategy before the other player, the follower, creates a best response strategy (Stackelberg, 2010).

In the traditional attacker-defender model of cybersecurity games, Stackelberg games will define the defender as the leader and the attacker as the follower (Tambe, 2011). Conceptually it follows that the defender of a system or an environment must first define the strategy that they are going to utilize before an attacker even considers making an attempt to identify a way to attack that system.

This thought generally holds, as the decisions made by the defender will influence the actions that an attacker has available to them.

The assumption made in this game, is the attacker will perform perfect reconnaissance against the target in order to ascertain the defenses used. This means that the attacker could see the strategy being used by the defender, and for any time period would be able to define the best response (Bagwell, 1995).

While the attacker has knowledge of the defenders strategy, this does not mean that they know the exact move that the defender will play at any given moment. Despite the concept defining that the game is given the description leader-follower, the players still act simultaneously.

In the case of the simple security game, the defender may prefer to protect t_1, which has been observed by the attacker, so they would, as a response, prefer to attack t_2. This means that while for any given time slot, the attacker would predict that the defender would protect t_1, the attacker must still be aware that the defender may select t_2 to defend. As such, the attacker does not get perfect information as to the target to attack based on the decision of the defender, but more how to predict the defender.

The equilibrium conditions for a Stackelberg equilibrium can be defined as:

$$u_d(\sigma_d, g(\sigma_a)) \geq u_d(\sigma_d', g(\sigma_a)), \forall \sigma_d', \quad (11.13)$$

where

$$u_a(g(\sigma_a), \sigma_d) \geq u_a(g(\sigma_a'), \sigma_d), \forall \sigma_d, g', \quad (11.14)$$

where $g(\sigma_a)$ is the follower best response to the defender strategy σ_d

When calculating the strategies of the players, both players define their strategy with the idea that one of the players has perfect observability of the other player's strategy. This means that a defender will make the assumption that their strategy is known, and play to maximize their own payoff given the optimal response to their strategy.

It is this that makes a Stackelberg game a specific form of extensive form game. In this formulation, all of the defender's moves are represented in the tree before any of the attacker's moves.

The effective method for solving Stackelberg games requires the calculation of the sub-game payoffs for the attacker, before solving the game for the defender. This formulation identifies that the attacker strategy will be the best response to any defender strategy. The defender, therefore, aims to create a scenario where they are indifferent knowing the strategy of the attacker.

Reconnaissance over an infinite time scale would determine the strategy that the defender is using. While in reality, this is not possible, observation over a long enough time period would give an approximation to the strategy, as the attacker would see the defensive decisions made at each time period and account for how often each different strategy occurs. It is with this capability, that the attacker can define their strategy to maximize the chances of success.

It is important to note, that the strategy decision on the part of the defender needs to be a random selection distributed from their optimal strategy (Rubenstein, 1991). This is such, that if a predictable pattern can be established, then the game theoretic solution no longer holds. In this instance, the attacker would only need to discern the pattern in order to identify their own optimal strategy. While this concept is true regardless of the formulation of the game, due to the leader-follower nature of Stackelberg games, a pattern-based approach would undermine the perceived risk mitigation of the solution.

Should an attacker not perform reconnaissance to identify the defenders strategy, then any strategy they play will be suboptimal. Based on the assumption that the attacker could build complete knowledge of the defender strategy, then any decision would be made with the information associated with a Nash game.

11.5.1 Strong Stackelberg Equilibria

When solving a Stackelberg game, there is a common occurrence that there may be multiple viable equilibria for the player acting as a follower. The strength of the equilibria is attributed to the utility of the leader. In a strong equilibrium, the utility of the leader is the best that it can be of the possible solutions that give the same expected utility to the follower.

In addition to the two main conditions of the Stackelberg equilibrium given previously, the strong Stackelberg equilibrium has the additional condition, that represents the tiebreaking property.

$$u_d(\sigma_d, g(\sigma_a)) \geq u_d(\sigma'_d, \tau(\sigma_a)), \forall \sigma'_d, \quad (11.15)$$

where $\tau(\sigma_a)$ is the set of follower best responses to σ_d.

It is generally held that the strong Stackelberg equilibria would be the solution played. Given the indifference that the attacker has to the solutions they have no reason to consider that the leader would play anything but the strategy that would maximize their utility.

11.5.2 Weak Stackelberg Equilibria

In the previous section, the equilibrium that the follower is indifferent to, but is most advantageous to the other player is the strong equilibrium. Other possible equilibria are then termed as weak equilibria. Each of these solutions provides the same level of utility to the follower but are less beneficial to the leader.

While the idea that a strong Stackelberg equilibrium is the expected outcome to security games, there are scenarios that have been raised where a weak equilibrium would be the optimal choice for an attacker (Korzhyk et al., 2011).

The general scenario under which this is true, is when there is a resource allocation that impacts more than one target. The reason that this is applicable is that often in cybersecurity decision making, a single resource will cover multiple targets or vulnerabilities. For example, utilization of strong perimeter firewall rules will protect a wide variety of targets on a network, not just a single device.

11.6 PRACTICAL OPERATION

While a predominantly theoretical concept, there is a need to obtain tangible benefits from utilizing game theory for security (Hamilton et al., 2002).

Within cybersecurity, the utilization of game theory has best been established as a form of decision support for complex interactions. The complexity of decisions for cybersecurity scenarios, lend themselves to the abstractions associated with game theory and other modelling techniques. Specifically game theory has found use in the space of risk assessment and evaluation of network security.

The realization of a game theoretic model into practical application is underpinned by the applicability to the real-world scenario. It is this understanding of the requirements of translating from theory that has a number of issues that must be addressed to achieve usable solutions.

11.6.1 Data

Practical applicability to a scenario for cybersecurity requires relevant data to support the decision-making process. Much of the theoretical underpinning has been performed on parameterized rationalization of the space (Liang and Xiao, 2012). Parameterization is used in this case, as there is often a lack of suitably detailed data to support empirical validation. Within the field of cybersecurity, there has been a long-standing issue of effectively representing the cost and impact of cyberattacks (Smith, 2004). The inability to effectively value data and reputation along with unknowable costs prior to an attack, indicate that the true cost of attacks can vary significantly (Abubakar et al., 2015).

One of the other major issues for defining accurate game theoretic decision support methods is the identification of appropriate efficacy measures (Wang and Shroff, 2017). It is often unknowable the extent to which any given mechanism supports the mitigation of a threat. In addition to this, the compound impact of controls is not well represented. In a study by Khouzani et al. (2016), they identify three different mechanisms for combining the efficacy of security controls: best of, additive, and multiplicative. Each of these methods provides different optimal security decision strategies because of the impact they have on the utility calculation, and the number of controls to implement.

One area that has been a topic for significant discussion has surrounded the modelling of capturing the state of threat for an attacker. One of the guiding principles of game theory is that the agents in a system act perfectly rationally. The difficulty in defining utility for a defender is in being able to adequately understand losses from a cyberattack as well as the effectiveness of defensive mechanisms (Romanosky, 2016).

Likewise for an attacker, understanding the motivation for an attack, and therefore, the utility of the attack as well as the effectiveness of their capabilities to deliver such attacks limits the capability to define appropriate rational models of attackers (Holmgren et al., 2007).

It is for this reason that many cybersecurity problems are represented as zero-sum games. The assumption made is that an attacker seeks only to maximize the damage to a defender (Manshaei et al., 2013), although there have been attempts to rationalize the problem as non-zero sum games (Luo et al., 2010).

11.6.2 Assumptions

The capability to build a perfectly accurate model to the real world, is generally not achievable. Assumptions and abstractions provide a core foundation for making models work. In order to make practical the tools that are being designed, there is a need to reduce the complexity (Mahajan et al., 2004). This is often to make the problem tractable in both time and computational feasibility.

Game theoretic models already implement a number of assumptions. The most common is the perfect rationality of players. This conjecture simply states that players will always act to make the most optimal play for them, ignoring sub-optimal strategies. In reality, it has been studied that attackers act irrationally (Basu, 1990; Bachmann, 2010).

Much like having data to accurately represent the utility the attacker receives; a realistic game theoretic model of cybersecurity should take into account a model of the attacker (Choi et al., 2014). An attacker model should seek to capture any of the irrationality that the available data is unable to represent. Applying an attacker model to represent a players preference will change the irrationality of a generalized attacker into a rational one.

Defining a good set of assumptions, the objective is to capture only the factors required to solve the problem. Acknowledgement of the assumptions and the limitations that they provide is essential as this defines the operational bounds of the model.

11.6.3 Uncertainty

While briefly mentioned previously, there is a concern regarding the accuracy of the data. The difficulty to create relevant metrics for cybersecurity is a well noted problem, where relevant and accurate data is often difficult to obtain (Black et al., 2008). This lack of trust in the data impacts the reliability of the models and by extension the trust in the support to a decision being made. Uncertainty dictates the degree to which the state or values that underpin the decision making can be trusted. Uncertainty in the game and actions can change the optimality of solutions.

In cybersecurity, uncertainty can occur in a variety of ways. Two general forms of uncertainty are seen in the type of attacker, and the state of the game.

Thus far, there has been an assumption that there is a single type of attacker, either generic or represented by a threat profile. In some cases, it is not viable to consider a single generic profile and so there can be a distribution amongst the profile of potential attackers (Seebruck, 2015). Games of this type are called incomplete. In this case, the player with incomplete information needs to reason over the distribution of attacker types and provide a solution that is optimal based on uncertainty of the type of attacker they will face.

The capability to build an understanding of the variety of threats faced, and a rational for their preferences over the space of possible objectives requires detailed specific domain knowledge. As with incorrect or incomplete data, decision making on incorrect distributions or preferences of attacker type can result in advice that doesn't represent the real world (Rass et al., 2015).

Uncertainty is used frequently in games where stealth is an issue. These games represent scenarios where the players do not strictly know what state the game is in. The most representative example is the FLIPIT game, where two players compete over a single resource (Van Dijk et al., 2013). The game considers a similar attacker-defender model that has been consistent with security games. In this case, the two players attempt to hold control of a cryptographic key. The game dismisses the pretense of being able to secure the key and instead opts that if a player wants to take control they can do so. The game works by restricting the time between requests and knowledge of if the player actually has taken control of the key.

In this game players define a strategy for when they attempt to take control of the resource, where each player aims to maximize the time they have control of the resource subject to a limitation on when and how they can act. Players do not know when they lose control of the resource, only the time at which they believe they gain control of the resource. This uncertainty means that players cannot define responsive actions and must instead set out a strategy that defines their plan proactively.

11.6.4 Timing

One assumption that exists when using game theory in cybersecurity contexts is that there is some consistency to the timing of decisions being made. An example would be that for every security decision made, an attacker only has a single opportunity to attack. However, many decisions and operations do not work in that kind of discrete time-step. An attacker might have numerous opportunities to attack utilizing a variety of methods before any kind of action can be taken by a defender in response.

In an example of advanced persistent threat attacks (APTs), some stages of the attack may involve an attacker sitting on a network for an extended period of time with little to no activity (Ghafir and Prenosil, 2014). Representing this in a scenario where other components of the attack might happen within a very short period of time, such as malware deployment, requires specific design steps (Rass et al., 2017).

Likewise, decisions that a defender makes over a long period might not be reflected by the time frame of the attacker. Designs of practical decision support systems need to operate inline with the time frame of the requirements for the advice. If the advice from a game theoretic tool is needed in real time, then the design needs to reflect this decision (Reddy, 2009).

Similarly, even with a reduction in complexity through assumptions, the potential number of solutions that are viable can make the problem computationally infeasible to solve. Approaches that can efficiently approximate an optimal solution are generally used (Fox, 2010).

11.6.5 Solution Design

When defining the decision support that the game theoretic methodology will provide, there needs to be a clear definition of the meaning of both pure and mixed strategies. Fundamentally, this is about ensuring that the solutions can be clearly actioned by either the attacker or defender. The design of the game needs to be performed in a way that either findings are not trivialized or would be impossible to perform.

To create non-trivial problems, appropriate bounds are needed. These bounds will create constraints within which the players must devise their strategies. The main purpose is to give suitable reason for choice in the games. Allowing an option to select all actions would appear to be an optimal move in security. However, factoring in the costs of these actions or other negative consequences of actions means that trade-off may need to be made between security with regards to the constraints.

For cybersecurity, these bounds are generally seen as some restriction in available resourcing, such as the number of staff available (Fielder et al., 2014), availability of actions (Lye and Wing, 1947), or budget (Nagurney et al., 2017).

In some cases there will be multiple different constraining factors. In a paper by Panaousis et al. (2014), there are two constraining factors for cybersecurity decision making, the budget and indirect costs. The budget dictated the maximum amount of money that was available to purchase security controls and set a fixed maximum for the amount of controls that could be used to solve the security problem. The novel introduction of indirect cost demonstrated that other factors outside of price and efficiency might impact the uptake of a control. Factors such as the training requirements and productivity impact were factored in to ensure that the optimal solutions were also optimal for the user base.

The solution design is coupled closely with an understanding of the timing of activities and decision making. The representation of the outputs and information flows need to be representative of the real world. Doing this correctly means that there is a better connection between the solutions and any practical implementations.

Appropriate design for decision making does not always guarantee that mixed strategies will be viable. For example, it is a reasonable design decision to have a solution implement a perimeter firewall. As a pure strategy this makes sense, since the strategy involves setting up and running a firewall. However, as part of a mixed strategy this makes less sense. If the optimal strategy only calls for using a firewall with $p = 0.7$, then there needs to be a clear definition for this. Depending on context, this could involve operating it only for 70% of time steps, or running it at only 70% capacity.

11.6.6 Scalability

Many of the examples of game theory assess problems at a small scale, either reducing the problem to a minimal decision set or simplified utility structure.

To build a model that is representative of the real world, a wider range of actions, decisions, and targets need to be taken into consideration, much more than the literature tests on. Predominantly, the issue with scalability is not that the theory fails at a large scale, more that the techniques and methods used to solve the problems do not effectively scale.

An example from physical security, the Federal Air Marshal Service was not capable of analyzing all possible flight paths under the original designs due to issues scaling the game solvers (Tsai et al., 2009; Jain et al., 2010; Guo et al., 2019). New designs for solvers were needed to scale to the appropriate size under the time constraint required for creating schedules for the air marshals.

The complexity of scenarios necessitates the need for abstractions to be tractable. When scaling game theoretic problems, any abstractions and assumptions still need to hold when scaling the game to contain the real world. Much like the scalability of tools and techniques reduce the capability to ensure optimal decisions, then the scalability of design decisions can introduce unchecked uncertainty that will impact the results in a similar manner.

Many game theoretic problems suffer from the state explosion problem. As new decision points or actions are introduced into the game, the solution space can increase exponentially. In work using states of network compromise, the number of states of the game are limited to only those of critical importance. This might be known bad states that a network can reach or interim states with the highest number of effective decisions from either player.

In the case of work by Caulfield and Fielder (2015), the model developed considered multiple states across only three interconnected devices. This was done to facilitate the state explosion that came from the probabilistic creation of vulnerabilities across the network. The state explosion limited the capability to reason on larger networks, as identifying the optimal strategy could only be done on predicted states, and still with high resource overhead for processing.

11.7 RESEARCH EXTENSIONS

There are predominantly three broad areas in which game theory is being used for cybersecurity: operational security, privacy, and machine learning (Do et al., 2017; Wang et al., 2016).

11.7.1 Operational Security

Operational security is defined here as any aspect of security that relates to the capability to maintain desired operational functionality. This is an intentionally broad category, as there are a wide range of approaches and implementations. However, all approaches at their core seek to improve understanding of risk to security decision makers (Schatz and Bashroush, 2017).

Closest to the early economic decision making associated with game theory, the aspect of understanding return on security investment is where game theory is most commonly implemented (Anderson and Moore, 2006). The fundamental question is how do organizations make strategic investments in security when the tangible benefits might not be understood (Gordon and Loeb, 2002).

Conventionally, the approach to representing investment decision making has been as a function of loss minimization (Fielder et al., 2016). A typical formulation is to create a set of security controls for a network defender, and a set of feasible attacks for the attacker.

Supporting the incentivization of security, game theory has also begun being applied to cyber insurance markets (Marotta et al., 2017). A specific area of use is in the design of mechanisms, where the aim of the provider is to minimize the risk that they take from their customers (Panda et al., 2019). In this paradigm, the consumers rationality is linked to minimizing the amount of combined cost for security and insurance that ensure a desired level of risk coverage.

While many of the approaches have considered traditional network security, with the increasing uptake of managed services and cloud computing, security for these devices needs to be thought of in the same way (Furuncu and Sogukpinar, 2015). Many of the same game theoretic techniques can be used for cloud computing, as secure design, resource allocation, and malicious users are still relevant in new domains (Kamhoua et al., 2014; Maghrabi et al., 2016). Likewise, consideration of cyber-physical domains such as those related to critical national infrastructure, can also gain extensive insights from game theoretic approaches (Ferdowsi et al., 2017).

Deception techniques provide a way for defenders to gather intelligence on attackers by providing realistic benign environments to capture attacker behavior. Defining how to deploy such environments across a network has long been a topic of research (Kiekintveld et al., 2015; Wagener et al., 2009), with work focusing on strategically identifying means to deploy honeypots (Boumkheld et al., 2019).

This is also consistent with creating moving target defenses for networks (Jajodia et al., 2012). The research aims to create a means for probabilistically deciding how to configure a network in order to be able to confuse potential attackers against the network. The concept is that by using game theory, the optimal configuration for any time period can be calculated.

11.7.2 Privacy

Privacy is an area where game theory can better support the rationale for designs. The idea is to create a set of preferences for the development of better privacy controls. There is a wide range of use-cases of game theory for enhancing or maintaining privacy covering the whole spectrum from technical cryptographic design to business decisions on the availability of data.

One of the early uses of game theory in the privacy space has concerned information sharing methods. The scope of game theory in this space has covered aspects such as measuring the collective and individual benefit to entities providing information to a shared threat aspect (Fuchsbauer et al., 2010). This kind of approach seeks to identify socially optimal designs where all players have an interest in inputting into a system, attempting to address the freeloader problem. However, even within these systems, game theory has been used to identify ways to support the protection of sensitive data from malicious users.

Similarly, a number of cryptographic studies have also been performed; examples of this are in multiparty computation and secret sharing protocols. In these instances, game theory is used to identify desirable properties of the design of cryptographic methods (Katz, 2008). Game theory provides a means of understanding the method by which the malicious actors can manipulate and distort the outputs of the mechanisms. Likewise, game theory can be used to define optimal properties for implementation.

In some cases such as anonymity, the competitive design can be surrounding the conflicting desires of service providers and users (Karimi Adl et al., 2012). This removes the convention of a malicious adversary that has been seen in classic security games, replacing it with more traditional economic models of supply and demand of services (Liu et al., 2013).

Some approaches have sought to derive aspects of human behavior and potential vulnerabilities using game theory to explore the risks that are exposed (Ryutov et al., 2015). An example of password security balancing strength of passwords with the cognitive load required to study them identifies the types of trade-off that are sometimes needed in usable security design (Rass and Konig, 2018).

11.7.3 Machine Learning

Game theory has previously been used in deriving understanding of decisions being made in machine learning-based classifiers (Kononenko et al., 2010). A more recent growing use of game theory is in understanding vulnerabilities in machine learning systems.

With increased emphasis on the use of machine learning to support a wide variety of activities including security, there is a need to ensure that those systems operate correctly.

Strategic decisions between a classification algorithm and an adversary can be represented as a game. In this kind of game, an adversary is attempting to disrupt the training algorithm by contaminating a data set (Bruckner and Scheffer, 2011). The concept is that the adversary is attempting to find effective ways to disrupt the training of the classifier without being discovered. This is being performed against a classifier that is aiming to minimize the error rate and identify adversarial actions. An example of this type of game is in spam filtering, where this approach has been used to create a model of training data generation for malicious email misclassification (Bruckner et al., 2012).

The ability to manipulate the manner in which training is performed by neural networks can be used to cause significant shifts in the reliability of the algorithms to perform their tasks correctly. The addition of small perturbations across the whole of the input space has been shown to decrease the effectiveness of the training algorithms (Goodfellow et al., 2014).

11.8 CONCLUSIONS

Game theory provides a powerful tool for furthering the capability to understand and support security concerns. The approach utilizes the interactions and payoffs between a number of different agents who act rationally to optimize their behavior to their benefit.

The range of methods for representing the problem and interactions allows for complex scenarios to be represented and reasoned on. Understanding where and how to build an appropriate model for the problem is a necessity for gaining a useful actionable output.

While game theory certainly provides support for understanding, a number of issues exist that can inhibit the uptake and effectiveness of the approach. It is important to take the findings as a means of providing insights into the decisions surrounding security. Game theoretic approaches of security should be used to aid the decision making of those tasked with cyber-related responsibilities and not to replace them.

With further advancements in methods for attack and defense, there will be new ways for utilizing game theory to provide decision support for currently unknown cyber scenarios.

REFERENCES

Abubakar, A., Chiroma, H., Muaz, S.A., and Ila, L.B. (2015). A review of the advances in cyber security benchmark datasets for evaluating data-driven based intrusion detection systems. In *SCSE*, 221–227.

Adl, R.K., Askari, M., Barker, K., and Safavi-Naini, R. (2012). Privacy concensus in anonymization systems via game theory. In *IFIP Annual Conference on Data and Applications Security and Privacy,* Springer, 74–89.

Alpcan, T., and Basar, T. (2003). A game theoretic approach to decision and analysis in network intrusion detection. In *42nd IEEE International Conference on Decision and Control (IEEE Cat. No. 03CH37475)*, IEEE, 2595–2600.

Anderson, R., and Moore, T. (2006). The economics of information security. *Science*, 314(5799): 610–613.

Armstrong, R., Mayo, J., and Siebenlist, F. (2009). Complexity science challenges in cybersecurity. *Sandia National Laboratories SAND Report.*

Axelrod, R. (1987). The evolution of strategies in the iterated prisoner's dilemma. *The Dynamics of Norms*, 1–16.

Bachmann, M. (2010). The risk propensity and rationality of computer hackers. *International Journal of Cyber Criminology*, 4(1-2), 643–656.

Bagwell, K. (1995). Commitment and observability in games. *Games and Economic Behavior*, 8(2), 271–280.

Bakker, C., Bhattacharya, A., Chatterjee, S., and Vrabie, D.L. (2019). Learning and information manipulation: repeated hypergames for cyber-physical security. *IEEE Control Systems Letters*, 4(2), 295–300.

Basu, K. (1990). On the non-existence of a rationality definition for extensive games. *International Journal of Game Theory*, 19(1), 33–34.

Ben-Porath, E. (1997). Rationality, nash equilibrium and backwards induction in perfect-information games. *The Review of Economic Studies*, 64(1), 23–46.

Bennett, P.G. (1980). Hypergames: developing a model of conflict. *Futures*, 12(6), 489–507.

Black, P.E., Scarfone, K., and Souppaya, M. (2008). Cyber security metrics and measures. *Wiley Handbook of Science and Technology for Homeland Security,* 1–15.

Boumkheld, N., Panda, S., Rass, S., and Panaousis, E. (2019). Honeypot type selection games for smart grid networks. In *International Conference on Decision and Game Theory for Security,* Springer, 85–96.

Bruckner, M., Kanzow, C., and Scheffer, T. (2012). Static prediction games for adversarial learning problems. *Journal of Machine Learning Research,* 13, 2617–2654.

Bruckner, M., and Scheffer, T. (2011). Stackelberg games for adversarial prediction problems. In *Proceedings of the 17th ACM SIGKDD International Conference on Knowledge Discovery and Data Mining,* 547–555.

Caulfield, T., and Fielder, A. (2015). Optimizing time allocation for network defense. *Journal of Cybersecurity,* 1(1), 37–51.

Choi, S., Song, J., Kim, S., and Kim, S. (2014). A model of analyzing cyber threats trend and tracing potential attackers based on darknet traffic. *Security and Communication Networks,* 7(10), 1612–1621.

Cox, L.A. Jr. (2009). Game theory and risk analysis. *Risk Analysis: An International Journal,* 29(8), 1062–1068.

Do, C.T., Tran, N.H., Hong, C., Kamhoua, C.A., Kwiat, K.A., Blasch, E., Ren, S., Pissinou, N., and Iyengar, S.S. (2017). Game theory for cyber security and privacy. *ACM Computing Surveys (CSUR),* 50(2), 1–37.

Ferdowsi, A., Sanjab, A., Saad, W., and Mandayam, N.B. (2017). Game theory for secure critical interdependent gas-power-water infrastructure. In *2017 Resilience Week (RWS),* IEEE, 184–190.

Fielder, A., Panaousis, E., Malascaria, P., Hankin, C., and Smeraldi, F. (2014). Game theory meets information security management. In *IFIP International Information Security Conference,* Springer, 15–29.

Fielder, A., Panaousis, E., Malascaria, P., Hankin, C., and Smeraldi, F. (2016). Decision support approaches for cyber security investment. *Decision Support Systems,* 86, 13–23.

Fox, W.P. (2010). Teaching the applications of optimization in game theory's zero sum and non-zero sum games. *International Journal of Data Analysis Techniques and Strategies,* 2(3), 258–284.

Fuchsbauer, G., Katz, J., and Naccache, D. (2010). Efficient rational secret sharing in standard communication networks. In *Theory of Cryptography Conference,* Springer, 419–436.

Furuncu, E., and Sogukpinar, I. (2015). Scalable risk assessment method for cloud computing using game theory (ccram). *Computer Standards and Interfaces*, 38, 44–50.

Ghafir, I., and Prenosil, V. (2014). Advanced persistent threat attack detection: an overview. *International Journal of Advanced Computer Network Security*, 4(4), 5054.

Goodfellow, I.J., Shlens, J., and Szegedy, C. (2014). Explaining and harnessing adversarial examples. ArXiv Preprint 1412.6572.

Gordon, L.A., and Loeb, M.P. (2002). The economics of information security investment. *ACM Transactions on Information and System Security (TISSEC)*, 5(4), 438–457.

Guan, S., Wang, J., Yao, H., Jiang, C., Han, Z., and Ren, Y. (2019). Colonel blotto games in network systems: models, strategies, and applications. *IEEE Transactions on Network Science and Engineering*.

Guo, Q., Gan, J., Fang, F., Tran-Thanh, L., Tambe, M., and An, B. (2019). On the inducibility of stackelberg equilibrium for security games. In *Proceedings of the AAAI Conference on Artificial Intelligence*, 33, 2020–2028.

Gupta, A., Basar, T., and Schwartz, G.A. (2014). A three-stage colonel blotto game: when to provide more information to an adversary. In *International Conference on Decision and Game Theory for Security*, Springer, 216–233.

Hamilton, S.N., Miller, W.L., Ott, A., and Saydjari, O.S. (2002). Challenges in applying game theory to the domain of information warfare. In *Information Survivability Workshop (ISW)*, Citeseer.

Holmgren, A.J., Jenelius, E., and Westin, J. (2007). Evaluating strategies for defending electric power networks against antagonistic attacks. *IEEE Transactions on Power Systems*, 22(1), 76–84.

Jain, M., Tsai, J., Pita, J., Kiekintveld, S.R., Tambe, M., and Ordonez, F. (2010). Software assistants for randomized patrol planning for the lax airport police and the federal air marshal service. *Interfaces*, 40(4), 267–290.

Jajodia, S., Ghosh, A.K., Subrahmanian, V.S., Swarup, V., Wang, C., and Wang, X.S. (2012). *Moving Target Defense II: Application of Game Theory and Adversarial Modeling*, 100.

Kamhoua, C.A., Kwiat, L., Kwiat, K.A., Park, J.S., Zhao, M., and Rodriguez, M. (2014). Game theoretic modeling of security and interdependency in a public cloud. In *2014 IEEE 7th International Conference on Cloud Computing*, IEEE, 514–521.

Katz, J. (2008). Bridging game theory and cryptography: recent results and future directions. In *Theory of Cryptography Conference*, Springer, 251–272.

Khouzani, M.H.R., Malascaria, P., Hankin, C., Fielder, A., and Smeraldi, F. (2016). Efficient numerical frameworks for multi-objective cyber security planning. In *European Symposium on Research in Computer Security,* Springer, 179–197.

Kiekintveld, C., Jain, M., Tsai, J., Pita, J., Ordonez, F., and Tambe, M. (2009). Computing optimal randomized resource allocations for massive security games. In *Proceedings of the 8th International Conference on Autonomous Agents and Multiagent Systems,* 1, 689–696.

Kiekintveld, C., Lisy, V., and Pibil, R. (2015). Game-theoretic foundations for the strategic use of honeypots in network security. In *Cyber Warfare,* Springer, 81–101.

Kononenko, I., and Strumbelj, E. (2010). An efficient explanation of individual classifications using game theory. *Journal of Machine Learning Research,* 11, 1–18.

Korzhyk, D., Yin, Z., Kiekintveld, C., Conitzer, V., and Tambe, M. (2011). Stackelberg vs. nash in security games: an extended investigation of interchangeability, equivalence, and uniqueness. *Journal of Artificial Intelligence Research,* 41, 297–327.

Liang, X., and Xiao, Y. (2012). Game theory for network security. *IEEE Communications Surveys and Tutorials,* 15(1), 472–486.

Liu, X., Liu, K., Guo, L., Li, X., and Fang, Y. (2013). A game-theoretic approach for achieving k-anonymity in location based services. In *2013 Proceedings IEEE INFOCOM,* IEEE, 2985–2993.

Luo, Y., Szidarovszky, F., Al-Nashif, Y., and Hariri, S. (2010). Game theory based network security. *Journal of Information Security,* 1(1), 41–44.

Lye, K.-W., and Wing, J.M. (2005). Game strategies in network security. *International Journal of Information Security,* 4(1-2), 71–86.

Maghrabi, L., Pfluegel, E., and Noorji, S.F. (2016). Designing utility functions for game-theoretic cloud security assessment: a case for using the common vulnerability scoring system. In *2016 International Conference on Cyber Security and Protection of Digital Services (Cyber Security),* IEEE, 1–6.

Mahajan, R., Rodrig, M., Wetherall, D., and Zahorjan, J. (2004). Experiences applying game theory to system design. In *Proceedings of the ACM SIGCOMM Workshop on Practice and Theory of Incentives in Networked Systems,* 183–190.

Manshaei, M.H., Zhu, Q., Alpcan, T., Bacsar, T., and Hubaux, J.-P. (2013). Game theory meets network security and privacy. *ACM Computing Surveys (CSUR),* 45(3), 1–39.

Marotta, A., Martinelli, F., Nanni, S., Orlando, A., and Yautsiukhin, A. (2017). Cyber-insurance survey. *Computer Science Review,* 24, 35–61.

Min, M., Xiao, L., Xie, C., Hajimirsadeghi, M., and Mandayam, N.B. (2017). Defense against advanced persistent threats: a colonel blotto game approach. In *2017 IEEE International Conference on Communications (ICC)*, 1–6.

Nagurney, A., Daniele, P., and Shukla, S. (2017). A supply chain network game theory model of cybersecurity investments with nonlinear budget constraints. *Annals of Operations Research,* 248(1-2), 405–427.

Nowak, M., and Sigmund, K. (1993). A strategy of win-stay, lose-shift that outperforms tit-for-tat in the prisoner's dilemma game. *Nature,* 364(6432), 56–58.

Panaousis, E., Fielder, A., Malascaria, P., Hankin, C., and Smeraldi, F. (2014). Cybersecurity games and investments: a decision support approach. In *International Conference on Decision and Game Theory for Security,* Springer, 266–286.

Panda, S., Woods, D.W., Laszka, A., Fielder, A., and Panaousis, E. (2019). Post-incident audits on cyber insurance discounts. *Computers and Security,* 87, 101593.

Rass, S., and Konig, S. (2018). Password security as a game of entropies. *Entropy,* 20(5), 312.

Rass, S., Konig, S., and Schauer, S. (2015). Uncertainty in games: using probability-distributions as payoffs. In *International Conference on Decision and Game Theory for Security,* Springer, 346–357.

Rass, S., Konig, S., and Schauer, S. (2017). Defending against advanced persistent threats using game theory. *PloS One,* 12(1).

Reddy, Y.B. (2009). A game theory approach to detect malicious nodes in wireless sensor networks. In *2009 Third International Conference on Sensor Technologies and Applications,* IEEE, 462–468.

Roberson, B. (2006). The colonel blotto game. *Economic Theory,* 29(1), 1–24.

Romanosky, S. (2016). Examining the costs and causes of cyber incidents. *Journal of Cybersecurity,* 2(2), 121–135.

Roy, S., Ellis, C., Shiva, S., Dasgupta, D., Shandilya, V., and Wu, Q. (2010). A survey of game theory as applied to network security. In *2010 43rd Hawaii International Conference on System Services,* IEEE, 1–10.

Rubinstein, A. (1991). Comments on the interpretation of game theory. *Econometrica: Journal of the Econometric Society,* 900–924.

Ryutov, T., Orosz, M., Blythe, J., and von Winterfeldt, D. (2015). A game theoretic framework for modeling adversarial cyber security game among attackers, defenders, and users. In *International Workshop on Security and Trust Management,* Springer, 274–282.

Schatz, D., and Bashroush, R. (2017). Economic valuation for information security investment: a systematic literature review. *Information Systems Frontiers,* 19(5), 1205–1228.

Seebruck, R. (2015). A typology of hackers: classifying cyber malfeasance using a weighted arc circumplex model. *Digital Investigation,* 14, 36–45.

Smith, G.S. (2004). Recognizing and preparing loss estimates from cyber attacks. *Information Systems Security,* 12(6), 46–57.

Tambe, M. (2011). *Security and Game Theory: Algorithms, Deployed Systems, Lessons Learned.* Cambridge University Press.

Tsai, J., Rathi, S., Kiekintveld, C., Ordonez, F., and Tambe, M. (2011). Iris: a tool for strategic security allocation in transportation networks. *AAMAS (Industry Track),* 37–44.

Van Dijk, M., Juels, A., Oprea, A., and Rivest, R.L. (2013). Flipit: the game of stealty takeover. *Journal of Cryptology,* 26(4), 655–713.

Von Neumann, J., and Morgenstern, O. (1947). *Theory of Games and Economic Behavior.* Princeton University Press.

Von Stackelberg, H. (2010). *Market Structure and Equilibrium.* Springer Science and Business Media.

Wagener, G., State, R., Dulaunoy, A., and Engel, T. (2009). Self adaptive high interaction honeypots driven by game theory. In *Symposium on Self-Stabilizing Systems,* Springer, 741–755.

Wang, S., and Shroff, N. (2017). Security game with non-additive utilities and multiple attacker resources. *Proceedings of the ACM on Measurement and Analysis of Computing Systems,* 1(1), 1–32.

Wang, Y., Wang, Y., Liu, J., Huang, Z., and Xie, P. (2016). A survey of game theoretic methods for cyber security. In *2016 IEEE First International Conference on Data Science in Cyberspace (DSC),* IEEE, 631–636.

CHAPTER 12

Number Theory

Dane Skabelund

CONTENTS

12.1	Introduction	393
12.2	The Integers	394
	12.2.1 Divisibility and Modular Arithmetic	394
	12.2.2 The Euclidean Algorithm	395
	12.2.3 Prime Numbers	397
	12.2.4 ISBN Codes	398
12.3	Units and Multiplicative Functions	399
	12.3.1 Units and Modular Inverses	399
	12.3.2 Affine Ciphers	400
	12.3.3 The Chinese Remainder Theorem	400
	12.3.4 Multiplicative Functions and the Euler Phi Function	402
	12.3.5 The RSA Cryptosystem	403
12.4	Polynomials, Bits, and Error Correction	404
	12.4.1 Polynomials	404
	12.4.2 Binary and Hexadecimal	405
	12.4.3 Cyclic Redundancy Checks	406
12.5	Primitive Roots	408
	12.5.1 Multiplicative Order and Primitive Roots	408
	12.5.2 Discrete Logarithms	409
	12.5.3 Diffie-Hellman Key Exchange	410
	12.5.4 Baby-Step Giant-Step	410
12.6	Primality Testing	411
	12.6.1 Trial Division and the Sieve of Eratosthenes	411
	12.6.2 The Miller-Rabin Test	412
12.7	Integer Factorization	413
	12.7.1 Pollard's Rho Algorithm	414
	12.7.2 The Quadratic Sieve	416
12.8	Conclusion	418

12.1 INTRODUCTION

Number theory concerns the study of the integers and their properties. Since the integers are so fundamental to mathematics, one should not be surprised when number theory comes up in unexpected places. It is particularly useful in the cyber realm, where it forms the basis of large parts of modern cryptography and has many applications to error-correction. In this chapter we give a brief introduction to several fundamental topics in number theory and its applications.

Here is a brief outline of the chapter. We will begin in Sections 12.2 and 12.3 by exploring fundamental properties of the integers. In these sections we discuss divisibility, factorization, units modulo n, and multiplicative functions, and as applications of these ideas we introduce ISBN codes, affine ciphers, and the RSA cryptosystem. Then in Section 12.4 we will briefly discuss polynomials in one variable over a field, and how these can be used to create cyclic redundancy checks. In Section 12.5 we introduce primitive roots modulo n, along with the discrete logarithm problems and its application to the Diffie-Hellman key exchange. To finish the chapter, we discuss in Sections 12.6 and 12.7 the problems of primality testing and integer factorization, and introduce some algorithms for solving these problems.

12.2 THE INTEGERS

12.2.1 Divisibility and Modular Arithmetic

One of the most fundamental notions in number theory is that of divisibility. Given integers m and n, we say that n divides m, written $n \mid m$, if $m = nk$ for some integer k. In this case, we call n a divisor of m and m a multiple of n. We list a few of the basic properties of divisibility:

1. If $a \mid b$ and $b \mid c$, then $a \mid c$.

2. If $a \mid b$ and $a \mid c$, then $a \mid (b+c)$.

3. $a \mid b$ if and only if $ac \mid bc$.

Theorem 12.1 (Division algorithm). *Let n be a positive integer. Then for any integer a, there are unique integers q and r with $0 \le r < n$ such that $a = qn + r$.*

Proof: Consider the set of integer multiples of n which are less than or equal to a. This set has a maximal element $b = qn$. Moreover, the maximality of b implies that

$$b + d = n(q+1) > a = b + (a - b),$$

so that the nonnegative number $r = a - b$ satisfies $r < n$. □

The numbers q and r in the theorem are called the quotient and remainder of the division of a by n. Sometimes r is written as $a \bmod n$.

For any positive integer n, the set of integers $\mathbb{Z}$ is a disjoint union of the sets

$$[r] = \{\ldots, r - n,\ r,\ r + n,\ r + 2n, \ldots\}, \qquad r = 0, \ldots, n-1,$$

called residue classes modulo n, composed of all integers that leave the same remainder when divided by n. We write $\mathbb{Z}/n\mathbb{Z}$ for the set of these residue classes.

We can indicate that two integers s and t are in the same residue class modulo n by writing $s \equiv t \bmod n$. This means that $s = t + kn$ for some integer k, or that the difference $s - t = kn$ is divisible by n. Thus, with this notation, writing $m \equiv 0 \bmod n$ is the same as saying that n divides m.

If two integers are added or multiplied, then the residue class of the result only depends on the residue classes of the original numbers. It follows that addition and multiplication can be defined on the set of residue classes by

$$[r] + [s] = [r_1 + r_2], \qquad [r_1] \cdot [r_2] = [r_1 r_2].$$

This is called modular arithmetic. The addition and multiplication on $\mathbb{Z}/n\mathbb{Z}$ inherit the properties from the integers that make $\mathbb{Z}/n\mathbb{Z}$ a commutative ring with unity.

Example 12.1.

1. Working modulo 6, we have $4 + 4 = 8 \equiv 2 \bmod 6$, and $2 \cdot 3 = 6 \equiv 0 \bmod 6$, and $5^2 = 25 \equiv 1 \bmod 6$.

2. When working in $\mathbb{Z}/10\mathbb{Z}$, we can simply forget all decimal digits of a number except for the units: $142 \cdot 77 \equiv 2 \cdot 7 = 14 \equiv 4 \bmod 10$.

12.2.2 The Euclidean Algorithm

The greatest common divisor (GCD) of two nonzero integers m and n, is the largest positive integer $d = \gcd(m, n)$ with $d \mid m$ and $d \mid n$. If one or both of m and n are zero, we set $\gcd(m, n) = 0$. Similarly, the least common multiple (LCM) of m and n is the least nonnegative integer divisible by both m and n.

The following lemma contains an idea that will allow us to efficiently compute the GCD of two integers by repeated use of the division algorithm above.

Lemma 12.1. If m, n, and q are any integers, then

$$\gcd(m, n) = \gcd(n, m - qn).$$

Proof: If d divides both m and n, then d also divides $m - qn$. Conversely, if d divides n and $m - qn$, then d also divides m. □

Suppose that we want to compute the GCD of two nonnegative integers m and n. Provided that $n \neq 0$, we have $m = q_0 n + r_0$ with $0 \leq r_0 < n$. Then by the lemma,

$$\gcd(m, n) = \gcd(n, m - qn) = \gcd(n, r_0).$$

If $r_0 = 0$, then n is the desired GCD and we are done. Otherwise, we we can repeat this process with n and r_0 in the place of m and n, to obtain $n = q_1 r_0 + r_1$ and $\gcd(n, r_0) = \gcd(r_0, r_1)$.

We continue similarly for $i \geq 2$, letting

$$r_{i-2} = q_i r_{i-1} + r_i, \qquad 0 \leq r_i < r_{i-1},$$

until $r_i = 0$, at which point we have $r_{i-1} = \gcd(r_{i-1}, 0) = \gcd(m, n)$. This must eventually happen because the r_i are nonnegative and get smaller at each step.

Example 12.2. Let us compute the GCD of 2470 and 123 using the Euclidean algorithm. We have

$$270 = 2 \cdot 123 + 24$$
$$123 = 5 \cdot 24 + 3$$
$$24 = 8 \cdot 3 + 0.$$

Thus, $\gcd(270, 123) = 3$.

Notice from the second line of the preceding example that $3 = 1 \cdot 123 - 5 \cdot 24$ is an linear combination of 123 and 24 with integer coefficients. Similarly, the first line shows that $24 = 1 \cdot 270 - 2 \cdot 123$ is an integer linear combination of 270 and 123. We can piece these together to obtain

$$3 = -5 \cdot 270 + 11 \cdot 123.$$

This is an instance of the following very useful fact about the greatest common divisor.

Lemma 12.2 (Bézout's Lemma). Let m and n be nonzero integers. Then

$$\gcd(m, n) = am + bn$$

for some integers a and b. Moreover, $\gcd(m, n)$ is the smallest positive integer that can be expressed in this manner.

Proof: Let $d = am + bn$ be the smallest positive integer linear combination of m and n. Then d is dividible by $\gcd(m, n)$, since $\gcd(m, n)$ divides both m and n.

We claim that $d = \gcd(m, n)$. Suppose to the contrary that $d > \gcd(m, n)$. Then d must not divide one of m or n. Without loss of generality, we assume that d does not divide m, and write $m = qd + r$ with $0 < r < d$. But then

$$r = m - qd = (1 - qa)m - qbn$$

is a positive integer linear combination of m and n which is strictly smaller than d, contradicting our definition of d. □

Integers a and b as in the lemma are often called Bézout coefficients. As demonstrated in the example above, these coefficients can be computed by first performing the Euclidean algorithm, storing the intermediate quotients and remainders along the way, and then backtracking to reconstruct the coefficients a and b. We call this process, which takes m and n and returns the tuple (d, a, b) with $d = \gcd(m, n) = am + bn$, the extended Euclidean algorithm. Alternate ways of implementing this algorithm require much less storage, at the cost of adding a few more computations to the intial pass of the Euclidean algorithm.

Integers m and n are called relatively prime, or coprime, if $\gcd(m, n) = 1$. By Bézout's Lemma, any integer may be written as an integer linear combination of coprime m and n. This fact often proves useful.

12.2.3 Prime Numbers

The prime numbers $2, 3, 5, 7, \ldots$ are those integers $p \geq 2$ whose only positive divisors are 1 and p. An integer $n \geq 2$ which is not prime is called a composite number. A composite number n admits a proper factorization $n = de$, that is, one such that neither of the factors is equal to n. Prime numbers form the building blocks for all integers. We collect some of their fundamental properties in this section.

Theorem 12.2 (Fundamental Theorem of Arithmetic). *Every positive integer can be factored as a product of prime numbers, and this factorization is unique up to re-ordering of the prime factors.*

Proof: The result is surely true for all prime numbers, including $n = 2$. If n is any composite number, then it is a factorization $n = de$ with both d and e strictly between 1 and n. By induction, each of d and e has a factorization into primes, and concatenating these yields one for n.

The uniqueness of the factorization follows from the following alternate characterization of prime numbers. □

Lemma 12.3 (Euclid's Lemma). *An integer $p \geq 2$ is prime if and only if p has the property that whenever $p \mid mn$, either $p \mid m$ or $p \mid n$.*

Proof: Suppose that p divides mn, but does not divide m. Then $\gcd(p, m) = 1$, and by Bézout's Lemma there are integers a and b satisfying $ap + bm = 1$. Multiplying through by n yields

$$apn + bmn = n.$$

Then since p divides both terms on the left-hand side, it must also divide n.

On the other hand, if p is not prime then it admits a factorization $p = de$ with d, e both strictly smaller than p. Thus p divides de, but p divides neither d nor e. □

The following fact has been known since ancient times.

Theorem 12.3 (Euclid). *There are infinitely many prime numbers.*

Proof: Let $S = \{p_1, \ldots, p_t\}$ be any finite set of prime numbers, and consider the number $N = p_1 p_2 \cdots p_t + 1$. Since $N \geq 2$, it is divisible by some prime q. But q is not in S, since when N is divided by any of the p_i, it leaves a remainder of 1. Thus N is not divisible by any of the p_i. We conclude that no finite set contains all the prime numbers, so there must be infinitely many of them. □

The least common multiple and greatest common divisor of two integers are easily expressed in terms of prime factorization.

Theorem 12.4. *Let m and n be positive integers, written as the product of prime powers as $m = p_1^{a_1} \cdots p_s^{a_s}$ and $n = p_1^{b_1} \cdots p_s^{b_s}$. Then*

$$\gcd(m, n) = p_1^{\min(a_1, b_1)} \cdots p_s^{\min(a_s, b_s)}$$

and

$$\operatorname{lcm}(m, n) = p_1^{\max(a_1, b_1)} \cdots p_s^{\max(a_s, b_s)}.$$

Consequently, $mn = \gcd(m, n) \operatorname{lcm}(m, n)$.

Proof: The proposed GCD certainly divides both m and n, so the power of p_i that divides $\gcd(m,n)$ is at least $\min(a_i, b_i)$. On the other hand, suppose that p_i^c divides m with $c > a_i$, so that $m = p_i^c d$ for some integer d. Then

$$\prod_{j \neq i} p_j^{a_j} = m/p_i^{a_i} = p_i^{c-a_i} d.$$

But then Euclid's Lemma implies that $p_i = p_j$ for some $j \neq i$, a contradiction.

A similar argument gives the formula for the least common multiple. □

12.2.4 ISBN Codes

Most books published around the world, be they in print form or digital form as ebooks or audiobooks, are assigned an International Standard Book Number (ISBN). This ten or thirteen digit code is a unique identifier for the book, including its version and edition. Built into this code is a method of error correction that makes use of modular arithmetic. We will see a similar but more complicated method of error-correction used for storing and transferring data when we discuss cyclic redundancy checks in Section 12.4.

Books published before 2007 received a ten digit ISBN. For example, the ISBN-10 for the hardcover version of *Rational Points on Elliptic Curves* by Joseph Silverman and John Tate, is

$$0 - 387 - 97825 - 9.$$

The first nine digits of the string contain information about the book, and are often separated into groups of various length by spaces or dashes. The first two groups 0 and 387 indicate the geographical or language area (English) and identity of the publisher (Springer), while the third group 97825 indicates the title. The last digit 9 is called the check digit, and is used to detect some of the most common types of single errors that occur when reading and transcibing strings of numbers.

If the nine digit string containing the book information is $a_{10} a_9 \ldots a_2$, then the check digit a_1 is chosen to satisfy

$$a_1 + 2a_2 + 3a_3 + \cdots 10 a_{10} \equiv 0 \bmod 11.$$

There will always be an a_1 satisfying this equation, namely $a_1 = -(2a_2 + 3a_3 + \cdots + 10 a_{10})$, reduced modulo 11. If it happens that $a_1 = 10$, then the digit a_1 is represented by a capital X. In the example above, we have

$$2 \cdot 5 + 3 \cdot 2 + 4 \cdot 8 + 5 \cdot 7 + 6 \cdot 9 + 7 \cdot 7 + 8 \cdot 8 + 9 \cdot 3 + 10 \cdot 0 = 277 \equiv 2 \bmod 11,$$

so that $a_1 = 11 - 2 = 9$ is indeed the right check digit.

There are two types of errors that the ISBN-10 check digit helps detect: misread digits and transposition errors. A misread digit happens when one of the digits a_n is mis-entered as a'_n. A single misread digit will always yield an invalid ISBN. To see this, we perform the check above to obtain

$$a_1 + 2a_2 + \cdots + n a'_n + c \cdots + 10 a_{10} \equiv n(a'_n - a_n) \bmod 11.$$

Then since neither n nor $a'_n - a_n$ are divisible by 11 and 11 is prime, the result is nonzero, indicating that the code entered is not a valid ISBN. A transposition error occurs when two adjacent digits a_n and a_{n+1} are accidentally permuted (as often happens when typing). In this case, performing the check yields

$$a_1 + \cdots + na_{n+1} + (n+1)a_n + \cdots + 10a_{10} \equiv a_n - a_{n+1} \bmod 11,$$

so this type of error can be detected as well. Although the check digit allows either of these types of errors to be detected, it does not help to correct the errors, nor can it be used to reliably detect multiple errors.

The set of all ISBN-10 codes lies inside of a 9-dimensional vector space defined by the ISBN check equation. This space is a simple example of a linear error-correcting code over the finite field $\mathbb{F}_{11}$, and the ISBN codes inherit their error-correcting properties from this space. More robust error-correcting codes are designed to provide reliable communication over noisy channels, where better error-correction capability is necessary.

As 10-digit ISBNs began to run out, a newer standard was needed to replace it. The modern ISBN-13 standard, which was adopted in 2007, is similar to its 10-digit cousin, in that it also has a check digit, but uses a check equation of the form

$$a_1 + 3a_2 + a_3 + 3a_4 + \cdots + 3a_{12} + a_{13} \equiv 0 \bmod 10.$$

The check-digit for ISBN-13 allows for the detection of most of the types of errors mentioned above. However, because of the increased length relative to the modulus, and because 10 is not a prime number, it cannot detect all errors. For example, if digits a_2 and a_3 of a valid ISBN-13 $a_{13} \ldots a_3 a_2 a_1$ differ by 5, then transposing them causes the check to yield

$$a_1 + 3a_3 + a_2 \cdots + 3a_{12} + a_{13} \equiv 2(a_3 - a_2) = \pm 10 \equiv 0 \bmod 10,$$

so that transposing digits a_2 and a_3 yields another valid ISBN.

12.3 UNITS AND MULTIPLICATIVE FUNCTIONS

12.3.1 Units and Modular Inverses

An integer r is invertible modulo n if there is an integer a such that $ar \equiv 1 \bmod n$. In this case, the residue class of a is called the inverse of the residue class of r. The set $(\mathbb{Z}/n\mathbb{Z})^\times$ of invertible residue classes forms a group under multiplication, called the group of units modulo n.

If r is invertible modulo n, then multiplying through by the inverse of r modulo n shows that the congruence $rs \equiv rt \bmod n$ implies $s \equiv t \bmod n$. In other words, $(\mathbb{Z}/n\mathbb{Z})^\times$ consists of those classes that one is allowed to divide by when dealing with congruences modulo n.

Lemma 12.4. The group $(\mathbb{Z}/n\mathbb{Z})^\times$ consists of those residue classes $[r]$ with $\gcd(n, r) = 1$.

Proof: If $\gcd(r,n) = 1$, then there are Bézout coefficients a and b such that

$$ar + bn = 1.$$

Reducing modulo n, this says that $ar \equiv 1 \bmod n$, so we have constructed an inverse a for $r \bmod n$.

On the other hand, if $\gcd(r,n) = d > 1$, then

$$r \cdot \frac{n}{d} = \frac{r}{d} \cdot n \equiv 0 \bmod n.$$

If r had an inverse a modulo n, then multiplying through by a would yield $n/d \equiv 0 \bmod n$, which is not the case. □

The most common use of the extended Euclidean algorithm is the computation of inverses modulo n. Note, however, that only one of the Bézout coefficients is necessary for this purpose, as demonstrated by the proof of the lemma.

If p is prime, then every integer r that is not divisible by p satisfies $\gcd(r,p) = 1$. This motivates the term 'relatively prime' introduced earlier. By Lemma 12.4, every nonzero element of $\mathbb{Z}/p\mathbb{Z}$ is invertible, so that $\mathbb{Z}/p\mathbb{Z}$ forms a field, often denoted $\mathbb{F}_p$. As a result, the tools of linear algebra apply when solving systems of linear congruences modulo p.

12.3.2 Affine Ciphers

An affine cipher is a classical cipher that uses modular arithmetic. To describe it, let us fix an association of the alphabet $\{A, B, \ldots, Z\}$ with the residue classes modulo 26:

$$A = 0, \quad B = 1, \quad C = 2, \quad \ldots, \quad Z = 25.$$

To encrypt a message, consisting of a string of letters, we first choose a polynomial of the form $f(x) = ax + b$, where $a, b \in \mathbb{Z}/26\mathbb{Z}$ with a invertible, called an affine function. Then we apply the function f to each letter in our message. For example, if $f(x) = 5x + 6$, then the word GRANARY encrypts to KNGTGNW.

Someone who knows the inverse of $f(x)$ can then decrypt the message. The inverse of f is also an affine function, namely $f^{-1}(x) = a'(x - b)$, where $aa' \equiv 1 \bmod 26$ (this is why we needed a to be invertible). In our example, we can use the extended Euclidean algorithm to find that the inverse of $a = 5 \bmod 26$ is $a' = 21$, and so $f(x) = 5x + 6$ has inverse

$$f^{-1}(x) = 21(x - 6) \equiv 21x + 4 \bmod 26.$$

Since there are only 12 units modulo 26, namely 1, 3, 5, 7, 9, 11, 15, 17, 19, 21, 23, and 25, there are only $12 \cdot 26 = 312$ different functions that f^{-1} could be. For this reason, among others, an affine cipher is easily broken (i.e., it is not a fine cipher).

12.3.3 The Chinese Remainder Theorem

Just as there is a unique straight line through two given points on the plain, a polynomial function $f(x)$ of degree d is determined by its values at any $d + 1$

points. Suppose that $(x_0, y_0), \ldots, (x_d, y_d)$ are points on the graph $y = f(x)$ with distinct x-coordinates. Then Lagrange Interpolation may be used to recover f as follows. For each $i = 0, \ldots, d$, construct the polynomial

$$g_i(x) = \prod_{j \neq i} \frac{x - x_j}{x_i - x_j},$$

which has degree d, and takes the value 1 at x_i and the value 0 at all x_j with $j \neq i$. Then

$$f(x) = y_0 g_0(x) + \cdots + y_d g_d(x).$$

The same idea may be used to solve systems of congruences. Suppose we wish to find an integer x satisfying the system of congruences

$$x \equiv a_1 \bmod m_1, \quad x \equiv a_2 \bmod m_2, \quad \ldots, \quad x \equiv a_t \bmod m_t,$$

where the moduli m_i are pairwise coprime. Let m be the product of the m_i. Then since the m_i are pairwise coprime, m_i and m/m_i are relatively prime for each i, so m_i has an inverse m_i' modulo m/m_i, which we can compute using the Euclidean algorithm. The integers $u_i = m_i' m/m_i$ are analogous to the polynomials $g_i(x)$ above, in that they satisfy $u_i \equiv 1 \bmod m_i$ and $u_i \equiv 0 \bmod m_j$ for $j \neq i$. The integer $x = a_1 u_1 + \cdots a_t u_t$ is the desired solution, since for each i we have

$$x \equiv a_1 \cdot 0 + \cdots + a_i \cdot 1 + \cdots + a_t \cdot 0 \equiv a_i \bmod m_i.$$

We have proved the following important theorem.

Theorem 12.5 (Chinese Remainer Theorem). *Let $m_1, \ldots, m_t$ be pairwise coprime integers, all greater than 1, with product $m = m_1 m_2 \cdots m_t$. Then for any integers $a_1, \ldots, a_t$, there is a unique solution x modulo m of the system of congruences*

$$x \equiv a_1 \bmod m_1, \quad x \equiv a_2 \bmod m_2, \quad \ldots, \quad x \equiv a_t \bmod m_t.$$

Note that our proof not only shows the existence of a solution, but provides a practical method of finding one. A nice way to think of this theorem is that the ring $\mathbb{Z}/m\mathbb{Z}$ decomposes as a direct product

$$\mathbb{Z}/m\mathbb{Z} \cong \mathbb{Z}/m_1\mathbb{Z} \times \mathbb{Z}/m_2\mathbb{Z} \times \cdots \times \mathbb{Z}/m_t\mathbb{Z},$$

so that congruences modulo the various m_i are independent of one another. Of particular interest is when the m_i are the prime powers in the prime factorization of m.

Note that although we used Lagrange interpolation to motivate the Chinese Remainder Theorem, it itself is a special case of a more general version of the theorem.

Example 12.3. The Chinese Remainder Theorem can be useful when one wants to perform a large computation involving arithmetic operations. Depending on the situation, it may be more efficient to perform the computation modulo several primes or prime powers, and then use the process outlined in the proof of the theorem to reconstruct the answer.

For example, suppose that we wanted to find all square roots of unity modulo m, where $m = p_1 p_2 \cdots p_t$ is the product of several distinct primes. Then the Chinese Remainder Theorem says that the equation $x^2 \equiv 1 \bmod m$ is equivalent to a simultaneous solution of

$$x^2 \equiv 1 \bmod p_1, \quad x^2 \equiv 1 \bmod p_2, \quad \ldots \quad x^2 \equiv 1 \bmod p_t.$$

But each of these has exactly two solutions $\pm 1 \bmod p_i$, corresponding to the factorization of the polynomial $x^2 - 1 = (x-1)(x+1)$ modulo p_i. Therefore, there are exactly 2^t solutions of $x^2 \equiv 1 \bmod m$, each corresponding to t choices of solutions $\epsilon_i = \pm 1 \bmod p_i$, one for each prime modulus. We can reconstruct x from $(\epsilon_1, \ldots, \epsilon_t)$ by first finding u_i as in the proof of the theorem, and then taking

$$x = \epsilon_1 u_1 + \cdots + \epsilon_t u_t.$$

12.3.4 Multiplicative Functions and the Euler Phi Function

A complex-valued function $f \colon \mathbb{Z} \to \mathbb{C}$ is called multiplicative if

$$f(mn) = f(m)f(n)$$

whenever m and n are relatively prime. A multiplicative function f is determined by its values on prime powers, since if $n = p_1^{a_t} \cdots p_t^{a_t}$, then

$$f(n) = f(p_1^{a_1}) \cdots f(p_t^{a_t}).$$

The quintessential example of a multiplicative function is the Euler phi function

$$\varphi(n) = \#(\mathbb{Z}/n\mathbb{Z})^\times = \#\{1 \le k \le n-1 \mid \gcd(n,k) = 1\}.$$

Why is this function multiplicative? If $n = p_1^{a_t} \cdots p_t^{a_t}$, then the Chinese Remainder Theorem says that the function

$$\Psi \colon \mathbb{Z}/n\mathbb{Z} \longrightarrow \mathbb{Z}/p_1^{a_1}\mathbb{Z} \times \cdots \times \mathbb{Z}/p_t^{a_t}\mathbb{Z}$$

taking $k \bmod n$ to $(k \bmod p_1^{a_1}, \ldots, k \bmod p_t^{a_t})$ is bijective. But an integer k is coprime to n if and only if k is coprime to each of the prime powers dividing n. Therefore, the bijection Ψ identifies $(\mathbb{Z}/n\mathbb{Z})^\times$ with the product $(\mathbb{Z}/p_1^{a_1}\mathbb{Z})^\times \times \cdots \times (\mathbb{Z}/p_t^{a_t}\mathbb{Z})^\times$. The size of the former is $\varphi(n)$, while the size of the latter is $\varphi(p_1^{a_1}) \cdots \varphi(p_t^{a_t})$.

Since there are exactly p^{a-1} multiples of p in the interval from 0 to $p^a - 1$, it follows that $\varphi(p^a) = p^a - p^{a-1} = p^a \left(1 - \frac{1}{p}\right)$. Therefore, the Euler phi function may be expressed as the product

$$\phi(n) = n \prod_{p \mid n} \left(1 - \frac{1}{p}\right).$$

Theorem 12.6 (Euler's theorem). *For any integer a coprime to n,*
$$a^{\varphi(n)} \equiv 1 \bmod n.$$

Proof: The number $\varphi(n)$ is the order of the group $(\mathbb{Z}/n\mathbb{Z})^\times$. Therefore, by Lagrange's theorem in group theory, any element of this group becomes the identity when raised to the power $\varphi(n)$. □

In the case that $n = p$ is prime, we have $\varphi(p) = p - 1$, and Euler's theorem implies that
$$a^p \equiv a \bmod p$$
for any integer a. This fact is called Fermat's little theorem.

A multiplicative function closely related to Euler's phi function is the Charmichael lambda function $\lambda(n)$, defined as the exponent of the group of units modulo n, i.e., the smallest number e such that $a^e \equiv 1 \bmod n$ for all a with $\gcd(a, n) = 1$. By Euler's theorem, it follows that $\varphi(n)$ is a multiple $\lambda(n)$ for any n. Theorem 12.8 in the next section will show that in fact $\lambda(p^a) = \varphi(p^a)$ for all prime powers p^a other than powers of 2 which are at least 8. In the exceptional cases, $\lambda(2^a) = 2^{a-2} = \frac{1}{2}\varphi(2^a)$.

Other important multiplicative functions that come up often include:

- The sum of divisor functions $\sigma_k(n) = \sum_{d|n} d^k$;

- The Möbius function $\mu(n)$, defined by $\mu(n) = (-1)^t$ if n is the product of t distinct primes, and $\mu(n) = 0$ otherwise;

- For a fixed prime p, the Legendre symbol
$$\left(\frac{n}{p}\right) = \begin{cases} 1, & n \text{ a nonzero square mod } p \\ -1, & n \text{ not a square mod } p \\ 0, & n \equiv 0 \bmod p. \end{cases}$$

12.3.5 The RSA Cryptosystem

The RSA cryptosystem is a widely used asymmetric public-key cryptosystem based on the difficulty of the problem of integer factorization. It's setup involves three integers n, e, and d. The pair (n, e) forms the public key, and the remaining integer d forms the private key. The modulus $n = pq$ is taken as the product of two large prime numbers (this factorization is not disclosed), and the encryption and decryption exponents e and d are chosen to be inverses modulo $\lambda(n)$, so that $de \equiv 1 \bmod \lambda(n)$.

To encrypt a message, represented by an integer m, one looks up the public key (n, e) of the intended recipient and creates the ciphertext $c = m^e \bmod n$. Only the intended recipient knows the decryption exponent and can recover m as
$$c^d \equiv m^{de} \equiv m^{1+k\lambda(n)} \equiv m \bmod n.$$

If the factorization $n = pq$ is discovered, then the security of this system is completely broken. For then $\varphi(n) = (p-1)(q-1)$ is known, and the extended

Euclidean algorithm can be used to quickly compute an inverse d' for $e \bmod \varphi(n)$. Euler's theorem assures that exponentiating by d' will work for decryption. Determining whether breaking RSA is theoretically as difficult as factoring numbers of the form $n = pq$ remains a difficult open problem.

Since RSA was created in the 1970s, attacks against it have been developed that work against certain ranges of parameters. Care must be taken when implementing RSA to avoid these. Because RSA is relatively slow, in practice it is commonly used to initially exchange shared keys for a symmetric-key cryptosystem such as AES. Given the fact that there exist quantum algorithms, such as Shor's algorithm, which can factor integers in polynomial time, if at some point sufficiently large quantum computers are able to be built, then RSA and other crytosystems that rely on the difficulty of factoring may become obsolete.

12.4 POLYNOMIALS, BITS, AND ERROR CORRECTION

12.4.1 Polynomials

Most of the properties of the integers mentioned in Section 12.2 are also enjoyed by polynomials over the rational numbers $\mathbb{Q}$ or a field $\mathbb{F}_p$ of prime order. Since these are so important, we mention some of them here. In this section, F will denote either $\mathbb{Q}$ or $\mathbb{F}_p$ (or any other field, if you wish). Of particular interest will be the case when $F = \mathbb{F}_2$ is the binary field. The ring of polynomials $a_0 + a_1 x + \cdots + a_n x^n$ over F is denoted by $F[x]$. Addition and multiplication of polynomials in $F[x]$ is done as in elementary algebra, by collecting terms with the same power of x.

Divisibility and modular arithmetic work the same with polynomials as they do with integers. We say that a polynomial $f(x)$ divides $g(x)$ if there is another polynomial $h(x)$ such that $f(x)h(x) = g(x)$. We also write $f(x) \equiv g(x) \bmod h(x)$ to indicate that $f(x)$ and $g(x)$ differ by a multiple of $h(x)$, i.e., that there is a polynomial $k(x)$ such that $f(x) - g(x) = h(x)k(x)$. Since the degree of the product of two polynomials $f(x)$ and $g(x)$ is equal to the product of their degrees, the only units in $F[x]$, i.e., polynomials $f(x)$ such that $f(x)g(x) = 1$ for some $g(x)$, are the nonzero constant polynomials.

The residue of one polynomial modulo another can be computed using polynomial long division, yielding a division algorithm similar to Theorem 12.1. This may be iterated to compute greatest common divisors of polynomials, or to find inverses modulo a given polynomial.

Theorem 12.7 (Division algorithm for polynomials). *Let $f(x)$ and $g(x)$ be a polynomial in $F[x]$, with $g(x) \neq 0$. Then there are unique polynomials $q(x)$ and $r(x)$*

$$f(x) = q(x)g(x) + r(x)$$

and such that either $r(x) = 0$ and $\deg r(x) \leq \deg g(x)$.

The polynomials in $F[x]$ that play the role that prime numbers do in the integers are called irreducible polynomials. These are polynomials which cannot be factored as the product of two non-constant polynomials. Up to re-ordering the factors or multiplying factors nonzero elements of F, any polynomial in $F[x]$ admits a unique factorization into irreducible polynomials, similar to Theorem 12.2.

12.4.2 Binary and Hexadecimal

> *"Believe me, my young friend, there is nothing—absolutely nothing—half so much worth doing as simply messing about in boats. Simply messing [...] about in boats—or with boats."* [1]
> – Kenneth Grahame, *The Wind in the Willows*

Data is represented and manipulated by computers as sequences of bits, this is, zeros and ones. The standard unit of data, the byte, is typically 8 bits long; an individual byte can represent a total of $2^8 = 256$ different values.

Because long strings of 0's and 1's can be difficult to parse by sight, binary data is often represented for humans to read by strings of hexadecimal (base 16) digits. Each byte is made up of two hexadecimal digits, which take the place of four bits each. Hexadecimal uses the digits 0 through 9, along with the letters `a` through `f` to represent the digits 10 through 15. This is summarized in Table 12.1.

TABLE 12.1 Conversion between bases

Dec	Hex	Bin	Dec	Hex	Bin
0	0	0000	8	8	1000
1	1	0001	9	9	1001
2	2	0010	10	a	1010
3	3	0011	11	b	1011
4	4	0100	12	c	1100
5	5	0101	13	d	1101
6	6	0110	14	e	1110
7	7	0111	15	f	1111

For example, the three bytes `00011111 10001011 00001000` often used to indicate the beginning of a file compressed with `gzip` are represented in hexadecimal as `1f 8b 08`.

A given string of bits can be interpreted very differently, depending on the context. Understanding this context is important. For example, the byte `01001011` could be interpreted as the integer 147 written in base 2:

$$2^7 + 2^4 + 2^1 + 2^0 = 147.$$

The same byte could also be interpreted as the ASCII character K (ASCII is a 7 bit encoding—the first bit is usually set to 0, or used for error correction), or could represent the vector $(0, 1, 0, 0, 1, 0, 1, 1)$ in a vector space over $\mathbb{F}_2$, or the polynomial $x^7 + x^4 + x + 1$ over $\mathbb{F}_2$.

---

[1] If, unlike the water rat, you expect to find yourself frequently messing about with *bits*, it is worth getting comfortable converting quickly between numbers in binary and hexadecimal form.

If bits are thought of as elements of the binary field $\mathbb{F}_2$, then the operations of addition and multiplication in the field are the same as the logical XOR and AND operations on bits. These operations are often done bitwise, as in the following example:

$$
\begin{array}{r}
10101110 \\
\text{XOR } 01110100 \\
\hline
11011010
\end{array}
$$

This is the same as addition of vectors or polynomials over $\mathbb{F}_2$.

Example 12.4. In the context of polynomials over $\mathbb{F}_2$, adding or removing a zero from the right-hand side of a binary string (shifting left or right) corresponds to multiplication or division by x. The division algorithm in Theorem 12.7 can be performed using a combination of bitwise XOR's and shifting of the modulus as follows:

1. Shift the modulus f so that the leading 1's of f and g line up.

2. Replace g with the bitwise XOR of f and g.

3. Repeat until remainder is shorter than the modulus.

We demonstrate this in Figure 12.1, where we reduce the polynomial $g(x) = x^{12} + x^{11} + x^9 + x^6 + x^5 + x^3 + 1$ modulo $f(x) = x^4 + x^3 + 1$ over $\mathbb{F}_2$.

1101001101001	g
11001	f shifted left 8 bits
1101101001	
11001	f shifted left 5 bits
1001001	
11001	f shifted left 2 bits
101101	
11001	f shifted left 1 bit
11111	
11001	f itself
110	$g \bmod f$

FIGURE 12.1 Dividing two polynomials over $\mathbb{F}_2$.

This shows that $g(x) = (x^8 + x^5 + x^2 + x + 1)f(x) + (x^2 + x)$.

12.4.3 Cyclic Redundancy Checks

A cyclic redundancy check, or CRC, is a method used to protect against corruption or data loss when storing information in memory or sending it across a network. CRCs are used frequently in packet data or transmitted data because they are easy to implement, and because they are particularly good at detecting short bursts of errors, which often occur in this setting.

The simplest kind of CRC is called a parity bit. This is a bit which is added to a string of data to ensure that the number of 1's in the data is even or odd. In our discussion, we will focus on even parity bits, which ensure that the data transmitted has an even number of 1's.

For example, suppose Sam wants to send the 7-bit string 1100100 to Theresa over a noisy channel, and is worried that one or more of the bits might get flipped. Since the desired payload has an odd number of 1's, Sam appends a 1 to the data, and sends 11001001. When Theresa receives the data, however, it reads 10001001! Since this has an odd number of 1's, she knows that something was wrong with the transmission, so she asks Sam to send it again. In general, a parity bit will detect whenever an odd number of bits is flipped. It will not tell us which bits were flipped, however, nor will it detect an even number of flipped bits.

A single parity bit is sometimes incorporated as the most significant digit of bytes representing ASCII data, which is a 7-bit encoding. For example, instead of using

01001000 01001001 00100000 01000010 01001111 01000010

to represent "HI BOB" (note that all the first bits in these bytes are 0, as the characters are encoded in the last 7 bits), we could change the first bit of each individual byte as a parity bit, and use

01001000 11001001 10100000 01000010 11001111 01000010.

This contains all the same information, but if any given character is received incorrectly, causing it to have the wrong parity, then we will notice.

Now let us interpret a string of binary data as a polynomial $g(x)$ over $\mathbb{F}_2$. If we evaluate $g(x)$ at $x = 1$, we end up taking the sum of several 1's, one for each term of $g(x)$, so that $g(1) = 0$ if there is an even number of terms, and $g(1) = 1$ if there is an odd number of terms. But $g(1)$ is precisely the remainder when $g(x)$ is divided by $f(x) = x - 1$ (more commonly written $x + 1$, since we are working over $\mathbb{F}_2$). This observation is what allows us to generalize parity bits to CRCs which are many bits long.

More generally, an n-bit CRC is a sequence of n bits which are added to data to ensure that it is divisible by a fixed polynomial $f(x)$ of degree n, when thought of as a polynomial over $\mathbb{F}_2$. The choice of polynomial $f(x)$ has a large effect on the error-correction capabilities of a CRC. We will not get into the details here, as it would lead us too far afield but will mention that an appropriately chosen $f(x)$ is capable of detecting any two bit errors within a block of data of size $2^n - 1$, or any two bit errors or any odd-number of errors in a block of size $2^{n-1} - 1$.

As an example, let us add a 4-bit CRC to 1101001101001 by using $f(x) = x^4 + x^3 + 1$, which is the polynomial we used as a modulus in Figure 12.1. We want to know which 4 bits, when appended to the data, yields a string corresponding with a polynomial divisible by $f(x)$. If $g(x)$ be the 12-degree polynomial we wish to encode, then we want to find $h(x)$ of degree ≤ 3 such that

$$x^4 g(x) + h(x) \equiv 0 \bmod f(x).$$

In other words, we want to compute $x^4 g(x) \bmod f(x)$. We can do this by first appending 0000 to the data (multiplying by x^4), and proceeding as in Figure 12.1.

$$
\begin{array}{r}
1101001101001\ 0000 \\
11001 \\
\ddots \\
11111\ 0000 \\
11001 \\
110\ 0000 \\
110\ 01 \\
\hline
0100
\end{array}
$$

FIGURE 12.2 Computing a 4-bit CRC.

The resulting CRC, as seen in Figure 12.2, is 0100, which corresponds to the polynomial $h(x) = x^2$.

12.5 PRIMITIVE ROOTS

12.5.1 Multiplicative Order and Primitive Roots

For any integer a coprime to n, its (multiplicative) order modulo n, written as $\text{ord}_n(a)$, is the smallest positive integer k such that $a^k \equiv 1 \bmod n$. Euler's theorem implies that such an integer exists. In fact, $\text{ord}_n(a)$ is always a divisor of $\varphi(n)$, since $\text{ord}_n(a)$ is the order of a in the group $(\mathbb{Z}/n\mathbb{Z})^\times$, which has order $\varphi(n)$.

If a is such that $\text{ord}_n(a) = \varphi(n)$, then a is called a primitive root modulo n.

Lemma 12.5. *If a is a primitive root modulo n, then $(\mathbb{Z}/n\mathbb{Z})^\times$ consists of the distinct residue classes*
$$1,\ a,\ a^2,\ \ldots,\ a^{\varphi(n)-1}.$$

Proof: There are $\varphi(n)$ of these residue classes in this list, so we only need to show that they are distinct. Suppose that $a^j \equiv a^k \bmod n$ for some j, k with $0 \leq j \leq k < \varphi(n)$. Then multiplying through by a^{-j} yields $a^{k-j} \equiv 1 \bmod n$. Since $k - j \leq \varphi(n) = \text{ord}_a(n)$, it follows that $k - j = 0$. □

In other words, n has a primitive root a if and only if the group $(\mathbb{Z}/n\mathbb{Z})^\times$ is cyclic, with a as a generator. It is not the case that every n has a primitive root. The following theorem, for which we omit a proof, gives a classification of those n that do.

Theorem 12.8. *Let n be a positive integer. Then there exists a primitive root modulo n only in the following cases:*

- $n = 1, 2, 4$

- $n = p^a$ or $2p^a$ with p an odd prime.

Proof: A proof of this theorem may be found in many introductory number theory texts. See, for example, Apostol (1976, Chapter 10). □

In particular, the theorem states that $(\mathbb{Z}/p^a\mathbb{Z})^\times$ is cyclic for p^a an odd prime power or if $p = 2$ and $a \leq 2$. For higher powers of the prime 2, one has $(\mathbb{Z}/2^a\mathbb{Z})^\times \cong \mathbb{Z}/2\mathbb{Z} \times \mathbb{Z}/2^{a-2}\mathbb{Z}$.

Theorem 12.9. *If n has a primitive root, then there are exactly $\varphi(\varphi(n))$ primitive roots modulo n.*

Proof: Suppose that n has a primitive root a. Then every element of $(\mathbb{Z}/n\mathbb{Z})^\times$ is of the form a^k for some k. Now a^k is a primitive root if and only if there is k' such that
$$a \equiv (a^k)^{k'} \equiv a^{kk'} \mod n.$$
But this is the case if and only if $kk' \equiv 1 \mod \varphi(n)$, since $\varphi(n) = \text{ord}_n(a)$. In other words, the primitive roots are of the form a^k where k is invertible modulo $\varphi(n)$. There are exactly $\varphi(\varphi(n))$ such k. □

In particular, every prime p has a primitive root, and there are $\varphi(p-1)$ such.

12.5.2 Discrete Logarithms

Fix an integer n such that n has a primitive root a. Then every $b \in (\mathbb{Z}/n\mathbb{Z})^\times$ can be written in the form
$$b \equiv a^k \mod n$$
for unique $k \mod \varphi(n)$. This exponent k is referred to as the discrete logarithm of b with respect to a, and we will write it as $\text{Log}_a(b)$. The function Log_a satisfies properties analogous to the ordinary logarithm, such as
$$\text{Log}_a(bc) \equiv \text{Log}_a(b) + \text{Log}_a(c) \mod \varphi(n)$$
and
$$\text{Log}_a(b^m) \equiv m \text{Log}_a(b) \mod \varphi(n).$$
The first property means that Log_a defines an isomorphism
$$(\mathbb{Z}/n\mathbb{Z})^\times \to \mathbb{Z}/\varphi(n)\mathbb{Z}$$
of cyclic groups. Discrete logarithms can also be defined for any cyclic group, and in particular for the multiplicative group of a finite field.

The problem of determining $\text{Log}_a(b)$ given a and b, is called the discrete logarithm problem. There are currently no efficient general solutions to this problem, and several systems in cryptography base their security around it. To finish this section, we present one such system, along with one method for computing discrete logarithms.

12.5.3 Diffie-Hellman Key Exchange

Diffie-Hellman key exchange is a system used when two parties want to agree on a shared secret key to use, for example, to exchange messages using a symmetric-key cryptosystem such as AES. It is built around the difficulty of the following problem:

$$\text{Given } a^j \text{ and } a^k \text{ mod } p, \text{ compute } a^{jk} \text{ mod } p.$$

Suppose that our two parties, which we call Alice and Bob, wish to establish a shared private key using Diffie-Hellman key exchange. Then they begin by agreeing on a large prime number p and a primitive root a mod p. Each of them then chooses a secret exponent modulo $p-1$. Alice uses the exponent j to compute a^j mod p, which she sends to Bob; Bob uses the exponent k to compute a^k mod p, which he sends to Alice. With these, they can both calculate the shared key a^{jk} mod p. Alice computes it as $(a^k)^j$ mod p and Bob computes it as $(a^j)^k$ mod p.

The hope is that even if a third party eavesdrops on Alice and Bob, and is able to obtain both a^j and a^k, that it will be prohibitively difficult to reconstruct the key a^{jk}. An efficient solution to the discrete problem would break this sytem, since it would allow the eavesdropper to obtain j from a^j, and use it to compute the key. Therefore, the discrete logarithm problem is at least as difficult as the one mentioned above. Whether there is an attack against Diffie-Hellman that do not require the full strength of the discrete logarithm problem, however, is unknown.

12.5.4 Baby-Step Giant-Step

The baby-step giant-step algorithm is a general-purpose algorithm that solves the discrete logarithm problem in any cyclic group. It is a meet-in-the-middle type algorithm, which seeks to find a collision between two sequences.

The main idea behind the algorithm is as follows. Let a be a primitive root modulo p, and b coprime to p. Then for any integer $N \geq \sqrt{p}$, we can write $\text{Log}_a(b)$ in base N, as

$$\text{Log}_a(b) = j + kN, \quad \text{with} \quad 0 \leq j, k < N.$$

Then $a^{j+kN} \equiv b \mod p$, or in other words

$$a^j \equiv ba^{-kN} \mod p.$$

This suggests the following. First we compute and store the values of a^j for $0 \leq j < N$ in a data structure that provides efficient lookup. This setup is the "baby-step" part of the algorithm. Then we go through the sequence ba^{-kN}, taking "giant steps" of size a^{-N}, and checking whether any of these values have already appeared as one of the a^j. As soon as a collision is found, we have obtained $\text{Log}_a(B) = j + kN$. This algorithm requires enough memory to store $\sqrt{p}$ integers of size p, and takes between $\sqrt{p}$ and $2\sqrt{p}$ steps to run. Note that nothing in the above description depended on p being prime. As such, it is an exponential-time algorithm (exponential in the number of bits of p).

Other exponential-time algorithms for discrete logarithms include the Pohlig-Hellman algorithm and Pollard's rho algorithm for logarithms. Pohlig-Hellman works very well for primes p such that $p - 1$ is the product of small prime numbers. Pollard's rho algorithm has runtime approximately $O(\sqrt{p})$, but requires much less storage than the baby-step giant-step algorithm. It's main idea is the same as Pollard's rho algorithm for factorization, which we discuss in Section 12.7.1. Subexponential algorithms include the index calculus algorithm and the number field sieve for logarithms, which are similar to the quadratic sieve algorithm for integer factorization. The function field sieve is a subexponential algorithm that computes discrete logarithms in finite fields.

12.6 PRIMALITY TESTING

The Prime Number Theorem states that the number $\pi(x)$ of primes up to x is approximately

$$\int_2^x \frac{1}{\log t}\, dt \approx \frac{x}{\log x}.$$

If one thinks of the prime numbers as occurring randomly, this can be interpreted roughly as saying that the probability that a given number n is prime is about $1/\log n$. But for a specific n, how can we determine whether n is prime or composite? It turns out, perhaps surprisingly, that answering this question is quite a bit easier than the problem of finding the prime factorization of n.

There are several very efficient methods for testing primality. For example, there is a deterministic algorithm created by Adelman, Pomerance, and Rumely (1983), later refined by Cohen and Lenstra (1984), which has runtime $(\log n)^{O(\log \log \log n)}$. More recently, Agrawal, Kayal, and Saxena have even proven that there is a deterministic polynomial-time algorithm for primality testing, although their method of proof does not lend itself to the creation of an efficient algorithm (Agrawal et al., 2004). There are also several probabilistic primality tests such as the Miller-Rabin test, which gain speed at the expense of introducing a small chance of false positives.

12.6.1 Trial Division and the Sieve of Eratosthenes

The most basic (and perhaps worst) method of testing whether an integer n is prime is trial division, i.e., attempting to divide n by each prime p up to $\sqrt{n}$. This works because every composite n has such a prime divisor. This method requires about $\frac{1}{2}\sqrt{n}/\log n$ trial divisions after the task of enumerating the primes up to $\sqrt{n}$ is accomplished.

The Sieve of Eratosthenes is an ancient and surprisingly efficient algorithm for generating a list of all primes up to a given bound n. The Sieve of Eratosthenes requires $O(n)$ storage and has time-complexity $O(n \log n \log \log n)$. It proceeds as follows:

1. Begin with the array of all integers from 2 to n.

2. For primes $p = 2, 3, \ldots$, mark as composite all multiples of p between p^2 and n.

3. Stop when $p^2 > n$.

By the principle of trial division, all the numbers that have not been marked composite after this process are prime. The reason for the sieve's efficiency is that the multiples of a given prime are equally spaced in the integers, and so no integer that is not a multiple of p has to be considered or tested during step p of the process.

For example, in Figure 12.3 we have shaded all the multiples of 2, 3, 5, and 7. Since these are all the primes up to $\sqrt{50}$, the boxes left unshaded constitute the primes up to 50.

	2	3	4	5	6	7	8	9	10
11	12	13	14	15	16	17	18	19	20
21	22	23	24	25	26	27	28	29	30
31	32	33	34	35	36	37	38	39	40
41	42	43	44	45	46	47	48	49	50

FIGURE 12.3 Sieve of Eratosthenes for $n = 50$.

12.6.2 The Miller-Rabin Test

In this section we describe the Miller-Rabin test. This test identifies composite numbers with high probability. Thus if it fails, the number tested is prime with high likelihood.

Because of its efficiency, the Miller-Rabin test is particularly useful for generating large random prime numbers. To do this, we would first select an interval of length proportional to $\log n$, which likely contains a prime. Then after removing multiples of small primes, we test the remaining integers for probable primality using the Miller-Rabin test. Once a probable prime has been found, a deterministic test can be used to rigorously verify that it is prime.

Let n be a positive odd integer to be tested for primality, and let a be an integer satisfying $1 \leq a < n$. Then since n is odd, we have the factorization

$$a^{n-1} - 1 = (a^{(n-1)/2} - 1)(a^{(n-1)/2} + 1)$$

as a difference of squares. Moreover, if we write $n - 1 = 2^v m$ with m odd, then continued factoring yields

$$a^{n-1} - 1 = (a^m - 1) \prod_{w=0}^{v-1} (a^{2^w m} + 1).$$

If n is prime, then Fermat's little theorem implies that this product is divisible by n, and hence that at least one of the factors is divisible by n. In such a case, one of the following conditions is satisfied:

- $a^m \equiv 1 \bmod n$,
- $a^{2^w m} \equiv -1 \bmod n$ for some $w = 0, 1, \ldots, v-1$.

On the other hand, if none of these are satisfied for a particular a, then this demonstrates that n is composite, and a is called a Miller-Rabin witness for (the compositeness of) n. The Miller-Rabin test consists of checking the conditions above for several a.

This test was originally developed by Miller, who proved, under the assumption of an important open problem in number theory called the Generalized Riemann Hypothesis, that testing all a up to $2(\log n)^2$ suffices to show that n is prime (Miller, 1976). The following was indepependently proved in Monier (1980) and Rabin (1980).

Theorem 12.10. *Let n be a composite odd integer. Then the proportion of integers a in $\{1, 2, \ldots, n-1\}$ which are Miller-Rabin witnesses for n is at least $3/4$.*

It follows that if n is composite, that there is less than a $(1/4)^k$ chance that the Miller-Rabin test fails to produce a witness with k randomly chosen integers a in $\{2, \ldots, n-1\}$. Using $1/\log n$ as a heuristic probability for n being prime, it follows from Bayes' Theorem in probability that the probability that an odd number n is composite after failing k iterations of the Miller-Rabin test is at most

$$\frac{(\frac{1}{4})^k \cdot (1 - \frac{1}{\log n})}{(\frac{1}{4})^k \cdot (1 - \frac{1}{\log n}) + 1 \cdot \frac{1}{\log n}} = \frac{\log n - 1}{\log n - 1 + 4^k} \leq \frac{\log n}{4^k}.$$

Therefore, the number of iterations k should be chosen to satisfy $k > \frac{\log \log n}{\log 4}$, and that an n which fails this many iterations of the Miller-Rabin test may be considered as being prime with probability at least $1 - (\log n)/4^k$.

Example 12.5. We apply the Miller-Rabin test to show that $n = 1373653$ is composite. We have $n - 1 = 2^2 \cdot 343413$, so $v = 2$ and $m = 343413$. For $a = 2$ we have

$$2^m \equiv 890592 \not\equiv \pm 1 \bmod n$$
$$2^{2m} \equiv -1 \bmod n$$

so that 2 is not a witness. Similarly, $3^m \equiv 1 \bmod n$, so that 3 is not a witness for n. Moving on to $a = 5$ we find that

$$5^m \equiv 1199564 \not\equiv \pm 1 \bmod n$$
$$5^{2m} \equiv 73782 \not\equiv -1 \bmod n,$$

so that 5 is a witness for n, showing that n is composite.

12.7 INTEGER FACTORIZATION

The problem of efficient integer factorization is not only a fundamental problem in computational number theory, but one that has large practical ramifications for cryptography and cyber security. As large-scale computing has become common, the factorization of large integers has even become somewhat of a benchmark for computing itself.

It may seem strange that so basic a problem as factoring an integer could be difficult, especially with our successful experiences of factoring small integers by hand with simple algorithms such as trial division. But as soon as the numbers get sufficiently large, the inefficiency of such methods become painfully obvious. In this section we describe two factorization algorithms to give a flavor for the various options available for factoring.

12.7.1 Pollard's Rho Algorithm

Pollard's rho algorithm is a probabilistic factorization algorithm that tends to work well on integers with a small prime factor. Although it does not always succeed at producing a factorization, its heuristic runtime is $O(\sqrt{p})$, where p is the smallest prime factor of the input. This algorithm was created by Pollard (1975), and improved versions have been introduced in by Brent (1980) and Brent and Pollard (1981). We will only discuss the original version. This algorithm was also introduced in the chapter on Combinatorics, but we will go into slightly more detail here, and include some examples.

Pollard's rho algorithm takes as input a choice of polynomial $f(x)$ and an integer seed c_0. With these, it generates a pseudo-random sequence of congruence classes $\{c_i \bmod n\}$ by setting $c_i = f(c_{i-1}) \bmod n$ for $i \geq 1$. Since there are finitely many congruence classes modulo n, this sequence eventually repeats itself in a cycle. This is illustrated in Figure 12.4, which depicts, as a directed graph, the orbit of the intial value c_0 under the function f modulo n. The ρ-shaped appearance of this graph is the reason for the name of the algorithm.

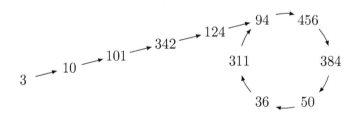

FIGURE 12.4 Orbit of $c_0 = 3$ under $f(x) = x^2 + 1$, modulo $n = 493$.

By the following lemma, we will likely see a repetition in the sequence when its length reaches a multiple of $\sqrt{n}$.

Lemma 12.6 (Birthday Problem). Suppose that k objects are chosen uniformly at random, and with replacement, from a set of size n. Then the probability that at least two of the objects chosen are equal is at least $1 - e^{-k(k-1)/2n}$.

Proof: Let P_k be the probability that no two of the objects chosen are equal, and consider what happens when we choose the k objects one at a time. The probability that the second object is not equal to the first is $1 - 1/n$. Then given that the first two objects are distinct, the probability that the third object is not equal to the first two is $1 - 2/n$. Continuing in this manner yields

$$P_k = \left(1 - \frac{0}{n}\right)\left(1 - \frac{1}{n}\right)\left(1 - \frac{2}{n}\right)\cdots\left(1 - \frac{k-1}{n}\right).$$

Now using the fact that $\log(1 - x) \geq -x$ for any $x \geq 0$, we have

$$\log P_k = \sum_{j=1}^{k-1} \log\left(1 - \frac{j}{n}\right) \geq \sum_{j=1}^{k-1} \frac{-j}{n} = -\frac{k(k-1)}{2n}.$$

Therefore, $1 - P_k \leq 1 - e^{-k(k-1)/2n}$, as desired. $\square$

This is often referred to as the Birthday Problem, or Birthday Paradox, because it implies that in a room with $k = 23$ people, it is more likely than not that two of them share a birthday, since

$$1 - e^{-23 \cdot 22/(2 \cdot 365)} = 0.50000175\ldots > 1/2.$$

This is not actually a paradox, but may seem counterintuitive at first glance. Note that the probability in the theorem can be made as close to 1 as desired by taking k to be a multiple of $\sqrt{n}$.

If p is a factor of n, then the sequence $\{c_i \bmod p\}$ also eventually cycles back on itself, but in expected time $O(\sqrt{p})$ instead of $O(\sqrt{n})$. Since we do not know any divisors of n beforehand, we cannot observe directly when such a cycle occurs. However, if there are c_i and c_j which are distinct modulo n, but satisfy $c_i \equiv c_j \bmod p$ for some divisor p of n, this will be detected by computing $\gcd(|c_i - c_j|, n)$. The goal of Pollard's algorithm is to find such c_i and c_j.

Instead of inefficiently computing $\gcd(|c_i - c_j|, n)$ each pair (i, j), we make use of Floyd's cycle finding algorithm, which finds cycles in the directed graph pictured above by using two subsequences of $\{c_i\}$ that move at different speeds. Namely, we take $j = 2i$, and successively compute $g_i = \gcd(|c_i - c_{2i}|, n)$ until we get a value with $g_i \neq 1$. If $g_i \neq n$, then g_i is a proper factor of n, and we are done. If $g_i = n$, then the algorithm has failed to factor n using the inputs f and c_0, and trying again with a different c_0 or f may possibly succeed.

Example 12.6. We factor $n = 493 = 17 \cdot 29$ using Pollard's rho algorithm. Using $f(x) = x^2 + 1$ and $c_0 = 3$ as in Figure 12.4, we collect the relevant values of c_i and g_i in Table 12.2. This yields only the trivial divisor 493, so our first attempt fails. Hidden in the background of this calculation are the two graphs corresponding to reduction modulo 17 and 29, shown in Figure 12.5. When $i = 6$, cycles are found simultaneously modulo 17 and 29, which is the reason why we did not find a proper factor using $c_0 = 3$.

TABLE 12.2 $c_0 = 3$

i	1	2	3	4	5	6
c_i	10	101	342	124	94	456
c_{2i}	101	124	456	50	311	456
g_i	1	1	1	1	1	493

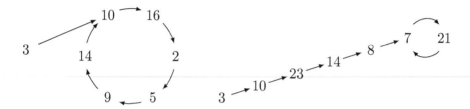

FIGURE 12.5 Orbits of 3 under $x^2 + 1$, modulo 17 and 29.

Let us now try with $c_0 = 2$ instead. We collect the values of c_i and g_i in Table 12.3. From this we obtain the divisor 17 of 493, giving us the factorization $493 = 17 \cdot 29$.

TABLE 12.3 $c_0 = 2$

i	1	2	3	4	5	6
c_i	5	26	184	333	458	240
c_{2i}	26	333	240	485	28	427
g_i	1	1	1	1	1	17

12.7.2 The Quadratic Sieve

A simple idea that goes back to Fermat is that any factorization $n = de$ can be realized by a difference of squares, namely

$$n = de = a^2 - b^2,$$

where $a = \frac{d+e}{2}$ and $b = \frac{d-e}{2}$. Thus, one (inefficient) way of trying to factor n is to go through the values of the polynomial $x^2 - n$ one-by-one, checking whether they are square. In this section we discuss the quadratic sieve, invented in 1981 by Carl Pomerance, which is one in a series of factorization algorithms which generalize this idea.

Although several conjectures remain unproven regarding the performance of the quadratic sieve, it works remarkably well in practice. With a heuristic runtime of $\exp((1 + o(1))\sqrt{\log n \log \log n})$, it is currently the fastest general-purpose algorithm for factoring integers of up to about 100 decimal digits. For numbers with many more than 100 digits, a more sophisticated relative of the quadratic sieve called the general number field sieve becomes more efficient.

Instead of searching for solutions of $n = a^2 - b^2$, the quadratic sieve relaxes this a bit and attempts to find a and b satisfying

$$a^2 \equiv b^2 \mod n \quad \text{and} \quad a \not\equiv \pm b \mod n.$$

Then $(a-b)(a+b) = kn$ for some k and neither of $a-b$ nor $a+b$ is divisible by n, so that $\gcd(a+b, n)$ and $\gcd(a-b, n)$ are both proper factors of n.

The main idea behind the algorithm is to find several squares a_i^2 which are equivalent modulo n to smooth numbers, i.e., numbers s_i whose prime factors are all small. Linear algebra can then be used on the lists of exponents in the prime factorizations of the s_i to yield a product of some subset of the a_i^2 which is a square modulo n. This gives us our solution $a^2 \equiv b^2 \bmod n$. Moreover, assuming that n is odd and is not a prime power, the undesirable situation $a \equiv \pm b \bmod n$ happens only with probability $1/2^{t-1} \leq 1/2$, where t is the number of distinct prime divisors of n. This is because each pair (a, b) is of the form $(a, \mu b)$ with $\mu^2 \equiv 1 \bmod n$, and there are exactly 2^t such μ.

We demonstrate the approach by factoring the number $n = 10441$. First we choose the set $S = \{2, 3, 5, 7\}$ of small primes, and then begin listing those integers $a_i \geq \lceil n \rceil$ such that $a_i^2 \bmod n$ have all their prime factors in S:

$$103^2 \equiv 2^3 \cdot 3 \cdot 7 \bmod n$$
$$104^2 \equiv 3 \cdot 5^3 \bmod n$$
$$107^2 \equiv 2^4 \cdot 3^2 \cdot 7 \bmod n$$
$$109^2 \equiv 2^5 \cdot 3^2 \cdot 5 \bmod n$$

We wish to multiply a subset of these congruences so that the powers of the primes on the right occur with even exponent, yielding a square on both sides. Note that taking a product on the right-hand side is the same as adding the integer vectors formed by the exponents of the primes in S, and that for our purposes, we only care about the parity of the exponents. Thus we arrange the exponents in a matrix and reduce modulo 2, to obtain

$$\begin{bmatrix} 3 & 1 & 0 & 1 \\ 0 & 1 & 3 & 0 \\ 4 & 2 & 0 & 1 \\ 5 & 2 & 1 & 0 \end{bmatrix} \equiv \begin{bmatrix} 1 & 1 & 0 & 1 \\ 0 & 1 & 1 & 0 \\ 0 & 0 & 0 & 1 \\ 1 & 0 & 1 & 0 \end{bmatrix} \bmod 2.$$

We are left with a linear algebra problem over the binary field $\mathbb{F}_2$. The (left) nullspace of this matrix is spanned by the vector $(1, 1, 1, 1)$, and this solution corresponds to the congruence

$$(103 \cdot 104 \cdot 107 \cdot 109)^2 \equiv (2^6 \cdot 3^3 \cdot 5^2 \cdot 7)^2 \bmod n.$$

Letting $a = 103 \cdot 104 \cdot 107 \cdot 109$ and $b = 2^6 \cdot 3^3 \cdot 5^2 \cdot 7$, we then find that $\gcd(a+b, n) = 53$, yielding the factorization $n = 53 \cdot 197$.

Thus far, the factorization strategy we have outlined is common to several algorithms which preceded the quadratic sieve, including the approaches of Kraitchik and Dixon. Indeed, we have elided the *sieve* part of the quadratic sieve, which is what makes it so effective. The step in the above description where sieving comes into play when we need to find many a_i with smooth squares modulo n. Searching for these in a naive way would involve doing trial division by each prime in S for each potential a_i, which is inefficient. We describe now how sieving provides a more better way to accomplish this task.

Recall that the sieve of Eratosthenes produces a list of primes $p \leq L$ by starting with the list $1, 2, \ldots, L$ and successively crossing off the multiples of each prime up to $\sqrt{L}$. Those numbers which are left over are the desired primes. A similar sieving technique can be used to produce smooth numbers. Suppose we want a list of all integers in $\{1, 2, \ldots, L\}$ whose prime factors are in S. Then we start with the array $1, 2, \ldots, L$, and for each prime power $p^a \leq L$ with $p \in S$, we run over the array, dividing those numbers whose positions are multiples of p^a by p. When this is finished, the desired smooth numbers are those that have been changed into 1's.

This sieving technique works over any interval, and for the values of any polynomial $f(x)$. The main observation is that the consecutive values of $f(x)$ are periodic modulo any modulus. In particular, the quadratic sieve uses the polynomial $f(x) = x^2 - n$.

If n is a square modulo p, then there are exactly two congruence classes x mod p^a satisfying $x^2 - n \equiv 0$ mod p^a. Moreover, these can be efficiently computed using algorithms such as Tonelli-Shanks or a combination of the Cippola-Lehmer algorithm with Hensel's lemma. See Cohen (2010) for more details on these algorithms. If the range to be sieved is divided into blocks of length p^a, then these two congruence classes occur in the same positions of each block. On the other hand, primes p such that n is not a square modulo p have no solutions of $x^2 - n \equiv 0$ mod p^a, so these are excluded from the set S. These account for approximately half of the primes up to a given bound, and can be distinguished easily by using properties of the Legendre symbol.

The set S is generally chosen to include all primes p up to some bound B such that n is a square modulo p. The larger B is, the more dense our set of smooth numbers is, and the smaller B is, the fewer smooth numbers we need to find before the linear algebra problem is guaranteed to have a solution. The parameter B is chosen to optimize between this tradeoff. Without going into a detailed analysis, we mention that $B = \exp(\frac{1}{2}\sqrt{\log n \log \log n})$ is the choice that yields the heuristic runtime mentioned above, although in practice is often chosen to be smaller for various implementation reasons.

12.8 CONCLUSION

Number theory is incredibly vast, perhaps due to the fact that it is defined by very simple objects of study, yet uses quite varied tools to understand them. We have only been able to give a brief introduction here, but hopefully it has become clear how ubiquitous number theory is both within mathematics and in the cyber realm.

We conclude by mentioning some sources for further study and exploration of the field. For a more thorough introduction to standard topics in number theory in a few different flavors, we recommend Apostol (1976), Marcus (2018), and Ireland and Rosen (1990). A topic that we did not have space to discuss, but which is fundamental to number theory and is used extensively in cryptography, is the study of elliptic curves. An engaging introduction to this topic may be found in

Silverman and Tate (1994). For more about the connections between number theory and cryptography, we recommend Koblitz (1994) and Trappe and Washington (2006). For those interested in learning more about coding theory, we recomment van Lint (1999), or the encyclopedic treatment by MacWilliams and Sloane (1977). For more details about the history and creation of the quadratic sieve, we recommend Pomerance's article (Pomerance, 1996). Crandall and Pomerance have also written a very in-depth reference for those interested in prime factorization (Crandall and Pomerance, 2005). Finally, for those interested in programming or implementing number theoretic algorithms, Cohen (2010) is an excellent reference.

REFERENCES

Adleman, L. M., Pomerance, C., and Rumely, R. S. (1983). On distinguishing prime numbers from composite numbers. *Ann. of Math. (2)*, 117(1):173–206.

Agrawal, M., Kayal, N., and Saxena, N. (2004). PRIMES is in P. *Ann. of Math. (2)*, 160(2):781–793.

Apostol, T. M. (1976). *Introduction to analytic number theory*. Springer Verlag.

Brent, R. P. (1980). An improved Monte Carlo factorization algorithm. *BIT*, 20(2):176–184.

Brent, R. P. and Pollard, J. M. (1981). Factorization of the eighth Fermat number. *Math. Comp.*, 36(154):627–630.

Cohen, H. (2010). *A course in computational algebraic number theory*. Graduate Texts in Mathematics. Springer.

Cohen, H. and Lenstra, Jr., H. W. (1984). Primality testing and Jacobi sums. *Math. Comp.*, 42(165):297–330.

Crandall, R. and Pomerance, C. (2005). *Prime numbers: A computational perspective*. Springer, New York, second edition.

Ireland, K. and Rosen, M. (1990). *A classical introduction to modern number theory*. Graduate texts in mathematics. Springer, 2nd edition.

Koblitz, N. (1994). *A course in number theory and cryptography*, volume 114 of Graduate Texts in Mathematics. Springer-Verlag, second edition.

MacWilliams, F. J. and Sloane, N. J. A. (1977). *The theory of error-correcting codes. II*. North-Holland Publishing Co., Amsterdam-New York-Oxford. North-Holland Mathematical Library, Vol. 16.

Marcus, D. A. (2018). *Number fields*. Universitext. Springer.

Miller, G. L. (1976). Riemann's hypothesis and tests for primality. *J. Comput. System Sci.*, 13(3):300–317.

Monier, L. (1980). Evaluation and comparison of two efficient probabilistic primality testing algorithms. *Theoret. Comput. Sci.*, 12(1):97–108.

Pollard, J. M. (1975). A Monte Carlo method for factorization. *Nordisk Tidskr. Informationsbehandling (BIT)*, 15(3):331–334.

Pomerance, C. (1996). A tale of two sieves. *Notices Amer. Math. Soc.*, 43(12):1473–1485.

Rabin, M. O. (1980). Probabilistic algorithm for testing primality. *J. Number Theory*, 12(1):128–138.

Silverman, J. H. and Tate, J. (1994). *Rational points on elliptic curves*. Undergraduate Texts in Mathematics. Springer.

Trappe, W. and Washington, L. C. (2006). *Introduction to cryptography with coding theory*. Pearson Prentice Hall, 2nd edition.

van Lint, J. H. (1999). *Introduction to coding theory*, volume 86 of Graduate Texts in Mathematics. Springer-Verlag, 3rd edition.

CHAPTER 13

Quantum Theory

Travis B. Russell

CONTENTS

13.1	Introduction	421
13.2	Quantum Information	422
	13.2.1 Hilbert Spaces and Their Operators	423
	13.2.2 Postulates of Quantum Mechanics	425
	13.2.3 Properties of Quantum Information	427
13.3	Quantum Communication	429
	13.3.1 Quantum Teleportation	429
	13.3.2 Superdense Coding	431
	13.3.3 Quantum Key Distribution	432
13.4	Quantum Graph Theory	433
	13.4.1 Classical Channels	434
	13.4.2 Quantum Channels	436
	13.4.3 Quantum Graphs	438
13.5	Nonlocal Games	440
	13.5.1 A Quantum Game	440
	13.5.2 Quantum Correlations	442
	13.5.3 Device Independent Cryptography	444
13.6	Conclusion	447

13.1 INTRODUCTION

Historians afford groups of people the label of 'civilization' upon meeting a number of requirements. One of these requirements is the development of some system of symbolic communication, usually a form of writing. By converting information—ideas, events, observations, and so on—into abstract symbols and embedding those symbols into physical media, it becomes possible for civilizations to communicate across vast distances in both space and time.

However, with the advent of abstract means of communication come a number of obstacles. How is information efficiently stored? How can we prevent information from being manipulated, intentionally or unintentionally? How can we prevent unwanted persons from viewing sensitive information? These questions, which are central to the study of information theory, have persisted from the earliest days

DOI: 10.1201/9780429354649-13

of abstract communication and their relevance has increased dramatically in the modern age.

Progress along these lines has historically coincided with the creation of more and more sophisticated physical media upon which abstract communication can be encoded. The creation of new information-carrying media has occasionally prompted changes in the actual abstract symbolic forms we use to express language and ideas. For example, the creation of the computer, which manipulates clusters of low and high voltage electrical currents, ushered in the creation of the bit—the unit of the two-letter alphabet $\{0, 1\}$.

Quantum information theory seeks to address the usual questions of information theory with the assumption that information is now stored in a device which is quantum—one that harnesses the fundamental quantum principles of superposition, entanglement and measurement. What is astonishing about quantum information theory is not just that a new alphabet is being used, qubits instead of bits. Rather it is that information can be processed on quantum devices in a provably faster and more secure manner than on their non-quantum or 'classical' counterparts. In the past, increases in speed or security were due to the addition of faster mechanics and more durable materials. Quantum devices owe their efficiency not to greater horsepower but to the exercise of distinct principles of physics never before employed in information-carrying routines.

In this chapter, we seek to introduce the reader to the notion of quantum information, explain some of its fascinating properties, and survey a few cutting-edge topics within the field, namely quantum graph theory and non-local games. We seek to do this as efficiently as possible, so we will skip many details that veterans to the subject might consider essential—for example, the density operator picture of quantum mechanics. That is because our purpose is only to offer an invitation to this exciting subject—for a thorough survey on all that quantum information theory has to offer, we refer the reader to the textbook of Nielsen and Chuang (2000).

The chapter is outlined as follows. In Section 13.2 we provide the basic mathematical notions which define the arena of quantum mechanics. In Section 13.3, we describe some standard communication and cryptographic protocols that employ quantum information to motivate the subject. In Section 13.4, we introduce the subject of quantum graph theory. In Section 13.5, we introduce the subject of non-local games. We finish with a conclusion in Section 13.6 which provides references and commentary for readers interested in pursuing the topic further.

13.2 QUANTUM INFORMATION

In this section we introduce the medium in which quantum information is stored (Hilbert space) and the mechanisms for transferring quantum information (linear operators).

13.2.1 Hilbert Spaces and Their Operators

We first provide a quick summary of the mathematical structures we will employ. The most fundamental mathematical structure in quantum mechanics is the Hilbert space, which we will define next. For brevity, we will assume all Hilbert spaces are finite dimensional, although infinite dimensional Hilbert spaces exist and play an important role in quantum physics.

Let $\mathbb{R}$ denote the set of real numbers and let $\mathbb{C}$ denote the set of complex numbers, i.e., numbers of the form $a + bi$ where $a, b \in \mathbb{R}$ and $i^2 = -1$. For $\alpha = a + bi \in \mathbb{C}$, we let $\alpha^* = a - bi$ denote the complex conjugate. By a **vector space** over $\mathbb{C}$, we mean a set V, whose elements we shall denote as $|x\rangle \in V$, upon which the operations of addition and scalar multiplication are well defined. Hence, for each $|x\rangle, |y\rangle \in V$ and $\alpha \in \mathbb{C}$, there exists a vector $|x\rangle + \alpha |y\rangle \in V$. There is also a special zero vector 0 satisfying $|x\rangle - |x\rangle = 0$ for every vector $|x\rangle$, where $-|x\rangle$ means $(-1)|x\rangle$. Other predictable properties (e.g. the distributive property $\alpha(|x\rangle + |y\rangle) = \alpha|x\rangle + \alpha|y\rangle$) hold, but we will not list them all here.

A set of vectors $\{|0\rangle, |1\rangle, \ldots, |n-1\rangle\}$ are called **linearly independent** if the equation $\sum_{j=0}^{n-1} \alpha_j |j\rangle = 0$ implies that each $\alpha_j = 0$. The set of vectors of the form $\sum_{j=0}^{n-1} \alpha_j |j\rangle$ is called the **span** of the set $\{|j\rangle\}$. If $\{|0\rangle, |1\rangle, \ldots, |n-1\rangle\}$ is linearly independent and spans the entire vector space, then we say the dimension of the vector space is n, and we call the vector space **finite dimensional**. It is a nice exercise to prove that whenever n exists, it is unique.

We now formally define a Hilbert space, which is essentially a vector space in which we can multiply vectors to form a scalar.

Definition 13.1 (Hilbert space). *A **Hilbert space** is a finite dimensional vector space H over the complex numbers together with an operation $(|x\rangle, |y\rangle) \mapsto \langle x|y\rangle \in \mathbb{C}$ called the **inner product** satisfying the following properties:*

1. *$\langle x|(|y\rangle + \alpha |z\rangle) = \langle x|y\rangle + \alpha \langle x|z\rangle$ for all $|x\rangle, |y\rangle, |z\rangle \in H$ and $\alpha \in \mathbb{C}$.*

2. *$\langle x|y\rangle^* = \langle y|x\rangle$ for all $|x\rangle, |y\rangle \in H$.*

3. *$\langle x|x\rangle > 0$ for all non-zero $|x\rangle \in H$.*

The notation above is called 'bra-ket' notation. Vectors such as $|x\rangle$ are called 'kets". Given a ket $|x\rangle$, we can define its dual 'bra' to be the function $\langle x| : H \to \mathbb{C}$ defined by $\langle x|(|y\rangle) = \langle x|y\rangle$, the 'bra-ket' of $|x\rangle$ and $|y\rangle$.

It is convenient to express vectors in Hilbert spaces in terms of orthonormal bases. For an n-dimensional Hilbert space, an **orthonormal basis** is a collection of vectors $\{|0\rangle, |1\rangle, \ldots |n-1\rangle\}$ with the properties

- for each $i < n$, $\langle i|i\rangle = 1$ and

- for each $i, j < n$ with $i \neq j$, $\langle i|j\rangle = 0$.

It is a straightforward exercise to prove that every n-dimensional Hilbert space has an orthonormal basis and that such a basis is indeed a linearly independent spanning set for the Hilbert space (though this basis is far from unique). Thus, for each n-dimensional Hilbert space H and each vector $|\psi\rangle \in H$, there exist scalars $\alpha_0, \alpha_1, \ldots, \alpha_{n-1} \in \mathbb{C}$ such that

$$|\psi\rangle = \alpha_0 |0\rangle + \alpha_1 |1\rangle + \cdots + \alpha_{n-1} |n-1\rangle.$$

With this in mind, we often think of vectors as column matrices via the identification

$$|\psi\rangle \mapsto \begin{pmatrix} \alpha_0 \\ \alpha_1 \\ \vdots \\ \alpha_{n-1} \end{pmatrix} \in \mathbb{C}^n.$$

Thus we may identify any n-dimensional Hilbert space with the vector space $\mathbb{C}^n$. A vector $|\psi\rangle \in \mathbb{C}^2$ is called a **qubit**, while a vector $|\psi\rangle \in \mathbb{C}^d$ (for an arbitrary integer d) is called a **qudit**.

In quantum mechanics, we are often more interested in linear operators on Hilbert spaces rather than Hilbert spaces themselves. Linear operators describe the dynamics of a quantum system over time. A linear operator $T : H \to H$ is a function with the property that, for all $|\psi\rangle, |\phi\rangle \in H$ and $\alpha \in \mathbb{C}$,

$$T(\alpha |\psi\rangle + |\phi\rangle) = \alpha T |\psi\rangle + T |\phi\rangle.$$

To each linear operator T on H, there exists a corresponding linear operator $T^\dagger$ satisfying the relation

$$\langle\psi| (T |\phi\rangle) = (\langle\psi| T^\dagger), |\phi\rangle$$

for all $|\psi\rangle, |\phi\rangle \in H$ (in bra-ket notation, this allows us to write $\langle\psi| T |\phi\rangle$ without ambiguity - we assume T acts to the right as T and acts to the left as $T^\dagger$). Any operator satisfying $T = T^\dagger$ is called **self-adjoint** or **hermitian**. If a linear operator U satisfies $UU^\dagger = U^\dagger U = I$ (where I is the identity operator $I |\psi\rangle = |\psi\rangle$) then U is called **unitary**. If a linear operator P is hermitian and satisfies $P^2 = P$ then P is called a **projection**.

We should mention two useful ways to regard linear operators. First, if we fix an orthonormal basis $\{|0\rangle, |1\rangle, \ldots, |n-1\rangle\}$ for H, then we may associate to each linear operator T a matrix M_T whose i, j-th entry is given by $(M_T)_{i,j} = \langle i| T |j\rangle$. With this identification, we can perform the operation $T |\psi\rangle$ by multiplying the $n \times n$ matrix M_T by the $n \times 1$ column matrix corresponding to $|\psi\rangle$. The adjoint $T^\dagger$ has matrix $M_{T^\dagger} = (M_T)^\dagger$, where $(M_T)^\dagger$ is the conjugate transpose of the matrix M_T. It follows from all of this that every linear operator can be written as a linear combination of operators of the form $|\psi\rangle \langle\phi|$ which operate on H via $|\psi\rangle \langle\phi| (|\gamma\rangle) = \langle\phi|\gamma\rangle |\psi\rangle$. Moreover, we have $(|\psi\rangle \langle\phi|)^\dagger = |\phi\rangle \langle\psi|$.

Tensor products of Hilbert spaces play an important role in distinguishing operations occurring at distinct locations in space. Let H and K be Hilbert spaces. Then we define $H \otimes K$ to be the vector space of all linear combinations of elementary tensors $|x\rangle \otimes |y\rangle$, where $|x\rangle \in H$ and $|y\rangle \in K$, subject to the linearity conditions

$|x\rangle \otimes (|y\rangle + \alpha |z\rangle) = |x\rangle \otimes |y\rangle + \alpha |x\rangle \otimes |z\rangle$ and $(|y\rangle + \alpha |z\rangle) \otimes |x\rangle = |y\rangle \otimes |x\rangle + \alpha |z\rangle \otimes |x\rangle$. We define an inner product on this vector space as

$$(|x\rangle \otimes |y\rangle, |x'\rangle \otimes |y'\rangle) \mapsto \langle x|x'\rangle \langle y|y'\rangle$$

and extending this relation to all of $H \otimes K$ by properties (1) and (2) of the inner product. It is straightforward to verify that if $\{|i\rangle\}$ is a basis for H and $\{|j\rangle\}$ is a basis for K then $\{|i\rangle \otimes |j\rangle\}$ is a basis for $H \otimes K$. For simplicity, we often use the abbreviation $|i\rangle \otimes |j\rangle = |ij\rangle$.

In general, there are many different ways to write a vector in $H \otimes K$ as a linear combination of elementary tensors $|x\rangle \otimes |y\rangle$ for $|x\rangle \in H$ and $|y\rangle \in K$. Any vector $h \in H \otimes K$ that can be written as $h = |x\rangle \otimes |y\rangle$ is called **separable**. Any vector that cannot be written this way is called **entangled**. For example, the reader can verify that $|00\rangle + |01\rangle + |10\rangle + |11\rangle \in \mathbb{C}^2 \otimes \mathbb{C}^2$ is separable, whereas $|00\rangle + |11\rangle \in \mathbb{C}^2 \otimes \mathbb{C}^2$ is entangled.

Finally, suppose we have linear operators S on H and T on K, where H and K are Hilbert spaces. Then the formula

$$S \otimes T (\sum_{i,j} \alpha_{i,j} |i\rangle \otimes |j\rangle) = \sum_{i,j} \alpha_{i,j} (S |i\rangle) \otimes (T |j\rangle)$$

defines a linear operator $S \otimes T$ on $H \otimes K$. In general, a linear operator on $H \otimes K$ can always be written as a linear combination of operators of the form $S \otimes K$.

13.2.2 Postulates of Quantum Mechanics

We are ready to define the rules governing quantum mechanics. These are captured in four postulates codified by John von Neumann in Von Neumann (1927).

Postulate 13.1. *Each physical system is associated to a Hilbert space H. The state of the system is given by a unit vector $|\phi\rangle \in H$ unique up to phase. That is, the vectors $|\phi\rangle$ and $|\psi\rangle$ correspond to the same state if and only if $|\phi\rangle = \alpha |\psi\rangle$ where $\alpha \in \mathbb{C}$ is a scalar with $|\alpha| = 1$.*

Postulate 13.2. *The evolution of a closed physical system from time t to time t' is implemented by a unitary transformation $U = U_{t,t'}$ depending only on the times t and t'. If the state of the system at time t was $|\phi\rangle$, then the state of the system at time t' is $U |\phi\rangle$.*

Postulate 13.3. *If two distinct physical systems are associated to Hilbert spaces H and K, then the joint system is associated to the Hilbert space $H \otimes K$. If the systems do not interact, then the state of the system is a separable state $|\phi\rangle \otimes |\psi\rangle \in H \otimes K$. Otherwise, the state is entangled.*

Postulate 13.4. *An outside observer can perform measurements on a system associated to Hilbert space H as follows. Each possible observation corresponds to a linear operator M_k. If the initial state of the system was $|\phi\rangle$, then the new state of the system will be $|\psi_k\rangle := \langle \phi | M_k^\dagger M_k | \phi \rangle^{-1/2} M_k |\phi\rangle$ with probability $\langle \phi | M_k^\dagger M_k | \phi \rangle$. If the possible observations are indexed by $k = 0, 1, \ldots, m-1$, then the operators M_k must satisfy $\sum_{k=0}^{m-1} = M_k^\dagger M_k = I$ where I is the identity operator on H.*

The fourth postulate requires some explanation. First, the condition $\sum_{k=0}^{m-1} M_k^\dagger M_k = I$ implies that, for any state $|\phi\rangle$, the mapping $k \mapsto \langle\phi| M_k^\dagger M_k |\phi\rangle$ specifies a discrete probability distribution on the set $\{0, 1, \ldots, m-1\}$. After the measurement, the factor $\langle\phi| M_k^\dagger M_k |\phi\rangle^{-1/2}$ appearing in the definition of $|\psi_k\rangle$ ensures that $|\psi_k\rangle$ is a unit vector but serves no other purpose. Note that $\langle\phi| M_k^\dagger M_k |\phi\rangle^{-1/2}$ is undefined when $\langle\phi| M_k^\dagger M_k |\phi\rangle = 0$. However, the new state will be $|\psi_k\rangle$ with probability zero in that case, so this is not really a problem.

The fourth postulate leads to the unhappy realization that measurements may change the state of a physical system and the state of the system after measurement may be unknown. To illustrate this, consider a system associated the the two-dimensional Hilbert space $\mathbb{C}^2$ with orthonormal basis $\{|0\rangle, |1\rangle\}$. First, suppose Alice prepares the system in the state $|i\rangle$ for either $i = 0$ or $i = 1$. If Bob measures with operators $\{|0\rangle\langle 0|, |1\rangle\langle 1|\}$, then the system will remain in state $|i\rangle$ after measurement and Bob can conclude which state Alice prepared (assuming Bob knew that the state was either $|0\rangle$ or $|1\rangle$). If instead Alice prepares the system in either state $|+\rangle = \frac{1}{\sqrt{2}}(|0\rangle + |1\rangle)$ or $|-\rangle = \frac{1}{\sqrt{2}}(|0\rangle - |1\rangle)$, then the new state will be $|0\rangle$ with probability $1/2$ or $|1\rangle$ with probability $1/2$. In either case, Bob's measurement yields no information about which state Alice chose! In other words, Alice has securely encrypted her state from the measurement $\{|0\rangle\langle 0|, |1\rangle\langle 1|\}$.

In fact, the third postulate leads us to consider an even more dire scenario. Suppose two qubit systems are each in the state $|0\rangle$, so that their joint system is in the state $|00\rangle \in \mathbb{C}^2 \otimes \mathbb{C}^2$. Suppose that these systems interact and, after some time has passed, evolve to the state $\frac{1}{\sqrt{2}}(|00\rangle + |11\rangle)$. Because this state cannot be written in the form $|\phi\rangle \otimes |\psi\rangle$ for qubits $|\phi\rangle$ and $|\psi\rangle$, it is not clear what the state of the first system is at all. If we choose an orthonormal basis $\{|\phi\rangle, |\psi\rangle\}$ for the first system, then we could perform corresponding measurements $\{|\phi\rangle\langle\phi|\otimes I, |\psi\rangle\langle\psi|\otimes I\}$ to determine the state of the first system. However, it is easy to check that this measurement yields the state $|\phi\rangle$ with probability $1/2$ and $|\psi\rangle$ with probability $1/2$, regardless of the selection of basis $\{|\phi\rangle, |\psi\rangle\}$! It seems the state of the first qubit is entirely hidden.

Some of the difficulties above can be alleviated by considering a different notion of 'state'. By an **ensemble** on a Hilbert space H, we mean a finite set of pairs $\{(p_k, |\phi_k\rangle)\}$ such that, for each k, $p_k \in [0, 1]$, $|\phi_k\rangle \in H$ is a unit vector, and $\sum_k p_k = 1$. We associate to each ensemble an operator $\rho := \sum_k p_k |\phi_k\rangle\langle\phi_k|$ on H, called the **density operator** for the ensemble. For example, in the case when two qubits enter the entangled state $\frac{1}{\sqrt{2}}(|00\rangle + |11\rangle)$, as described in the previous paragraph, the state of the first qubit can be associated with any ensemble of the form $\{(1/2, |\phi\rangle), (1/2, |\psi\rangle)\}$ where $\langle\phi|\psi\rangle = 0$. The corresponding density operator is $\frac{1}{2}I$ regardless of which orthonormal basis $\{|\phi\rangle, |\psi\rangle\}$ is chosen.

In general, any hermitian operator ρ with unit trace and positive eigenvalues can be realized as the density operator for some ensemble. The four postulates described previously can be rephrased in terms of density operators, though we omit that description for the sake of brevity.

13.2.3 Properties of Quantum Information

We conclude this section by defining what we mean by quantum information, then pointing out some basic differences between quantum and classical information.

The basic building blocks of classical information are **alphabets**, finite sets composed of symbols called **letters**. Information is encoded by combining letters into **strings**, finite sequences of letters from some fixed alphabet.

Definition 13.2. *We define a **quantum alphabet** to be a finite dimensioanl Hilbert space H. We define a **quantum letter** to be a state (unit vector) in H. We define a **quantum string** to be a state in $H^{\otimes n} = H \otimes \cdots \otimes H$, the n-fold tensor product of H, for some positive integer n.*

Suppose we are working in a d-dimensional Hilbert space with canonical orthonormal basis $\{|0\rangle, \ldots |d-1\rangle\}$. We will think of these basis vectors as classical letters in a classical alphabet. Thus, the possible quantum letters include not only classical letters such as $|i\rangle$, but also superpositions of classical letters such as $\sum \alpha_i |i\rangle$. Likewise, quantum strings include classical strings (the elementary tensors) but also superpositions of classical strings. For example, if we take our quantum alphabet to be $\mathbb{C}^2$, we can use classical letters ($|0\rangle$ and $|1\rangle$) to encode classical strings like $|000\rangle$ or $|010\rangle$, though many other possibilities are available to us (for example, $\frac{1}{\sqrt{2}}(|000\rangle + |010\rangle)$).

If we agree to only work with classical letters and classical strings (i.e. basis vectors), then ordinary classical string manipulations can be encoded as quantum processes (unitary matrix multiplication). However, these operations cannot be generalized to all quantum states! As an example, consider the process of duplicating strings. Suppose Alice would like to duplicate the classical string 0110. This can be done by first encoding the string as 01100000 (i.e. padding with zeros), then applying a classical transformation that replaces the last four strings with the sum of the first four strings and the last four strings (modulo two), yielding 01100110. Indeed, we could design a unitary to accomplish this task for computational basis states - i.e. there exists a unitary U on $(\mathbb{C}^2)^{\otimes 8}$ such that $U|abcdefgh\rangle = |abcd\rangle \otimes |a+e\rangle \otimes |b+f\rangle \otimes |c+g\rangle \otimes |d+h\rangle$ for all $a,b,c,d,e,f,g,h \in \{0,1\}$. Naturally, one might wonder if a process like this could be used to duplicate arbitary states. In other words, for some fixed H, is there a unitary U on $H \otimes H$ such that $U|\phi\rangle \otimes |0\rangle = |\phi\rangle \otimes |\phi\rangle$ for all $|\phi\rangle \in H$? In fact, this is impossible.

Theorem 13.1 (No cloning theorem). *Let H be a finite dimensional Hilbert space and $|0\rangle \in H$ some fixed unit vector. Then there does not exist a unitary U on $H \otimes H$ with the property that $U|\phi\rangle \otimes |0\rangle = |\phi\rangle \otimes |\phi\rangle$ for all $|\phi\rangle \in H$.*

Proof: We'll check this for the Hilbert space $H = \mathbb{C}^2$. Suppose such a unitary U exists. Then $U|10\rangle = |11\rangle$ and $U|\phi\rangle \otimes |0\rangle = |\phi\rangle \otimes |\phi\rangle$ for $|\phi\rangle = \frac{1}{\sqrt{2}}(|0\rangle + |1\rangle)$. Notice that

$$|\phi\rangle \otimes |\phi\rangle = \frac{1}{2}(|00\rangle + |01\rangle + |10\rangle + |11\rangle).$$

Thus $\langle 10| (|\phi\rangle \otimes |0\rangle) = \frac{1}{\sqrt{2}}$ while $\langle 11| (|\phi\rangle \otimes |\phi\rangle) = \frac{1}{2}$. However, this is a contradiction since

$$\langle 11| (|\phi\rangle \otimes |\phi\rangle) = \langle 10| U^\dagger U (|\phi\rangle \otimes |0\rangle)$$
$$= \langle 10| (|\phi\rangle \otimes |0\rangle).$$

for any unitary U. □

These arguments show that quantum information is more general than classical information, yet has some apparent limitations. So what is the advantage to replacing classical information with quantum information? Our goal for the remainder of this chapter is to explain just that.

Exercise 13.1. *Let H be an n-dimensional Hilbert space. Prove that H has an orthonormal set of n vectors $\{|0\rangle, |1\rangle, \ldots, |n-1\rangle\}$. Show that this set is linearly independent and thus spans H.*

Exercise 13.2. *Fix an orthonormal basis for an n-dimensional Hilbert space H, and let T be a linear operator on H. Prove that $M_{T^\dagger} = M_T^\dagger$, where we define $A^\dagger$ to be the matrix whose i,j-th entry is $a_{j,i}^*$ (where $a_{i,j}$ is the i,j-th entry of A).*

Exercise 13.3. *Prove that $|00\rangle + |01\rangle + |10\rangle + |11\rangle$ is separable and that $|00\rangle + |11\rangle$ is entangled in $\mathbb{C}^2 \otimes \mathbb{C}^2$.*

Exercise 13.4. *Let ρ be an operator on an n-dimensional Hilbert space, and let $\{|0\rangle, \ldots, |n-1\rangle\}$ and $\{|0'\rangle, \ldots, |n-1'\rangle\}$ be two orthonormal sets in H. Prove that*

$$\sum_{k=0}^{n-1} \langle k| \rho |k\rangle = \sum_{k=0}^{n-1} \langle k'| \rho |k'\rangle.$$

Moreover, show that this quantity is equal to $Tr(M_\rho)$ where Tr is the usual matrix trace (sum of the diagonal entries).

Exercise 13.5. *Redefine the state of a quantum system with Hilbert space H to be a density operator, i.e., a positive-semidefinite[1] operator ρ with $Tr(\rho) = 1$, where Tr is the usual matrix trace. Reformulate postulates 13.2, 13.3, and 13.4 in this setting. For example, in postulate 13.2, you must describe how a closed system in state ρ evolves to a new state ρ'. These postulates should be consistent with the original postulates, so that if $|\phi\rangle$ is the original state and $|\phi'\rangle$ is the new state in the 'state picture', then whenever $|\phi\rangle \langle\phi|$ is the original state in the 'density operator picture', we should have final state $|\phi'\rangle \langle\phi'|$.*

---

[1] An operator is positive-semidefinite if it is self-adjoint and has only non-negative eigenvalues.

Exercise 13.6. Let $\gamma = \frac{1}{\sqrt{2}}(|00\rangle + |11\rangle) \in \mathbb{C}^2 \otimes \mathbb{C}^2$, and let I be the identity operator on $\mathbb{C}^2$. Prove that if $\{|\phi\rangle, |\psi\rangle\}$ is an orthonormal set in $\mathbb{C}^2$, then for $M_0 = |\phi\rangle\langle\phi| \otimes I$ and $M_1 = |\psi\rangle\langle\psi| \otimes I$ we have $\langle\gamma| M_a^\dagger M_a |\gamma\rangle = 1/2$ for $a = 0, 1$. Furthermore, prove that $(1/2)(|\phi\rangle\langle\phi| + |\psi\rangle\langle\psi|) = (1/2)I$.

Exercise 13.7. Prove that there exists a unitary U on $(\mathbb{C}^2)^{\otimes 8}$ such that $U|abcdefgh\rangle = |abcd\rangle \otimes |a+e\rangle \otimes |b+f\rangle \otimes |c+g\rangle \otimes |d+h\rangle$ for all $a, b, c, d, e, f, g, h \in \{0, 1\}$.

Exercise 13.8. Prove Theorem 13.1 for $H = \mathbb{C}^n$ for $n > 2$.

13.3 QUANTUM COMMUNICATION

In this section we introduce some important quantum communication protocols, namely quantum teleportation (see Bennett et al., 1993), superdense coding (see Bennett and Wiesner, 1992), and quantum key distribution (see Bennett and Brassard, 1984). These protocols are not entirely quantum. They involve the sharing of both classical and quantum information. Sometimes the goal of the protocol is to communicate quantum information, and sometimes the goal is to communicate classical information. Moreover, this is by no means an exhaustive list of quantum communication protocols. Instead, we hope to leave the reader with a taste of what is possible when we employ quantum technology to communicate information.

13.3.1 Quantum Teleportation

We will first consider the problem of transmitting a single qubit from Alice to Bob. To perform the operation, we will require that Alice and Bob share an additional pair of qubits - thus, three qubits will be employed in the protocol. The shared qubits are old friends from the previous section, namely the entangled state $|\phi\rangle = \frac{1}{\sqrt{2}}(|00\rangle + |11\rangle)$. We will assume that Alice has access to the first qubit and Bob has access to the second. Physically, such a scenario can be achieved by entangling a pair of particles then sending one to Alice and the other to Bob.

Now suppose that Alice would like to send Bob the qubit $|\psi\rangle = \alpha|0\rangle + \beta|1\rangle$. Initially the qubit $|\psi\rangle$, possessed by Alice, is kept separate from the shared qubit. Thus we may consider the entire system to be initialized in the tripartite state $|\psi\rangle|\phi\rangle \in \mathbb{C}^2 \otimes \mathbb{C}^2 \otimes \mathbb{C}^2$. Since Alice has access to the first two qubits, she may perform measurements of the form $\{M_k \otimes I_2\}$ where M_k acts on the first two qubits, while Bob can perform measurements of the form $\{I_4 \otimes N_k\}$ with each N_k acting on the final qubit. The protocol will proceed as follows. First, Alice will perform a measurement on the first two qubits. Then, using classical communication, Alice will transmit two classical bits of information to Bob. Finally, Bob will perform a measurement based on the information received from Alice. After the final measurement, the system will be in the state $|\rho\rangle|\psi\rangle$ where $|\rho\rangle \in \mathbb{C}^2 \otimes \mathbb{C}^2$.

We first describe Alice's measurements. To do so, we need to introduce a basis on $\mathbb{C}^2 \otimes \mathbb{C}^2$ called the Bell[2] basis. The Bell basis consists of the following vectors:

$$|\phi_0\rangle = \frac{1}{\sqrt{2}}(|00\rangle + |11\rangle)$$

$$|\phi_1\rangle = \frac{1}{\sqrt{2}}(|00\rangle - |11\rangle)$$

$$|\phi_2\rangle = \frac{1}{\sqrt{2}}(|01\rangle + |10\rangle)$$

$$|\phi_3\rangle = \frac{1}{\sqrt{2}}(|01\rangle - |10\rangle).$$

With these definitions, we define Alice's measurements by setting $M_k := |\phi_k\rangle\langle\phi_k| \otimes I_2$ for $k = 0, 1, 2, 3$. Thus the new state of the system will be $p_k^{-1/2} M_k |\psi\rangle |\phi\rangle$ with probability $p_k = \langle\psi|\langle\phi| M_k |\psi\rangle |\phi\rangle$ (where we have used the fact that M_k is a projection). The reader can verify that $p_k = 1/4$ for each k regardless of the state $|\psi\rangle$ being transmitted.

After measurement, Alice will transmit to Bob two bits of information representing the outcome of her measurement: a 00, 01, 10, or 11 representing outcome 0, 1, 2, or 3 (respectively). Based on the information received from Alice, Bob will apply a unitary operator to the state. Specifically, if Bob receives the bit string ij, then Bob will apply the unitary $X^i Z^j$ to his qubit, where

$$X = \begin{bmatrix} 0 & 1 \\ 1 & 0 \end{bmatrix} \text{ and } Z = \begin{bmatrix} 1 & 0 \\ 0 & -1 \end{bmatrix}.$$

Now that we have explained how the protocol is carried out, let's check that it works. Recall that the initial state of the system was $|\psi\rangle |\phi\rangle$ where $|\psi\rangle = \alpha |0\rangle + \beta |1\rangle$ is the state we intend to transmit to Bob and $|\phi\rangle = \frac{1}{\sqrt{2}}(|00\rangle + |11\rangle)$ is the state shared by Alice and Bob - Alice possessing the first qubit and Bob the second. Observe that

$$\begin{aligned} |\psi\rangle |\phi\rangle &:= (\alpha |0\rangle + \beta |1\rangle) \otimes \frac{1}{\sqrt{2}}(|00\rangle + |11\rangle) \\ &= \frac{\alpha}{\sqrt{2}}(|000\rangle + |011\rangle) + \frac{\beta}{\sqrt{2}}(|100\rangle + |111\rangle). \end{aligned}$$

Let us consider the case when the result of Alice's measurement is 2. In this case, the state of the system after measurement will be

$$(1/4)^{-1/2} M_2 \otimes I |\psi\rangle |\phi\rangle = |\phi_2\rangle (\alpha |1\rangle + \beta |0\rangle).$$

Thus, after measurement, Bob's qubit has been changed to the state $(\alpha |1\rangle + \beta |0\rangle)$. Since Alice's measurement resulted in a two, she transmits the bits 10 to Bob, who then applies the unitary operator X to his qubit, yielding the final state $|\phi_2\rangle |\psi\rangle$.

---

[2]Named for John Bell whose research resolved the famous Einstein-Rosen-Podolsky paradox.

It is important to note that the process just described is, in some sense, secure (assuming Alice and Bob have successfully shared the state $|\phi\rangle$). Since Alice and Bob do not physically transfer qubits after the initial distribution of the entangled state $|\phi\rangle$, it is not possible for an eavesdropper to observe or otherwise manipulate the state in any way. The only information vulnerable to eavesdropping is the pair of bits ij transferred from Alice to Bob. However, the values $\{00, 01, 10, 11\}$ are random, occurring with equal probability, and reveal no information whatsoever about the state $|\psi\rangle$. Of course, manipulating the classical bits Bob receives will result in Bob ending up with the wrong state—so this protocol is vulnerable to some attacks. This can be alleviated by using classical cryptography and error correction to encode the classical bits ij that Alice must send to Bob. We will discuss error correction briefly in Section 13.4.

13.3.2 Superdense Coding

We now consider the problem of using quantum technology to transmit classical information. The protocol we will follow is called superdense coding. It is dual to quantum teleportation in the following sense: while quantum teleportation allows us to transmit a single qubit using only entanglement and the transmission of two classical bits, superdense coding allows us to transmit two classical bits using only entanglement and the transmission of a single qubit.

The superdense coding protocol proceeds as follows. Initially Alice and Bob share the two-qubit state $|\phi\rangle = \frac{1}{\sqrt{2}}(|00\rangle + |11\rangle)$ as in quantum teleportation. Alice will apply one of four unitary operations to her qubit depending on which message she would like Bob to receive. Then Alice physically sends her qubit to Bob. Once Bob has possession of both qubits, he will perform unitary operations to the two qubit states to uncover a classical two-bit message.

We now describe the protocol in more detail. Assume Alice would like to transmit the classical two-bit string ij with $i, j \in \{0, 1\}$. Alice will encode this message by applying the unitary $X^i Z^j$ to her qubit. For example, let's assume she intends to transmit the message 10. Then she will apply X to the first qubit, so that the new state of the system is

$$X \otimes I_2 |\phi\rangle = |\phi_2\rangle.$$

Indeed, it is easily checked that the new state of the system is $|\phi_0\rangle, |\phi_1\rangle, |\phi_2\rangle$, or $|\phi_3\rangle$ when the message Alice encodes is $00, 01, 10$, or 11, respectively. Alice then sends her qubit to Bob by some physical means. Upon receiving the qubit, Bob applies the measurement $\{M_k\}$ where $M_k := |\phi_k\rangle \langle \phi_k|$. It is easily verified that when Bob receives $|\phi_k\rangle$ from Alice, the result of his measurement is k.

As in quantum teleportation, we should check that this process is secure. Suppose an eavesdropper, Eve, intercepts Alice's edited qubit as it is being transmitted to Bob. Suppose Eve performs a measurement of the form $\{|a\rangle \langle a|, |b\rangle \langle b|\}$ where $\{|a\rangle, |b\rangle\}$ is some basis for $\mathbb{C}^2$. Then Eve will obtain outcome 0 with probability $1/2$ and outcome 1 with probability $1/2$. Indeed, without access to Bob's qubit, no information about Alice's qubit can be obtained by measurement. Though Eve learns nothing about the message Alice sends, it is possible that Eve's measurement changes the final classical message that Bob receives from Alice—so the protocol is vulnerable to some manipulation. In Section 13.4 we will learn about quantum coding and quantum error correction, which provides us with some tools for dealing with random errors like the ones inflicted by an eavesdropper.

13.3.3 Quantum Key Distribution

We conclude with one final protocol that has achieved considerable attention in recent years. Quantum key distribution allows two parties (Alice and Bob as usual) to generate a secret key by performing measurements on a quantum state and communicating publicly. Once the secret key is generated, Alice and Bob may use the key to encode and decode information, which is then shared via classical means. We will only consider an early variation of this protocol known as BB84, though other versions have since been developed.

We should first describe a classical cryptography protocol known as the one time pad. For example, suppose Alice wants to share with Bob the four-bit message 0110. Before passing along the message, Alice and Bob agree on a randomly generated secret key, say 1011, for example. Keep in mind that Alice and Bob both know the key. Before Alice sends her message, she first encodes it by bit-wise binary addition. Thus the message 0110 becomes $1101 = 0110 + 1011$. The encoded message 1101 is transmitted over a public channel. Upon receiving the encoded message, Bob then decodes by adding the key to the encoded message, yielding $1101 + 1011 = 0110$.

The one-time-pad is secure provided that the key is randomly generated, known only to Alice and Bob, and only used once. The main difficulty is securely transmitting the key to Alice and Bob. This task is the goal of quantum key distribution. To achieve it we proceed as follows.

First, Alice and Bob each generate binary strings, a and b for Alice and a single string c for Bob, each of length N (where N is much larger than the length of the message we intend to transmit). Alice then prepares N qubits $\{|\phi_i\rangle\}$ as follows. For each $i = 1, \ldots N$, Alice defines $|\phi_i\rangle = H^{b_i}|a_i\rangle$ where a_i and b_i are the i-th letters of the words a and a', respectively, and

$$H = \frac{1}{\sqrt{2}} \begin{bmatrix} 1 & 1 \\ 1 & -1 \end{bmatrix}.$$

When b_i is 0, we obtain either $|0\rangle$ or $|1\rangle$, and when $b_i = 1$ we obtain either $|+\rangle = \frac{1}{\sqrt{2}}(|0\rangle + |1\rangle)$ or $|-\rangle = \frac{1}{\sqrt{2}}(|0\rangle - |1\rangle)$.

After the states are prepared, Alice transmits the quantum string $|\phi_1\rangle \otimes \cdots \otimes |\phi_N\rangle$ to Bob. Then, for each $i = 1, \ldots, N$, Bob applies the measurement $\{|0\rangle\langle 0| H^{c_i}, |1\rangle\langle 1| H^{c_i}$ to ϕ and records the outcome of the measurement as d_i. Notice that whenever $b_i = c_i$, Bob will obtain outcome $d_i = a_i$ with certainty. However, if $b_i \neq c_i$, then the outcome of Bob's measurement will be either 0 or 1 with probability $1/2$. After the measurements are concluded, Alice and Bob share the values of the strings b and c over a public channel. Any bits i where $b_i \neq c_i$ are discarded. For the remaining bits, Alice and Bob know $d_i = a_i$ provided that no errors (random more intentional) occurred in the state transmission. To test for errors, a random subset called the testing set of the remaining bits can be identified (again over the public channel) and Alice and Bob can publicly compare a_i with d_i for these bits. If the values a_i and d_i are sufficiently correlated, then Alice and Bob conclude that the key is secure, discard the testing set, and use the remaining bits of a and d for their one-time pad.

The security of this protocol depends on the fact that an eavesdropper who intercepts the state $|\phi_1\rangle \otimes \cdots \otimes |\phi_N\rangle$ cannot determine the identity of the states $|\phi_i\rangle$ without knowing whether it belongs to the basis $\{|0\rangle, |1\rangle\}$ or $\{|+\rangle, |-\rangle\}$. This is because these bases are mutually unbiased bases. For example, if Eve uses measurement operators $|0\rangle\langle 0|$ and $|1\rangle\langle 1|$ when the state is either $|+\rangle$ or $|-\rangle$, she will obtain the outcome 0 or 1 with equal probability, yielding no information whatsoever about the original state. Moreover, the state will be transformed to $|0\rangle$ or $|1\rangle$ with equal probability, so that if Bob then measures this modified state using $|+\rangle\langle +|$ and $|-\rangle\langle -|$ the outcome will be 0 or 1 with probability $1/2$. Many such modifications will result in detectable errors so that Alice and Bob can conclude that the key is not secure.

Exercise 13.9. *Verify the details of the quantum teleportation protocol.*

Exercise 13.10. *Generalize the quantum teleportation to arbitrary qudits $|\psi\rangle \in \mathbb{C}^d$. Hint: Consider the operators X' and Z', which act on an orthonormal basis $\{|0\rangle, |1\rangle, \ldots, |d-1\rangle\}$ via $X'|k\rangle = |k+1\rangle$ (using arithmetic modulo d) and $Z|k\rangle = e^{2\pi i k/d}|k\rangle$ for each k, and the state $|\phi\rangle = \frac{1}{\sqrt{d}}(\sum_{k=0}^{d-1}|kk\rangle)$.*

Exercise 13.11. *For the superdense coding protocol, verify the claim that measurements of the form*

$$\{M_0 = |a\rangle\langle a|, M_1 = |b\rangle\langle b|\}$$

on the first qubit of $|\phi_k\rangle$, where $\{|a\rangle, |b\rangle\}$ is an arbitrary orthonormal basis for $\mathbb{C}^2$, result in outcome 0 or outcome 1 with equal probability. In other words, verify that superdense coding is secure.

Exercise 13.12. *'Dualize' the protocol developed in Exercise 10 to obtain a generalized superdense coding for transmitting arbitrary two-letter strings ij for $i, j \in \{0, 1, \ldots, d-1\}$.*

13.4 QUANTUM GRAPH THEORY

In the previous section, we saw instances where a quantum protocol could be disrupted by a malicious third party, Eve, who randomly corrupts classical or quantum information being transferred from Alice to Bob. In fact, whenever information (classical or quantum) is transferred from one party to another, it is subject to manipulation due to random error or technological deficiencies. It is, therefore, necessary to have some means of correcting these kinds of errors. For that reason we will briefly turn our attention away from security and focus on the problem of clear communication and error correction. We will describe in rough terms how this is done in classical communication for the purpose of introducing concepts that will be important later on. We then define the notion of a quantum channel and explain how error correction can be performed in that setting. Finally, we will use these ideas to introduce the recent notion of 'quantum graph theory' and a surprising result this new area of research has produced.

13.4.1 Classical Channels

Before describing quantum channels in more detail, we need to introduce the notion of a **classical channel**[3]. These kinds of channels were studied extensively by Claude Shannon in Shannon (1948).

Suppose Alice wishes to send a message to Bob. For simplicity, let us assume that the message is a single letter from a finite alphabet, and that after transmission Bob will receive a single letter from the same alphabet. We will not, however, assume that the channel is perfect - it may be subject to noise, whether environmental or malevolent in nature. We will model the noise with a probability distribution. Let $N(y|x)$ denote the probability that Bob receives the letter y given that Alice transmitted the letter x, where x and y are arbitrary letters in our finite alphabet. More specifically, we require that for each fixed x we have $\sum_y N(y|x) = 1$, and any distribution satisfying this condition is considered a classical channel.

For classical channels, we are concerned with the question of whether or not Bob can discern what letter x Alice has attempted to send upon receiving some letter y. If Alice is only permitted a single use of the channel, then we can answer this question by considering the matrix of probabilities $(N(y|z))_{y,z}$ where y is the (fixed) letter Bob receives as z runs over all possible letters in the alphabet. If both $N(y|x_1) > 0$ and $N(y|x_2) > 0$, then it is not possible for Bob to know with certainty whether Alice intended to send the letter x_1 or the letter x_2. Notice that this determination does not depend on the particular values of $N(y|x_1)$ and $N(y|x_2)$, only the observation that they are both positive.

In light of the preceding example, we make the following definition. First, recall that a **simple graph** is a set V of vertices together with a set $E \subseteq V \times V$ of ordered pairs of vertices with the property that $(i,j) \in E$ implies $(j,i) \in E$.

Definition 13.3 (Confusability Graph). *Let $\mathcal{F} = \{x_1, x_2, \ldots, x_N\}$ be a finite alphabet and let $N(y|x)$ be a classical channel over $\mathcal{F}$. The confusability graph for N, denoted G_N, is the simple graph with vertex set $V = \mathcal{F}$ (one vertex for each letter in the alphabet $\mathcal{F}$) and with edge set*

$$\{(x_i, x_j) : N(y|x_i)N(y|x_j) > 0 \text{ for some } y \in \mathcal{F}\}.$$

The confusability graph is a kind of certificate produced from the channel that encodes whether or not the channel may confuse certain letters. While every channel has a unique confusability graph, it is easy to see that many different channels may produce the same confusability graph.

---

[3]What we call 'classical channels' are more commonly known as discrete memoryless channels.

To make these ideas more concrete, consider the following example. Suppose Alice and Bob share the alphabet $\{a, b, c, d, e, f\}$. They communicate via a channel N described in the following table.

TABLE 13.1 Communication protocol

Input x	Possible output y
a	b or c
b	b or d
c	d or e
d	e or f
e	a or d
f	a

The possible outputs in the table tell us, for example, that $N(a|e)$ is positive, though it does not tell us what the particular value of $N(a|e)$ is. Notice that our channel mixes up letters in our alphabet significantly! However, if these properties are already known to Alice and Bob, they may yet be able to use the channel effectively to communicate. To see this, consider the confusability graph G_N depicted in Figure 13.1. Because the letters $\{b, c, e\}$ all may be changed to d by the channel, it is not possible for Bob to distinguish them. However, the letters $\{a, d, f\}$ are all mapped to distinct subsets of the alphabet. Therefore, it is possible for Bob to distinguish when Alice is attempting to transmit these letters, if he knows in advance that the letter sent by Alice is one of these three.

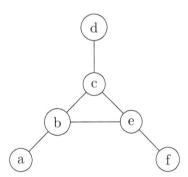

FIGURE 13.1 Confusability graph.

We conclude with a definition and a theorem from graph theory, which is relevant to the above discussion. Let $G = (V, E)$ be a simple graph. By a **clique**, we mean a subset $C \subseteq V$ such that for every $x, y \in C$ there exists a corresponding edge $(x, y) \in E$. In other words, the subgraph formed by G is a complete graph. By an **anticlique**, we mean a subset $A \subset V$ such that for every pair $x, y \in A$ with $x \neq y$ we have $(x, y) \notin E$. In other words, the subgraph formed by A is completely disconnected. A clique of size k is called a k-clique and an anticlique of size k is called a k-anticlique.

When G is the confusability graph of a channel, the cliques of G represent letters, which are indistinguishable, while the anticliques represent letters, which are completely distinguishable. When Alice's intent is to transmit messages to Bob with certainty, she should encode her messages using letters from an anticlique. When her intent is to keep the message hidden from Bob, she should encode the message using letters from a clique. These properties motivate the following terminology. Let C be a subset of an alphabet used by Alice and Bob to communicate over some classical channel N. If C is an anticlique for the confusability graph G_N then we call C an **error-correcting code**, and if C is a clique for G_N, we call C a **private code**.

The following classic theorem will be important for us later. It implies random channels on sufficiently large alphabets necessarily contain large cliques or anticliques, though it does not specify which.

Theorem 13.2 (Ramsey, 1930). *Let k be a positive integer. Then there exists a positive integer N_k such that for all $N \geq N_k$, every graph with at least N vertices contains either a k-clique or a k-anticlique (or both).*

13.4.2 Quantum Channels

In this section we wish to describe the notion of a quantum channel. In particular we would like to understand when it is possible for Bob to identify a quantum message sent by Alice through a given channel. In fact, we have already seen quantum channels before - they are the same as quantum measurements! The difference will be that, for a quantum channel, we are not concerned with the outcome of the quantum measurement - rather, we are interested in the state of the system after measurement.

The following definition is from Kraus et al. (1983).

Definition 13.4 (Quantum Channel). *By a quantum channel, we mean a set $\{E_k\}$ of linear operators $E_k : H \to H$ on a Hilbert space H satisfying the completeness relation $\sum_k E_k^\dagger E_k = I$. When the input of the quantum channel is a state $|x\rangle \in H$, its output is the state $|y_k\rangle = \langle x| E_k^\dagger E_k |x\rangle^{-1/2} E_k |x\rangle$ with probability $\langle x| E_k^\dagger E_k |x\rangle$. The operators E_k are called the **noise operators** for the channel.*

In the following we offer a brief proof of a key observation that we have made use of already. Whenever two states $|x_1\rangle$ and $|x_2\rangle$ are orthogonal (so $\langle x_1|x_2\rangle = 0$) there exists a measurement that distinguishes them with certainty. Conversely, no measurement can distinguish $|x_1\rangle$ from $|x_2\rangle$ with certainty when this fails to be the case. By 'distinguish with certainty' we mean that our measurement includes operators M_1 and M_2 such that $\langle x_1| M_1^\dagger M_1 |x_1\rangle = 1$ (the probability of outcome 1 given input $|x_1\rangle$ is 100%) and $\langle x_2| M_2^\dagger M_2 |x_2\rangle = 1$ (the probability of outcome 2 given input $|x_2\rangle$ is 1).

Proposition 13.1. *Let $|x_1\rangle, |x_2\rangle \in H$ be unit vectors. Then there exists a measurement $\{M_k\}$ such that $\langle x_1| M_1^\dagger M_1 |x_1\rangle = 1$ and $\langle x_2| M_2^\dagger M_2 |x_2\rangle = 1$ if and only if $\langle x_1|x_2\rangle = 0$.*

Proof: If $\langle x_1|x_2\rangle = 0$, we may use measurement operators $M_0 = I - |x_1\rangle\langle x_1| - |x_2\rangle\langle x_2|$, $M_1 = |x_1\rangle\langle x_1|$ and $M_2 = |x_2\rangle\langle x_2|$. On the other hand, suppose that M_1 and M_2 have the properties $\langle x_1| M_1^\dagger M_1 |x_1\rangle = 1$ and $\langle x_2| M_2^\dagger M_2 |x_2\rangle = 1$. By the completeness relation, $M_k |x_i\rangle = 0$ whenever $k \neq i$ for each $i = 1, 2$, since $\| M_k |x_i\rangle \|^2 = \langle x_i| M_k^\dagger M_k |x_i\rangle = 0$. From this it follows that

$$\langle x_1|x_2\rangle = \sum_k \langle x_1| M_k^\dagger M_k |x_1\rangle = 0.$$

So $|x_1\rangle$ and $|x_2\rangle$ must be orthogonal in that case. □

We now turn our attention to the question of how Bob is to distinguish quantum states sent to him from Alice through some quantum channel. In the classical case, we had to consider the case when two distinct letters x_1 and x_2 could be mapped by a channel to the same letter y. Proposition 13.1 shows that two states can be considered distinct if and only if they are orthogonal. Thus we must consider whether a quantum channel maps orthogonal states $|x_1\rangle, |x_2\rangle$ to non-orthogonal states $|y_1\rangle, |y_2\rangle$. If the input of the quantum channel is $|x_i\rangle$, then its output will be a scalar multiple of $E_k |x_i\rangle$ for some k. Bob can distinguish $|x_1\rangle$ from $|x_2\rangle$ with certainty after application of the quantum channel if and only if every possible output state $E_k |x_1\rangle$ is orthogonal to every possible output state $E_{k'} |x_2\rangle$. We summarize this idea in the following corollary.

Corollary 13.1. *Let $|x_1\rangle$ and $|x_2\rangle$ be orthogonal states in a Hilbert space H, and let $\{E_k\}$ be a quantum channel. Then $|x_1\rangle$ and $|x_2\rangle$ are distinguishable after application of the channel if and only if $\langle x_1| E_i^\dagger E_j |x_2\rangle = 0$ for all i, j.*

We now have the tools we need to give a suitable definition for quantum error-correcting codes. By a **quantum code**, we mean a linear subspace C of a Hilbert space H. Let $\{E_k : H \to H\}$ be a quantum channel. Then we call C a **quantum error-correcting code** if all orthogonal pairs $|x_1\rangle, |x_2\rangle \in C$ are distinguishable after application of the channel.

Given a code $C \subset H$, there exists a unique projection P with the property that $P|\phi\rangle = |\phi\rangle$ for all $|\phi\rangle \in C$ and $P|\psi\rangle = 0$ whenever $\langle\psi|\phi\rangle = 0$ for all $|\phi\rangle \in C$. This is called the **projection onto** C. Using projection operators, we can explain exactly how Bob is to identify messages passed to him from Alice through the channel. Suppose Bob knows that Alice will encode her message in the code C by choosing letters from a known basis $\{|0\rangle, |1\rangle, \ldots, |d-1\rangle\}$ of C. For each $k = 0, 1, \ldots, d-1$, let P_k be the projection onto the linear span of the operators $E_j |k\rangle$. Set $P_d = I - (\sum_{k=0}^{d-1} P_k)$. Then $\{P_k\}_{k=0}^d$ describes a new quantum channel. Now, suppose Alice sends the letter $|j\rangle$ to Bob. Then Bob will receive a vector of the form $\alpha E_i |j\rangle$ where α is a positive scalar. Bob can perform the measurement associated with the channel $\{P_k\}$. Since $P_j \alpha E_i |j\rangle = \alpha E_i |j\rangle$, there is some non-zero probability of measuring outcome j. Moreover, since C is error correcting, $E_i |j\rangle$ is orthogonal to $E_l |k\rangle$ for all l and all $k \neq j$. It follows that $P_k E_i |j\rangle = 0$ for all $j \neq k$. Therefore, Bob's measurement yields outcome j with probability 1.

13.4.3 Quantum Graphs

In this section we would like to define a quantum analog of the confusability graph. Along with the quantum graph, we will define quantum versions of cliques and anticliques. These notions will allow us to rephrase the ideas from the previous sections in graph-theoretic terms. It will also provide us with a suitable notion of a quantum private code.

Definition 13.5 (Quantum graph, Duan et al., 2013). *Let $\mathcal{E} = \{E_k : H \to H\}$ be the noise operators for a quantum channel. We define the **quantum confusability graph** to be the linear span of the operators $\{E_i^\dagger E_j\}$. We denote this vector space $V(\mathcal{E})$.*

The quantum confusability graph $V(\mathcal{E})$ is a vector space of operators on a Hilbert space. There is no obvious connection to the vertices and edges of classical graphs we introduced earlier, so we will rely on a few theorems to motivate this definition. We should point out some key properties of the quantum confusability graph. First, it is closed under the dagger operation. Indeed, this follows from the fact that $(E_i^\dagger E_j)^\dagger = E_j^\dagger E_i \in V(\mathcal{E})$. Also, $V(\mathcal{E})$ contains the identity operator I since $\sum_k E_k^\dagger E_k = I$. In general, any subspace V of the operators on a Hilbert space that contains I and is closed under the dagger operation is called a **quantum graph**[4]. Like classical graphs, every quantum graph can be realized as the quantum confusability graph of some quantum channel, though many different quantum channels can have the same quantum confusability graph.

In the following definition, note that because a projection operator P is hermitian, we have $P = P^\dagger$.

Definition 13.6 (Quantum cliques and anticliques, Weaver, 2017). *Let $\mathcal{E}$ be a quantum channel with noise operators $\{E_k : H \to H\}$ and let $V(\mathcal{E})$ be its quantum confusability graph. Let $C \subset H$ be a quantum code, and let P be the projection onto C. We call C a **quantum anticlique** for $V(\mathcal{E})$ if $dim(PV(\mathcal{E})P) = 1$. We call C a **quantum clique** for $V(\mathcal{E})$ if $dim(PV(\mathcal{E})P) = (dim(C))^2$. If $dim(C) = k$, then we call C a quantum k-clique or quantum k-anitclique whenever it is a quantum clique or quantum anticlique, respectively.*

We will motivate the definitions of quantum cliques and anticliques further in a moment, but we can start with some superficial observations. For classical graphs, cliques are subsets of vertices, which are maximally connected, in the sense that they contain as many edges as possible, whereas anticliques are subsets of vertices, which are minimally connected. Similarly quantum cliques are maximal in the sense that the vector space $PV(\mathcal{E})P$ is as large as possible, and quantum anticliques are minimal in the sense that the vector space $PV(\mathcal{E})P$ is as small as possible. Indeed, any operator of the form PTP can be regarded as a linear operator on the Hilbert space C, and the set of all linear operators on C is isomorphic to the algebra of

---

[4] The more appropriate name for these spaces is 'operator system.' Operator systems were defined by William Arveson in Arveson (1969) and are of great importance in both quantum mechanics and operator algebras.

$dim(C) \times dim(C)$ matrices, which has dimension $dim(C)^2$. On the other hand, since $I \in V(\mathcal{E})$, the operator $PIP = P$ is always an element of $PV(\mathcal{E})P$. Hence

$$1 \leq dim(PV(\mathcal{E})P) \leq dim(C)^2.$$

In the classical case, anticliques correspond to error-correcting codes. The next theorem, a major result from Knill and Laflamme (1997), shows that this is also the case for quantum codes.

Theorem 13.3 (Knill and Laflamme, 1997). *Let $\mathcal{E}$ be a quantum channel on a Hilbert space H and $C \subset H$ a quantum code. Then C is a quantum error-correcting code if and only if C is a quantum anticlique.*

We leave the proof as an exercise to the reader. We turn our attention instead to quantum cliques. We showed earlier that classical cliques corresponded to private codes. Letters chosen from a private code cannot be distinguished with certainty after the application of a classical channel. We will now show that the same is true of codes from a quantum clique.

Definition 13.7. *Let $\mathcal{E} = \{E_i\}$ be a quantum channel on a Hilbert space H. A code $C \subset H$ is called a **quantum private code** if for every pair of orthogonal vectors $|x_1\rangle, |x_2\rangle \in C$ there exist indices i, j such that $\langle x_1| E_i^\dagger E_j |x_2\rangle \neq 0$ and $\langle x_1| E_i^\dagger E_i |x_1\rangle \langle x_2| E_j^\dagger E_j |x_2\rangle > 0$.*

When C is a quantum private code, there is always some non-zero probability that orthogonal vectors $|x_1\rangle$ and $|x_2\rangle$ are mapped to non-orthogonal vectors. Thus it is possible that Bob cannot distinguish between the output of the channel applied to $|x_1\rangle$ and $|x_2\rangle$. The next theorem shows that quantum cliques are examples of quantum private codes.

Theorem 13.4. *Let $\mathcal{E}$ be a quantum channel on a Hilbert space H and let $C \subset H$ be a quantum clique. Then C is a quantum private code.*

Proof: Let us assume that C is a quantum clique. Thus, $dim(PV(\mathcal{E})P) = \dim(C)^2$. Let $L(C)$ denote the set of all linear operators on C. Given $T \in V(\mathcal{E})$ we can identify PTP with a linear operator on C by just restricting its domain to C and observing that the range of PTP is a subset of the range of P, which equals C. Since $dim(L(C)) = dim(C)^2 = dim(PV(\mathcal{E})P)$, this allows us to identify $PV(\mathcal{E})P$ with $L(C)$. Now pick orthogonal $|x_1\rangle$ and $|x_2\rangle$. Then there exists $T \in V(\mathcal{E})$ such that $PTP = |x_1\rangle \langle x_2|$. Since $V(\mathcal{E}) = span\{E_i^\dagger E_j\}$, we must have $T = \sum_{i,j} \alpha_{i,j} E_i^\dagger E_j$. If $\langle x_1| E_i^\dagger E_j |x_2\rangle = 0$ for all i, j, then $\langle x_1| T |x_2\rangle = 0$. But this is impossible, since

$$\begin{aligned} \langle x_1| T |x_2\rangle &= \langle x_1| PTP |x_2\rangle \\ &= \langle x_1| (|x_1\rangle \langle x_2|) |x_2\rangle \\ &= 1. \end{aligned}$$

So there exist some i, j such that $\langle x_1| E_i^\dagger E_j |x_2\rangle \neq 0$. Since $\langle x_1| E_i^\dagger E_i |x_1\rangle = \|E_i |x_1\rangle\|^2 > 0$ and $\langle x_2| E_j^\dagger E_j |x_2\rangle = \|E_j |x_2\rangle\|^2 > 0$, and since $|x_1\rangle$ and $|x_2\rangle$ were chosen arbitrarily, it follows that C is a quantum private code. $\square$

We conclude with an interesting recent discovery about quantum graphs from Weaver (2017). Its proof is much too difficult to include in this paper, so we will only state the result. It is the quantum analogue of the classical Ramsey theorem.

Theorem 13.5 (Weaver, 2017). *Let k be a positive integer. Then there exists a positive integer N_k such that for every Hilbert space H of dimension at least N_k and every quantum graph V of operators acting on H either V has a quantum k-clique or V has a quantum k-anticlique.*

In terms of quantum codes, this theorem tells us that random quantum channels on sufficiently large Hilbert spaces necessarily possess either a quantum error-correcting code or a quantum private code.

Exercise 13.13. *Let $U_1, U_2, \ldots, U_n$ be unitaries on a Hilbert space H, and let p_i be a finite probability distribution over $\{1, 2, \ldots, n\}$. Show that $\{p_i^{1/2} U_i\}$ defines a quantum channel.*

Exercise 13.14. *Consider the Pauli operators X, Y, Z, where X and Z are as defined in subsection 3.1 and $Y := iXZ$. Let $U_1 = X \otimes I \otimes I$, $U_2 = I \otimes Y \otimes I$, and $U_3 = I \otimes I \otimes Z$. Find a quantum 2-anticlique $C \subset \mathbb{C}^2 \otimes \mathbb{C}^2 \otimes \mathbb{C}^2$ for the channel with error operators $\{p_i^{1/2} U_i\}$ where p_i is a finite probability distribution over $\{1, 2, 3\}$.*

Exercise 13.15. *Consider the Pauli operators X, Y, Z. Let $U_1 = X \otimes I$, $U_2 = Z \otimes I$. Find a quantum 2-clique $C \subseteq \mathbb{C}^2 \otimes \mathbb{C}^2 \otimes \mathbb{C}^2$ for the channel with error operators $\{p_i^{1/2} U_i\}$ where p_1 and p_2 are positive real numbers satisfying $p_1 + p_2 = 1$.*

Exercise 13.16. *Prove the Knill-LaFlamme theorem (Theorem 13.3). Hint: Use Corollary 13.1 and Definition 13.6.*

13.5 NONLOCAL GAMES

In previous sections, we introduced some interesting applications of quantum technology. However, we have not attempted to quantify how quantum technologies outperform classical technologies. In this section, we will give an example of a task for which quantum technology outperforms all possible classical technologies. This example leads naturally to the concept of quantum correlations, a field whose body of literature is extremely recent and growing. We conclude by using these concepts to devise an even more secure method of quantum key distribution and introduce the concept of device-independent quantum cryptography.

13.5.1 A Quantum Game

Consider the following scenario. Alice and Bob are going to play a game. During the game, Alice and Bob are sent to separate rooms where they are unable to communicate with one another. At each round of the game, a referee will give a random question x to Alice and another random question y to Bob. The questions x and y are elements of $\{0, 1\}$. Separately (perhaps simultaneously), Alice and Bob will each reply to the question with an answer, a for Alice and b for Bob. Again, the answers are bits, so $a, b \in \{0, 1\}$. Alice and Bob win the game if $xy = a + b$ (using modular arithmetic, of course). Alice and Bob know the rules of the game in advance and can decide on any strategy they want, as long as they agree not to communicate during the game.

TABLE 13.2 Rules of the quantum game

xy	a	b	Outcome
0	0	0	win
0	0	1	lose
0	1	0	lose
0	1	1	win
1	0	0	lose
1	0	1	win
1	1	0	win
1	1	1	lose

Let's first consider 'classical' strategies. One classical strategy would be for Alice and Bob to simply agree which answers they will provide given certain questions. This is a deterministic strategy. It is not hard to see that there is no deterministic strategy that will guarantee that Alice and Bob will win the game. However, there is a deterministic strategy that is fairly successful. Since $xy = 1$ only 25% of the time, Alice and Bob could choose to always provide the answer 0. Then they will win with probability 75% (assuming the referee chooses questions at random). Another strategy Alice and Bob could employ is a probabilistic strategy. For example, upon receiving the question from the referee, Alice and Bob could separately roll dice to determine which of several deterministic strategies they could follow. It turns out that this kind of strategy provides no advantage for this particular game—the highest win probability is still 75%.

Let us now consider a 'quantum' strategy. Before the game begins, Alice and Bob prepare the entangled state $|\phi\rangle = \frac{1}{\sqrt{2}}(|00\rangle + |11\rangle)$. Alice keeps the first qubit for herself, and Bob keeps the second qubit for himself. Upon receiving question x, Alice will perform measurement $\{A_{x,a}, A_{x,b}\}$ and provide answer a or b depending on the outcome of this measurement. Likewise, Bob will perform measurement $\{B_{y,c}, B_{y,d}\}$ and provide answer c or d depending on the outcome of his measurement. Could such a strategy provide an advantage?

Consider the measurement operators depicted in the table above. These operators indicate which measurements Alice and Bob will perform upon receiving questions x and y, respectively. The outcome of the measurement tells us which pair a and b Alice and Bob will return to the referee. For example, if the referee provides Alice with question 1 and Bob with question 0, the probability of winning the game will be the probability $p(0, 0|1, 0)$ that Alice and Bob both return the answer 0 plus the probability $p(1, 1|1, 0)$ that Alice and Bob both return the answer 1. These probabilities can be calculated by performing an inner product. For

TABLE 13.3   Measurements for the quantum game

Measurement	Operators	Vectors
$A_{0,0}$	$\lvert 0\rangle\langle 0\rvert$	$\lvert +\rangle = \frac{1}{\sqrt{2}}(\lvert 0\rangle + \lvert 1\rangle)$
$A_{0,1}$	$\lvert 1\rangle\langle 1\rvert$	$\lvert -\rangle = \frac{1}{\sqrt{2}}(\lvert 0\rangle - \lvert 1\rangle)$
$A_{1,0}$	$\lvert +\rangle\langle +\rvert$	$\lvert \rightarrow\rangle = \cos\frac{\pi}{8}\lvert 0\rangle + \sin\frac{\pi}{8}\lvert 1\rangle$
$A_{1,1}$	$\lvert -\rangle\langle -\rvert$	$\lvert \leftarrow\rangle = \cos\frac{\pi}{8}\lvert 1\rangle - \sin\frac{\pi}{8}\lvert 0\rangle$
$B_{0,0}$	$\lvert \rightarrow\rangle\langle \rightarrow\rvert$	$\lvert \uparrow\rangle = \sin\frac{\pi}{8}\lvert 0\rangle + \cos\frac{\pi}{8}\lvert 1\rangle$
$B_{0,1}$	$\lvert \leftarrow\rangle\langle \leftarrow\rvert$	$\lvert \downarrow\rangle = \cos\frac{\pi}{8}\lvert 0\rangle - \sin\frac{\pi}{8}\lvert 1\rangle$
$B_{1,0}$	$\lvert \downarrow\rangle\langle \downarrow\rvert$	
$B_{1,1}$	$\lvert \uparrow\rangle\langle \uparrow\rvert$	

example,
$$p(0,0\lvert 1,0) = \langle \phi\rvert (\lvert +\rangle\langle +\rvert \otimes \lvert \rightarrow\rangle\langle \rightarrow\rvert) \lvert \phi\rangle.$$

Astonishingly, this strategy yields a win probability of over 85%! In fact, this is the win probability for any question asked by the referee. Hence, it is not even necessary that the referee draw the questions at random. As long as the rules of the game remain the same, Alice and Bob will win with the same probability. In fact, it was shown in Tsirelson (1980) that the quantum strategy provided here provides the greatest possible win probability. In other words, Alice and Bob have achieved the highest success rate allowed by the laws of physics!

13.5.2   Quantum Correlations

The quantum game in the previous section motivates the study of quantum correlations - joint probability distributions $p(i,j\lvert x,y)$ arising from joint measurements of quantum states. We will first provide a rigorous definition of these probability distributions and state a few of their properties. Then we will consider some related distributions.

While we have defined measurements to be linear operators $\{M_k\}$ on a Hilbert space satisfying the completeness relation $\sum_k M_k^\dagger M_k = I$, many of the examples we have seen involved projection operators $M_k = P_k$ where $P_k^\dagger = P_k$ and $P_k^2 = P_k$. Such measurements are called **projection-valued measures**. We will work exclusively with these kinds of measurements, although more general measurements could be considered instead.

Definition 13.8 (Correlation). *Let n and k be positive natural numbers. A tuple $(p(i,j\lvert x,y))_{x,y \leq n, i,j \leq k}$ of positive numbers is called a **correlation** if it satisfies the property*
$$\sum_{i,j} p(i,j\lvert x,y) = 1$$
for all $x,y \leq n$. The set of all correlations is denoted $C(n,k)$.

Correlations are candidate joint probability distributions. Given a correlation $(p(i,j\lvert x,y))$, we think of each value $p(i,j\lvert x,y)$ as representing the probability that

Alice produces answer i and Bob produces answer j given that Alice is asked question x and Bob is asked question y. The condition $\sum_{i,j} p(i,j|x,y) = 1$ for all x, y simply ensures that the probability of Alice and Bob providing some answer is always 1.

We can consider the geometry of the set $C(n, k)$ as a subset of the vector space $\mathbb{R}^{n^2 k^2}$. It is not too hard to check that, in this context, the set $C(n, k)$ is convex - i.e., given $p_1, p_2 \in C(n, k)$ and a positive real number $\lambda \le 1$, $\lambda p_1 + (1 - \lambda) p_2 \in C(n, k)$. It is also evident that $C(n, k)$ is a simplex - i.e., there exist finitely many points $p_1, p_2, \ldots, p_N \in C(n, k)$ such that every point in $C(n, k)$ is a convex combination of $p_1, p_2, \ldots, p_N$.

The set $C(n, k)$ turns out to be too large to be of interest to us. In fact, it includes correlations that can only be obtained in scenarios where Alice and Bob are allowed to communicate. We can remove this flaw by slightly restricting the set of correlations.

Definition 13.9 (Non-signalling correlations). *Let n and k be positive integers. A tuple $\{p(i, j|x, y)\} \in C(n, k)$ is called **non-signalling** if the marginal density values*
$$p_a(i|x) := \sum_j p(i,j|x,y), \quad p_b(j|y) := \sum_i p(i,j|x,y)$$
are well defined, i.e., $p_a(i|x)$ is independent of the value y in the above definition, and $p_b(j|y)$ is independent of the value x in the above definition. We let $C_{ns}(n, k)$ denote the set of all non-signalling correlations.

Again considering $C_{ns}(n, k)$ as a subset of $\mathbb{R}^{n^2 k^2}$, it is not too hard to check that $C_{ns}(n, k)$ is convex—in fact, a simplex.

We have yet to invoke quantum mechanics in our definitions, so it is not clear if the non-signalling correlations are actually achievable by the laws of physics. It is a surprising result of Tsirelson that some of them are not! To understand what this result means, we must identify those correlations which can be achieved using quantum strategies. We call these correlations 'quantum' correlations.

Definition 13.10 (Quantum correlations). *Let n and k be positive integers. A tuple $\{p(i,j|x,y) \in C(n,k)\}$ is called a **quantum correlation** if there exist finite dimensional Hilbert spaces H_a and H_b, projection-valued measures $\{A_{x,i}\}_{i=1}^k$ on H_a for each $x \in \{1, 2, \ldots, n\}$ and $\{B_{y,j}\}_{j=1}^k$ on H_b for each $y \in \{1, 2, \ldots, n\}$, and a state $|\phi\rangle \in H_a \otimes H_b$ such that*
$$p(i,j|x,y) = \langle \phi | A_{x,i} \otimes B_{y,j} | \phi \rangle.$$

The set of all quantum correlations is denoted $C_q(n, k)$.

It is straightforward to check that every quantum correlation is a non-signalling correlation. It is a little harder, though not too much so, to check that the set $C_q(n, k)$ is convex for every choice of n and k. However, it turns out that $C_q(n, k)$ is not a simplex! In fact, the geometry of $C_q(n, k)$ for arbitrary n and k is quite mysterious. For example, it was only recently discovered that for large values of n

and k, the set $C_q(n,k)$ is not topologically closed - in other words, it has boundary points which are not quantum correlations. Even in small cases, much remains open. For example, there is a considerable body of literature on the question of whether or not $C_q(3,2)$ is topologically closed!

One could perhaps simplify the problem by considering instead the geometry of the closure of $C_q(n,k)$. The geometry of this set is also quite mysterious. In fact, there is a nearly 50-year-old problem in mathematics, known as Connes' embedding problem, which is equivalent to a question about the closure of $C_q(n,k)$, though it is beyond the scope of this chapter to discuss the details here (see Section 13.6 for more details).

We conclude this section by introducing one final correlation set, the local correlations.

Definition 13.11 (Local correlations). *Let n and k be positive integers. By a **deterministic correlation**, we mean a correlation $(p(i,j|x,y)) \in C_{ns}(n,k)$ such that for every i,j,x,y, $p(i,j|x,y) \in \{0,1\}$. A correlation is called a **local correlation** if it is a convex combination of deterministic correlations. The set of all local correlations is denoted by $C_{loc}(n,k)$.*

The set of local correlations were considered implicitly in the famous paper Einstein et al. (1935). These correlations were proposed as part of an alternate theory of quantum mechanics called the local hidden variable theory. In short, Einstein, Rosen and Podolsky were disturbed by the strange properties which the mathematics of quantum mechanics seemed to endow upon entangled particles, and thus proposed an alternate theory which could, in principle, be compared with quantum mechanics. Many years later, the physicist John Bell devised a test to check which of the two theories accurately predicted the outcomes of a laboratory experiments. These experiments have since been carried out (see Aspect et al., 1981), suggesting that quantum correlations do indeed arise in nature. Thus we can prove, both mathematically and physically, the following theorem.

Theorem 13.6 (Bell, 1964). *The sets $C_{loc}(n,k)$ and $C_q(n,k)$ do not coincide.*

Bell's theorem has a very interesting application. It can be shown that whenever the state $|\phi\rangle$ in the definition of a quantum correlation is of the form $|\phi_a\rangle \otimes |\phi_b\rangle$ where $|\phi_a\rangle \in H_a$ and $|\phi_b\rangle \in H_b$ (i.e. $|\phi\rangle$ is separable), then the correlation given by $p(i,j|x,y) = \langle \phi | A_{x,i} \otimes B_{y,j} | \phi \rangle$ is in $C_{loc}(n,k)$. The contrapositive of this statement is that whenever a correlation p is in $C_q(n,k)$ but not $C_{loc}(n,k)$, then the state $|\phi\rangle$ in its definition must be entangled.

13.5.3 Device Independent Cryptography

We conclude with an application which is slightly more serious than playing games - quantum key distribution. We have already seen how the BB84 protocol allows two parties to generate a one-time pad so that a classical message can be encoded, shared, and decoded securely. We did not explain in great detail how Alice and Bob verify that their key is secure. We will now modify the BB84 protocol by allowing

Alice and Bob to share a string of entangled states rather than Alice sending Bob a string of states. We will not assume very much about the string of entangled states - we even allow for the case when a malicious third party, Eve, is the one who generated and distributed these states! Even under these extreme conditions, it is possible for Alice and Bob to generate a secure one-time pad. This is done by playing quantum games.

We begin by allowing an untrustworthy third party, Eve, to generate a string of qubits. She will generate at least $2N$ qubits, though she may generate even more that are kept secret. Next, N of the qubits are distributed to Alice and N are distributed to Bob - any remaining qubits are kept by Eve.

Next, Alice will generate a random string of N bits $\vec{b}$. Unlike the BB84 protocol, Alice will generate the string of bits $\vec{a}$ (from which we will obtain our secret key) by measuring qubits. If b_i is a 0, Alice will apply the measurement $\{|0\rangle\langle 0|, |1\rangle\langle 1|\}$ to the i-th qubit. Otherwise, she will apply the measurement $\{|+\rangle\langle +|, |-\rangle\langle -|\}$. She records a 0 for a_i if the outcome of the measurement was 0 and a 1 otherwise. In either case, the probability that $a_i = 0$ is $1/2$ and the probability that $a_i = 1$ is $1/2$.

Separately, Bob will perform slightly different operations on his qubit. First, he will generate a random string $\vec{c}$ as in the BB84 protocol, except that his string will consist of letters from a four-letter alphabet - so $c_i \in \{0, 1, 2, 3\}$. For each qubit, Bob will look at his random bit c_i. If it is a 0, Bob will apply the measurement $\{|0\rangle\langle 0|, |1\rangle\langle 1|\}$, and if it is a 1, he will apply the measurement $\{|+\rangle\langle +|, |-\rangle\langle -|\}$, just as Alice did. However, if the bit is a 2, he will apply the measurement $\{|\rightarrow\rangle\langle\rightarrow|, |\leftarrow\rangle\langle\leftarrow|\}$, and if the bit is a 3, he will apply the measurement $\{|\downarrow\rangle\langle\downarrow|, |\uparrow\rangle\langle\uparrow|\}$, as in the quantum game. Just as Alice did, Bob will record the outcome of his measurements in a string $\vec{d}$.

After all measurements are complete, Alice and Bob publicly share the randomly generated strings $\vec{b}$ and $\vec{c}$, initially keeping $\vec{a}$ and $\vec{d}$ secret. As a security check, Alice and Bob compare the values of $\vec{a}$ and $\vec{d}$ on those bits for which $c_i \in \{2, 3\}$. Notice that on these bits, the measurements performed by Alice and Bob are the same as the measurements used in the quantum game. If the qubits used in these measurements were each in the state $|\phi\rangle = \frac{1}{\sqrt{2}}(|00\rangle + |11\rangle)$ then Alice and Bob should find that $a_i + d_i = b_i(c_i - 2)$ (the '−2' just converts $\{2, 3\}$ to $\{0, 1\}$ here) for about 85% of their bits. If this is the case, they agree that the key is secure, choose a random subset of the remaining bits where their measurements agree, and take this subset to be their secret key.

If the state shared by Alice and Bob was indeed $|\phi\rangle \otimes \cdots \otimes |\phi\rangle$ as we have supposed, then it can be checked that whenever $b_i = c_i$ it must be that $a_i = d_i$— that is, Alice and Bob's measurements yield the same outcome. However, the state was not prepared by Alice or Bob, but rather by Eve, who may not be trustworthy. So how can we assume that the states have any particular form? The answer is in Tsirelson's theorem. We mentioned that Tsirelson proved that the quantum game has a maximum win probability of about 85%, and that this win probability can be realized using the entangled state $|\phi\rangle$ and employing the measurements $\{A_{i,j}\}$ and $\{B_{i,j}\}$ described earlier. In fact, Tsirelson's theorem tells us a bit more. The win probability of 85% indicates that the correlation $\{p(i,j|x,y)\}$ produced by the measurements $\{A_{i,j}\}$ and $\{B_{i,j}\}$ and the state $|\phi\rangle$ is an extreme point of the set of quantum correlations $C_q(2,2)$. Moreover, this extreme point lies outside the set of local correlations $C_{loc}(2,2)$. In fact, since Alice and Bob are measuring qubits, they can consider the question of what states $|\phi'\rangle$ produce the same win probability. It turns out that the state $|\phi\rangle$ is unique (up to multiplication by a scalar phase). In fact, it can be shown that if the overall state is of the form $|\psi\rangle \in \mathbb{C}^2 \otimes \mathbb{C}^2 \otimes H_e$, where the first qubit is Alice's, the second is Bob's, and the space H_e belongs to Alice, then $|\psi\rangle = |\phi\rangle \otimes |\rho\rangle$ where $|\rho\rangle \in H_e$. This means that by just testing win probabilities, Alice and Bob can guarantee not only the identity of their state but that it is not entangled with Eve's state - and hence it is secure!

The protocol which we just described is an example of device-independent quantum cryptography. The idea behind device-independence is that a quantum state prepared in a lab can be treated as a black box producing an entirely unknown state—however, by performing measurements and considering the correlations produced by those measurements, it is possible to learn something about the identity of the state produced by the black box. In some cases, it is not necessary to assume any prior knowledge of the measurements being performed or the size of the Hilbert space controlled by Alice and Bob - all that matters is what kind of correlations are produced from the black box. The implications of these ideas for practical cryptography are quite astonishing!

Exercise 13.17. *Verify that 75% is the highest possible classical win probability in the quantum game described in section 13.5.1.*

Exercise 13.18. *Use the table to verify that the quantum strategy provided yields the same win probability regardless of which question the referee asks, and that this probability is greater than 85%.*

Exercise 13.19. *Prove that for every pair of positive integers n, k, $C(n, k)$ is convex.*

Exercise 13.20. *Prove that for every pair of positive integers n, k, $C(n, k)$ is a simplex. Hint: Consider correlations where $p(i,j|x,y) \in \{0,1\}$ for every x, y, i, and j.*

Exercise 13.21. *Identify the extreme points of the set of non-signalling correlations $C_{ns}(2,2)$.*

Exercise 13.22. *Decide whether or not there exists a non-signalling correlation $p(i,j|x,y) \in C_{ns}(2,2)$ such that the probability of winning the game described in section 13.5.1 is greater than the quantum strategy.*

Exercise 13.23. *Prove that every quantum correlation is a non-signalling correlation.*

Exercise 13.24. *Show that the set $C_q(n,k)$ is convex. Hint: If p_1 is implemented with $|\phi_1\rangle \in H_a \otimes H_b$ and p_2 is implemented with $|\phi_2\rangle \in K_a \otimes K_b$, consider states of the form $\alpha|\phi_1\rangle \oplus \beta|\phi_2\rangle$ in $(H_a \oplus K_a) \otimes (H_b \oplus K_b)$. Here, $H \oplus K$ is* the $\dim(H) + \dim(K)$-dimensional Hilbert space defined by $H \oplus K = \{(h, k) : h \in H, k \in K\}$. Addition and scalar multiplication are entry-wise, and the inner product of $(|h_1\rangle, |k_1\rangle)$ with $(|h_2\rangle, |k_2\rangle)$ is $\langle h_1|h_2\rangle + \langle k_1|k_2\rangle$.

Exercise 13.25. *Prove that every local correlation is a quantum correlation. In other words, given a local correlation, produce corresponding quantum states and measurements. Hint: consider the deterministic correlations first, then use the previous exercise.*

Exercise 13.26. *Suppose that $|\phi\rangle = |\phi_a\rangle \otimes |\phi_b\rangle$ where $|\phi_a\rangle \in H_a$ and $|\phi_b\rangle \in H_b$, and let $\{A_{x,i}\}_{i=1}^{k}$ and $\{B_{y,j}\}_{j=1}^{k}$ be projection-valued measures on H_a and H_b, respectively. Prove that the correlation defined by $p(i,j|x,y) = \langle\phi| A_{x,i} \otimes B_{y,j} |\phi\rangle$ is an element of the set $C_{loc}(n, k)$.*

Exercise 13.27. *Prove the claim that Alice and Bob will always obtain the same outcome when performing the same measurement, whether $\{|0\rangle\langle 0|, |1\rangle\langle 1|\}$ or $\{|+\rangle\langle +|, |-\rangle\langle -|\}$, on the state $|\phi\rangle = \frac{1}{\sqrt{2}}(|00\rangle + |11\rangle)$, where Alice measures the first qubit and Bob measures the second. Check that this is true regardless of the order in which the measurements are performed (even when the measurement is simultaneous).*

13.6 CONCLUSION

We conclude this chapter with a few notes on the origins of the ideas described in this chapter and some other ideas that were left out for the sake of brevity.

The theory of Hilbert spaces and their operators can be attributed to the work of many different mathematicians, mostly from the nineteenth and early twentieth century. However, this field owes a great debt of gratitude to John von Neumann who initiated its rigorous study in Von Neumann (1927). His motivation was the notion of 'matrix mechanics' introduced by Werner Heisenberg to explain quantum mechanics. Thus, von Neumann used his theory of Hilbert spaces to write down rigorous axioms of quantum mechanics. These axioms are now the basis of modern quantum mechanics, though alternate axioms exist for dealing with things like relativistic quantum physics (for example, the Haag-Kastler axioms of quantum field theory from Haag and Kastler, 1964). The study of Hilbert space operators is known to mathematicians as 'operator theory' or 'operator algebras', depending on whether you care more about the properties of single linear operators or the collective properties of families of linear operators. After von Neumann's initial contributions, these subjects were studied by mathematicians for many years mostly independently of physicists. This trend has reversed recently as mathematicians and physicists have begun to discover a great deal of common ground in their work.

The protocols explained in Section 13.3 were all introduced in papers co-authored by physicist Charles Bennett (Bennett and Brassard, 1984; Bennett et al., 1993; and Bennett and Wiesner, 1992). For an interesting recounting of the events that led to the development of these algorithms, see Brassard (2005). Quantum teleportation has received considerable attention in popular science circles, often misinterpreted as an example of faster-than-light communication (the need for classical communication in this protocol implies that this is not the case). Quantum key distribution is an area of active research in both theoretical and practical domains. It has been shown that this protocol can be achieved not only in a laboratory, but also across great distances using fiber-optic cables (see Zhang et al., 2019) or satellite communication (see Yin et al., 2017). Indeed, private companies like ID Quantique have been producing commercial implementations of quantum key distribution since the turn of the twenty-first century.

The theory of quantum channels was developed independently by Choi and Kraus (see Kraus et al., 1983). While we have defined quantum channels as multi-valued probabilistic functions, they are more commonly defined as linear operators mapping density matrices to density matrices. Such operators can be equivalently defined as completely positive trace-preserving maps. Completely positive maps were defined and studied by Stinespring in Stinespring (1955). Given an initial density operator ρ, the output of a channel $\mathcal{E}$ is $\mathcal{E}(\rho) = \sum_k E_k^\dagger \rho E_k$. This was proven independently by Kraus and Choi based on Stinespring's theorem. The theory of completely positive maps is closely connected with the theory of operator systems developed by William Arveson. The quantum confusability graph, first introduced in Duan et al. (2013), is an example of an operator system. The notion of quantum cliques and anticliques is due to Weaver (2017). It is a surprising coincidence of history that these ideas have only recently converged into a single active area of research. While we chose to introduce the idea of quantum cliques, anticliques, and quantum Ramsey theory to illustrate some features of this new research area, there are many other very interesting applications that were left unmentioned. For example, using the idea of a quantum graph, one can define a quantum analogue of the Lovasz theta function, an important parameter in graph theory which bounds the capacity of a classical channel. The quantum Lovasz theta function plays a similar role for quantum channels.

The quantum game introduced in Section 13.5.1 is a modification of the argument used by John Bell to prove that the quantum correlations are strictly larger than the local correlations. The study of the geometry of the quantum correlations was initiated by Boris Tsirelson who discovered the upper bound used in the quantum game (see Tsirelson, 1980). Tsirelson's work opened many new questions, collectively known as Tsirelson's problems. Some of these problems have been answered, though only in the past few years. Perhaps the biggest open question raised by Tsirelson was the question of whether or not the quantum correlations are dense in the so-called quantum-commuting correlations. These are correlations which would arise in a universe governed by the Haag-Kastler axioms of quantum mechanics used in relativistic physics. In a surprising turn of events, it was shown in Ozawa (2013), building on the previous work of Junge et al. (2011) and Fritz (2012), that this question is equivalent to Connes' embedding conjecture (see Connes (1976)), an open problem from the mathematical theory of operator algebras. This problem remained unsolved for nearly fifty years. These connections have attracted the attention of mathematical physicists who have produced some surprising observations at the intersection of physics, mathematics and computer science.

In the first draft of this chapter, submitted in December of 2019, the previous paragraph included one additional prophetic sentence suggesting that Connes' long-standing problem may very well be solved by quantum physicists and computer scientists. In fact, only a few weeks after the submission of this chapter, a group of computer scientists working on quantum algorithms submitted the preprint Ji et al. (2020) proving Connes' embedding conjecture to be false. While the author did not know the paper would appear so rapidly (if at all!), this prophetic statement was not entirely coincidental. The author had attended a number of talks by the authors of Ji et al. (2020) explaining their progress, and had heard rumors that Connes' problem was solved—so this 'prophecy' was more akin to a placeholder for a development that seemed inevitable. In any case, it remains a remarkable coincidence that such a long standing conjecture would fall within a decade of being connected to problems of practical significance in quantum communication.

We have avoided introducing the notion of quantum computing since this is beyond the scope of what we have set out to discuss. However, we should mention a few important connections between quantum computing and the ideas discussed above. While the idea of using quantum technology to perform encryption is indeed very attractive, the recent push towards developing quantum computers is related to a very different task—breaking classical cryptography. In the early 1990s, Peter Shor developed an algorithm for factoring large integers in polynomial time on a quantum computer in Shor (1994). A working quantum computer would thus be able to break popular encoding schemes such as RSA encryption (see Rivest et al., 1978). For this reason, there is also significant research going into the development of new classical cryptographic protocols which are not vulnerable to decryption by Shor's algorithm (see Bernstein et al. (2008) for an introduction to this topic).

Perhaps the biggest impediment to the construction of large scale quantum computers is the problem of error correction. Because qubits decohere upon interacting with any outside system, they are extremely fragile. Developing machines which can store and use many qubits without introducing too much error has proven to be quite difficult. However, it is not an entirely mysterious problem. Errors are modeled by quantum channels, as described in Section 13.4, so the ideas of quantum error-correction, particularly the Knill-Laflamme theorem, apply in this setting as well.

Of course, there are many other important open problems and connection which we have left untouched—the PPT-square conjecture, the SIC-POVM existence problem, etc (see Werner (2020) for a list of these and other open problems of interest). We hope that this introduction at least leaves the reader with some excitement about the promise of quantum technologies and the importance of mathematical research in this field.

ACKNOWLEDGMENTS

The author thanks the editors for the invitation to submit this chapter. He also thanks the referee for many insightful comments which contributed to significant improvements in the exposition.

REFERENCES

Arveson, W. B. (1969). Subalgebras of C^*-algebras. *Acta Math.*, 123:141–224.

Aspect, A., Grangier, P., and Roger, G. (1981). Experimental tests of realistic local theories via bell's theorem. *Phys. Rev. Lett.*, 47:460–463.

Bell, J. S. (1964). On the Einstein Podolsky Rosen paradox. *Physics Physique Fizika*, 1:195–200.

Bennett, C. and Brassard, G. (1984). Quantum cryptography: Public key distribution and coin tossing. *Theoretical Computer Science*, 560:7-11.

Bennett, C. H., Brassard, G., Crépeau, C., Jozsa, R., Peres, A., and Wootters, W. K. (1993). Teleporting an unknown quantum state via dual classical and Einstein-Podolsky-Rosen channels. *Phys. Rev. Lett.*, 70:1895–1899.

Bennett, C. H. and Wiesner, S. J. (1992). Communication via one- and two-particle operators on Einstein-Podolsky-Rosen states. *Phys. Rev. Lett.*, 69:2881–2884.

Bernstein, D. J., Buchmann, J., and Dahmen, E. (2008). *Post Quantum Cryptography*. Springer, 1st edition.

Brassard, G. (2005). Brief history of quantum cryptography: a personal perspective. In *IEEE Information Theory Workshop on Theory and Practice in Information-Theoretic Security*, pages 19–23.

Connes, A. (1976). Classification of injective factors cases II_1, II_∞, III_λ, $\lambda \neq 1$. *Annals of Mathematics*, 104(1):73–115.

Duan, R., Severini, S., and Winter, A. (2013). Zero-error communication via quantum channels, noncommutative graphs, and a quantum Lovasz number. *IEEE Transactions on Information Theory*, 59:1164–1174.

Einstein, A., Podolsky, B., and Rosen, N. (1935). Can quantum-mechanical description of physical reality be considered complete? *Phys. Rev.*, 47:777–780.

Fritz, T. (2012). Tsirelson's problem and Kirchberg's conjecture. *Reviews in Mathematical Physics*, 24(05):1250012.

Haag, R. and Kastler, D. (1964). An algebraic approach to quantum field theory. *J. Math. Phys.*, 5(7):848–861.

Ji, Z., Natarajan, A., Vidick, T., Wright, J., and Yuen, H. (2020). MIP*= RE. arXiv:2001.04383.

Junge, M., Navascues, M., Palazuelos, C., Perez-Garcia, D., Scholz, V., and Werner, R. (2011). Connes' embedding problem and Tsirelson's problem. *J. Math. Phys.*, 52(1):012102.

Knill, E. and Laflamme, R. (1997). Theory of quantum error-correcting codes. *Physical Review A*, 55(2):900–911.

Kraus, K., Böhm, A., Dollard, J. D., and Wootters, W. H. (1983). *States, Effects, and Operations Fundamental Notions of Quantum Theory*, volume 190.

Nielsen, M. A. and Chuang, I. L. (2000). *Quantum computation and quantum information*. Cambridge University Press, Cambridge.

Ozawa, N. (2013). About the Connes embedding conjecture: algebraic approaches. *Jpn. J. Math.*, 8(1):147–183.

Ramsey, F. P. (1930). On a problem of formal logic. *Proceedings of the London Mathematical Society*, s2-30(1):264–286.

Rivest, R. L., Shamir, A., and Adleman, L. (1978). A method for obtaining digital signatures and public-key cryptosystems. *Commun. ACM*, 21(2):120126.

Shannon, C. E. (1948). A mathematical theory of communication. *Bell System Technical Journal*, 27(3):379–423.

Shor, P. W. (1994). Algorithms for quantum computation: discrete logarithms and factoring. In *Proceedings 35th Annual Symposium on Foundations of Computer Science*, pages 124–134.

Stinespring, W. F. (1955). Positive functions on C^*-algebras. *Proc. Amer. Math. Soc.*, 6:211–216.

Tsirelson, B. (1980). Quantum generalizations of Bell's inequality. *Letters in Mathematical Physics*, 4:93–100.

Von Neumann, J. (1927). Mathematische begrundung der quantenmechanik (mathematical foundation of quantum mechanics). *Mathematisch-Physikalische Klasse*, (1):1–57.

Weaver, N. (2017). A "quantum" Ramsey theorem for operator systems. *Proc. Amer. Math. Soc.*, 145:4595–4605.

Werner, R. (2020). Open quantum problems. *https://oqp.iqoqi.univie.ac.at/open-quantum-problems*.

Yin, J., Cao, Y., Li, Y.-H., Liao, S.-K., Zhang, L., Ren, J.-G., Cai, W.-Q., Liu, W.-Y., Li, B., Dai, H., Li, G.-B., Lu, Q.-M., Gong, Y.-H., Xu, Y., Li, S.-L., Li, F.-Z., Yin, Y.-Y., Jiang, Z.-Q., Li, M., Jia, J.-J., Ren, G., He, D., Zhou, Y.-L., Zhang, X.-X., Wang, N., Chang, X., Zhu, Z.-C., Liu, N.-L., Chen, Y.-A., Lu, C.-Y., Shu, R., Peng, C.-Z., Wang, J.-Y., and Pan, J.-W. (2017). Satellite-based entanglement distribution over 1200 kilometers. *Science*, 356(6343):1140–1144.

Zhang, Y., Li, Z., Chen, Z., Weedbrook, C., Zhao, Y., Wang, X., Huang, Y., Xu, C., Zhang, X., Wang, Z., Li, M., Zhang, X., Zheng, Z., Chu, B., Gao, X., Meng, N., Cai, W., Wang, Z., Wang, G., Yu, S., and Guo, H. (2019). Continuous-variable QKD over 50 km commercial fiber. *Quantum Science and Technology*, 4(3):035006.

CHAPTER 14

Group Theory

William Cocke

Meng-Che 'Turbo' Ho

CONTENTS

14.1	Introduction	453
	14.1.1 Motivation of Group Theory	455
14.2	Introduction to Group Theory	456
	14.2.1 Group Actions	458
	14.2.2 Further Examples	460
14.3	When are Two Graphs the Same?	462
	14.3.1 Everything is a Graph	465
14.4	Cryptography and the Diffie-Hellman Problem	468
	14.4.1 Diffie-Hellman Key Exchange	469
	14.4.2 An Attack on Diffie-Hellman Key Exchange	469
14.5	Conclusion	471
	14.5.1 Cryptography	471
	14.5.2 Graph Isomorphism	471
	14.5.3 Group Theory and Computational Group Theory	472

14.1 INTRODUCTION

This chapter is on group theory and its relationship to the cyber realm. We will introduce the reader to group theory as the study of symmetry. It was in this vein that group theory originates and that its major applications emerge. Groups help us understand the sameness of objects or structures. Many cryptographic systems and coding schemes utilize group theoretic constructions, where the structure of the group allows one to manipulate equations. In addition, many scientists (including computer scientists) find themselves doing group theory when handling large data sets or search spaces that have some structure. One area of particular interest in the cyber realm is the use of group theory to distinguish objects from one another. Indeed, when defining a data structure or object, researchers must be careful to capture the appropriate concept of equality for their given structures.

Group theory is part of a broader subject known as abstract algebra, also called modern algebra, or often just algebra. Much like the algebra you learn in grade school, the theme of abstract algebra is to study systems with enough structure to solve equations for unknown variables, i.e., find the value of 'x'. One somewhat technical definition of a group is an associative algebraic system G such that the equation $x * y = z$ has a unique solution for any single unknown variable. In this section, we focus on the motivation for group theory and will save the technical definitions for Section 14.2.

Groups, the object of study in group theory, arise naturally from the study of symmetry. Here symmetry does not just refer to the geometric symmetry of various objects from grade school, but to any operation that preserves the object itself. How an object is defined or accepted determines what symmetries are allowed. Hence the symmetries of an object are often part of the proper description of the object. In the cyber realm, the symmetry problem occurs with both hardware and data. For example, how do we uniquely choose a switch, router, or server to fulfill a given role, i.e., given a collection of switches or routers, how do we select one to serve a unique purpose? Similarly, how do we determine if a dataset is the same up to some equivalence? The reader has probably heard the terms symmetric key cryptography (private key) and asymmetric key cryptography (public key). The 'symmetry" in symmetric key reflects the fact that in many symmetric key cryptosystems the encryption and decryption keys are related via some group, the symmetry of the keys being part of the problem. The symmetry of an object helps us define the object.

We now provide an outline of the rest of the chapter. The next subsection begins with an example from the first author's daily life of how the definition of an object can incorporate or fail to incorporate symmetry.[1] The section continues with some more traditional examples of groups as the set of symmetries of an object. We also discuss how group theory is often implicitly used to narrow the size of a search space. In Section 14.2 we introduce the traditional mathematical notation used for groups.

Section 14.3 explores McKay's graph isomorphism algorithm, which is a beautiful example of how groups capture the symmetries of an object and how group theory can be used to distinguish two objects.

Section 14.4 discusses one use of group theory in cryptography, i.e., the Diffie-Hellman key exchange. We will introduce the Diffie-Hellman key exchange and the related discrete log problem. We will also discuss an interesting paper showing how the implementation of cryptographic systems by mathematicians, computer scientists, and engineers can add new exploit opportunities and security challenges.

Finally, the paper concludes in Section 14.5 with a brief discussion of some other potential topics of study wherein group theory intersects with the cyber realm.

---

[1] The designation of first and second author follows alphabetic convention and does not reflect any ordering among the coauthors. Equivalently, the full symmetric group of order 2 acts on the set of authors.

14.1.1 Motivation of Group Theory

Example 14.1. Most mornings I get breakfast ready for my three oldest children, named Max, Min, and Pi. I hand out three bowls and then get the cereal down from the shelf. Since I prefer to pour the cereal in one bowl at a time, I must choose an ordering of bowls to pour the cereal. In my mind, the problem is simple: I have a set of three empty bowls and I want to have a set of three full bowls. The order I pour the cereal in doesn't matter. There are six different orders in which I can pour the cereal, i.e., Max, Min, then Pi or Max, Pi, then Min etc. However, the order I choose is irrelevant to the function of filling the three bowls. Choosing a different order to pour the bowls would be equivalent—at least equivalent from my point of view.

However, to my children, the order I pour the cereal matters greatly. They view the problem as one of sequences. The order I pour the cereal matters in a sequence and the interpretation of the order matters. In my children's view, I pour the cereal to my favorite child first, then to my second favorite child, and finally to the last child. There are still a total of six ways to do this, but the different orders have different interpretations. The sequence (Max, Min, Pi) is not the same as the sequence (Min, Max, Pi). While as sets {Max, Min, Pi} and {Min, Max, Pi} are equivalent.

In the language of group theory, we would say that there are no nontrivial isomorphisms on a sequence of distinct elements. However, for a set of distinct elements, switching two of the elements preserves the set itself. (See more in Example 14.9.)

The origin of modern group theory comes from E. Galois's study of the symmetries of solutions to equations (Stillwell, 2002). Galois used group theory as a tool to address field extensions of the rational numbers. The now-named Fundamental Theorem of Galois Theory gives a correspondence between subfields of certain field extensions and subgroups of a group known as Galois group of the extension. The question of which groups occur as the Galois groups of rational field extensions is known as the Inverse Galois Problem, and is a major open problem in modern number theory. While it is unknown if all groups occurs in the original context introduced by Galois, all groups do occur as the group of symmetries of certain graphs. In Section 14.2, we give some formal examples of groups.

Almost all structures we are interested in have symmetries—humans are not good at randomness and symmetry is the antithesis to randomness. For example, consider the set {Max, Min, Pi} from the story described earlier. In writing the set down, say as input into a computer, we are required to write one element at a time. Hence, the data we type into the computer is actually ordered. The computer processes, stores, and accesses the list in an order. But, even if we typed the data into the computer differently we would want to know that the two objects correspond to the same set. Somehow our handling of the data needs to take this into account. For an unordered list, sorting the list in advance is a widespread method of producing a canonical form of the list. There are different ways to order a list and our preference for one over another is more cultural than mathematical. There are also lists where the order matters, e.g., if we were announcing runners crossing the finish line.

When dealing with an object, both in the physical world and within cyber space, understanding the symmetries inherent in the definition of that object can increase our understanding of the object itself. The symmetries of an object are captured in a corresponding group known as the automorphism group of the object. We give the following example to demonstrate how automorphisms are often implicitly used to assist in problem solving.

Example 14.2. Consider the following problem. Find the largest integer below 1,000,000 that can be expressed as the sum of two squares. Using elementary number theory it is not hard to show that neither 999,999 nor 999,998 are candidates. A simple computational approach would be to search over all pairs of integers i, j in the range 1 to 1,000 and ask if $i^2 + j^2$ is less than 1,000,000 and what the value is. However, the reader has probably already noticed that testing the pair $i = 4$ and $j = 366$ is equivalent to testing the pair $i = 366$ and $j = 4$. This is a simple symmetry of the search space, i.e., the indices i and j can be switched without changing the value of $i^2 + j^2$. This almost halves the size of the naive search space. There are, of course, other ways to reduce the search space as well that do not readily relate to group theory.

Of course, this example is very simplistic, but we hope it helps the reader understand some of the places where symmetry appears in a problem. For those wondering the largest number below 1,000,000 that can be written as a sum of two squares is 999,997 as demonstrated by $999,997 = 194^2 + 981^2$. For the interested reader, see also *Fermat's theorem on sum of two squares*, which gives a way to test whether a number is a sum of two squares by dividing by 4, but also requires much more mathematical background for its proof.

We now move to the formal mathematical treatment of group theory.

14.2 INTRODUCTION TO GROUP THEORY

Let G be a set. Suppose $*$ is a *binary operation* on G, i.e., $*$ is a function from G^2 to G. We will write $x * y$ for the value of $*$ evaluated on x and y. Some texts use $\circ$ instead of $*$ for the binary operation. We will reserve the $\circ$ symbol for group actions introduced later in this section. The actual symbol used is not mathematically relevant. We say G is a *group* if the binary operation $*$ satisfies the following conditions:

1. **Associative**: For all $x, y, z \in G$ we have
$$x * (y * z) = (x * y) * z.$$

2. **Identity**: There is an element $e \in G$ such that for all $x \in G$ we have that
$$x * e = e * x = x.$$

3. **Inverse**: For every $x \in G$ there is a $y \in G$ such that
$$x * y = y * x = e.$$

From these definitions we can infer many more properties of a group. For example, many group theory texts contain the following problem:

Example 14.3. Show that in a group there is a unique identity element. Here an *identity element* of a group is an element f such that $f*x = x*f = x$ for all $x \in G$.

The answer to this problem is an exercise in symbol manipulation. Suppose that f is an identity element of a group G. We will show that $f = e$. Consider the term $f*e$. We see that $f*e = e$, since f is an identity element and $f*e = f$ by the definition of G. Hence $f = f*e = e$ and we conclude that $f = e$.

Of course, there are many more properties of groups that we could explore. For the interested reader we recommend (Isaacs (2008)). We now give some examples of different groups.

Example 14.4. Consider the set $G = \{1, -1\}$ under multiplication. Clearly, multiplication is a binary operation on this set (meaning the product of two elements is contained in the set). Multiplication of integers is associative, the element 1 serves as the identity, and the inverse of each element is itself. Thus, G is a group.

Example 14.5. Fix a natural number $n \in \mathbb{N}$. Consider $G = \{0, 1, 2, \cdots, n-1\}$ under the operation $*$ which is *addition* modulo n, meaning we associate each summation with its residue modulo n. This forms a group, which is often denoted as $\mathbb{Z}/n\mathbb{Z}$ or C_n.

As an example, consider $n = 17$. Then $G = \{0, 1, 2, \cdots, 16\}$, and $1*2 = ((1+2) \mod 17) = 3$ while $15*10 = ((15+10) \mod 17) = (25 \mod 17) = 8$. The reader should convince themselves that this operation is associative, $e = 0$ is the identity, and the inverse of k is $17 - k$.

Example 14.6. Fix a prime p. Consider $G = \{1, 2, \cdots, p-1\}$ under the operation $*$ which is *multiplication* and then modulo p. This forms a group, which is often denoted as $(\mathbb{Z}/p\mathbb{Z})^\times$.[2]

As an example, consider $p = 17$. Then $G = \{1, 2, 3, \cdots, 16\}$, and $1*2 = ((1 \cdot 2) \mod 17) = 2$ while $7*8 = ((7 \cdot 8) \mod 17) = 5$. This operation is associative and $e = 1$ is the identity. An inverse does exist for every element, but it is harder to describe than in the previous example.

Given $k \in G$, to find the inverse of k, we need to find r such that $k*r = 1$. This means that $(k*r \mod 17) = 1$, so there is some s such that $kr + ps = 1$. Such r and s can be found by using the Euclidean algorithm. For instance, to find the inverse of $7 \in G$, we solve $7r + 17s = 1$. The Euclidean algorithm provides a solution $r = 5$, $s = -2$. From this we see that $7*5 = (35 \mod 17) = 1$, so the inverse of 7 is 5.

---

[2] An analogous construction still works when p is a natural number not necessarily a prime, but it is more complicated and we will omit it here.

This example is very closely related to the previous one.[3] One difference is that 0 is an element of $\mathbb{Z}/p\mathbb{Z}$ but not $(\mathbb{Z}/p\mathbb{Z})^\times$. This is because 0 does not have an inverse in $(\mathbb{Z}/p\mathbb{Z})^\times$, since $0 * x = 0$ for all $x \in (\mathbb{Z}/p\mathbb{Z})^\times$. We have also seen that finding the inverse in $\mathbb{Z}/p\mathbb{Z}$ is easy, while finding the inverse in $(\mathbb{Z}/p\mathbb{Z})^\times$ is harder. In fact, there are many computation problems that are easy in $\mathbb{Z}/p\mathbb{Z}$ but are difficult in $(\mathbb{Z}/p\mathbb{Z})^\times$. The discrete log problem is such a problem, and is the basis of the Diffie-Hellman Key Exchange discussed in Section 14.4.

14.2.1 Group Actions

The real power of group theory in applications comes when groups act on another object.

We say that a group G acts on a set X if there is a function $\circ : G \times X \to X$ that satisfies the following properties. We will write $g \circ x$ for the value of $\circ$ on g and x.

1. For all $g, h \in G$ and $x \in X$ we have

$$g \circ (h \circ x) = (g * h) \circ x.$$

2. For all $x \in X$ we have $e \circ x = x$.

Often, we are interested not just in sets, but in sets (or collections of sets) with some structure attached. In these cases the group acts not just on the set, but on the attached structure as well. An *automorphism* of an object X is a function from X to itself that preserves all of the properties of X. For example, if X is a graph, i.e., a set of nodes and edges, then an automorphism of X is a function that maps nodes to nodes such that an edge between two nodes is mapped to an edge between the images of the two nodes.

The automorphism group $\mathrm{Aut}(X)$ of an object X is the group whose elements are the automorphisms of X. Returning to the comparison of lists and sets, the set {Max, Min} and the set {Min, Max} are the same. There is a function that replaces the occurrence of Max with Min and the occurrence of Min with Max. There is another function that returns each name without altering it, i.e., it replaces Max with Max and Min with Min (this function is much less interesting.) These two functions capture the 'sameness' of the two sets. However, the alphabetical list [Max, Min] can only be written one way. The only automorphism of the alphabetical list is the function that simply returns each name.

Example 14.7. The *Caesar code* is the simple coding scheme where one shifts the letters in plain text by a certain number of letters to obtain a cipher text. Julius Caesar famously used the three-letter shift. For instance, the letter A becomes D, B becomes E, etc., so the text PRINT becomes SULQW. To decipher, one simply has to shift the letters backward by 3 (or equivalently, forward by 23).

---

[3]Indeed, if we combine the two examples together, we may obtain another algebraic structure called a *field*. Also, $(\mathbb{Z}/p\mathbb{Z})^\times$ is *isomorphic* to $\mathbb{Z}/(p-1)\mathbb{Z}$.

Consider the group C_{26} from Example 14.5. This group acts on the English letters by the same shifting as in the Caesar code, for instance, $3 \circ A = D, 5 \circ B = G$, and $0 \circ X = X$. In this view, the Caesar code is just choosing a fixed element in C_{26} and looking at its action on the letters one by one, e.g., $3 \circ \text{PRINT} = \text{SULQW}$.

We may further identify the English letters as numbers, i.e., A as 1, B as 2, ..., Y as 25, and Z as 0. Thus, we are really identifying the letters with elements in C_{26}, and we may consider the action of C_{26} on the letters as an action of C_{26} on itself, where the action is just the group operation, addition. This is a very common form of group action, i.e. a group acting on itself by group operation.

Example 14.8. Another class of group actions naturally arises when we consider the symmetries of certain geometric objects. We will show that a square has 8 symmetries and these symmetries form a group called the dihedral group of order 8. Consider the square in figure 14.8 with vertices labeled A, B, C, and D.

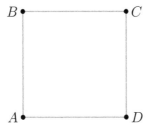

FIGURE 14.1 A square with labeled vertices.

Imagine we can pick up the square, spin it around, flip it in the air, and then return it to the page. In how many different ways could we return the square? We claim that there are 8 possible ways in total, and these actions form a group under the operation, which is doing two actions in a row.

First, we consider the ways that do not involve flipping. Since we need to make sure the square, as a whole shape, is still the same (while the vertices could be at a different place), we can only rotate the square by 90, 180, or 270 degrees. We can also choose to not rotate at all (0 degree). We shall denote the action 'rotating by 90 degrees counterclockwise' as a, so rotating by 180, 270, and 0 degrees are a^2, a^3, $a^0 = e$ correspondingly. Notice that $a * a^2 = a^3$, i.e. rotating by 90 degrees then 180 degrees amounts to just rotating by 270 degrees. Furthermore, $a^3 * a = a^4 = a^0 = e$, as rotating by 270 degrees then 90 degrees returns the vertices to their original positions, and is the same as not rotating at all.

Now we consider the actions that only involves flipping. We consider flipping as 'reflection over a certain line', and the only possible choices for these lines are the horizontal line at exactly the middle of the square, the vertical line at exactly the middle of the square, the diagonal connecting B and D, and the diagonal connecting A and C. We will write the action corresponding to reflecting over these lines as h (horizontal), v (vertical), d (diagonal), and d' (the other diagonal), respectively.

Now let us study the interaction between these two kinds of actions. In particular, we will show that these are all the possible symmetries of the square, i.e., doing a rotation and then a flip does not produce any new actions. Mathematically, we are saying that the group product of any two of these operations is still an operation we have already defined, i.e., this set is *closed under group product*. For instance, consider first the action of rotating by 90 degrees (a) then reflecting over the horizontal line (h). We claim that this is equivalent to reflecting over the diagonal connecting A and C. To see this, we examine the vertices - A is sent to B by a, and then sent back to A by h; B is sent to C by a, which is then sent to D by h, etc. This is exactly what the action d' does to the square: sending A and C back to themselves, while switching B and D. Mathematically, we verified that $h * a = d'$.[4] The reader should try to convince themselves that any product of two of $e, a, a^2, a^3, h, v, d, d'$ is still in this set.

So now we have the set $G = \{e, a, a^2, a^3, h, v, d, d'\}$, and since this set is closed under group product, $*$ is indeed a function from G^2 to G, i.e., a binary operation. This set G is called the *dihedral group of order 8*. We will write D_8 for the dihedral group of order 8.[5] To show that D_8 is indeed a group, we need to show it is associative, has an identity element, and every element has an inverse. This may be checked manually by the arguments in the previous paragraph. However, a more general fact is that the composition of functions is associative — and we have seen previously that we may think of the symmetries really as functions from the set of vertices to itself. The element e is the identity for this operation, as it corresponds to the action of 'not moving'. The inverse of a^i is a^{4-i} as rotating by 360 degrees amounts to not moving at all, and the inverse of the reflections h, v, d, d' are themselves, respectively. Thus, we have shown that D_8 is indeed a group.

14.2.2 Further Examples

There are many more examples of groups. Here, we will show a few more examples that are important in many applications.

Example 14.9. In the dihedral group example (Example 14.8), one key argument was to understand the group elements as bijections from the vertices to the vertices. If we consider the vertices as a set, then we say the dihedral group is a *permutation group* on the set of vertices. More generally, we may consider a set of bijections on a certain set S that is closed under function composition and includes the identity and inverses. Then this set of bijections form a group under the operation of function composition, and these groups are called *permutation groups*.[6]

A special case of this is the *symmetric group*. Fix a set $S = \{1, 2, \cdots, n\}$, and let G consists of all bijections from S to S. We claim that G forms a group under the operation of function composition. Indeed, the composition of any two bijections is still a bijection, so $*$ is a binary operation. As we have noted function composition is

---

[4]Note the order of the group multiplication: when we write $(h * a) \circ A$, we apply the a action on A first, which is what we did in the example. Thus we are saying that $(h * a) \circ X = d' \circ X$ for every $X = A, B, C, D$.

[5]Some authors prefer D_4, since the group acts on a 4-sided polygon, while others prefer $D_{2 \cdot 4}$.

[6]Actually, Cayley's Theorem says that every group is isomorphic to a permutation group.

associative, so $*$ is an associative binary operation. The identity function $f(k) = k$ serves as the identity of the group. Lastly, recall that a bijection always has an inverse function, and this serves as the inverse element in the group. Thus, G is a group, and G is often called the *symmetric group*, denoted as S_n.

Let us now discuss the ways of writing down an element of S_n. For simplicity, we will suppose $n = 3$. Every element g of S_3 is a function from $\{1, 2, 3\}$ to $\{1, 2, 3\}$ and can be described in a few different, albeit equivalent, ways. One way to describe a function is by describing its value for every element in the domain, for instance, $g(1) = 2$, $g(2) = 1$, and $g(3) = 3$. However, this is often cumbersome, and we introduce two standard notations here.

The first one is to write in the *two-line notation*, which is to write $g = \begin{pmatrix} 1 & 2 & 3 \\ 2 & 1 & 3 \end{pmatrix}$. The first line is the input, and the second line is the output. So to find out what the image $g(k)$ is in this notation, we find k in the first line and the number right below it is $g(k)$. Thus, we may retrieve that $g(1) = 2$, $g(2) = 1$, and $g(3) = 3$.

The other common notation is the *cycle notation*. This notation utilizes the fact that g is a bijection. In this notation, we will write $g = (12)(3)$.[7] Each of the full parentheses ((12) and (3)) are called *cycles*. To find $g(k)$ in this notation, we first locate k, then the next number in the cycle is $g(k)$. If k is the last number in a cycle, then $g(k)$ is the first number in the cycle (as we consider cycles to be 'cyclic".) So, again we can get $g(1) = 2$, $g(2) = 1$, and $g(3) = 3$. On the other hand, to find the cycle presentation of an element, one typically starts with 1, and proceeds by finding the image of 1, and the image of the image of 1, etc., until it comes back to 1 at which point we close the parentheses; then start with the smallest number not used so far, find the image of that number, and continue in this manner. In this example, one would construct the cycle permutation of g in the following steps:

(1	(Start with 1)
(12	($g(1) = 2$)
(12)	($g(2) = 1$, so close the parentheses)
(12)(3	(Start with the next unused number)
(12)(3)	($g(3) = 3$, so close the parentheses)

In Examples 14.4, 14.5, and 14.6, the group operations are the usual addition or multiplication. Thus, the various group operations are commutative, i.e., $a*b = b*a$ for all a, b in the group. Groups in which the operation is commutative are called *abelian groups*. We now show that the group S_3 is not abelian. Let $g = (12)(3)$ and $h = (13)(2)$. We will compute that $g * h \neq h * g$.

First, we compute $g * h = (12)(3)(13)(2)$. To find the cycle notation of gh, we proceed as in the previous paragraph. One computes $g * h(1) = g(h(1)) = g(3) = 3$, $g * h(3) = g(1) = 2$, and $g * h(2) = g(2) = 1$.[8] Thus $g * h = (132)$. Again, the

---

[7]Some texts omit the parentheses with only one number in it.
[8]The careful reader will notice that we are overloading the notations here: we use (3) for cycles in the cycle notation, but also (1) for 'evaluating at 1'.

operation written on the right is the one that we apply first. Similarly, the reader is invited to check that $h * g = (123)$. Thus, $g * h \neq h * g$. In particular, $g * h(1) = 3$ but $h * g(1) = 2$. So, the group S_3 is not abelian.

Although all the groups we have seen so far are finite, there are many infinite groups.

Example 14.10. Let $\mathbb{Z}$ be the set of integers. When equipped with the usual addition operation, this is a group. The reader is encouraged to verify this (the operation is associative, and the set contains identity and inverses.) This is often called the *additive group of integers*.

One may wonder what happens if we consider multiplication instead of addition. If we consider the set of integers together with multiplication as the binary operation, then one can check that the operation is associative and contains an identity, but does not (always) contain inverses. So a set can be a group under one natural operation but not another.

Example 14.11. For this example, we will try to construct a group G whose elements are 2×2 matrices, and the operation is matrix multiplication. We know matrix multiplication is associative, so we are good there. We know from linear algebra that the only matrix satisfying $AI = A = IA$ for every matrix A is the identity matrix $I = \begin{bmatrix} 1 & 0 \\ 0 & 1 \end{bmatrix}$, so we must have $I \in G$. Lastly, we know the inverse in G must be the inverse of a matrix in the sense of linear algebra. This means that every matrix in G must actually be invertible, so equivalently, must have non-zero determinant.

To include as many elements in G as possible, we may, for instance, consider G to be the set of all 2×2 matrices with determinant $\neq 0$. From linear algebra, we know that this set is closed under matrix multiplication as $\det(AB) = \det(A)\det(B)$. Thus, if we define $G = \left\{ A = \begin{bmatrix} a & b \\ c & d \end{bmatrix} \mid \det(A) = ad - bc \neq 0 \right\}$, then G equipped with matrix multiplication forms a group.

More generally, any set G of $n \times n$ matrices forms a group under matrix multiplication if:

- G is closed under matrix multiplication.

- G contains the identity matrix.

- Every matrix in G is invertible and the inverse is still in G.

14.3 WHEN ARE TWO GRAPHS THE SAME?

One problem of significant cyber importance connected to group theory is when two graphs are the same. Equivalently, a graph can be displayed in a number of ways; is there a best way to display it? The answer, perhaps surprisingly is no. Recall that when given a list, we can readily decide on an easy to describe canonical form, i.e., alphabetical. However, there are other well-known and well-desired forms such as reverse-alphabetical, by numerical order in ASCII, etc. For a list, a canonical form of the list can be created by sorting the list using some predefined rules. We can easily recognize if a list contains multiple copies of a single entry. We want to do something similar for graphs. The first problem is to determine an ordering of the vertices. The second problem, which is much more complicated, is how to determine when two vertices are 'the same" or represent the same entry.

Recall that a graph $\mathcal{G}$ is an ordered tuple of sets $\mathcal{G} = (V, E)$ where V is the set of vertices and E is the set of edges, i.e., an element $\{v_1, v_2\} \in E$ is a pair of vertices. As mentioned in the earlier discussion of group actions, isomorphisms between two graphs must respect the relationship between edges and nodes. Explicitly, an *isomorphism* between two graphs $\mathcal{G} = (V, E)$ and $\mathcal{G}' = (V', E')$ is a bijective function $f : V \to V'$ such that $\{v_1, v_2\} \in E$ if and only if $\{f(v_1), f(v_2)\} \in E'$. If there is an isomorphism from one graph to another, we say they are isomorphic. Thus, an *automorphism* of a graph $\mathcal{G}$ is an isomorphism from $\mathcal{G}$ to itself, and the automorphism group of $\mathcal{G}$ is the set of automorphisms of $\mathcal{G}$ with function composition as the group operation. For instance, if we view the square in Example 14.8 as a graph, then D_8 is its automorphism group.

We now present some examples of graphs that are isomorphic.

Example 14.12. Are the two graphs in Figure 14.2 isomorphic?

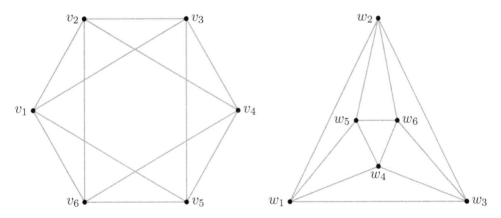

FIGURE 14.2 Two equivalent graphs.

If the two graphs are isomorphic, there should be a bijection from the set

$$f : \{v_1, \ldots, v_6\} \to \{w_1, \ldots, w_6\},$$

such that two vertices share an edge in the first graph if and only if their images under f share an edge.

The graphs are isomorphic, despite appearing different. The function

$$f(v_1) = w_1$$
$$f(v_2) = w_5$$
$$f(v_3) = w_2$$
$$f(v_4) = w_6$$
$$f(v_5) = w_3$$
$$f(v_6) = w_4$$

is an isomorphism of the two graphs. The example serves to demonstrate how seemingly different arrangements of vertices can encode the same graph.

Example 14.13. Consider the following graph $\mathcal{G}$.

FIGURE 14.3   A graph with three vertices.

The vertices of G are v_1, v_2, and v_3 and the edges are (v_1, v_2) and (v_2, v_3). No automorphism of $\mathcal{G}$ can move v_2 because it is the only vertex of degree 2. Informally, we will say that such a vertex is distinguishable from all other vertices. The automorphism group of $\mathcal{G}$ contains two functions: the function that doesn't move any of the vertices and the function that swaps v_1 and v_3. Graphically, the reader can readily see that swapping the vertices v_1 and v_3 amounts to 'spinning the graph' on the page.

Example 14.14. Now consider the following graph $\mathcal{H}$.

FIGURE 14.4   A simple graph.

As with the graph $\mathcal{G}$ from the above example, the automorphism group of $\mathcal{H}$ has two elements (which we leave to the reader to determine.) However, unlike $\mathcal{G}$ there are no vertices of $\mathcal{H}$ that are distinguishable. But, if we declare or mark a single vertex, it can be used to distinguish the other vertices. For example, if we distinguish w_2 the other vertices are distinguished in the following way:

1. w_1 is one vertex away from w_2 and does not connect to any other vertices.

2. w_2 is the distinguished vertex.

3. w_3 is one vertex away from w_2 and connects to another vertex.

4. w_4 is the only vertex that is two away from w_2.

The above two examples contain the key insights needed for McKay's graph isomorphism algorithm. First, if an automorphism takes one vertex to another, then we cannot tell them apart; this is because an automorphism of the graph corresponds exactly to the properties of the graph that we cannot distinguish. Second, if we arbitrarily distinguish a vertex of an object, then we can use that information to distinguish more vertices. The McKay algorithm distinguishes two graphs by attempting to build an isomorphism between the two graphs.

In general, determining if two graphs are isomorphic can be a difficult problem. Indeed, the process of building an isomorphism between the graphs is computationally equivalent to computing the automorphism group of a graph. We will not discuss the details of the McKay algorithm here, but instead refer the interested reader to (Hartke and Radcliffe, 2009) for an introduction of the McKay algorithm.

For the cyber researcher, graphs play a particularly important role. Every data structure you can encode in a computer can be represented as a graph.

14.3.1 Everything is a Graph

In this section we demonstrate how different data structures can be represented as graphs. We will use colors for our vertices. Coloring vertices does not increase the complexity of the problem, but serves to simplify potentially unwieldy arguments. An automorphism on a graph with colored vertices must preserve the color of the vertex. Here colors are represented by different shapes at the vertices.

Example 14.15. We make two graphs corresponding to the list (Max, Min, Pi) as follows. Notice, how the automorphism group of both graphs is trivial. There are no ways to permute distinct entries in a list.

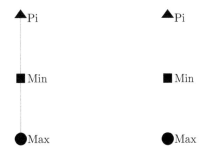

FIGURE 14.5 Two labeled graphs representing the same list.

The two graphs above are not the only two graphs that encode the list and indeed there are other ways to encode a list.

While it was easy to make graphs corresponding to the list, it is unclear what we actually gain from the encoding process. The reader might wonder if encoding structures as graphs is simply one of the weird games mathematicians play. The next example introduces the concept of an unordered table and will hopefully help illuminate why we would want to encode structures as graphs.

Example 14.16. We will call a table unordered if the rows and columns can be swapped independently of each other while preserving the underlying data. Notice, how the accepted automorphisms play a very important role in the definition of this very mathematical object. In practice, given two representatives of unordered tables, we want to know when the tables are the same. For example, are the following two tables the same as unordered tables?

$$\begin{pmatrix} 1 & 1 & 1 \\ 1 & -1 & 1 \\ 2 & 0 & -1 \end{pmatrix}, \begin{pmatrix} 0 & 2 & -1 \\ -1 & 1 & 1 \\ 1 & 1 & 1 \end{pmatrix}.$$

As unordered tables, they are the same. The row containing the 2 is easy to identify in both tables. From that row, we can pivot off of the entries to check that the columns are the same (even if they occur in different orders)

$$\begin{pmatrix} 1 & 1 & 1 \\ 1 & -1 & 1 \\ \mathbf{2} & \mathbf{0} & \mathbf{-1} \end{pmatrix}, \begin{pmatrix} 0 & \mathbf{2} & \mathbf{-1} \\ -1 & 1 & 1 \\ 1 & 1 & 1 \end{pmatrix}.$$

We now ask if the following tables are the same as unordered tables?

$$\begin{pmatrix} 1 & -1 & 1 & -1 \\ -1 & 1 & -1 & 1 \\ 2 & -1 & 1 & -1 \\ -2 & 1 & -1 & 1 \end{pmatrix}, \begin{pmatrix} 1 & -1 & 1 & -1 \\ -1 & 1 & -1 & 1 \\ -2 & -1 & 1 & -1 \\ 2 & 1 & -1 & 1 \end{pmatrix}$$

As in the previous example, we will try to pivot off of an easy to identify row.

$$\begin{pmatrix} 1 & -1 & 1 & -1 \\ -1 & 1 & -1 & 1 \\ \mathbf{2} & \mathbf{-1} & \mathbf{1} & \mathbf{-1} \\ -2 & 1 & -1 & 1 \end{pmatrix}, \begin{pmatrix} 1 & -1 & 1 & -1 \\ -1 & 1 & -1 & 1 \\ -2 & -1 & 1 & -1 \\ \mathbf{2} & \mathbf{1} & \mathbf{-1} & \mathbf{1} \end{pmatrix}$$

The tables are not the same since in each matrix the boldfaced row is the only row containing a 2, and in one table the row contains two -1's and in the other table it does not.

As one can imagine, there are many tables that appear extremely close making the process of telling them apart very labyrinthine. Encoding the unordered table as a graph allows us to run McKay's graph isomorphism algorithm, which operates in a manner similar to our heuristic approach above. To encode the table as a graph, we do the following:

Suppose that M is an n-by-n table with k distinct entries. Then our graph $\mathcal{G}$ of M will have vertex set

$$V = \{r_1, \ldots, r_n\} \cup \{c_1, \ldots, c_n\} \cup \{e_{i,j} : 1 \leq i, j \leq n\} \cup \{v_1, \ldots, v_k\},$$

where the divisions represent different colors as follows: the r-vertices are all one color, the c-vertices are a different color, and the e-vertices are a third color; moreover each of the v-vertices is a distinct color, meaning that the entire vertex set V is colored by $3 + k$ colors. The r-vertices will correspond to rows of M, the c-vertices to columns, the e-vertices to entries, and the v-vertices to distinct values of M. The edge set of $\mathcal{G}$ is

$$E = \left(\bigcup_{1 \leq i \leq n} \{\{r_i, e_{i,j}\} : 1 \leq j \leq n\} \right) \cup \left(\bigcup_{1 \leq j \leq n} \{\{c_j, e_{i,j}\} : 1 \leq i \leq n\} \right)$$
$$\cup \left(\bigcup_{1 \leq \ell \leq k} \{\{v_\ell, e_{i,j}\} : M[i,j] = v_\ell, 1 \leq i, j \leq n\} \right).$$

Using

$$M = \begin{pmatrix} 1 & 1 & 1 \\ 1 & -1 & 1 \\ 2 & 0 & -1 \end{pmatrix},$$

the graph $\mathcal{G}$ is:

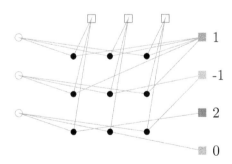

FIGURE 14.6 A graph representing the matrix M.

We note that the graph $\mathcal{G}$ does not depend on the ordering of the rows or columns of M. Hence the graph $\mathcal{G}$ exactly captures what we meant in the definition of M. This is a subtle and important idea: when defining a data structure, or any object, the explicit or implicit description of what constitutes the 'same' object is necessary.

14.4 CRYPTOGRAPHY AND THE DIFFIE-HELLMAN PROBLEM

In this section we introduce the Diffie-Hellman Key Exchange and the discrete log problem. The Diffie-Hellman key exchange protocol was the first explicit example of an asymmetric or public key cryptosystem and the security of the Diffie-Hellman key exchange protocol is related to the computationally difficult discrete log problem. Such computationally difficult 'one-way' functions were described long before computers. For example, William Stanley Jevons, a British economist once asked:

> 'What two numbers multiplied together will produce 8616460799? I think it unlikely that anyone but myself will ever know.'
> - William Stanley Jevons (Jevons, 1866)

Researchers at Government Communications Headquarters, GCHQ, discovered the foundations of public key cryptography about six years before the first publication on public key cryptography by Diffie and Hellman. In 1976, Bailey Whitfield Diffie and Martin Edward Hellman (Diffie and Hellman, 1976) published the now-named Diffie-Hellman Key Exchange protocol that allows two users to establish a shared secret, i.e., a private key, in a public channel. Before explaining the mathematics of their protocol in Section 14.4.1, we wish to remind the reader of the two groups $\mathbb{Z}/17\mathbb{Z}$ and $(\mathbb{Z}/17\mathbb{Z})^\times$ in Example 14.5 and 14.6. Both groups are cyclic, meaning they can be generated by repeatedly composing a single element by itself. For example, in $\mathbb{Z}/17$ all elements can be obtained by composing 5 with itself, i.e., adding 5 to itself to get 10, then adding 5 to 10 to get 15, etc. In the group $(\mathbb{Z}/17\mathbb{Z})^\times$ each element can be obtained by composing 5 with itself, i,e., $5*5 = 8$ and $8*5 = 6$. For convenience we list out the sequence so obtained in tables 14.2 and 14.1.

TABLE 14.1 Table showing the value of $5 \cdot i$ in $\mathbb{Z}/17\mathbb{Z}$

i	1	2	3	4	5	6	7	8	9	10	11	12	13	14	15	16	17
$5 \cdot i$	5	10	15	3	8	13	1	6	11	16	4	9	14	2	7	12	0

TABLE 14.2 Table showing the value of 5^i in $(\mathbb{Z}/17\mathbb{Z})^\times$

i	1	2	3	4	5	6	7	8	9	10	11	12	13	14	15	16
5^i	5	8	6	13	14	2	10	16	12	9	11	4	3	15	7	1

The reader should examine the second row of each table. For Table 14.1, the second row is very easy to compute. For Table 14.2 the second row appears more random, and this apparent randomness is the key to the discrete log problem.

The discrete log problem for a group G is the following: Fix a group G and element $g \in G$. Given g^a, find a. For example, fix $G = (\mathbb{Z}/17\mathbb{Z})^\times$ and $g = 5$. Then if $g^a = 10$, we consult Table 14.2 to see that $a = 7$. In general, without a precomputed table or other information the problem *appears* difficult. We emphasize appears, because determining which problems are difficult is tricky and lies in the realm of computational complexity. With specific regards to the discrete log problem, see Odlyzko's survey article (Odlyzko, 2000).

14.4.1 Diffie-Hellman Key Exchange

The Diffie-Hellman Key Exchange protocol is used to establish a shared secret between two parties. The secret is established over a public or open medium, i.e., the two parties do not have a private channel of communication and may not have met in person.

Initial Parameters: Two parties, known as Alice and Bob. A fixed group G and an element $g \in G$.

End Product: Alice and Bob have secretly agreed upon an element $h \in G$.

The protocol works as follows, Alice wants to establish a shared secret with Bob and they publicly agree on a group G and an element $g \in G$. Alice then privately chooses an integer a and computes g^a. Alice then sends g^a to Bob. Bob similarly privately chooses an integer b and computes g^b and sends g^b to Alice. When sending information to each other, Alice and Bob **do not** assume that their correspondence is secure.

To recap, the group G, the element g, and the elements g^a and g^b are assumed to be public information, or at least publicly available. The secret key Alice and Bob will use is g^{ab}, which Alice can compute from g^b by raising it to the a and Bob can compute from g^a by raising it to the b. An attacker or eavesdropper would only know the public information and cannot deduce g^{ab} from g, g^a, and g^b.

We return to our example where $G = (\mathbb{Z}/17\mathbb{Z})^\times$ and $g = 5$. Alice and Bob want to communicate with each other. Alice chooses $a = 5$ and computes $5^5 \equiv 14 \pmod{17}$. Bob chooses $b = 12$ and computes $5^{12} \equiv 4 \pmod{17}$. Alice sends 14 to Bob and Bob sends 4 to Alice. An eavesdropper would have to solve for x in $5^x \equiv 14$ to solve for a or $5^x \equiv 4$ to solve for b. Alice and Bob then use $5^{ab} = (14)^{12} \equiv (4)^5 \equiv 4 \pmod{17}$. So their shared secret is the element 4 in G.

Of course, there are good and bad choices for the parameters G, g, a, and b. For example, if we worked over the group $G = \mathbb{Z}/17\mathbb{Z}$, the discrete log problem amounts to division. The next section examines how for even good choices of groups, the Diffie-Hellman Key Exchange protocol can be mitigated by nation-state-level attacks.

14.4.2 An Attack on Diffie-Hellman Key Exchange

This section discusses major weaknesses of implementation for the Diffie-Hellman Key Exchange protocol as documented by the Weak Diffie-Hellman group (the WeakDH group) (Adrian et al., 2015). The WeakDH group did not mathematically solve the discrete log problem, but presented a highly efficient 'engineering' level solution. At its core, their attack involves a large pre-computation using the number field sieve for a fixed prime modulus, and an analysis that many servers use the same prime. From this pre-computed database, one can quickly break instances of Diffie-Hellman. The attack itself is highly effective when a large number of networks utilize the same prime. Of note, the WeakDH paper argues that the NSA has exploited these flaws in the Diffie-Hellman Key Exchange protocols. The authors call for greater cooperation between cryptographers and developers in implementing secure protocols.

The WeakDH group is comprised of 14 researchers from various countries working for universities (all located in the US), government institutes (all located in France), and Microsoft Research. The group consists of mathematicians and computer scientists broadly interested in security research. Most of the group members are highly supportive of data privacy and this motivated their collaboration. The senior researchers are well-known as academics in cybersecurity and cyber-related policy.

The group's formation was a direct response to the NSA leaks as revealed by Edward Snowden, in which Snowden claimed the NSA had the ability to decrypt a large amount of SSH traffic. As academics, the group responded with an overwhelming desire to identify how the NSA was accomplishing this and to prevent further surveillance by the NSA. The group is overwhelming committed to preserving internet privacy for average users with research interests of members including the importance of data privacy and the use of technology in international affairs, discovering vulnerabilities via internet-wide scanning, and how nation-state actors can compromise security protocols.

As with many cryptosystems, choosing weak parameters will compromise the security of the Diffie-Hellman Key Exchange. To circumvent this, there is a published list of safe primes. However, many of the safe primes are reused in numerous applications. Other safe primes were preloaded with various software packages as the default value. By reusing primes, targets encouraged attackers to perform expensive attacks that break the cryptosystem for a single prime. The attack utilized by the WeakDH group focused on pre-computation of information about a fixed prime.

Pre-computation is essentially how we solved the inverse logarithm problem in $(\mathbb{Z}/17\mathbb{Z})^{\times}$. There are faster and better techniques than simply computing powers of g in advance. The modern pre-computation step is computationally expensive, but can be parallelized. We do not discuss the mathematics behind the pre-computations here. Instead we will treat the pre-computation step as an expensive way to compute and store information that will enable a quick solution to the discrete log problem for a fixed prime modulus. In general, the information secured should be less expensive than the computation needed to break it. Pre-computations for a 512-bit prime currently take about a week over a moderately robust distributed computing network. The computations are done once for the specific prime and then stored. Hence, the use of the same prime by multiple targets allows an attacker to use the same pre-computed information repeatedly.

Another attack mentioned in the WeakDH paper involves the use of an inherent flaw in the Transport Layer Security Protocol, called the Logjam attack. We will not comment on this attack, other than that it depends on the ability to pre-compute information about 512-bit prime numbers. The reuse of certain 512-bit prime numbers provides an attacker with numerous targets. In addition to the Logjam attack, the WeakDH group analyzed the prospects of attacks by nation-states against Diffie-Hellman protocols that use 1024-bit prime numbers. Primes of the

1024-bit size would require exorbitantly expensive equipment to break, in the order of magnitude of one hundred million dollars. The WeakDH group estimates that it would require nation-state level resources to perform such calculations. However, by performing the calculations for a handful of the right primes, a state could launch large scale attacks against various targets.

When conducting their research, the WeakDH group estimate that a pre-computed attack on roughly 10 1024-bit primes would render over a quarter of all publicly accessible SSH servers subject to eavesdropping by a nation-state. The WeakDH group also estimates that over 170,000 of the top 1 million https-enabled web sites utilized the same single 1024-bit prime number in their Diffie-Hellman implementations.

The WeakDH group encourage users to regularly switch the prime field in their public keys by utilizing random prime numbers. If a large number of organizations utilize the same Diffie-Hellman prime, then the prime becomes a high-value target for a precomputation attack. The precomputation attack on Diffie-Hellman highlights an important and recurring theme in cybersecurity: strong systems, implemented poorly, are vulnerable. Cybersecurity teams need to implement safe, secure encryption on their networks, but also need to ensure that those networks do not share a single vulnerability. Regarding analysis groups, the WeakDH group encouraged cybersecurity professionals to more closely coordinate with cryptographers to ensure strong and up-to-date systems.

14.5 CONCLUSION

As we hope the reader has seen, group theory is ubiquitous in the physical and cyber worlds. We naturally try to catalogue when two objects are the same and when they differ for various applications. Because we input data into a computer in a linear way, we are forced to consider how different orderings of the input could correspond to the same structure. This sameness of structures is always present when we are forced to encode an object into a machine readable format. More blatantly, modern cryptosystems use group theoretic ideas to encrypt information and rely on group theoretic problems for their security. In this concluding section we will point the interested reader towards other resources in the broad category of group theory in the cyber realm.

14.5.1 Cryptography

We highly recommend Koblitz's (Koblitz, 1994) as an introduction to cryptography and the mathematical background, including groups and fields. More advanced students interested in group theoretic cryptography should consult the text by Myasnikov, Shpilrain, and Ushakov (Myasnikov et al., 2011).

14.5.2 Graph Isomorphism

The work of Lázló Babai on graph isomorphism represents one of the largest and most important uses of group theory within computer science (Babai, 2016).

McKay's nauty algorithm (McKay, 1981) and the newer traces are both worth further study (Piperno, 2008; McKay and Piperno, 2014).

14.5.3 Group Theory and Computational Group Theory

Group theory occupies a special place within mathematics. It is typically the first topic discussed in any undergraduate abstract algebra class. The interested reader is encouraged to learn more from any undergraduate abstract algebra or group theory textbook that they find easy to read. In particular, we recommend (Isaacs, 2008) and (Serre, 2016) for studying finite group theory. For infinite groups, combinatorial group theory takes a more algorithmic approach (Lyndon and Schupp, 2001). A more modern approach to infinite groups is geometric group theory (de la Harpe, 2000; Clay and Margalit, 2017).

Furthermore, computational questions regarding group theory are abound. Many of these questions range from theoretic complexity-type questions to practical implementation of algorithms. GAP (GAP, 2019) and Magma (Bosma et al., 1997) are two programming languages with extensive ability to perform computations involving groups. A comprehensive treatment of this subject can be found in Holt et al., 2005.

REFERENCES

Adrian, D., Bhargavan, K., Durumeric, Z., Gaudry, P., Green, M., Halderman, J. A., Heninger, N., Springall, D., Thomé, E., Valenta, L., et al. (2015). Imperfect forward secrecy: How diffie-hellman fails in practice. In *Proceedings of the 22nd ACM SIGSAC Conference on Computer and Communications Security*, pages 5–17. ACM.

Babai, L. (2016). Graph isomorphism in quasipolynomial time. In *Proceedings of the Forty-Eighth Annual ACM Symposium on Theory of Computing*, pages 684–697.

Bosma, W., Cannon, J., and Playoust, C. (1997). The magma algebra system i: The user language. *Journal of Symbolic Computation*, 24(3-4):235–265.

Clay, M. and Margalit, D., editors (2017). *Office hours with a geometric group theorist*. Princeton University Press, Princeton, NJ.

de la Harpe, P. (2000). *Topics in geometric group theory*. Chicago Lectures in Mathematics. University of Chicago Press, Chicago, IL.

Diffie, W. and Hellman, M. (1976). New directions in cryptography. *IEEE Transactions on Information Theory*, 22(6):644–654.

GAP (2019). *GAP – Groups, Algorithms, and Programming, Version 4.10.2*. The GAP Group.

Hartke, S. G. and Radcliffe, A. J. (2009). McKay's canonical graph labeling algorithm. In *Communicating mathematics*, volume 479 of *Contemp. Math.*, pages 99–111. Amer. Math. Soc., Providence, RI.

Holt, D. F., Eick, B., and O'Brien, E. A. (2005). *Handbook of computational group theory*. Chapman & Hall/CRC, Boca Raton, FL.

Isaacs, I. M. (2008). *Finite group theory*, volume 92 of Graduate Studies in Mathematics. American Mathematical Society, Providence, RI.

Jevons, W. S. (1866). Brief account of a general mathematical theory of political economy. *Journal of the Royal Statistical Society*, 29:282–87.

Koblitz, N. (1994). *A course in number theory and cryptography*, Springer Science & Business Media.

Lyndon, R. C. and Schupp, P. E. (2001). *Combinatorial group theory*. Classics in Mathematics. Springer-Verlag, Berlin. Reprint of the 1977 edition.

McKay, B. D. (1981). Practical graph isomorphism. *Congr. Numer.*, 30:45–87.

McKay, B. D. and Piperno, A. (2014). Practical graph isomorphism, II. *J. Symbolic Comput.*, 60:94–112.

Myasnikov, A. G., Shpilrain, V., and Ushakov, A. (2011). *Non-commutative cryptography and complexity of group-theoretic problems*. American Mathematical Soc.

Odlyzko, A. (2000). Discrete logarithms: The past and the future. In *Towards a Quarter-Century of Public Key Cryptography*, pages 59–75. Springer.

Piperno, A. (2008). Search space contraction in canonical labeling of graphs. arXiv preprint arXiv:0804.4881.

Serre, J.-P. (2016). *Finite groups: an introduction*, volume 10 of Surveys of Modern Mathematics. International Press, Somerville, MA; Higher Education Press, Beijing. With assistance in translation provided by Garving K. Luli and Pin Yu.

Stillwell, J. (2002). Mathematics and its history. *The Australian Mathem. Soc*, page 168.

CHAPTER 15

Ring Theory

Lindsey-Kay Lauderdale

CONTENTS

15.1	Introduction	475
15.2	Ring Theory Background	477
	15.2.1 Basic Definitions	477
	15.2.2 Polynomial and Factor Rings	478
15.3	The NTRU Cryptosystem	485
	15.3.1 Notational Conventions and Definitions	485
	15.3.2 Key Generation Algorithm	487
	15.3.3 Encryption Algorithm	488
	15.3.4 Decryption Algorithm	489
	15.3.5 Summary	491
15.4	Attacking the NTRU Cryptosystem	491
	15.4.1 General Attacks of the Spy	491
	15.4.2 The NTRU Key Recovery Problem	493
15.5	Conclusion	494

15.1 INTRODUCTION

Cryptography is the study of the systems that send secure messages; this area of study lies in an intersection of the fields of mathematics and computer science. Cryptographic advancement is vital to national security as the ability to send secure communications is required. Secure communications are also required for internet security and firewalls, digital signatures and authentication, and data compression, among others (Batten, 2013; Mollin, 2007).

In its most basic form, cryptography is based on the following situation, which is illustrated in Figure 15.1. A person, termed the *sender*, wants to relay a confidential message to another person, termed the *receiver*. Since this message contains secret or sensitive information, the sender does not want a third party, termed the *spy*, to be able to intercept or modify the message in any way. In an attempt to keep the message confidential, the sender will take the original message and obfuscate its contents using a process called *encryption*. This encryption process produces an unreadable message called *ciphertext* that is then sent to the receiver. Once the receiver obtains the ciphertext message, they use a process called *decryption* to

DOI: 10.1201/9780429354649-15

turn the ciphertext back into a readable message. The encryption and decryption process must work together if the receiver is to obtain the original message at the end of this process.

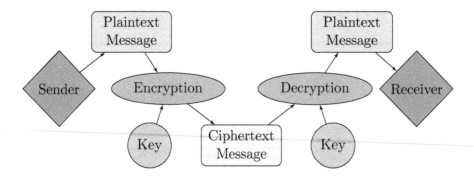

FIGURE 15.1 General situation on which cryptography is based.

A *cryptosystem* is a mathematical function that comprises an encryption algorithm and a decryption algorithm. All cryptosystems require mathematical objects called *keys*, which are usually represented by extremely large numbers or polynomials with integer coefficients. If the decryption key and the encryption key fit together for a given message, then the original message can be properly recovered from the ciphertext message.

The sender and receiver will never rely on securing a cryptosystem through obscurity; rather, when evaluating the security of a given cryptosystem, one usually takes on the perspective of the spy. It is assumed that the spy knows both the encryption algorithm and the decryption algorithm. Consequently, security is not based on the specific details of said cryptosystem and is based on the secrecy of the key(s).

The security spectrum for cryptosystems is vast. A cryptosystem is *insecure* if the spy can learn something about the plaintext message from the ciphertext message. On the other end of the security spectrum is perfect secrecy, which was introduced by Shannon (1949). A cryptosystem is *perfectly secret* provided that all plaintext messages have the same probability of occurring for a given ciphertext message. Thus, even with unlimited computing power, a spy would not be able to successfully attack a cryptosystem that has perfect secrecy assuming they had seen the ciphertext message.

There are two main types of cryptosystems that require keys, namely symmetric-key algorithms and public-key algorithms. The collection of symmetric-key algorithms has a wide range of security, varying from insecure to perfectly secret. As the name suggests, in cryptosystems with symmetric-key algorithms the encryption key and the decryption key are either the same or can be calculated easily from one another. Since both the sender and receiver need to know the same key, they must have prior contact. For example, the key might be sent by a courier prior to the sender relaying the confidential message to the receiver. In any case, this key must be kept secret; if the spy were to intercept the key, then they could read any confidential messages.

The idea of public-key cryptography was introduced by Diffie and Hellman (1976). While perfect secrecy is not attainable for public-key cryptosystems, they are designed around the computational hardness assumptions for computers. In this case, the sender and the receiver do not require a courier because the encryption key and the decryption key are different. Nevertheless, private-key communication is often established by first using public-key cryptography to exchange or create keys. In fact, the encryption key is often made public because the decryption key cannot be calculated from the encryption key in any reasonable amount of time using a modern computer. However, the invention of quantum computers (exceptionally powerful computers that use superposition and entanglement of bits to perform computations) will render many public-key cryptosystems ineffective, which will threaten our national security.

Modern cryptosystems, and cryptography in general, use many areas of mathematics and computer science. One way to create new cryptographic algorithms or to improve the security of known cryptographic algorithms is through the study of mathematical objects called rings; informally, a ring can be thought of as a set of objects on which both addition and multiplication can be performed.

In this chapter, we will consider an area of mathematics, called ring theory, and discuss its applications to cryptography at an introductory level. We will focus on a public-key cryptosystem called the N-th Degree Truncated Polynomial Ring (NTRU), which was first introduced by Hoffstein et al. (1998). The major advantages of the NTRU (pronounced *en-trū*) cryptosystem are that it is based on certain lattice problems, and it is thought to be resistant to all quantum-based computer attacks (Hoffstein et al., 2010). Additionally, it is considerably faster than other public-key cryptosystems like RSA (Rivest-Shamir-Adleman) and ECC (elliptic-curve cryptography), which are susceptible to quantum-based computer attacks (Hoffstein et al., 1998). The application of the NTRU cryptosystem requires some basic ring theory, and thus we proceed with the mathematical content about rings that will be required for this cryptosystem.

15.2 RING THEORY BACKGROUND

The NTRU cryptosystem was first introduced using polynomial rings with integer coefficients. Consequently, we will discuss some basic terminology and concepts for these rings in this section. This material is standard and more details can be found in most abstract algebra textbooks; see, for example, Dummit and Foote (2004).

15.2.1 Basic Definitions

Definition 15.1. *A* binary operation *on the set S is a function from $S \times S$ to S.*

Definition 15.2. *A* ring *is an ordered triple $(R, +, \cdot)$, where R is a nonempty set and $+$ and $\cdot$ are two binary operations satisfying the following six conditions:*

1. $r + s = s + r$ for all $r, s \in R$;

2. $(r+s)+t = r+(s+t)$ for all $r,s,t \in R$;

3. there exists $0 \in R$ such that $0+r = r$ for all $r \in R$;

4. for each $r \in R$, there exists $s \in R$ such that $r+s = 0$ (the ring element s is also denoted by $-r$);

5. $(r \cdot s) \cdot t = r \cdot (s \cdot t)$ for all $r,s,t \in R$; and

6. $(r+s) \cdot t = r \cdot t + s \cdot t$ for all $r,s,t \in R$.

Each of the integers $\mathbb{Z}$, the rational numbers $\mathbb{Q}$, and the real numbers $\mathbb{R}$ form a ring under the usual arithmetic operations of addition and multiplication. Additionally, if n is a positive integer, then the set of integers modulo n, namely $\mathbb{Z}_n = \{0, 1, 2, \ldots, n-1\}$, form a ring under the operations of addition modulo n and multiplication modulo n. The natural numbers $\mathbb{N}$ do not form a ring under the usual arithmetic operations of addition and multiplication because the fourth condition of Definition 15.2 fails to hold. For example, there is no $s \in \mathbb{N}$ such that $2 + s = 0$.

Definition 15.3. *A ring R is commutative provided $r \cdot s = s \cdot r$ for all $r, s \in R$.*

Definition 15.4. *A ring R has an identity if there exists $1 \in R$ such that $r \cdot 1 = 1 \cdot r$ for all $r \in R$.*

Of course, not all rings are commutative or even contain an identity; for example, the ring of even integers $2\mathbb{Z}$ does not have an identity, and the set of 2×2 matrices with entries from $\mathbb{Z}$ is a ring that is not commutative.

Definition 15.5. *A commutative ring R with identity is a field if for each nonzero $r \in R$ there exists an $s \in R$ such that $r \cdot s = 1$. In this case, r and s are multiplicative inverses of one another.*

The ring $\mathbb{Z}$ is not a field, while both $\mathbb{Q}$ and $\mathbb{R}$ are fields. Additionally, the ring $\mathbb{Z}_n$ is a field if and only if n is a prime number.

15.2.2 Polynomial and Factor Rings

The rings considered for the remainder of this chapter will be commutative rings with identity. An important family of such rings that often arises in cryptography is polynomial rings.

Definition 15.6. *Let R be a commutative ring with identity. Define the set $R[x]$ to be the set of all polynomials in the indeterminate x of the form*

$$r_n x^n + r_{n-1} x^{n-1} + \cdots + r_2 x^2 + r_1 x + r_0,$$

where n is a non-negative integer and $r_i \in R$ for all $i \in \{0, 1, \ldots, n\}$.

It is easy to check that the set $R[x]$ satisfies all six conditions given in Definition 15.2 under the usual operations of polynomial addition and polynomial multiplication. Consequently, $R[x]$ is a ring, called the *polynomial ring* in the variable x with coefficients from R.

There were two types of rings that appeared in the original NTRU cryptosystem, namely polynomial rings with integer coefficients and factor rings with integer coefficients. In order to give the definition of a factor ring, we first recall the definition of an ideal. Ideals are special types of subsets of rings and can be used to construct so-called factor rings. Note that when the fourth condition stated in Definition 15.2 holds in a ring, then s is called the *additive inverse* for r and is denoted by $s = -r$.

Definition 15.7. *A nonempty subset I of a ring R is an ideal of R if $i - j \in I$ and $r \cdot i, i \cdot r \in I$ for all $i, j \in I$ and $r \in R$.*

When R is a ring and $r \in R$, we let $\langle r \rangle$ denote the smallest ideal of R containing r, and call $\langle r \rangle$ the *ideal generated* by r.

Lemma 15.1. *Let I be an ideal of a ring R. The set*

$$R/I = \{r + I : r \in R\}$$

forms a ring under the binary operations defined by

$$(r + I) + (s + I) = (r + s) + I$$

and

$$(r + I) \cdot (s + I) = (r \cdot s) + I,$$

where $r + I, s + I \in R/I$.

The elements of R/I are called *cosets* of I in R.

Definition 15.8. *Let I be an ideal of the ring R. The ring stated in Lemma 15.1 is called the factor ring of R by I.*

If n is an integer, then the set $n\mathbb{Z} = \{nk : k \in \mathbb{Z}\}$ is an ideal of the ring $\mathbb{Z}$. In fact, these sets are the only ideals of $\mathbb{Z}$ and can be used to construct a factor ring that is structurally the same as (i.e., isomorphic to) the ring $\mathbb{Z}_n$. In particular, this ideal can be used to construct the factor ring $\mathbb{Z}/n\mathbb{Z}$. The set of residue classes,

$$\mathbb{Z}/n\mathbb{Z} = \{[0], [1], [2], \ldots, [n-1]\},$$

form a ring under the operations of addition and multiplication of residue classes, where

$$[i] = \{k \in \mathbb{Z} : i \equiv k \pmod{n}\}$$

for all $i \in \{0, 1, \ldots, n-1\}$.

Example 15.1. *If $n = 13$, the factor ring $\mathbb{Z}/13\mathbb{Z}$ contains the following 13 residue classes:*
$$[0], [1], [2], [3], [4], [5], [6], [7], [8], [9], [10], [11], [12].$$
Addition and multiplication in this factor ring were defined in Lemma 15.1. For instance,
$$[5] + [7] = [5 + 7] = [12]$$
and
$$[5] \cdot [7] = [5 \cdot 7] = [35] = [9],$$
where the last equality holds because $35 \equiv 9 \pmod{13}$.

Let p be a prime number. The NTRU cryptosystem considers three types of rings, namely the polynomial ring $(\mathbb{Z}/p\mathbb{Z})[x]$, as well as the factor rings
$$\mathbb{Z}[x]/\langle x^n - 1 \rangle \quad \text{and} \quad (\mathbb{Z}/p\mathbb{Z})[x]/\langle x^n - 1 \rangle,$$
where n is a positive integer. Let us continue by investigating the later two factor rings more closely.

Definition 15.9. *A nonzero commutative ring is an integral domain provided the product of any two nonzero elements is also nonzero.*

For instance, the rings $\mathbb{Z}$ and $\mathbb{Z}/p\mathbb{Z}$, where p is a prime number, are integral domains. The ring $\mathbb{Z}/6\mathbb{Z}$ is not an integral domain because the product of the two nonzero elements [2] and [3] is $[2] \cdot [3] = [0]$ in $\mathbb{Z}/6\mathbb{Z}$. In general, if n is a composite number, then $\mathbb{Z}/n\mathbb{Z}$ is not an integral domain because there exists two nonzero elements whose product is zero. In particular, because n is composite it is possible to choose $a, b \in \mathbb{Z}$ such that $n = ab$ and $1 < a, b < n$; in this case,
$$[a] \cdot [b] = [a \cdot b] = [n] = [0],$$
where the last equality holds because $n \equiv 0 \pmod n$ and $\mathbb{Z}/n\mathbb{Z}$ is not an integral domain.

The polynomial ring $R[x]$, where R is a field, and the integer ring $\mathbb{Z}$ have more structure than arbitrary rings. In particular, they are Euclidean domains, and thus possess a division algorithm.

Definition 15.10. *Let R be an integral domain. A division algorithm on R is a map $\varphi : R \to \mathbb{N}$ such that for all $a, b \in R$ with $b \neq 0$, there exists $q, r \in R$ such that $a = bq + r$ and either $\varphi(r) < \varphi(b)$ or $r = 0$. A ring that can be endowed with a division algorithm is called a Euclidean domain.*

The repeated application of the division algorithm will be required for the NTRU cryptosystem in order to find the multiplicative inverse of a given polynomial. Before this concept is demonstrated in an example below, we will consider two rings as the same for ease of notation. Specifically, since the rings $\mathbb{Z}_n$ and $\mathbb{Z}/n\mathbb{Z}$ are isomorphic, we will identify them as follows: for each $i \in \{0, 1, \ldots, n-1\}$, identify the representative i and the residue class $[i] \in \mathbb{Z}/n\mathbb{Z}$.

Example 15.2. *Consider the polynomial ring $(\mathbb{Z}/3\mathbb{Z})[x]$. Since*

$$\mathbb{Z}/3\mathbb{Z} = \{[0], [1], [2]\}$$

is a field, this polynomial ring is a Euclidean domain. In this example, we will demonstrate how to use the Euclidean algorithm to compute the multiplicative inverse of a given polynomial in $(\mathbb{Z}/3\mathbb{Z})[x]$. For ease of notation, recall that the representatives 0, 1, and 2 have been identified with residues classes [0], [1], and [2], respectively; thus, we can consider the elements of $(\mathbb{Z}/3\mathbb{Z})[x]$ as polynomials whose coefficients are either 0, 1, or 2.

Let us compute the multiplicative inverse of

$$f(x) = 2x^5 + x^4 + 2x^2 + x + 2$$

in $(\mathbb{Z}/3\mathbb{Z})[x]$ modulo $g(x) = x^7 + 2$ with the Euclidean algorithm; note that the division algorithm used here for polynomials is polynomial long division where the arithmetic is preformed modulo 3.

First, divide $g(x)$ by $f(x)$ to obtain a quotient of $q_1(x) = 2x^2 + 2x + 2$ and a remainder of $r_1(x) = 2x^2 + 1$, so that

$$\underbrace{x^7 + 2}_{g(x)} = (\underbrace{2x^5 + x^4 + 2x^2 + x + 2}_{f(x)})(\underbrace{2x^2 + 2x + 2}_{q_1(x)}) + (\underbrace{2x^2 + 1}_{r_1(x)}). \tag{15.1}$$

Next, divide $f(x)$ by $r_1(x)$ to obtain a quotient of $q_2(x) = x^3 + 2x^2 + x$ and a remainder of $r_2(x) = 2$, so that

$$\underbrace{2x^5 + x^4 + 2x^2 + x + 2}_{f(x)} = (\underbrace{2x^2 + 1}_{r_1(x)})(\underbrace{x^3 + 2x^2 + x}_{q_2(x)}) + (\underbrace{2}_{r_2(x)}). \tag{15.2}$$

Third, divide $r_1(x)$ by $r_2(x)$ to obtain a quotient of $q_3(x) = x^2 + 2$ and a remainder of $r_3(x) = 0$, so that

$$\underbrace{2x^2 + 1}_{r_1(x)} = (\underbrace{2}_{r_2(x)})(\underbrace{x^2 + 2}_{q_3(x)}) + (\underbrace{0}_{r_3(x)}).$$

These computations imply that the greatest common divisor of $f(x)$ and $g(x)$ is the penultimate remainder $r_2(x) = 2$.

In order to find the multiplicative inverse of $f(x)$ modulo $g(x)$, we can use a back substitution process (i.e., the Extended Euclidean Algorithm) with Equations 15.1 and 15.2. Since $[-1] = [2]$ in $\mathbb{Z}/3\mathbb{Z}$, Equation 15.2 implies

$$\underbrace{2}_{r_2(x)} = \underbrace{2x^5 + x^4 + 2x^2 + x + 2}_{f(x)} + 2(\underbrace{2x^2 + 1}_{r_1(x)})(\underbrace{x^3 + 2x^2 + x}_{q_2(x)}), \tag{15.3}$$

and Equation 15.1 implies that

$$\underbrace{2x^2 + 1}_{r_1(x)} = \underbrace{x^7 + 2}_{g(x)} + 2(\underbrace{2x^5 + x^4 + 2x^2 + x + 2}_{f(x)})(\underbrace{2x^2 + 2x + 2}_{q_1(x)}). \tag{15.4}$$

Substitute the expression for $r_1(x)$ given by Equation 15.4 into Equation 15.3 to obtain

$$2 = f(x) + 2\big(\underbrace{g(x) + 2f(x)(2x^2 + 2x + 2)}_{r_1(x)}\big)(x^3 + 2x^2 + x)$$
$$= f(x)\big(1 + (2x^2 + 2x + 2)(x^3 + 2x^2 + x)\big) + 2g(x)(x^3 + 2x^2 + x)$$
$$= f(x)(2x^5 + 2x^3 + 2x + 1) + 2g(x)(x^3 + 2x^2 + x),$$

which implies

$$1 = f(x)(x^5 + x^3 + x + 2) + g(x)(x^3 + 2x^2 + x).$$

Therefore,

$$1 \equiv \big(f(x)(x^5 + x^3 + x + 2) + g(x)(x^3 + 2x^2 + x)\big)\,(\bmod\, g(x))$$
$$\equiv \big(f(x)(x^5 + x^3 + x + 2) + 0\big)\,(\bmod\, g(x))$$
$$\equiv f(x)(x^5 + x^3 + x + 2)\,(\bmod\, g(x))$$

and $x^5 + x^3 + x + 2$ is the multiplicative inverse of $f(x)$ in $(\mathbb{Z}/3\mathbb{Z})[x]$.

Observe that the polynomial

$$g(x) = x^7 + 2 \in (\mathbb{Z}/3\mathbb{Z})[x]$$

given in Example 15.2 could have been written as

$$g(x) = x^7 - 1 \in (\mathbb{Z}/3\mathbb{Z})[x]$$

because $-1 \equiv 2 \pmod{3}$. The element $n-1$ in $\mathbb{Z}/n\mathbb{Z}$ is often written simply as -1 in the implementation of the NTRU cryptosystem. Moreover, polynomials of the form $x^n - 1$, where n is a positive integer, are central to this cryptosystem. Thus, we continue by investigating them more closely.

Consider the ideal generated by the polynomial $x^n - 1$ in the ring $\mathbb{Z}[x]$; the ideal $\langle x^n - 1 \rangle$ comprises all polynomials of the form

$$f(x) \cdot (x^n - 1),$$

where $f(x) \in \mathbb{Z}[x]$. While the ring $\mathbb{Z}[x]$ is not a Euclidean domain, it does possess a division algorithm in certain situations. In this case, since $x^n - 1$ is monic (i.e., the leading coefficient is 1), dividing any polynomial $g(x) \in \mathbb{Z}[x]$ by $x^n - 1$ will result in two polynomials $q(x), r(x) \in \mathbb{Z}[x]$ such that

$$g(x) = (x^n - 1) \cdot q(x) + r(x),$$

where either the degree of $r(x)$ is less than n or $r(x)$ is the zero polynomial. In either case, we see that

$$g(x) + \langle x^n - 1 \rangle = \underbrace{\left((x^n - 1) \cdot q(x) + r(x) \right)}_{g(x)} + \langle x^n - 1 \rangle$$
$$= \left((x^n - 1) \cdot q(x) + \langle x^n - 1 \rangle \right) + \left(r(x) + \langle x^n - 1 \rangle \right)$$
$$= \left(0 + \langle x^n - 1 \rangle \right) + \left(r(x) + \langle x^n - 1 \rangle \right)$$
$$= \left(0 + r(x) \right) + \langle x^n - 1 \rangle$$
$$= r(x) + \langle x^n - 1 \rangle,$$

where the third equality holds because

$$(x^n - 1) \cdot q(x) \in \langle x^n - 1 \rangle.$$

Therefore, all elements of the factor ring $\mathbb{Z}[x]/\langle x^n - 1 \rangle$ have degree less than n (i.e., have the form

$$(r_{n-1}x^{n-1} + \cdots + r_2 x^2 + r_1 x + r_0) + \langle x^n - 1 \rangle,$$

where $r_i \in \mathbb{Z}$ for all $i \in \{0, 1, \ldots, n-1\}$). Addition in this ring is defined componentwise; that is, the sum of the elements

$$(r_{n-1}x^{n-1} + \cdots + r_2 x^2 + r_1 x + r_0) + \langle x^n - 1 \rangle \in \mathbb{Z}[x]/\langle x^n - 1 \rangle$$

and

$$(s_{n-1}x^{n-1} + \cdots + s_2 x^2 + s_1 x + s_0) + \langle x^n - 1 \rangle \in \mathbb{Z}[x]/\langle x^n - 1 \rangle$$

is equal to

$$((r_{n-1} + s_{n-1})x^{n-1} + \cdots + (r_1 + s_1)x + (r_0 + s_0)) + \langle x^n - 1 \rangle \in \mathbb{Z}[x]/\langle x^n - 1 \rangle.$$

Multiplication in $\mathbb{Z}[x]/\langle x^n - 1 \rangle$ is defined as multiplication of polynomials modulo $x^n - 1$ and can be thought of as a cyclic convolution product. In particular, the product of

$$r(x) = (r_{n-1}x^{n-1} + \cdots + r_2 x^2 + r_1 x + r_0) + \langle x^n - 1 \rangle \in \mathbb{Z}[x]/\langle x^n - 1 \rangle$$

and

$$s(x) = (s_{n-1}x^{n-1} + \cdots + s_2 x^2 + s_1 x + s_0) + \langle x^n - 1 \rangle \in \mathbb{Z}[x]/\langle x^n - 1 \rangle$$

in $\mathbb{Z}[x]/\langle x^n - 1 \rangle$ is equal to

$$t(x) = (t_{n-1}x^{n-1} + \cdots + t_2 x^2 + t_1 x + t_0) + \langle x^n - 1 \rangle \in \mathbb{Z}[x]/\langle x^n - 1 \rangle,$$

where

$$t_i = \sum_{j+k \equiv i \,(\mathrm{mod}\, n)} r_j s_k.$$

We demonstrate this cyclic convolution product of two elements in $\mathbb{Z}[x]/\langle x^n - 1 \rangle$ below.

Example 15.3. *Consider the factor ring $\mathbb{Z}[x]/\langle x^3 - 1\rangle$. Both of the polynomials*

$$r(x) = 2x^2 + 7x + 5 \in \mathbb{Z}[x]$$

and

$$s(x) = x^2 + 4x + 9 \in \mathbb{Z}[x]$$

are elements of this factor ring as their degrees are less than 3. To compute the product of $r(x)$ and $s(x)$ in $\mathbb{Z}[x]/\langle x^3 - 1\rangle$, say $t(x) = r(x) \cdot s(x)$, we first note that

$$r_2 = 2,\ r_1 = 7,\ r_0 = 5,\ s_2 = 1,\ s_1 = 4,\ \text{and}\ s_0 = 9.$$

The summation to compute the coefficient t_i of $t(x)$ for $i \in \{0, 1, 2\}$ is taken over all pairs (j, k) such that $j+k \equiv i \pmod 3$. Therefore, we obtain the following coefficients of $t(x)$:

$$t_0 = r_0 s_0 + r_1 s_2 + r_2 s_1 = (5)(9) + (7)(1) + (2)(4) = 60$$
$$t_1 = r_0 s_1 + r_1 s_0 + r_2 s_2 = (5)(4) + (7)(9) + (2)(1) = 85,$$

and

$$t_2 = r_0 s_2 + r_2 s_0 + r_1 s_1 = (5)(1) + (2)(9) + (7)(4) = 51.$$

Therefore, the product of $r(x)$ and $s(x)$ in $\mathbb{Z}[x]/\langle x^3 - 1\rangle$ is

$$t(x) = 51x^2 + 85x + 60.$$

In addition to the factor ring $\mathbb{Z}[x]/\langle x^n - 1\rangle$, the NTRU cryptosystem uses the factor ring

$$(\mathbb{Z}/p\mathbb{Z})[x]/\langle x^n - 1\rangle,$$

where p is an integer. Notice that the polynomial $x^n - 1 \in (\mathbb{Z}/p\mathbb{Z})[x]$ generates the ideal $\langle x^n - 1\rangle$ in $(\mathbb{Z}/p\mathbb{Z})[x]$; thus, we can form the factor ring

$$(\mathbb{Z}/p\mathbb{Z})[x]/\langle x^n - 1\rangle,$$

which has a similar structure to that of $\mathbb{Z}[x]/\langle x^n - 1\rangle$. This observation provides an alternate construction of the former ring, which can be seen through the concept of a ring homomorphism. A ring homomorphism is a map between rings that respects both the additive structure and multiplicative structure of these rings.

Definition 15.11. *Let R_1 and R_2 be rings. A ring homomorphism is a map*

$$\varphi : R_1 \to R_2$$

that satisfies the following two conditions:

1. $\varphi(r + s) = \varphi(r) + \varphi(s)$ for all $r, s \in R_1$; and

2. $\varphi(r \cdot s) = \varphi(r) \cdot \varphi(s)$ for all $r, s \in R_1$.

With this definition, we can establish the similarity between the factor rings $\mathbb{Z}[x]/\langle x^n - 1\rangle$ and
$$(\mathbb{Z}/p\mathbb{Z})[x]/\langle x^n - 1\rangle.$$

In particular, for each positive integer n, there is a ring homomorphism
$$\varphi : \mathbb{Z}[x]/\langle x^n - 1\rangle \to (\mathbb{Z}/p\mathbb{Z})[x]/\langle x^n - 1\rangle$$
defined by
$$\varphi(r_{n-1}x^{n-1} + \cdots + r_1 x + r_0 + \langle x^n - 1\rangle)$$
$$= (s_{n-1}x^{n-1} + \cdots + s_1 x + s_0) + \langle x^n - 1\rangle,$$
where $s_i \in \mathbb{Z}/p\mathbb{Z}$ and $s_i \equiv r_i \pmod{p}$ for all $i \in \{0, 1, \ldots, n-1\}$. Addition and multiplication are defined as in $\mathbb{Z}[x]/\langle x^n - 1\rangle$ with all coefficients reduced modulo p.

With these basics of ring theory in hand, we will proceed with the introduction of the NTRU cryptosystem.

15.3 THE NTRU CRYPTOSYSTEM

In the NTRU cryptosystem, the receiver will generate a private key and a public key in order to obtain the sender's confidential message. The public key is used to encrypt a message that contains secret or sensitive information into ciphertext, and the private key is used to decrypt the ciphertext message back into the original plaintext message. We require a few notational conventions and definitions in order to discuss the NTRU cryptosystem in greater detail.

15.3.1 Notational Conventions and Definitions

The NTRU cryptosystem is based on three positive integer parameters, namely n, p, and q, where

(i) n is prime number;

(ii) n and q are relatively prime;

(iii) $q \gg p$; and

(iv) p and q are relatively prime.

Note that there is no requirement for p and q to be prime numbers in this cryptosystem, although they are often chosen to be prime numbers. For ease of notation, we set $R = \mathbb{Z}[x]/\langle x^n - 1\rangle$,
$$R_p = (\mathbb{Z}/p\mathbb{Z})[x]/\langle x^n - 1\rangle,$$
and
$$R_q = (\mathbb{Z}/q\mathbb{Z})[x]/\langle x^n - 1\rangle.$$

We can identify any element of R, R_p, or R_q with a polynomial of degree at most n and coefficients in $\mathbb{Z}$, $\mathbb{Z}/p\mathbb{Z}$, or $\mathbb{Z}/q\mathbb{Z}$, respectively, in the natural way and do so now. For example, the polynomial

$$r(x) = (r_{n-1}x^{n-1} + \cdots + r_2 x^2 + r_1 x + r_0) + \langle x^n - 1 \rangle \in R$$

is identified with

$$r(x) = r_{n-1}x^{n-1} + \cdots + r_2 x^2 + r_1 x + r_0,$$

which effectively renders elements of R polynomials as opposed to equivalence classes of polynomials.

Finally, let a and b be non-negative integers. Define $\mathcal{L}(a,b)$ to be the set containing all polynomials $r(x) \in R$ such that

(i) a coefficients of r are equal to 1;

(ii) b coefficients of r are equal to -1; and

(iii) the remaining $n - a - b$ coefficients of r are equal to 0.

This set $\mathcal{L}(a,b)$ defines a subset of R whose elements have a small number coefficients (i.e., each polynomial in $\mathcal{L}(a,b)$ is ternary).

Example 15.4. *Assume that $n = 11$, so that $R = \mathbb{Z}[x]/\langle x^{11} - 1 \rangle$. If $a = 4$ and $b = 2$, then $\mathcal{L}(4,2)$ is the set of polynomials with degree less than 11 that have:*

(i) four coefficients equal to 1;

(ii) two coefficients equal to -1; and

(iii) five coefficients equal to 0.

For instance, the polynomials

$$r(x) = x^{10} + x^7 - x^6 + x^2 - x + 1$$

and

$$s(x) = -x^8 + x^7 - x^4 + x^3 + x^2 + x$$

are both elements of $\mathcal{L}(4,2)$, while

$$t(x) = x^{10} + x^8 + x^7 - x^6 + x^2 - x + 1$$

is not an element of $\mathcal{L}(4,2)$.

With these notational conventions and definitions in hand, we can discuss how keys are generated in the NTRU cryptosystem.

15.3.2 Key Generation Algorithm

To encrypt a plaintext message, the receiver will first generate a private key and a public key. Assume that a_f and a_g are positive integers, and set $\mathcal{L}_f = \mathcal{L}(a_f+1, a_f)$ and $\mathcal{L}_g = \mathcal{L}(a_g, a_g)$. (We note that the parameter a_f was chosen to equal a_g in the original implementation of the NTRU cryptosystem.) Next, the receiver will choose two polynomials $f \in \mathcal{L}_f$ and $g \in \mathcal{L}_g$; the polynomial f will be the private key, while the polynomial g will be used to calculate the public key.

Once f is identified, the receiver will find two polynomials $f_p, f_q \in R$ such that

$$f \cdot f_p \equiv 1 \pmod{p}$$

and

$$f \cdot f_q \equiv 1 \pmod{q};$$

in this case, f is invertible modulo in both R_p and R_q as f_p is the multiplicative inverse of f modulo p and f_q is the multiplicative inverse of f modulo q. If either f_p or f_q does not exist, then the receiver must choose a new polynomial $f \in \mathcal{L}_f$ that is invertible modulo p and q, and thus meets the aforementioned conditions. Finally, the receiver will use the polynomial defined by

$$h \equiv (f_q \cdot g) \pmod{q}$$

to be the public key; the parameters n, p, and q that were chosen at the start of the key generation process can also be made public.

The following example demonstrates this key generation process with specific parameters.

Example 15.5. *Assume that the receiver chooses the following parameters: $n = 11$, $p = 7$, and $q = 233$. In this case, $R = \mathbb{Z}[x]/\langle x^{11} - 1 \rangle$,*

$$R_7 = (\mathbb{Z}/7\mathbb{Z})[x]/\langle x^{11} - 1 \rangle,$$

and

$$R_{233} = (\mathbb{Z}/233\mathbb{Z})[x]/\langle x^{11} - 1 \rangle.$$

If the receiver chooses $a_f = 3$ and $a_g = 3$, then $\mathcal{L}_f = \mathcal{L}(4,3)$ and $\mathcal{L}_g = \mathcal{L}(3,3)$. Further, assume that the receiver chooses

$$f(x) = x^6 - x^5 + x^4 - x^3 + x^2 - x + 1 \in \mathcal{L}_f,$$

which will be the private key, and

$$g(x) = x^{10} + x^8 - x^7 + x^3 - x^2 - x \in \mathcal{L}_g,$$

which will be used to calculate the public key. Since f has a multiplicative inverse modulo 7 and a multiplicative inverse module 233, it represents a valid choice in the NTRU cryptosystem. In particular, the Euclidean algorithm establishes that

$$f_7(x) = 6x^8 + x^4 + x$$

is the multiplicative inverse of f modulo 7 and that

$$f_{233}(x) = 232x^8 + x^4 + x$$

is the multiplicative of f modulo 233. Therefore, the receiver can compute

$$\begin{aligned} h(x) &\equiv \bigl(f_{233}(x) \cdot g(x)\bigr) \pmod{233} \\ &\equiv (x^{10} + 2x^9 + 232x^8 + 232x^6 \\ &\quad + 231x^5 + 2x^4 + 232x^2 + x + 232) \pmod{233}, \end{aligned}$$

which will be used as the public key in this iteration NTRU cryptosystem.

We continue with a description of the encryption algorithm for the NTRU cryptosystem.

15.3.3 Encryption Algorithm

For the NTRU encryption process, the sender requires the sets $\mathcal{L}_\varphi$ and $\mathcal{M}$, which are defined as follows. First, choose a positive integer a_φ, and set $\mathcal{L}_\varphi = \mathcal{L}(a_\varphi, a_\varphi)$. The set $\mathcal{M}$ is defined to the the set of polynomials $m \in R$ with integer coefficients that lie in the interval $\left(-\frac{p}{2}, \frac{p}{2}\right]$, where p is the integer chosen for the NTRU cryptosystem as in Section 15.3.1. While the choice of other intervals is possible, we use $\left(-\frac{p}{2}, \frac{p}{2}\right]$ as it was used in the original implementation of the NTRU cryptosystem by Hoffstein et al. (1998). This chosen interval will capture the possible plaintext messages that the sender can relay to the receiver and correspond to a certain lift of polynomials in R_p.

Under the aforementioned assumptions, if the sender is encrypting the message $m \in \mathcal{M}$ with coefficients reduced modulo p, then the encryption function is

$$c \equiv (p\varphi \cdot h + m) \pmod{q},$$

where the polynomial $\varphi \in \mathcal{L}_\varphi$ is chosen at random. The output of this encryption function, namely c, is the ciphertext message that will be sent to the receiver.

In the example below, we will use the public key

$$\begin{aligned} h(x) \equiv (&x^{10} + 2x^9 + 232x^8 + 232x^6 \\ &+ 231x^5 + 2x^4 + 232x^2 + x + 232) \pmod{233} \end{aligned}$$

that was calculated in Section 15.3.2 to encrypt a plaintext message m using the NTRU cryptosystem.

Example 15.6. *As in Example 15.5, let* $n = 11$, $p = 7$, *and* $q = 233$. *Assume that* $a_\varphi = 2$ *and that the sender encodes their plaintext message using the polynomial*

$$m(x) = x^8 + 2x^7 + 3x^2 - 1.$$

This polynomial m is a value choice in $\mathcal{M}$ because its coefficients lie in the interval

$$\left(-\frac{p}{2}, \frac{p}{2}\right] = (-3.5, 3.5];$$

in other words, it is a polynomial whose coefficients are elements of $\{-3,-2,-1,0,1,2,3\}$. Suppose the sender chooses

$$\varphi(x) = x^{10} + x^7 - x^6 - x^3 \in \mathcal{L}(2,2)$$

randomly. Since the encryption function is

$$c \equiv (p\varphi \cdot h + m)(\bmod q),$$

the ciphertext message is

$$c(x) \equiv (212x^{10} + 7x^9 + 43x^8 + 200x^7 + 14x^6 + 7x^5 \\ + 191x^4 + 28x^3 + 222x^2 + 205x + 41)(\bmod 233).$$

In the forthcoming section, we will see how the receiver will decrypt the ciphertext message c to obtain the original plaintext message m.

15.3.4 Decryption Algorithm

Once the ciphertext message c is relayed to the receiver, it can be decrypted by the receiver as follows:

$$\bar{m} \equiv (f_p \cdot d)(\bmod p),$$

where

$$d \equiv (f \cdot c)(\bmod q).$$

The output of this decryption function, namely $\bar{m}$, is a plaintext message. In Theorem 15.1, we will establish the conditions that guarantee that the output message $\bar{m}$ from the decryption function is actually equal to the original plaintext message m. But first, let us apply this decryption algorithm to the ciphertext message created by the sender in Example 15.6 using the key that was generated in Example 15.5.

Example 15.7. As in Example 15.5, let $n = 11$, $p = 7$, $q = 233$,

$$f_7(x) = 6x^8 + x^4 + x,$$

and

$$f(x) = x^6 - x^5 + x^4 - x^3 + x^2 - x + 1.$$

Since the ciphertext message is

$$c(x) \equiv (212x^{10} + 7x^9 + 43x^8 + 200x^7 + 14x^6 + 7x^5 \\ + 191x^4 + 28x^3 + 222x^2 + 205x + 41)(\bmod 233),$$

(see Example 15.6), we have that

$$\begin{aligned} d(x) &\equiv \big(f(x) \cdot c(x)\big)(\bmod q) \\ &\equiv (13x^{10} + 227x^9 + 2x^8 + 13x^7 + 228x^6 + 231x^5 \\ &\quad + 16x^4 + 218x^3 + 10x^2 + 226x + 219)(\bmod 233) \\ &\equiv (13x^{10} - 6x^9 + 2x^8 + 13x^7 - 5x^6 - 2x^5 \\ &\quad + 16x^4 - 15x^3 + 10x^2 - 7x - 14)(\bmod 233). \end{aligned}$$

(*Observe that the coefficients of $d(x)$ were rewritten to lie in the set $\left(-\frac{q}{2}, \frac{q}{2}\right]$, where $q = 233$. Below, we explain why this is necessary in the NTRU cryptosystem.*) Therefore, the receiver will obtain the plaintext message

$$\bar{m}(x) \equiv (f_p(x) \cdot d(x)) \pmod{p}$$
$$\equiv (x^8 + 2x^7 + 3x^2 + 6) \pmod{7}$$
$$\equiv (x^8 + 2x^7 + 3x^2 - 1) \pmod{7},$$

as desired.

In any cryptosystem, including the NTRU cryptosystem described above, the sender takes the plaintext message m and uses an encryption function to obtain the ciphertext message c, which was then relayed to the receiver. Once the receiver obtains the ciphertext message c, they can use the decryption function to obtain the message $\bar{m}$. This process is only successful if the messages m and $\bar{m}$ are exactly the same. One of the disadvantages of the NTRU cryptosystem is that m and $\bar{m}$ are not guaranteed to be the same. Let us continue by investigating this situation further.

Theorem 15.1. *If the coefficients of the polynomial*

$$d \equiv (f \cdot c) \pmod{q}$$

are elements of $\left(-\frac{q}{2}, \frac{q}{2}\right]$, then the plaintext message m is equal to the output of the decryption function $\bar{m}$.

Proof: Assume that the coefficients of the polynomial d are elements of $\left(-\frac{q}{2}, \frac{q}{2}\right]$. Since

$$c \equiv (p\varphi \cdot h + m) \pmod{q},$$

we have that

$$d \equiv (f \cdot c) \pmod{q}$$
$$\equiv (f \cdot (p\varphi \cdot h + m)) \pmod{q}$$
$$\equiv (f \cdot p\varphi \cdot h + f \cdot m) \pmod{q}.$$

In this case, all coefficients of the polynomial

$$f \cdot p\varphi \cdot h + f \cdot m$$

are elements of the interval $\left(-\frac{q}{2}, \frac{q}{2}\right]$ and

$$d = f \cdot p\varphi \cdot h + f \cdot m.$$

Therefore,

$$\bar{m} \equiv (f_p \cdot d) \pmod{p}$$
$$= (f_p \cdot (f \cdot p\varphi \cdot h + f \cdot m)) \pmod{p}$$
$$= (f_p \cdot f \cdot m) \pmod{p}$$
$$= m \pmod{p},$$

where the last equality holds because f_p and f are inverses of one other modulo p.

Our work above shows that the readable messages m and $\bar{m}$ are the same provided that the coefficients of

$$f \cdot p\varphi \cdot h + f \cdot m$$

are elements of $\left(-\frac{q}{2}, \frac{q}{2}\right]$. In the case when the coefficients of this polynomial do not lie in the interval $\left(-\frac{q}{2}, \frac{q}{2}\right]$, the message m will not decrypt to the message $\bar{m}$. Additionally, the receiver will not realize that $\bar{m}$ is not the correct message relayed by the sender. However, Hoffstein et al. (1998) showed that this situation can be avoided if the difference between the greatest coefficient and the least coefficient of

$$p\varphi \cdot g + f \cdot m$$

is less than q. Notice that this condition is not a restrictive one as

$$q > (6\max\{a_f, a_g\} + 1)p$$

suffices.

15.3.5 Summary

The NTRU cryptosystem can be summarized in eight steps; see Table 15.1. These steps include the generation of the required parameters, key generation, the encryption algorithm, and the decryption algorithm.

15.4 ATTACKING THE NTRU CRYPTOSYSTEM

In this section, we will discuss how the spy might attack the NTRU cryptosystem.

15.4.1 General Attacks of the Spy

Naturally, when the sender relays the confidential message containing sensitive or secret information, the spy will try and attack this process. In cryptography, it is generally assumed (for example, see Washington and Trappe, 2006) that the spy's attack will consist of one of the following goals:

1. The spy aims to read the sender's confidential message;

2. The spy aims to discover the key and read all future messages the sender encrypts with this key;

3. The spy aims to alter the sender's confidential message in such a way that the receiver is not aware the message has been changed; or

4. The spy aims to create a new message and send it to the receiver in such a way that appear to be from the sender.

Moreover, when the spy is attacking a certain cryptosystem, we will assume Kerckhoff's Principle (Kerckhoffs, 1883).

TABLE 15.1 Summary of the NTRU cryptosystem

NTRU Cryptosystem
Parameter Generation
(1) The positive integer parameters n, q, p, a_f, and a_g are created such that: (i) n is prime number;   (ii) n and q are relatively prime;   (iii) p and q are relatively prime; and   (iv) $q > (6\max\{a_f, a_g\} + 1)p$.
Key Generation
(2) The receiver chooses the private key $f \in \mathcal{L}_f$ and polynomial f_p that is the multiplicative inverse of f modulo p.   (3) Next, the receiver chooses a polynomial f_q that is the multiplicative inverse of f modulo q, as well as the polynomial $g \in \mathcal{L}_g$.   (4) Finally, the receiver computes the public key $$h \equiv (f_q \cdot g)(\bmod q),$$ and then makes n, q, p, and h public.
Encryption Algorithm
(5) The sender chooses a positive integer a_φ and a polynomial $\varphi \in \mathcal{L}_\varphi$ at random.   (6) Next, the sender encrypts the plaintext message $m \in \mathcal{L}_m$ via with encryption function $$c \equiv (p\varphi \cdot h + m)(\bmod q),$$ and relays the ciphertext message c to the receiver.
Decryption Algorithm
(7) The receiver computes the polynomial $$d \equiv (f \cdot c)(\bmod q).$$ (8) Finally, the receiver computes $$\bar{m} \equiv (f_p \cdot d)(\bmod p),$$ and recovers the original plaintext message $m = \bar{m}$.

Principle 15.1 (Kerckhoff's Principle). *In evaluating the security of a given cryptosystem, it should be assumed that the spy knows the cryptosystem in use.*

In other words, the spy knows everything about the cryptosystem except the key. This is based on the idea that persons who know the cryptosystem in use can be captured and provide the spy with this information; it is also possible that the spy can gain access to the encryption algorithm or decryption algorithm for the cryptosystem currently in use.

15.4.2 The NTRU Key Recovery Problem

Let us suppose that the spy aims to discover the private key $f \in \mathcal{L}_f$ that is being used by the sender and receiver in an iteration of the NTRU Cryptosystem. Hoffstein et al. (1998) experimentally observed that the coefficients of the inverse of f in R_q, namely f_q, tend to be both randomly distributed modulo q and uniformly distributed modulo q. For example, assume that $n = 13$, $q = 89$, $a_f = 3$, and consider the randomly chosen polynomial

$$f(x) = x^{11} + x^8 - x^7 + x^5 - x^3 + x^2 - 1 \in \mathcal{L}_f = \mathcal{L}(a_f + 1, a_f).$$

The polynomial f is invertible module q, and its inverse in R_q is

$$\begin{aligned} f_q(x) = {} & 11x^{12} + 37x^{11} + 7x^{10} + 23x^9 + 38x^8 + 48x^7 \\ & + 68x^6 + 8x^5 + 76x^4 + 72x^3 + 70x^2 + 77. \end{aligned}$$

These coefficients of f_q appear to be randomly distributed modulo q and uniformly distributed modulo q. Moreover, the coefficients of public key

$$h \equiv (f_q \cdot g) \pmod{q}$$

also appear to be both randomly distributed modulo q and uniformly distributed modulo q. For instance, if $a_g = 3$ and

$$g(x) = x^{12} + x^{11} - x^8 - x^7 + x^3 - x^2,$$

then

$$\begin{aligned} h(x) \equiv {} & (9x^{12} + 44x^{11} + 5x^{10} + 83x^9 + 71x^8 + 41x^7 + 34x^6 \\ & + 70x^5 + 65x^4 + 11x^3 + 85x^2 + 52x + 53) \pmod{q}. \end{aligned}$$

Even though the coefficients of the public key h appear to be both randomly distributed modulo q and uniformly distributed modulo q, they are related to the private key f in the NTRU cryptosystem. In particular, we can multiply both sides of the congruence

$$h \equiv (f_q \cdot g) \pmod{q}$$

by the private key f to obtain

$$f \cdot h \equiv (f \cdot (f_q \cdot g)) \equiv ((f \cdot f_q) \cdot g) \equiv (1 \cdot g) \pmod{q}$$

or
$$f \cdot h \equiv g \pmod{q}.$$

Therefore, the spy can attack the NTRU cryptosystem by considering the following problem, called the *NTRU Key Recovery Problem*.

Problem 15.1 (NTRU Key Recovery Problem). *Given a polynomial h, find polynomials $f \in \mathcal{L}_f$ and $g \in \mathcal{L}_g$ such that*
$$f \cdot h \equiv g \pmod{q}.$$

While this problem is easy to state, it is believed a mathematical hard problem to solve. It particular, solving the NTRU Key Recovery Problem is thought to be equivalent to solving the Shortest Vector Problem in a certain class of lattices.

15.5 CONCLUSION

Although the NTRU cryptosystem does possess parameter choices for which decryption can fail without the receiver's knowledge, it offers many advantages. In particular, it is thought to be resistant to all quantum based computer attacks (Hoffstein et al., 2010). Moreover, the NTRU cryptosystem is faster than other public-key cryptosystems, and its security is built on the difficulty of certain lattice problems.

The NTRU cryptosystem has been modified to consider rings other than polynomial rings with integer coefficients. Such considerations have introduced faster algorithms and improved security of the NTRU public-key cryptosystem; for example, matrix rings with integer entries (Coglianese and Goi, 2005), the Eisenstein integers (Jarvis and Nevins (2015)), the Gaussian integers (Kouzmenko, 2006), and quaternion algebras (Malekian et al., 2009) have been considered in this context.

REFERENCES

Batten, L. M. (2013). *Public key cryptography: Applications and attacks.* John Wiley & Sons.

Coglianese, M. and Goi, B.-M. (2005). *MaTRU: A new NTRU-based cryptosystem*, pages 232–243. INDOCRYPT 2005, Lecture notes in computer science. Springer-Verlag.

Diffie, W. and Hellman, M. (1976). New directions in cryptography. *IEEE Transactions on Information Theory*, 22(6):644–654.

Dummit, D. and Foote, R. (2004). *Abstract algebra.* John Wiley & Sons, third edition.

Hoffstein, J., Howgrave-Graham, N., Pipher, J., and Whyte, W. (2010). *Practical lattice-based cryptography: NTRUEncrypt and NTRUSign*, pages 349–390. The LLL Algorithm: Survey and applications, Information security and cryptography. Springer-Verlag.

Hoffstein, J., Pipher, J., and Silverman, J. H. (1998). *NTRU: A ring-based public key cryptosystem*, pages 267–288. Algorithmic number theory, Lecture notes in computer science. Springer-Verlag.

Jarvis, K. and Nevins, M. (2015). ETRU: NTRU over the Eisenstein integers. *Des. Codes Cryptogr.*, 74:219–242.

Kerckhoffs, A. (1883). La cryptographie militaire. *Journal des sciences militaires*, IX:5–38.

Kouzmenko, R. (2005–2006). *Generalizations of the NTRU cryptosystem*. PhD thesis, École Polytechnique Federale de Lausanne.

Malekian, E., Zakerolhosseini, A., and Mashatan, A. (2009). QTRU: A lattice attack resistant version of NTRU. *IACR Cryptology ePrint Archive*, 2009.

Mollin, R. A. (2007). *An introduction to cryptography*. Chapman & Hall/CRC, second edition.

Shannon, C. E. (1949). Communication theory of secrecy systems. *The Bell System Technical Journal*, 28(4):656–715.

Washington, L. C. and Trappe, W. (2006). *Introduction to cryptography with coding theory*. Pearson, second edition.

Index

Note: Italicized pages refer to figures and **bold** pages refer to tables.

A

Abstract simplicial complexes, 135–136, 155–157; *see also* Topology
    interference complex, 155, *156*
    link complex, 155, 156–157
Activation sheaf, 157
    global sections, 157–159
    using activation patterns, 160–163
Actuarial science, 328
Affine ciphers, 400
Affinity diagrams, 241
Agent-based simulation, 271–272
Algebraic curves, 97–99, 101, *102*
    affine and projective varieties, 102–103
    common component of intersection, 104–105
    compact Riemann surface, 101
    coordinate rings, 105–107
    plane curves, 103–105
Algebraic geometry, 97

B

Baye's rule, 354–355, 413
Bézout's lemma, 396
Binomial theorem, 9–11, 19, 22, 30–32, 37, 39
    generalized, 10–11
Birthday attacks, 17–18
Birthday problem, 17–18, 121, 414–415
Brute force attacks, 13, 16
    expected time, 13
Bytes (or bits), 405–407
    binary, 405
    hexadecimal, 405

C

Chain complex, *137*, 138–139; *see also* Topology
Chinese remainder theorem, 400–402
Codes, 61
    Caesar codes, 458–459
    cosets, 75–77
    dimension, 61
    dual, 64
    information rate, 62
    Goppa codes, 70–71, 74, 78–79, 81
    Hamming codes, 56, 64, 73, *74–75*
    length, 61
    low-density parity-check (LDPC) codes, 71
    minimum distance, 61
    Moderate-Density Parity-Check (MDPC) codes, 71
    parity-check codes, 70–71
    puncturing and shortening, 65
    quantum codes, 437
    quantum private codes, 439
    Reed-Solomon codes, 67–71, 74, 79–82, 85–90
    repetition codes, 63
    simple parity-check codes, 63
    Singleton bound, 62–63, 68
    subfield subcode, 67
Codewords, 57, 61–64, 67–69, 75–77, 84
Coding theory, 54, **55**
Cohomological analysis, 163–164
Combinatorial classes, 3–4
    combining, 33–34
Combinatorial designs, 45–50
    black-box testing, 47–48
    Latin squares, 46

Combinatorics, 1
Correlation analysis, 321–322
Counting derangements, 22
Counting involutions, 26
Coupon collector's problem, 18–20
  expectation, 19–20
Cryptography, 53–54, **55**, *56*, 98–99, 109, 117, 119, 122, 128–131, 444–446, 468, 471, 475, *476*
  device independent cryptography, 444–447
  elliptic curve cryptography, 55, 77, 117
  isogeny-based cryptography, 119
  post-quantum cryptography, 54–55, 98, 119
  post-quantum hyperelliptic curve cryptography, 129–130
  public-key cryptography, 77, 477
Cryptosystems, 77, 476
  Berger-Loidreau-based cryptosystem, 86–88
  LDPC code-based cryptosystem, 71, 74–75, 90–92
  McEliece-based cryptosystem, 70, 78–82, 85–89, 92
  Niederreiter-based cryptosystem, 75, 79–82, 86–88, 92
  N-th Degree Truncated Polynomial Ring (NTRU) cryptosystem, 477, 479–480, 482, 484–494
  Random Linear Code Encryption (RLCE) cryptosystem, 88–90
  RSA cryptosystem, 54–55, 77, 98, 109, 119, 130, 403–404, 449, 477
Cyber attack, 219, 268
  on a voting process, 241–244
Cyber Attacker Model Profile (CAMP), 325
Cyberbullying, 307
Cyber defense, 268
  Kerckhoff's principle, 493
Cyber red teaming, 276–277

Cyber resilience, 255, 281, 304–305, 326–328
  assessment, 327
  definition and measures, 326
  enhancement, 327
Cyber risk measurement, 308
Cybersecurity model, 223, 309–310
Cyberspace, 219
Cyber terrorism, 209, 218–219
Cyber warfare, 268
Cyclic Redundancy Check (CRC), 406–407

D
Data analysis, 267
  classification methods, 283, **284**
  descriptive data analytic methods, 282
  dimension reduction methods, 283, **284**
  prediction methods, 283, **284**
  predictive data analytic methods, 282–283
  prescriptive data analytic methods, 282
  supervised learning, 283–284
  time series models, 284
  unsupervised learning, 283–284
Deception attacks, 188, 321
Decision analysis, 234–235
  Analytical Hierarchy Process (AHP), 240
  decision strategies, 235–236
  decision tree, 235–236, 287
  desirability functions, 239–240
  Expected Monetary Value (EMV), 235
  group decision making, 240–241
  utility functions, 237–239
  Value Focused Thinking (VFT), 241
  value functions, 239–240
Denial of Service (DoS) attack, 188–192, 272, 276, 308
Descriptive statistics, 305

Differential equations, 171
Diffie-Hellman key exchange, 77, 98, 118–121, 128, **129**, 394, 410, 454, 458, 468–471
    protocol, 469
    using elliptic curves, **118**, 119
    using hyperelliptic curves, 128, **129**
    using supersingular elliptic curves, 120
Digital viruses, 226–229
Discrete Event Simulation (DES), 271–272
Distinguisher attacks, 84–85
Dowker homology, 135, 141–142, *143*, 144, *145–147*; *see also* Topology

E
Eigenvalues, 108, 175, 177–178, 180, 182–183, 189–191, 198–200, 214, 240, 295–296, 426
Eigenvectors, 198, 214, 295–296
Elliptic curves, 101, 109–112; *see also* Algebraic curves
    division polynomials, 115–117
    Hasse's theorem, 112
    isogenies, 112–113
    modular polynomials, 114–115
    rational map, 114
    Velu's formula, 113–115
Enumeration, 2–3, *8*
    asymptotic enumeration, 29
    multiset enumeration, 8
    permutation enumeration, 5
    sequence enumeration, 4
    set partition, 24
    subset enumeration, 6–7
    surjections, 23–24
Equilibrium points, 172, 175–178, 180–185, 198, 200
    disease-free equilibrium points, 179–182
    stability, 175, 178–179, 181–185, 187, 198
    worm-free equilibrium points, 182–188

Error-correcting codes, 54–55, *56*, 63–64, 77, 92, 109, 399, 436–437, 439–440
    quantum, 437
Euler characteristic, 139, 144, *148*
Euler phi function, 402–403
Euler's theorem, 403–404, 408
Expectation (of a random variable), 12–15, 18–19, 44, 257, 352–353
    fixed point, 14
    linearity, 13

F
Fermat's little theorem, 403, 412
Fibonacci numbers, 32–33
Fields, 57–59, 99, 478
    algebraic closure, 100
    field extension, 99, *100*
    finite fields, 99–101
    perfect field, 100
    splitting field, 100–101
File integrity, 16–17
    checksum, 16
Floyd's cycle-finding algorithm, 27–28, 415

G
Games, 279, 371–379, 440–442
    extensive form games, 279, 373–377
    minimax, 280
    Nash games, 371–373
    normal form, 279
    players, 365–366
    prisoner's dilemma, 278–279
    quantum games, 440–442
    Stackelberg games, 279, 377–379
    utilities, 366–367
Game theory, 277, 279, 281, 363, 379
    assumptions, 380–381
    data, 380
    operational security, 384–385
    privacy, 385–386
    scalability, 383–384
    solution design, 382–383

timing, 382
uncertainty, 381–382
Generating functions, 29–37, 40–45
    bivariate generating functions, 42–45
    coefficients of rational generating functions, 40–41
    cycle and set operators, 35–36
    exponential generating function, 30, 35, 45
    ordinary generating function, 29–31, 34–35, 41–42
Genus 1 curves, *see* Elliptic curves
Genus 2 curves, *see* Hyperelliptic curves
Graphical displays, 309–313; *see also* Descriptive statistics
    bar graph, 310, *311*
    box-and-whisker plot, 312, *313*
    dot plot, 312, *313*
    frequency polygon graph, 312, *313*
    histogram, 310–311, *312*
    line graph, 310
    pie chart, 310, *311*
    Pareto chart, 310, *311*
    stem-and-leaf plot, 312, *313*
Graph metrics, 211–213
    average shortest path length, 212
    clustering coefficient, 212–213
    degree, 211
    degree distribution, 212
    density, 212
    diameter, 212
Graphs (or networks), 71, 107–109, 210–211, 462–467
    attack graphs (trees), 221–222, 236–237, *238*, *242*, *244*
    betweenness centrality, 150, 213–214, 220
    bipartite graphs, 71, *72*, 210
    centrality, 213–214, **220**, 229
    completed graph, 210
    component, 211
    cycle, 211
    degree centrality, 213
    directed graphs (or digraphs), 135, 149, *150*, 210–212, 226–227, 250, 414–415
    edges (or links), 71–74, 107–108, 133, 139, 149, 156, 193, 209–210, 435, 438, 458, 463–464, 467
    expander graphs, 108–109
    k-regular graphs, 211
    link graphs, 155
    isogeny graphs, 120–121
    iterated function graphs, 26
    nodes (or vertices), 36, 42–44, 72, 154–158, 160–161, 210–214, 225
    path, 211
    Ramanujan graphs, 108, 120–121
    random graph, 214, **215**
    shortest path, 107, 211–214, 222
    subgraphs, 210
    Tanner graphs, 72, *73–75*
    undirected graphs, 107–108, 148, 210–212
    weighted and unweighted graphs, 210
Graph theory, 2, 98, 107–109, 154, 209, 229, 422, 433–435, 448
Group, 109–110, 112–113, 117, 122–125, 408–410, 456
    group actions, 458–460
    identity element, 457
    permutation group, 460
    symmetric group, 460
Group theory, 403, 453, 472
    binary operation, 456
    computational group theory, 472
    fundamental theorem of Galois theory, 455

H

Hash functions, 15–18, 27, 45
    collision resistance, 17
    cryptographic, 15–16
    probability of collision, 16

Hilbert spaces, 423–427, 436–440,
    442–443, 446–447
  density operator, 426
  ensemble, 426
  no cloning theorem, 427–428
Hurwitz condition, 186–187,
    200–201
Hyperelliptic curves, 98, 102,
    122–123; *see also* Algebraic
    curves
  addition by interpolation, 126,
    127
  algebraic addition, 127–128
  discrete logarithm problem, 128
  Mumford representation, 125
  principal divisors, 123–126
  semi-reduced divisors, 124

I
Inferential statistics, 305, 318–321
  estimation methods, 318
  hypothesis testing, 318–321
  population, 306
  sample, 306
  variable, 306
Influence diagrams, 241
Information entropy, 8–9, 357–358
Information set decoding, 85–86
Input-to-State Stability (ISS), 201
Insider threat, 222
Integer programming, 248
Integers, 394–399
  characteristic, 99
  division and modular arithmetic,
    394–395, 399–400
  Euclidean algorithm, 395–396,
    400–401, 404, 457, 481, 487
ISBN codes, 398–399
Isomorphism, 58–59, 67, 111, 137, 139,
    438, 458, 460, 463–465,
    479–480

K
Kernel, 60, 68, 112–113, 120, 152–153,
    292

L
La Salle's invariance principle, 179, 184,
    187, 200
Linear optimization, 245–247
  constraints, 245
  cost and technological coefficients,
    245
  decision variables, 245
  dual, 249
  feasible solution or region, 245,
    246
  objective function, 245
  Simplex method, 247
Lyapunov equation, 190

M
Machine learning, 150, 165, 234, 262,
    268, 282, 297, 321, 327–328,
    386
Malware spread, 250, 252–253, 259–261
Markov chains, 252–258, 280–281,
    355–357
  absorbing state, 254
  aperiodic, 253, 280
  irreducible, 253, 280
  limiting distribution, 254
  memoryless property, 252, 355–356
  periodic, 253–254
  recurrent, 253, 280
  regular, 254
  state transition diagram, *253, 257,
    260*
  symmetric random walk, *253*
  transient, 253
  transition matrix, 252
Markov processes, 252–254
Mathematical optimization, 244
  convex hull, 249
Monte Carlo methods, 258–259, 261,
    271–272
Multisets, 7–8

N
Nash equilibrium, 277–279; *see also*
    Game theory

Network models, 214
    Barabási-Albert model, 217–218
    Erdös-Renyi model, 214–216
    Watts-Strogatz model, 216
Network interdiction, 250–251
Network science, 207
Nonlinear optimization, 247–248
Number theory, 50, 393, 455–456
Numerical descriptive measures, 313–315; *see also* Descriptive statistics
    mean (average), 313
    median, 313
    mode, 314
    quartiles, 314–315
    range, 314
    standard deviation, 314
    variance, 314

O

Operations research, 233
Orthogonal arrays, 46–50, 436–437, 439

P

Passwords, 2–3, 8–9, 13–16, 40–42
    asymptotics, 41
    password attack, 276
    password strength, 8–9, 386
Path homology, 147–150; *see also* Topology
Permutations, 4–6, 13–14, 20, 22, 25–26, 29–30, 36, 38, 42, 44–45
    cycle notation, 25–29, 461
    n-permutation, 5
Petri net, 272
Phase portrait, 198–200
Pollard's rho algorithm, 27–29, 410–411, 414–416
Polynomials, 404–405; *see also* Number theory
    division algorithm, 404–405
Preimage attacks, 26

Primality testing, 411–413
    Miller-Rabin test, 412–413
    Sieve of Eratosthenes, 411, *412*, 418
    trial division, 411
Prime numbers, 118, 128–129, 397, 399, 403–404, 410–412, 470–471, 478, 480, 485, 492; *see also* Integers
    fundamental theorem of arithmetic, 397
    Euclid's lemma, 397–398
    prime number theorem, 411
Primitive roots, 408–411; *see also* Number theory
    baby-step giant-step algorithm, 410–411
    discrete logarithms, 409
Principle of inclusion-exclusion, 20–22, 337–338
Probability, 12–20, 22, 25, 161, *162–163*, 212, 214–218, 235–239, 242–244, 252–254, 256–257, 260–261, 280, 285–287, 305, 319–320, 336–343, 345–348, 350, 354–357, 360–361, 366–367, 369, 371, 375, 411–415, 417, 425–426, 430–434, 436–437, 439–442, 445–446, 476
    axioms, 336–337
    conditional probability, 252, 339–340, 354, 356
    cumulative distribution function, 258, 341, *342*, *348*
    event, 336
    probability density function, 341, *348*
    probability mass function, 341, *342*
    sample space, 336
    Venn diagrams, *21*, 338, *339*
Probability distributions, 152, *153*, 255, 258, 320, 341, 347–349, 352, 366, 369
    Bernoulli distribution, 343
    binomial distribution, 215, 343–344
    chi-squared distribution, 351
    gamma distribution, 351

Index ■ 503

geometric distribution, 345
hypergeometric distribution, 344
negative binomial distribution, 345–346
normal distribution, 349–351
Poisson distribution, 215–216, 338, 346–347
standard normal distribution table, 362
uniform distribution, 349
Probability theory, 335, 358–359

Q

Quadratic sieve, 416–418
Qualitative data, 235, 306–307, 309–310, *311*
    nominal data, 306
    ordinal data, 307
Quantitative data, 293, 307–310, 312, *313*
    continuous data, 307
    discrete data, 307
Quantum communication, 429–433
    quantum key distribution, 432–433
    quantum teleportation, 429–431
    superdense coding, 431–432
Quantum correlations, 442–444
    correlation, 442
    local correlations, 444
    non-signalling correlations, 443
    projection-valued measures, 442
Quantum graph theory, 433–440
    classical channels, 434
    confusability graph, 434, *435*
    quantum channels, 436–438
    quantum cliques and anticliques, 438
    quantum graphs, 438–440
Quantum information, 427–429
    quantum alphabet, 427
    quantum letter, 427
Quantum mechanics (postulates), 425–426
Quantum theory, 421
Queueing models, 254–258
    Kendall notation, 255

R

Random functions, 23–24
Random variables, 12–15, 18, 249, 252, 258, 340–353, 355–358, 361
    discrete random variables, 341–342
    continuous random variables, 347–348
Recurrences, 31–32
Recursive structures, 36–38
    binary trees, 36–37
    Catalan trees, 38
    rooted trees, 36, 225
Relative homology, 159–160; *see also* Topology
Reliability engineering, 328
Ring, 401, 477–480, 482–485
    commutative property, 478
    factor ring, 479
    homomorphism, 484–485
    ideal, 479
    identity property, 478
    nonzero commutative ring, 480
    polynomial ring, 404, 478–481
Ring theory, 475
    binary operation, 477
Risk analysis, 328
Risk Management Frameworks (RMF), 273, *274*, 275–276

S

Sequences, 2–8, 17–18, *19*, 23, 25–36, 38–40, 48–49, 107–108, 138–139, 164, 221, 237, 253, 337, 355, 405, 407, 410, 414–415, 427, 455, 468
    asymptotic growth, 39
Sensitivity analysis, 246
Sidelnikov-Shestakov attack, 82–84
Simplicial homology, 137; *see also* Topology
Standard deviation (of a random variable), 255, 314, 321, 323, 349, 353
Statistical methods, *305*

Statistical process control charts, 316–317
    Cumulative Sum (CUSUM) chart, 317
    Exponentially Weighted Moving Average (EWMA) chart, 317
    Shewhart chart, 317
Statistical software packages, 315–317
Statistics, 281–282, 303, 328
Stirling numbers, 24
Stirling's approximation, 39
Stochastic process modeling, 234, 251–253, 259, 261–262, 280
Strategies, 367–371; *see also* Games
    dominant strategies, 369
    equilibria, 370
    in security games, 370–371
    mixed strategies, 368–369
    pure strategies, 367–368
Subsets, 5–7
Summary tables, 309; *see also* Descriptive statistics
    cumulative frequency, 309, **316**
    cumulative relative frequency, 309, **316**
    frequency, 309, **316**
    relative frequency, 309, **316**
Superelliptic curves, 102; *see also* Algebraic curves
Supervised clustering methods, 288–292
    discriminant analysis, 288–290
    Support Vector Machine (SVM), 290–292
Supervised predictive tools, 284–288; *see also* Data analysis; Statistics
    logistic regression, 285, *286*, 287–288
    regression analysis, 284–285, 322–324, *325*, 327
    tree methods, 287, *288*
Susceptible-Exposed-Infected-Quarantined-Recovered (SEIQR) model, 185, 188
Susceptible-Exposed-Infected-Recovered (SEIR) model, 176
Susceptible-Infected-Recovered (SIR) model, 171, 173–175, 179–180, 196
Syndrome decoding, 75–76

T

Test generation, 49–50
Topological data analysis, 152–153
    unimodal decomposition, 153
Topology, 133, 208, 228
Trace-determinant diagram, 199–200

U

Unsupervised clustering methods, 292–294
    k-means clustering, 294
    k-nearest neighbors, 293–294
    multidimensional scaling, 292–293
Unsupervised dimension reduction methods, 294–296
    Factor Analysis (FA), 295–296
    Principal Component Analysis (PCA), 294–296

V

Variance (of a random variable), 44, 282, 285, 289, 295–296, 314, 318, 321, 323, 353–354, 361
Vectors, 11, 60–64, 69, 75–76, 78, 80, 83–84, 86, 91, 99, 105, 135, 137–141, 149–150, 159, 163–164, 214, 224, 245–248, 254, 260, 281, 285, 357, 399, 405–406, 417, 423–428, 430, 436–439, 443,
    entangled, 425
    finite dimensional, 423
    inner product, 424
    linearly independent set, 423
    qubit, 424
    qudit, 424
    orthonormal basis, 423

separable, 425
span, 423
Vulnerable-Exposed-Infectious-Secured-Vulnerable (VEISV) model, 182–183
Visualization methods (for data), 269–271

dynamic dashboard visualization, *271*
static dashboard visualization, *270*

Y
Young's inequality, 191

Milton Keynes UK
Ingram Content Group UK Ltd.
UKHW050046270524
443136UK00004B/68